Universitext

For other titles in this series, go to
www.springer.com/series/223

Loring W. Tu

An Introduction to Manifolds

Second Edition

 Springer

Loring W. Tu
Department of Mathematics
Tufts University
Medford, MA 02155
loring.tu@tufts.edu

ISBN 978-1-4419-7399-3   ISBN 978-1-4419-7400-6 (eBook)
DOI 10.1007/978-1-4419-7400-6
Springer New York Dordrecht Heidelberg London

Library of Congress Control Number: 2010936466

Mathematics Subject Classification (2010): 58-01, 58Axx, 58A05, 58A10, 58A12

Printed on acid-free paper

Springer is part of Springer Science+Business Media (www.springer.com)

Dedicated to the memory of Raoul Bott

Preface to the Second Edition

This is a completely revised edition, with more than fifty pages of new material scattered throughout. In keeping with the conventional meaning of chapters and sections, I have reorganized the book into twenty-nine sections in seven chapters. The main additions are Section 20 on the Lie derivative and interior multiplication, two intrinsic operations on a manifold too important to leave out, new criteria in Section 21 for the boundary orientation, and a new appendix on quaternions and the symplectic group.

Apart from correcting errors and misprints, I have thought through every proof again, clarified many passages, and added new examples, exercises, hints, and solutions. In the process, every section has been rewritten, sometimes quite drastically. The revisions are so extensive that it is not possible to enumerate them all here. Each chapter now comes with an introductory essay giving an overview of what is to come. To provide a timeline for the development of ideas, I have indicated whenever possible the historical origin of the concepts, and have augmented the bibliography with historical references.

Every author needs an audience. In preparing the second edition, I was particularly fortunate to have a loyal and devoted audience of two, George F. Leger and Jeffrey D. Carlson, who accompanied me every step of the way. Section by section, they combed through the revision and gave me detailed comments, corrections, and suggestions. In fact, the two hundred pages of feedback that Jeff wrote was in itself a masterpiece of criticism. Whatever clarity this book finally achieves results in a large measure from their effort. To both George and Jeff, I extend my sincere gratitude. I have also benefited from the comments and feedback of many other readers, including those of the copyeditor, David Kramer. Finally, it is a pleasure to thank Philippe Courrège, Mauricio Gutierrez, and Pierre Vogel for helpful discussions, and the Institut de Mathématiques de Jussieu and the Université Paris Diderot for hosting me during the revision. As always, I welcome readers' feedback.

Paris, France
June 2010

Loring W. Tu

Preface to the First Edition

It has been more than two decades since Raoul Bott and I published *Differential Forms in Algebraic Topology*. While this book has enjoyed a certain success, it does assume some familiarity with manifolds and so is not so readily accessible to the average first-year graduate student in mathematics. It has been my goal for quite some time to bridge this gap by writing an elementary introduction to manifolds assuming only one semester of abstract algebra and a year of real analysis. Moreover, given the tremendous interaction in the last twenty years between geometry and topology on the one hand and physics on the other, my intended audience includes not only budding mathematicians and advanced undergraduates, but also physicists who want a solid foundation in geometry and topology.

With so many excellent books on manifolds on the market, any author who undertakes to write another owes to the public, if not to himself, a good rationale. First and foremost is my desire to write a readable but rigorous introduction that gets the reader quickly up to speed, to the point where for example he or she can compute de Rham cohomology of simple spaces.

A second consideration stems from the self-imposed absence of point-set topology in the prerequisites. Most books laboring under the same constraint define a manifold as a subset of a Euclidean space. This has the disadvantage of making quotient manifolds such as projective spaces difficult to understand. My solution is to make the first four sections of the book independent of point-set topology and to place the necessary point-set topology in an appendix. While reading the first four sections, the student should at the same time study Appendix A to acquire the point-set topology that will be assumed starting in Section 5.

The book is meant to be read and studied by a novice. It is not meant to be encyclopedic. Therefore, I discuss only the irreducible minimum of manifold theory that I think every mathematician should know. I hope that the modesty of the scope allows the central ideas to emerge more clearly.

In order not to interrupt the flow of the exposition, certain proofs of a more routine or computational nature are left as exercises. Other exercises are scattered throughout the exposition, in their natural context. In addition to the exercises embedded in the text, there are problems at the end of each section. Hints and solutions to selected exercises and problems are gathered at the end of the book. I have starred the problems for which complete solutions are provided.

This book has been conceived as the first volume of a tetralogy on geometry and topology. The second volume is *Differential Forms in Algebraic Topology* cited above. I hope that Volume 3, *Differential Geometry: Connections, Curvature, and Characteristic Classes*, will soon see the light of day. Volume 4, *Elements of Equivariant Cohomology*, a long-running joint project with Raoul Bott before his passing away in 2005, is still under revision.

This project has been ten years in gestation. During this time I have benefited from the support and hospitality of many institutions in addition to my own; more specifically, I thank the French Ministère de l'Enseignement Supérieur et de la Recherche for a senior fellowship (bourse de haut niveau), the Institut Henri Poincaré, the Institut de Mathématiques de Jussieu, and the Departments of Mathematics at the École Normale Supérieure (rue d'Ulm), the Université Paris 7, and the Université de Lille, for stays of various length. All of them have contributed in some essential way to the finished product.

I owe a debt of gratitude to my colleagues Fulton Gonzalez, Zbigniew Nitecki, and Montserrat Teixidor i Bigas, who tested the manuscript and provided many useful comments and corrections, to my students Cristian Gonzalez-Martinez, Christopher Watson, and especially Aaron W. Brown and Jeffrey D. Carlson for their detailed errata and suggestions for improvement, to Ann Kostant of Springer and her team John Spiegelman and Elizabeth Loew for editing advice, typesetting, and manufacturing, respectively, and to Steve Schnably and Paul Gérardin for years of unwavering moral support. I thank Aaron W. Brown also for preparing the List of Notations and the TeX files for many of the solutions. Special thanks go to George Leger for his devotion to all of my book projects and for his careful reading of many versions of the manuscripts. His encouragement, feedback, and suggestions have been invaluable to me in this book as well as in several others. Finally, I want to mention Raoul Bott, whose courses on geometry and topology helped to shape my mathematical thinking and whose exemplary life is an inspiration to us all.

Medford, Massachusetts   *Loring W. Tu*
June 2007

Contents

Chapter 4 Lie Groups and Lie Algebras

Chapter 5 Differential Forms

Appendices

A Brief Introduction

Undergraduate calculus progresses from differentiation and integration of functions on the real line to functions in the plane and in 3-space. Then one encounters vector-valued functions and learns about integrals on curves and surfaces. Real analysis extends differential and integral calculus from $\mathbb{R}^3$ to $\mathbb{R}^n$. This book is about the extension of calculus from curves and surfaces to higher dimensions.

The higher-dimensional analogues of smooth curves and surfaces are called *manifolds*. The constructions and theorems of vector calculus become simpler in the more general setting of manifolds; gradient, curl, and divergence are all special cases of the exterior derivative, and the fundamental theorem for line integrals, Green's theorem, Stokes's theorem, and the divergence theorem are different manifestations of a single general Stokes's theorem for manifolds.

Higher-dimensional manifolds arise even if one is interested only in the three-dimensional space that we inhabit. For example, if we call a rotation followed by a translation an affine motion, then the set of all affine motions in $\mathbb{R}^3$ is a six-dimensional manifold. Moreover, this six-dimensional manifold is not $\mathbb{R}^6$.

We consider two manifolds to be topologically the same if there is a homeomorphism between them, that is, a bijection that is continuous in both directions. A topological invariant of a manifold is a property such as compactness that remains unchanged under a homeomorphism. Another example is the number of connected components of a manifold. Interestingly, we can use differential and integral calculus on manifolds to study the topology of manifolds. We obtain a more refined invariant called the de Rham cohomology of the manifold.

Our plan is as follows. First, we recast calculus on $\mathbb{R}^n$ in a way suitable for generalization to manifolds. We do this by giving meaning to the symbols dx, dy, and dz, so that they assume a life of their own, as *differential forms*, instead of being mere notations as in undergraduate calculus.

While it is not logically necessary to develop differential forms on $\mathbb{R}^n$ before the theory of manifolds—after all, the theory of differential forms on a manifold in Chapter 5 subsumes that on $\mathbb{R}^n$, from a pedagogical point of view it is advantageous to treat $\mathbb{R}^n$ separately first, since it is on $\mathbb{R}^n$ that the essential simplicity of differential forms and exterior differentiation becomes most apparent.

Another reason that we do not delve into manifolds right away is so that in a course setting the students without a background in point-set topology can read Appendix A on their own while studying the calculus of differential forms on $\mathbb{R}^n$.

Armed with the rudiments of point-set topology, we define a manifold and derive various conditions for a set to be a manifold. A central idea of calculus is the approximation of a nonlinear object by a linear object. With this in mind, we investigate the relation between a manifold and its tangent spaces. Key examples are Lie groups and their Lie algebras.

Finally, we do calculus on manifolds, exploiting the interplay of analysis and topology to show on the one hand how the theorems of vector calculus generalize, and on the other hand, how the results on manifolds define new C^∞ invariants of a manifold, the de Rham cohomology groups.

The de Rham cohomology groups are in fact not merely C^∞ invariants, but also topological invariants, a consequence of the celebrated de Rham theorem that establishes an isomorphism between de Rham cohomology and singular cohomology with real coefficients. To prove this theorem would take us too far afield. Interested readers may find a proof in the sequel [4] to this book.

Chapter 1

Euclidean Spaces

The Euclidean space $\mathbb{R}^n$ is the prototype of all manifolds. Not only is it the simplest, but locally every manifold looks like $\mathbb{R}^n$. A good understanding of $\mathbb{R}^n$ is essential in generalizing differential and integral calculus to a manifold.

Euclidean space is special in having a set of standard global coordinates. This is both a blessing and a handicap. It is a blessing because all constructions on $\mathbb{R}^n$ can be defined in terms of the standard coordinates and all computations carried out explicitly. It is a handicap because, defined in terms of coordinates, it is often not obvious which concepts are intrinsic, i.e., independent of coordinates. Since a manifold in general does not have standard coordinates, only coordinate-independent concepts will make sense on a manifold. For example, it turns out that on a manifold of dimension n, it is not possible to integrate functions, because the integral of a function depends on a set of coordinates. The objects that can be integrated are differential forms. It is only because the existence of global coordinates permits an identification of functions with differential n-forms on $\mathbb{R}^n$ that integration of functions becomes possible on $\mathbb{R}^n$.

Our goal in this chapter is to recast calculus on $\mathbb{R}^n$ in a coordinate-free way suitable for generalization to manifolds. To this end, we view a tangent vector not as an arrow or as a column of numbers, but as a derivation on functions. This is followed by an exposition of Hermann Grassmann's formalism of alternating multilinear functions on a vector space, which lays the foundation for the theory of differential forms. Finally, we introduce differential forms on $\mathbb{R}^n$, together with two of their basic operations, the wedge product and the exterior derivative, and show how they generalize and simplify vector calculus in $\mathbb{R}^3$.

§1 Smooth Functions on a Euclidean Space

The calculus of C^∞ functions will be our primary tool for studying higher-dimensional manifolds. For this reason, we begin with a review of C^∞ functions on $\mathbb{R}^n$.

L. W. Tu, *An Introduction to Manifolds,* Universitext, DOI 10.1007/978-1-4419-7400-6_1,
© Springer Science+Business Media, LLC 2011

1.1 C^∞ Versus Analytic Functions

Write the coordinates on $\mathbb{R}^n$ as $x^1,\ldots,x^n$ and let $p = (p^1,\ldots,p^n)$ be a point in an open set U in $\mathbb{R}^n$. In keeping with the conventions of differential geometry, the indices on coordinates are *superscripts*, not subscripts. An explanation of the rules for superscripts and subscripts is given in Subsection 4.7.

Definition 1.1. Let k be a nonnegative integer. A real-valued function $f: U \to \mathbb{R}$ is said to be C^k *at* $p \in U$ if its partial derivatives

$$\frac{\partial^j f}{\partial x^{i_1} \cdots \partial x^{i_j}}$$

of all orders $j \le k$ exist and are continuous at p. The function $f: U \to \mathbb{R}$ is C^∞ *at* p if it is C^k for all $k \ge 0$; in other words, its partial derivatives $\partial^j f/\partial x^{i_1} \cdots \partial x^{i_j}$ of all orders exist and are continuous at p. A vector-valued function $f: U \to \mathbb{R}^m$ is said to be C^k *at* p if all of its component functions $f^1,\ldots,f^m$ are C^k at p. We say that $f: U \to \mathbb{R}^m$ is C^k *on* U if it is C^k at every point in U. A similar definition holds for a C^∞ function on an open set U. We treat the terms "C^∞" and "smooth" as synonymous.

Example 1.2.
  (i) A C^0 function on U is a continuous function on U.
 (ii) Let $f: \mathbb{R} \to \mathbb{R}$ be $f(x) = x^{1/3}$. Then

$$f'(x) = \begin{cases} \frac{1}{3}x^{-2/3} & \text{for } x \ne 0, \\ \text{undefined} & \text{for } x = 0. \end{cases}$$

  Thus the function f is C^0 but not C^1 at $x = 0$.
(iii) Let $g: \mathbb{R} \to \mathbb{R}$ be defined by

$$g(x) = \int_0^x f(t)\,dt = \int_0^x t^{1/3}\,dt = \frac{3}{4}x^{4/3}.$$

  Then $g'(x) = f(x) = x^{1/3}$, so $g(x)$ is C^1 but not C^2 at $x = 0$. In the same way one can construct a function that is C^k but not C^{k+1} at a given point.
(iv) The polynomial, sine, cosine, and exponential functions on the real line are all C^∞.

A *neighborhood* of a point in $\mathbb{R}^n$ is an open set containing the point. The function f is *real-analytic* at p if in some neighborhood of p it is equal to its Taylor series at p:

$$f(x) = f(p) + \sum_i \frac{\partial f}{\partial x^i}(p)(x^i - p^i) + \frac{1}{2!}\sum_{i,j} \frac{\partial^2 f}{\partial x^i \partial x^j}(p)(x^i - p^i)(x^j - p^j)$$

$$+ \cdots + \frac{1}{k!}\sum_{i_1,\ldots,i_k} \frac{\partial^k f}{\partial x^{i_1} \cdots \partial x^{i_k}}(p)(x^{i_1} - p^{i_1}) \cdots (x^{i_k} - p^{i_k}) + \cdots,$$

in which the general term is summed over all $1 \leq i_1, \ldots, i_k \leq n$.

A real-analytic function is necessarily C^∞, because as one learns in real analysis, a convergent power series can be differentiated term by term in its region of convergence. For example, if

$$f(x) = \sin x = x - \frac{1}{3!}x^3 + \frac{1}{5!}x^5 - \cdots,$$

then term-by-term differentiation gives

$$f'(x) = \cos x = 1 - \frac{1}{2!}x^2 + \frac{1}{4!}x^4 - \cdots.$$

The following example shows that a C^∞ function need not be real-analytic. The idea is to construct a C^∞ function $f(x)$ on $\mathbb{R}$ whose graph, though not horizontal, is "very flat" near 0 in the sense that all of its derivatives vanish at 0.

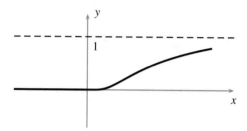

Fig. 1.1. A C^∞ function all of whose derivatives vanish at 0.

Example 1.3 (*A C^∞ function very flat at* 0). Define $f(x)$ on $\mathbb{R}$ by

$$f(x) = \begin{cases} e^{-1/x} & \text{for } x > 0, \\ 0 & \text{for } x \leq 0. \end{cases}$$

(See Figure 1.1.) By induction, one can show that f is C^∞ on $\mathbb{R}$ and that the derivatives $f^{(k)}(0)$ are equal to 0 for all $k \geq 0$ (Problem 1.2).

The Taylor series of this function at the origin is identically zero in any neighborhood of the origin, since all derivatives $f^{(k)}(0)$ equal 0. Therefore, $f(x)$ cannot be equal to its Taylor series and $f(x)$ is not real-analytic at 0.

1.2 Taylor's Theorem with Remainder

Although a C^∞ function need not be equal to its Taylor series, there is a Taylor's theorem with remainder for C^∞ functions that is often good enough for our purposes. In the lemma below, we prove the very first case, in which the Taylor series consists of only the constant term $f(p)$.

We say that a subset S of $\mathbb{R}^n$ is *star-shaped* with respect to a point p in S if for every x in S, the line segment from p to x lies in S (Figure 1.2).

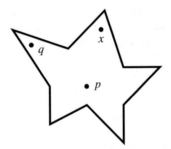

Fig. 1.2. Star-shaped with respect to p, but not with respect to q.

Lemma 1.4 (Taylor's theorem with remainder). *Let f be a C^∞ function on an open subset U of $\mathbb{R}^n$ star-shaped with respect to a point $p = (p^1, \ldots, p^n)$ in U. Then there are functions $g_1(x), \ldots, g_n(x) \in C^\infty(U)$ such that*

$$f(x) = f(p) + \sum_{i=1}^{n} (x^i - p^i) g_i(x), \quad g_i(p) = \frac{\partial f}{\partial x^i}(p).$$

Proof. Since U is star-shaped with respect to p, for any x in U the line segment $p + t(x - p)$, $0 \le t \le 1$, lies in U (Figure 1.3). So $f(p + t(x - p))$ is defined for $0 \le t \le 1$.

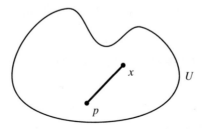

Fig. 1.3. The line segment from p to x.

By the chain rule,

$$\frac{d}{dt} f(p + t(x - p)) = \sum (x^i - p^i) \frac{\partial f}{\partial x^i}(p + t(x - p)).$$

If we integrate both sides with respect to t from 0 to 1, we get

$$f(p + t(x - p))]_0^1 = \sum (x^i - p^i) \int_0^1 \frac{\partial f}{\partial x^i}(p + t(x - p)) \, dt. \tag{1.1}$$

Let

$$g_i(x) = \int_0^1 \frac{\partial f}{\partial x^i}(p + t(x - p)) \, dt.$$

Then $g_i(x)$ is C^∞ and (1.1) becomes

$$f(x) - f(p) = \sum (x^i - p^i) g_i(x).$$

Moreover,

$$g_i(p) = \int_0^1 \frac{\partial f}{\partial x^i}(p) dt = \frac{\partial f}{\partial x^i}(p). \qquad \square$$

In case $n = 1$ and $p = 0$, this lemma says that

$$f(x) = f(0) + x g_1(x)$$

for some C^∞ function $g_1(x)$. Applying the lemma repeatedly gives

$$g_i(x) = g_i(0) + x g_{i+1}(x),$$

where g_i, g_{i+1} are C^∞ functions. Hence,

$$\begin{aligned}
f(x) &= f(0) + x(g_1(0) + x g_2(x)) \\
&= f(0) + x g_1(0) + x^2(g_2(0) + x g_3(x)) \\
&\qquad \vdots \\
&= f(0) + g_1(0)x + g_2(0)x^2 + \cdots + g_i(0)x^i + g_{i+1}(x)x^{i+1}.
\end{aligned} \qquad (1.2)$$

Differentiating (1.2) repeatedly and evaluating at 0, we get

$$g_k(0) = \frac{1}{k!} f^{(k)}(0), \quad k = 1, 2, \ldots, i.$$

So (1.2) is a polynomial expansion of $f(x)$ whose terms up to the last term agree with the Taylor series of $f(x)$ at 0.

Remark. Being star-shaped is not such a restrictive condition, since any open ball

$$B(p, \varepsilon) = \{x \in \mathbb{R}^n \mid \|x - p\| < \varepsilon\}$$

is star-shaped with respect to p. If f is a C^∞ function defined on an open set U containing p, then there is an $\varepsilon > 0$ such that

$$p \in B(p, \varepsilon) \subset U.$$

When its domain is restricted to $B(p, \varepsilon)$, the function f is defined on a star-shaped neighborhood of p and Taylor's theorem with remainder applies.

NOTATION. It is customary to write the standard coordinates on $\mathbb{R}^2$ as x, y, and the standard coordinates on $\mathbb{R}^3$ as x, y, z.

Problems

1.1. A function that is C^2 but not C^3
Let $g\colon \mathbb{R} \to \mathbb{R}$ be the function in Example 1.2(iii). Show that the function $h(x) = \int_0^x g(t)\,dt$ is C^2 but not C^3 at $x = 0$.

1.2.* A C^∞ function very flat at 0
Let $f(x)$ be the function on $\mathbb{R}$ defined in Example 1.3.

(a) Show by induction that for $x > 0$ and $k \geq 0$, the kth derivative $f^{(k)}(x)$ is of the form $p_{2k}(1/x)e^{-1/x}$ for some polynomial $p_{2k}(y)$ of degree $2k$ in y.
(b) Prove that f is C^∞ on $\mathbb{R}$ and that $f^{(k)}(0) = 0$ for all $k \geq 0$.

1.3. A diffeomorphism of an open interval with $\mathbb{R}$
Let $U \subset \mathbb{R}^n$ and $V \subset \mathbb{R}^n$ be open subsets. A C^∞ map $F\colon U \to V$ is called a *diffeomorphism* if it is bijective and has a C^∞ inverse $F^{-1}\colon V \to U$.

(a) Show that the function $f\colon\,]-\pi/2, \pi/2[\,\to \mathbb{R}$, $f(x) = \tan x$, is a diffeomorphism.
(b) Let a, b be real numbers with $a < b$. Find a linear function $h\colon\,]a,b[\,\to\,]-1,1[$, thus proving that any two finite open intervals are diffeomorphic.

The composite $f \circ h\colon\,]a,b[\,\to \mathbb{R}$ is then a diffeomorphism of an open interval with $\mathbb{R}$.

(c) The exponential function $\exp\colon \mathbb{R} \to\,]0,\infty[$ is a diffeomorphism. Use it to show that for any real numbers a and b, the intervals $\mathbb{R}$, $]a,\infty[$, and $]-\infty,b[$ are diffeomorphic.

1.4. A diffeomorphism of an open cube with $\mathbb{R}^n$
Show that the map

$$f\colon\, \left]-\frac{\pi}{2}, \frac{\pi}{2}\right[^n \to \mathbb{R}^n, \quad f(x_1,\dots,x_n) = (\tan x_1,\dots,\tan x_n),$$

is a diffeomorphism.

1.5. A diffeomorphism of an open ball with $\mathbb{R}^n$
Let $\mathbf{0} = (0,0)$ be the origin and $B(\mathbf{0},1)$ the open unit disk in $\mathbb{R}^2$. To find a diffeomorphism between $B(\mathbf{0},1)$ and $\mathbb{R}^2$, we identify $\mathbb{R}^2$ with the xy-plane in $\mathbb{R}^3$ and introduce the lower open hemisphere

$$S\colon x^2 + y^2 + (z-1)^2 = 1, \quad z < 1,$$

in $\mathbb{R}^3$ as an intermediate space (Figure 1.4). First note that the map

$$f\colon B(\mathbf{0},1) \to S, \qquad (a,b) \mapsto (a,b,1-\sqrt{1-a^2-b^2}),$$

is a bijection.

(a) The *stereographic projection* $g\colon S \to \mathbb{R}^2$ from $(0,0,1)$ is the map that sends a point $(a,b,c) \in S$ to the intersection of the line through $(0,0,1)$ and (a,b,c) with the xy-plane. Show that it is given by

$$(a,b,c) \mapsto (u,v) = \left(\frac{a}{1-c}, \frac{b}{1-c}\right), \quad c = 1 - \sqrt{1-a^2-b^2},$$

with inverse

$$(u,v) \mapsto \left(\frac{u}{\sqrt{1+u^2+v^2}}, \frac{v}{\sqrt{1+u^2+v^2}}, 1 - \frac{1}{\sqrt{1+u^2+v^2}}\right).$$

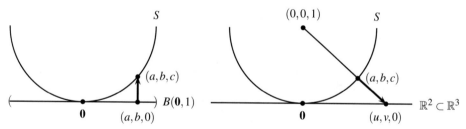

Fig. 1.4. A diffeomorphism of an open disk with $\mathbb{R}^2$.

(b) Composing the two maps f and g gives the map

$$h = g \circ f \colon B(\mathbf{0},1) \to \mathbb{R}^2, \qquad h(a,b) = \left(\frac{a}{\sqrt{1-a^2-b^2}}, \frac{b}{\sqrt{1-a^2-b^2}} \right).$$

Find a formula for $h^{-1}(u,v) = (f^{-1} \circ g^{-1})(u,v)$ and conclude that h is a diffeomorphism of the open disk $B(\mathbf{0},1)$ with $\mathbb{R}^2$.

(c) Generalize part (b) to $\mathbb{R}^n$.

1.6.* Taylor's theorem with remainder to order 2

Prove that if $f \colon \mathbb{R}^2 \to \mathbb{R}$ is C^∞, then there exist C^∞ functions g_{11}, g_{12}, g_{22} on $\mathbb{R}^2$ such that

$$f(x,y) = f(0,0) + \frac{\partial f}{\partial x}(0,0)x + \frac{\partial f}{\partial y}(0,0)y$$
$$+ x^2 g_{11}(x,y) + xy g_{12}(x,y) + y^2 g_{22}(x,y).$$

1.7.* A function with a removable singularity

Let $f \colon \mathbb{R}^2 \to \mathbb{R}$ be a C^∞ function with $f(0,0) = \partial f/\partial x(0,0) = \partial f/\partial y(0,0) = 0$. Define

$$g(t,u) = \begin{cases} \dfrac{f(t,tu)}{t} & \text{for } t \neq 0, \\ 0 & \text{for } t = 0. \end{cases}$$

Prove that $g(t,u)$ is C^∞ for $(t,u) \in \mathbb{R}^2$. (*Hint:* Apply Problem 1.6.)

1.8. Bijective C^∞ maps

Define $f \colon \mathbb{R} \to \mathbb{R}$ by $f(x) = x^3$. Show that f is a bijective C^∞ map, but that f^{-1} is not C^∞. (This example shows that a bijective C^∞ map need not have a C^∞ inverse. In complex analysis, the situation is quite different: a bijective holomorphic map $f \colon \mathbb{C} \to \mathbb{C}$ necessarily has a holomorphic inverse.)

§2 Tangent Vectors in $\mathbb{R}^n$ as Derivations

In elementary calculus we normally represent a vector at a point p in $\mathbb{R}^3$ algebraically as a column of numbers

$$v = \begin{bmatrix} v^1 \\ v^2 \\ v^3 \end{bmatrix}$$

or geometrically as an arrow emanating from p (Figure 2.1).

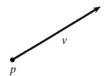

Fig. 2.1. A vector v at p.

Recall that a secant plane to a surface in $\mathbb{R}^3$ is a plane determined by three points of the surface. As the three points approach a point p on the surface, if the corresponding secant planes approach a limiting position, then the plane that is the limiting position of the secant planes is called the tangent plane to the surface at p. Intuitively, the tangent plane to a surface at p is the plane in $\mathbb{R}^3$ that just "touches" the surface at p. A vector at p is tangent to a surface in $\mathbb{R}^3$ if it lies in the tangent plane at p (Figure 2.2).

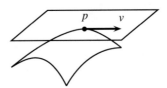

Fig. 2.2. A tangent vector v to a surface at p.

Such a definition of a tangent vector to a surface presupposes that the surface is embedded in a Euclidean space, and so would not apply to the projective plane, for example, which does not sit inside an $\mathbb{R}^n$ in any natural way.

Our goal in this section is to find a characterization of tangent vectors in $\mathbb{R}^n$ that will generalize to manifolds.

2.1 The Directional Derivative

In calculus we visualize the tangent space $T_p(\mathbb{R}^n)$ at p in $\mathbb{R}^n$ as the vector space of all arrows emanating from p. By the correspondence between arrows and column vectors, the vector space $\mathbb{R}^n$ can be identified with this column space. To distinguish between points and vectors, we write a point in $\mathbb{R}^n$ as $p = (p^1, \ldots, p^n)$ and a vector in the tangent space $T_p(\mathbb{R}^n)$ as

$$v = \begin{bmatrix} v^1 \\ \vdots \\ v^n \end{bmatrix} \quad \text{or} \quad \langle v^1, \ldots, v^n \rangle.$$

We usually denote the standard basis for $\mathbb{R}^n$ or $T_p(\mathbb{R}^n)$ by $e_1, \ldots, e_n$. Then $v = \sum v^i e_i$ for some $v^i \in \mathbb{R}$. Elements of $T_p(\mathbb{R}^n)$ are called *tangent vectors* (or simply *vectors*) at p in $\mathbb{R}^n$. We sometimes drop the parentheses and write $T_p\mathbb{R}^n$ for $T_p(\mathbb{R}^n)$.

The line through a point $p = (p^1, \ldots, p^n)$ with direction $v = \langle v^1, \ldots, v^n \rangle$ in $\mathbb{R}^n$ has parametrization

$$c(t) = (p^1 + tv^1, \ldots, p^n + tv^n).$$

Its ith component $c^i(t)$ is $p^i + tv^i$. If f is C^∞ in a neighborhood of p in $\mathbb{R}^n$ and v is a tangent vector at p, the *directional derivative* of f in the direction v at p is defined to be

$$D_v f = \lim_{t \to 0} \frac{f(c(t)) - f(p)}{t} = \frac{d}{dt}\bigg|_{t=0} f(c(t)).$$

By the chain rule,

$$D_v f = \sum_{i=1}^n \frac{dc^i}{dt}(0) \frac{\partial f}{\partial x^i}(p) = \sum_{i=1}^n v^i \frac{\partial f}{\partial x^i}(p). \tag{2.1}$$

In the notation $D_v f$, it is understood that the partial derivatives are to be evaluated at p, since v is a vector at p. So $D_v f$ is a number, not a function. We write

$$D_v = \sum v^i \frac{\partial}{\partial x^i}\bigg|_p$$

for the map that sends a function f to the number $D_v f$. To simplify the notation we often omit the subscript p if it is clear from the context.

The association $v \mapsto D_v$ of the directional derivative D_v to a tangent vector v offers a way to characterize tangent vectors as certain operators on functions. To make this precise, in the next two subsections we study in greater detail the directional derivative D_v as an operator on functions.

2.2 Germs of Functions

A *relation* on a set S is a subset R of $S \times S$. Given x, y in S, we write $x \sim y$ if and only if $(x, y) \in R$. The relation R is an *equivalence relation* if it satisfies the following three properties for all $x, y, z \in S$:

(i) (reflexivity) $x \sim x$,
(ii) (symmetry) if $x \sim y$, then $y \sim x$,

(iii) (transitivity) if $x \sim y$ and $y \sim z$, then $x \sim z$.

As long as two functions agree on some neighborhood of a point p, they will have the same directional derivatives at p. This suggests that we introduce an equivalence relation on the C^∞ functions defined in some neighborhood of p. Consider the set of all pairs (f, U), where U is a neighborhood of p and $f: U \to \mathbb{R}$ is a C^∞ function. We say that (f, U) is *equivalent* to (g, V) if there is an open set $W \subset U \cap V$ containing p such that $f = g$ when restricted to W. This is clearly an equivalence relation because it is reflexive, symmetric, and transitive. The equivalence class of (f, U) is called the *germ* of f at p. We write $C_p^\infty(\mathbb{R}^n)$, or simply C_p^∞ if there is no possibility of confusion, for the set of all germs of C^∞ functions on $\mathbb{R}^n$ at p.

Example. The functions

$$f(x) = \frac{1}{1-x}$$

with domain $\mathbb{R} - \{1\}$ and

$$g(x) = 1 + x + x^2 + x^3 + \cdots$$

with domain the open interval $]-1, 1[$ have the same germ at any point p in the open interval $]-1, 1[$.

An *algebra* over a field K is a vector space A over K with a multiplication map

$$\mu: A \times A \to A,$$

usually written $\mu(a, b) = a \cdot b$, such that for all $a, b, c \in A$ and $r \in K$,

  (i) (associativity) $(a \cdot b) \cdot c = a \cdot (b \cdot c)$,
 (ii) (distributivity) $(a + b) \cdot c = a \cdot c + b \cdot c$ and $a \cdot (b + c) = a \cdot b + a \cdot c$,
(iii) (homogeneity) $r(a \cdot b) = (ra) \cdot b = a \cdot (rb)$.

Equivalently, an algebra over a field K is a ring A (with or without multiplicative identity) that is also a vector space over K such that the ring multiplication satisfies the homogeneity condition (iii). Thus, an algebra has three operations: the addition and multiplication of a ring and the scalar multiplication of a vector space. Usually we omit the multiplication sign and write ab instead of $a \cdot b$.

A map $L: V \to W$ between vector spaces over a field K is called a *linear map* or a *linear operator* if for any $r \in K$ and $u, v \in V$,

 (i) $L(u + v) = L(u) + L(v)$;
(ii) $L(rv) = rL(v)$.

To emphasize the fact that the scalars are in the field K, such a map is also said to be *K-linear*.

If A and A' are algebras over a field K, then an *algebra homomorphism* is a linear map $L: A \to A'$ that preserves the algebra multiplication: $L(ab) = L(a)L(b)$ for all $a, b \in A$.

The addition and multiplication of functions induce corresponding operations on C_p^∞, making it into an algebra over $\mathbb{R}$ (Problem 2.2).

2.3 Derivations at a Point

For each tangent vector v at a point p in $\mathbb{R}^n$, the directional derivative at p gives a map of real vector spaces

$$D_v \colon C_p^\infty \to \mathbb{R}.$$

By (2.1), D_v is $\mathbb{R}$-linear and satisfies the Leibniz rule

$$D_v(fg) = (D_v f)g(p) + f(p)D_v g, \tag{2.2}$$

precisely because the partial derivatives $\partial/\partial x^i|_p$ have these properties.

In general, any linear map $D \colon C_p^\infty \to \mathbb{R}$ satisfying the Leibniz rule (2.2) is called a *derivation at p* or a *point-derivation* of C_p^∞. Denote the set of all derivations at p by $\mathcal{D}_p(\mathbb{R}^n)$. This set is in fact a real vector space, since the sum of two derivations at p and a scalar multiple of a derivation at p are again derivations at p (Problem 2.3).

Thus far, we know that directional derivatives at p are all derivations at p, so there is a map

$$\phi \colon T_p(\mathbb{R}^n) \to \mathcal{D}_p(\mathbb{R}^n), \tag{2.3}$$

$$v \mapsto D_v = \sum v^i \left. \frac{\partial}{\partial x^i} \right|_p.$$

Since D_v is clearly linear in v, the map ϕ is a linear map of vector spaces.

Lemma 2.1. *If D is a point-derivation of C_p^∞, then $D(c) = 0$ for any constant function c.*

Proof. Since we do not know whether every derivation at p is a directional derivative, we need to prove this lemma using only the defining properties of a derivation at p.

By $\mathbb{R}$-linearity, $D(c) = cD(1)$. So it suffices to prove that $D(1) = 0$. By the Leibniz rule (2.2),

$$D(1) = D(1 \cdot 1) = D(1) \cdot 1 + 1 \cdot D(1) = 2D(1).$$

Subtracting $D(1)$ from both sides gives $0 = D(1)$. $\qquad\square$

The *Kronecker delta* δ is a useful notation that we frequently call upon:

$$\delta_j^i = \begin{cases} 1 & \text{if } i = j, \\ 0 & \text{if } i \neq j. \end{cases}$$

Theorem 2.2. *The linear map $\phi \colon T_p(\mathbb{R}^n) \to \mathcal{D}_p(\mathbb{R}^n)$ defined in (2.3) is an isomorphism of vector spaces.*

Proof. To prove injectivity, suppose $D_v = 0$ for $v \in T_p(\mathbb{R}^n)$. Applying D_v to the coordinate function x^j gives

$$0 = D_v(x^j) = \sum_i v^i \left. \frac{\partial}{\partial x^i} \right|_p x^j = \sum_i v^i \delta_i^j = v^j.$$

Hence, $v = 0$ and ϕ is injective.

To prove surjectivity, let D be a derivation at p and let (f, V) be a representative of a germ in C_p^∞. Making V smaller if necessary, we may assume that V is an open ball, hence star-shaped. By Taylor's theorem with remainder (Lemma 1.4) there are C^∞ functions $g_i(x)$ in a neighborhood of p such that

$$f(x) = f(p) + \sum (x^i - p^i) g_i(x), \quad g_i(p) = \frac{\partial f}{\partial x^i}(p).$$

Applying D to both sides and noting that $D(f(p)) = 0$ and $D(p^i) = 0$ by Lemma 2.1, we get by the Leibniz rule (2.2)

$$Df(x) = \sum (Dx^i) g_i(p) + \sum (p^i - p^i) Dg_i(x) = \sum (Dx^i) \frac{\partial f}{\partial x^i}(p).$$

This proves that $D = D_v$ for $v = \langle Dx^1, \ldots, Dx^n \rangle$. $\qquad\square$

This theorem shows that one may identify the tangent vectors at p with the derivations at p. Under the vector space isomorphism $T_p(\mathbb{R}^n) \simeq \mathcal{D}_p(\mathbb{R}^n)$, the standard basis $e_1, \ldots, e_n$ for $T_p(\mathbb{R}^n)$ corresponds to the set $\partial/\partial x^1|_p, \ldots, \partial/\partial x^n|_p$ of partial derivatives. From now on, we will make this identification and write a tangent vector $v = \langle v^1, \ldots, v^n \rangle = \sum v^i e_i$ as

$$v = \sum v^i \left. \frac{\partial}{\partial x^i} \right|_p. \tag{2.4}$$

The vector space $\mathcal{D}_p(\mathbb{R}^n)$ of derivations at p, although not as geometric as arrows, turns out to be more suitable for generalization to manifolds.

2.4 Vector Fields

A *vector field* X on an open subset U of $\mathbb{R}^n$ is a function that assigns to each point p in U a tangent vector X_p in $T_p(\mathbb{R}^n)$. Since $T_p(\mathbb{R}^n)$ has basis $\{\partial/\partial x^i|_p\}$, the vector X_p is a linear combination

$$X_p = \sum a^i(p) \left. \frac{\partial}{\partial x^i} \right|_p, \quad p \in U, \quad a^i(p) \in \mathbb{R}.$$

Omitting p, we may write $X = \sum a^i \partial/\partial x^i$, where the a^i are now functions on U. We say that the vector field X is C^∞ on U if the coefficient functions a^i are all C^∞ on U.

Example 2.3. On $\mathbb{R}^2 - \{\mathbf{0}\}$, let $p = (x, y)$. Then

$$X = \frac{-y}{\sqrt{x^2 + y^2}} \frac{\partial}{\partial x} + \frac{x}{\sqrt{x^2 + y^2}} \frac{\partial}{\partial y} = \left\langle \frac{-y}{\sqrt{x^2 + y^2}}, \frac{x}{\sqrt{x^2 + y^2}} \right\rangle$$

is the vector field in Figure 2.3(a). As is customary, we draw a vector at p as an arrow emanating from p. The vector field $Y = x \partial/\partial x - y \partial/\partial y = \langle x, -y \rangle$, suitably rescaled, is sketched in Figure 2.3(b).

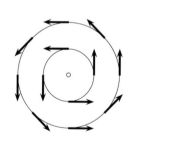

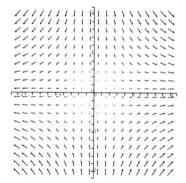

(a) The vector field X on $\mathbb{R}^2 - \{\mathbf{0}\}$ \qquad\qquad (b) The vector field $\langle x, -y \rangle$ on $\mathbb{R}^2$

Fig. 2.3. Vector fields on open subsets of $\mathbb{R}^2$.

One can identify vector fields on U with column vectors of C^∞ functions on U:

$$X = \sum a^i \frac{\partial}{\partial x^i} \quad \longleftrightarrow \quad \begin{bmatrix} a^1 \\ \vdots \\ a^n \end{bmatrix}.$$

This is the same identification as (2.4), but now we are allowing the point p to move in U.

The ring of C^∞ functions on an open set U is commonly denoted by $C^\infty(U)$ or $\mathcal{F}(U)$. Multiplication of vector fields by functions on U is defined pointwise:

$$(fX)_p = f(p)X_p, \quad p \in U.$$

Clearly, if $X = \sum a^i \partial/\partial x^i$ is a C^∞ vector field and f is a C^∞ function on U, then $fX = \sum (fa^i)\partial/\partial x^i$ is a C^∞ vector field on U. Thus, the set of all C^∞ vector fields on U, denoted by $\mathfrak{X}(U)$, is not only a vector space over $\mathbb{R}$, but also a *module* over the ring $C^\infty(U)$. We recall the definition of a module.

Definition 2.4. If R is a commutative ring with identity, then a (left) R-*module* is an abelian group A with a scalar multiplication map

$$\mu : R \times A \to A,$$

usually written $\mu(r,a) = ra$, such that for all $r, s \in R$ and $a, b \in A$,

  (i) (associativity) $(rs)a = r(sa)$,
 (ii) (identity) if 1 is the multiplicative identity in R, then $1a = a$,
(iii) (distributivity) $(r+s)a = ra + sa$, $r(a+b) = ra + rb$.

If R is a field, then an R-module is precisely a vector space over R. In this sense, a module generalizes a vector space by allowing scalars in a ring rather than a field.

Definition 2.5. Let A and A' be R-modules. An *R-module homomorphism* from A to A' is a map $f\colon A \to A'$ that preserves both addition and scalar multiplication: for all $a, b \in A$ and $r \in R$,

(i) $f(a+b) = f(a) + f(b)$,
(ii) $f(ra) = rf(a)$.

2.5 Vector Fields as Derivations

If X is a C^∞ vector field on an open subset U of $\mathbb{R}^n$ and f is a C^∞ function on U, we define a new function Xf on U by

$$(Xf)(p) = X_p f \quad \text{for any } p \in U.$$

Writing $X = \sum a^i \, \partial/\partial x^i$, we get

$$(Xf)(p) = \sum a^i(p) \frac{\partial f}{\partial x^i}(p),$$

or

$$Xf = \sum a^i \frac{\partial f}{\partial x^i},$$

which shows that Xf is a C^∞ function on U. Thus, a C^∞ vector field X gives rise to an $\mathbb{R}$-linear map

$$C^\infty(U) \to C^\infty(U),$$
$$f \mapsto Xf.$$

Proposition 2.6 (Leibniz rule for a vector field). *If X is a C^∞ vector field and f and g are C^∞ functions on an open subset U of $\mathbb{R}^n$, then $X(fg)$ satisfies the product rule (Leibniz rule):*

$$X(fg) = (Xf)g + fXg.$$

Proof. At each point $p \in U$, the vector X_p satisfies the Leibniz rule:

$$X_p(fg) = (X_p f)g(p) + f(p)X_p g.$$

As p varies over U, this becomes an equality of functions:

$$X(fg) = (Xf)g + fXg. \qquad \square$$

If A is an algebra over a field K, a *derivation* of A is a K-linear map $D\colon A \to A$ such that

$$D(ab) = (Da)b + aDb \quad \text{for all } a, b \in A.$$

The set of all derivations of A is closed under addition and scalar multiplication and forms a vector space, denoted by $\mathrm{Der}(A)$. As noted above, a C^∞ vector field on an open set U gives rise to a derivation of the algebra $C^\infty(U)$. We therefore have a map

$$\varphi\colon \mathfrak{X}(U) \to \mathrm{Der}(C^\infty(U)),$$
$$X \mapsto (f \mapsto Xf).$$

Just as the tangent vectors at a point p can be identified with the point-derivations of C_p^∞, so the vector fields on an open set U can be identified with the derivations of the algebra $C^\infty(U)$; i.e., the map φ is an isomorphism of vector spaces. The injectivity of φ is easy to establish, but the surjectivity of φ takes some work (see Problem 19.12).

Note that a derivation at p is not a derivation of the algebra C_p^∞. A derivation at p is a map from C_p^∞ to $\mathbb{R}$, while a derivation of the algebra C_p^∞ is a map from C_p^∞ to C_p^∞.

Problems

2.1. Vector fields
Let X be the vector field $x\,\partial/\partial x + y\,\partial/\partial y$ and $f(x,y,z)$ the function $x^2 + y^2 + z^2$ on $\mathbb{R}^3$. Compute Xf.

2.2. Algebra structure on C_p^∞
Define carefully addition, multiplication, and scalar multiplication in C_p^∞. Prove that addition in C_p^∞ is commutative.

2.3. Vector space structure on derivations at a point
Let D and D' be derivations at p in $\mathbb{R}^n$, and $c \in \mathbb{R}$. Prove that

(a) the sum $D + D'$ is a derivation at p.
(b) the scalar multiple cD is a derivation at p.

2.4. Product of derivations
Let A be an algebra over a field K. If D_1 and D_2 are derivations of A, show that $D_1 \circ D_2$ is not necessarily a derivation (it is if D_1 or $D_2 = 0$), but $D_1 \circ D_2 - D_2 \circ D_1$ is always a derivation of A.

§3 The Exterior Algebra of Multicovectors

As noted in the introduction, manifolds are higher-dimensional analogues of curves and surfaces. As such, they are usually not linear spaces. Nonetheless, a basic principle in manifold theory is the linearization principle, according to which every manifold can be locally approximated by its tangent space at a point, a linear object. In this way linear algebra enters into manifold theory.

Instead of working with tangent vectors, it turns out to be more fruitful to adopt the dual point of view and work with linear functions on a tangent space. After all, there is only so much that one can do with tangent vectors, which are essentially arrows, but functions, far more flexible, can be added, multiplied, scalar-multiplied, and composed with other maps. Once one admits linear functions on a tangent space, it is but a small step to consider functions of several arguments linear in each argument. These are the multilinear functions on a vector space. The determinant of a matrix, viewed as a function of the column vectors of the matrix, is an example of a multilinear function. Among the multilinear functions, certain ones such as the determinant and the cross product have an *antisymmetric* or *alternating* property: they change sign if two arguments are switched. The alternating multilinear functions with k arguments on a vector space are called *multicovectors of degree k*, or *k-covectors* for short.

It took the genius of Hermann Grassmann, a nineteenth-century German mathematician, linguist, and high-school teacher, to recognize the importance of multicovectors. He constructed a vast edifice based on multicovectors, now called the *exterior algebra*, that generalizes parts of vector calculus from $\mathbb{R}^3$ to $\mathbb{R}^n$. For example, the wedge product of two multicovectors on an n-dimensional vector space is a generalization of the cross product in $\mathbb{R}^3$ (see Problem 4.6). Grassmann's work was little appreciated in his lifetime. In fact, he was turned down for a university position and his Ph.D. thesis rejected, because the leading mathematicians of his day such as Möbius and Kummer failed to understand his work. It was only at the turn of the twentieth century, in the hands of the great differential geometer Élie Cartan (1869–1951), that Grassmann's

Hermann Grassmann

(1809–1877)

exterior algebra found its just recognition as the algebraic basis of the theory of differential forms. This section is an exposition, using modern terminology, of some of Grassmann's ideas.

3.1 Dual Space

If V and W are real vector spaces, we denote by $\mathrm{Hom}(V,W)$ the vector space of all linear maps $f\colon V \to W$. Define the *dual space* $V^\vee$ of V to be the vector space of all real-valued linear functions on V:

$$V^\vee = \mathrm{Hom}(V,\mathbb{R}).$$

The elements of $V^\vee$ are called *covectors* or *1-covectors* on V.

In the rest of this section, assume V to be a *finite-dimensional* vector space. Let $e_1,\ldots,e_n$ be a basis for V. Then every v in V is uniquely a linear combination $v = \sum v^i e_i$ with $v^i \in \mathbb{R}$. Let $\alpha^i\colon V \to \mathbb{R}$ be the linear function that picks out the ith coordinate, $\alpha^i(v) = v^i$. Note that α^i is characterized by

$$\alpha^i(e_j) = \delta^i_j = \begin{cases} 1 & \text{for } i = j, \\ 0 & \text{for } i \neq j. \end{cases}$$

Proposition 3.1. *The functions $\alpha^1,\ldots,\alpha^n$ form a basis for $V^\vee$.*

Proof. We first prove that $\alpha^1,\ldots,\alpha^n$ span $V^\vee$. If $f \in V^\vee$ and $v = \sum v^i e_i \in V$, then

$$f(v) = \sum v^i f(e_i) = \sum f(e_i)\alpha^i(v).$$

Hence,

$$f = \sum f(e_i)\alpha^i,$$

which shows that $\alpha^1,\ldots,\alpha^n$ span $V^\vee$.

To show linear independence, suppose $\sum c_i \alpha^i = 0$ for some $c_i \in \mathbb{R}$. Applying both sides to the vector e_j gives

$$0 = \sum_i c_i \alpha^i(e_j) = \sum_i c_i \delta^i_j = c_j, \quad j = 1,\ldots,n.$$

Hence, $\alpha^1,\ldots,\alpha^n$ are linearly independent. $\qquad\square$

This basis $\alpha^1,\ldots,\alpha^n$ for $V^\vee$ is said to be *dual* to the basis $e_1,\ldots,e_n$ for V.

Corollary 3.2. *The dual space $V^\vee$ of a finite-dimensional vector space V has the same dimension as V.*

Example 3.3 (*Coordinate functions*). With respect to a basis $e_1,\ldots,e_n$ for a vector space V, every $v \in V$ can be written uniquely as a linear combination $v = \sum b^i(v)e_i$, where $b^i(v) \in \mathbb{R}$. Let $\alpha^1,\ldots,\alpha^n$ be the basis of $V^\vee$ dual to $e_1,\ldots,e_n$. Then

$$\alpha^i(v) = \alpha^i\left(\sum_j b^j(v)e_j\right) = \sum_j b^j(v)\alpha^i(e_j) = \sum_j b^j(v)\delta^i_j = b^i(v).$$

Thus, the dual basis to $e_1,\ldots,e_n$ is precisely the set of coordinate functions $b^1,\ldots,b^n$ with respect to the basis $e_1,\ldots,e_n$.

3.2 Permutations

Fix a positive integer k. A *permutation* of the set $A = \{1,\ldots,k\}$ is a bijection $\sigma: A \to A$. More concretely, σ may be thought of as a reordering of the list $1, 2, \ldots, k$ from its natural increasing order to a new order $\sigma(1), \sigma(2), \ldots, \sigma(k)$. The *cyclic permutation*, $(a_1\ a_2\ \cdots\ a_r)$ where the a_i are distinct, is the permutation σ such that $\sigma(a_1) = a_2$, $\sigma(a_2) = a_3, \ldots, \sigma(a_{r-1}) = (a_r), \sigma(a_r) = a_1$, and σ fixes all the other elements of A. A cyclic permutation $(a_1\ a_2\ \cdots\ a_r)$ is also called a *cycle of length r* or an *r-cycle*. A *transposition* is a 2-cycle, that is, a cycle of the form $(a\ b)$ that interchanges a and b, leaving all other elements of A fixed. Two cycles $(a_1 \cdots a_r)$ and $(b_1 \cdots b_s)$ are said to be *disjoint* if the sets $\{a_1,\ldots,a_r\}$ and $\{b_1,\ldots,b_s\}$ have no elements in common. The *product* $\tau\sigma$ of two permutations τ and σ of A is the composition $\tau \circ \sigma: A \to A$, in that order; first apply σ, then τ.

A simple way to describe a permutation $\sigma: A \to A$ is by its matrix

$$\begin{bmatrix} 1 & 2 & \cdots & k \\ \sigma(1) & \sigma(2) & \cdots & \sigma(k) \end{bmatrix}.$$

Example 3.4. Suppose the permutation $\sigma: \{1,2,3,4,5\} \to \{1,2,3,4,5\}$ maps $1,2,3,4,5$ to $2,4,5,1,3$ in that order. As a matrix,

$$\sigma = \begin{bmatrix} 1\ 2\ 3\ 4\ 5 \\ 2\ 4\ 5\ 1\ 3 \end{bmatrix}. \tag{3.1}$$

To write σ as a product of disjoint cycles, start with any element in $\{1,2,3,4,5\}$, say 1, and apply σ to it repeatedly until we return to the initial element; this gives a cycle: $1 \mapsto 2 \mapsto 4 \to 1$. Next, repeat the procedure beginning with any of the remaining elements, say 3, to get a second cycle: $3 \mapsto 5 \mapsto 3$. Since all elements of $\{1,2,3,4,5\}$ are now accounted for, σ equals $(1\ 2\ 4)(3\ 5)$:

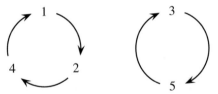

From this example, it is easy to see that any permutation can be written as a product of disjoint cycles $(a_1\ \cdots\ a_r)(b_1\ \cdots\ b_s)\cdots$.

Let S_k be the group of all permutations of the set $\{1,\ldots,k\}$. A permutation is *even* or *odd* depending on whether it is the product of an even or an odd number of transpositions. From the theory of permutations we know that this is a well-defined concept: an even permutation can never be written as the product of an odd number of transpositions and vice versa. The *sign* of a permutation σ, denoted by $\operatorname{sgn}(\sigma)$ or $\operatorname{sgn}\sigma$, is defined to be $+1$ or -1 depending on whether the permutation is even or odd. Clearly, the sign of a permutation satisfies

$$\operatorname{sgn}(\sigma\tau) = \operatorname{sgn}(\sigma)\operatorname{sgn}(\tau) \tag{3.2}$$

for $\sigma, \tau \in S_k$.

Example 3.5. The decomposition

$$(1\ 2\ 3\ 4\ 5) = (1\ 5)(1\ 4)(1\ 3)(1\ 2)$$

shows that the 5-cycle $(1\ 2\ 3\ 4\ 5)$ is an even permutation.

More generally, the decomposition

$$(a_1\ a_2\ \cdots\ a_r) = (a_1\ a_r)(a_1\ a_{r-1}) \cdots (a_1\ a_3)(a_1\ a_2)$$

shows that an r-cycle is an even permutation if and only if r is odd, and an odd permutation if and only if r is even. Thus one way to compute the sign of a permutation is to decompose it into a product of cycles and to count the number of cycles of even length. For example, the permutation $\sigma = (1\ 2\ 4)(3\ 5)$ in Example 3.4 is odd because $(1\ 2\ 4)$ is even and $(3\ 5)$ is odd.

An *inversion* in a permutation σ is an ordered pair $(\sigma(i), \sigma(j))$ such that $i < j$ but $\sigma(i) > \sigma(j)$. To find all the inversions in a permutation σ, it suffices to scan the second row of the matrix of σ from left to right; the inversions are the pairs (a, b) with $a > b$ and a to the left of b. For the permutation σ in Example 3.4, from its matrix (3.1) we can read off its five inversions: $(2, 1), (4, 1), (5, 1), (4, 3)$, and $(5, 3)$.

Exercise 3.6 (Inversions).* Find the inversions in the permutation $\tau = (1\ 2\ 3\ 4\ 5)$ of Example 3.5.

A second way to compute the sign of a permutation is to count the number of inversions, as we illustrate in the following example.

Example 3.7. Let σ be the permutation of Example 3.4. Our goal is to turn σ into the identity permutation $\mathbb{1}$ by multiplying it on the left by transpositions.

(i) To move 1 to its natural position at the beginning of the second row of the matrix of σ, we need to move it across the three elements $2, 4, 5$. This can be accomplished by multiplying σ on the left by three transpositions: first $(5\ 1)$, then $(4\ 1)$, and finally $(2\ 1)$:

$$\sigma = \begin{bmatrix} 1\ 2\ 3\ 4\ 5 \\ 2\ 4\ 5\ 1\ 3 \end{bmatrix} \xrightarrow{(5\ 1)} \begin{bmatrix} 2\ 4\ 1\ 5\ 3 \end{bmatrix} \xrightarrow{(4\ 1)} \begin{bmatrix} 2\ 1\ 4\ 5\ 3 \end{bmatrix} \xrightarrow{(2\ 1)} \begin{bmatrix} 1\ 2\ 4\ 5\ 3 \end{bmatrix}.$$

The three transpositions $(5\ 1), (4\ 1)$, and $(2\ 1)$ correspond precisely to the three inversions of σ ending in 1.

(ii) The element 2 is already in its natural position in the second row of the matrix.

(iii) To move 3 to its natural position in the second row, we need to move it across two elements $4, 5$. This can be accomplished by

$$\begin{bmatrix} 1\ 2\ 3\ 4\ 5 \\ 1\ 2\ 4\ 5\ 3 \end{bmatrix} \xrightarrow{(5\ 3)} \begin{bmatrix} 1\ 2\ 4\ 3\ 5 \end{bmatrix} \xrightarrow{(4\ 3)} \begin{bmatrix} 1\ 2\ 3\ 4\ 5 \end{bmatrix} = \mathbb{1}.$$

Thus,

$$(4\ 3)(5\ 3)(2\ 1)(4\ 1)(5\ 1)\sigma = \mathbb{1}. \tag{3.3}$$

Note that the two transpositions $(5\ 3)$ and $(4\ 3)$ correspond to the two inversions ending in 3. Multiplying both sides of (3.3) on the left by the transpositions $(4\ 3)$, then $(5\ 3)$, then $(2\ 1)$, and so on eventually yields

$$\sigma = (5\ 1)(4\ 1)(2\ 1)(5\ 3)(4\ 3).$$

This shows that σ can be written as a product of as many transpositions as the number of inversions in it.

With this example in mind, we prove the following proposition.

Proposition 3.8. *A permutation is even if and only if it has an even number of inversions.*

Proof. We will obtain the identity permutation $\mathbb{1}$ by multiplying σ on the left by a number of transpositions. This can be achieved in k steps.

(i) First, look for the number 1 among $\sigma(1), \sigma(2), \ldots, \sigma(k)$. Every number preceding 1 in this list gives rise to an inversion, for if $1 = \sigma(i)$, then $(\sigma(1), 1), \ldots,$ $(\sigma(i-1), 1)$ are inversions of σ. Now move 1 to the beginning of the list across the $i-1$ elements $\sigma(1), \ldots, \sigma(i-1)$. This requires multiplying σ on the left by $i-1$ transpositions:

$$\sigma_1 = (\sigma(1)\ 1) \cdots (\sigma(i-1)\ 1)\sigma = \Big[1\ \sigma(1)\ \cdots\ \sigma(i-1)\ \sigma(i+1)\ \cdots\ \sigma(k) \Big].$$

Note that the number of transpositions is the number of inversions ending in 1.

(ii) Next look for the number 2 in the list: $1, \sigma(1), \ldots, \sigma(i-1), \sigma(i+1), \ldots, \sigma(k)$. Every number other than 1 preceding 2 in this list gives rise to an inversion $(\sigma(m), 2)$. Suppose there are i_2 such numbers. Then there are i_2 inversions ending in 2. In moving 2 to its natural position $1, 2, \sigma(1), \sigma(2), \ldots$, we need to move it across i_2 numbers. This can be accomplished by multiplying σ_1 on the left by i_2 transpositions.

Repeating this procedure, we see that for each $j = 1, \ldots, k$, the number of transpositions required to move j to its natural position is the same as the number of inversions ending in j. In the end we achieve the identity permutation, i.e, the ordered list $1, 2, \ldots, k$, from $\sigma(1), \sigma(2), \ldots, \sigma(k)$ by multiplying σ by as many transpositions as the total number of inversions in σ. Therefore, $\operatorname{sgn}(\sigma) = (-1)^{\#\text{ inversions in } \sigma}$. $\qquad\square$

3.3 Multilinear Functions

Denote by $V^k = V \times \cdots \times V$ the Cartesian product of k copies of a real vector space V. A function $f\colon V^k \to \mathbb{R}$ is k-*linear* if it is linear in each of its k arguments:

$$f(\ldots, av + bw, \ldots) = af(\ldots, v, \ldots) + bf(\ldots, w, \ldots)$$

for all $a, b \in \mathbb{R}$ and $v, w \in V$. Instead of 2-linear and 3-linear, it is customary to say "bilinear" and "trilinear." A k-linear function on V is also called a k-*tensor* on V. We will denote the vector space of all k-tensors on V by $L_k(V)$. If f is a k-tensor on V, we also call k the *degree* of f.

Example 3.9 (*Dot product on* $\mathbb{R}^n$). With respect to the standard basis $e_1, \ldots, e_n$ for $\mathbb{R}^n$, the *dot product*, defined by

$$f(v, w) = v \bullet w = \sum_i v^i w^i, \quad \text{where } v = \sum v^i e_i, \ w = \sum w^i e_i,$$

is bilinear.

Example. The determinant $f(v_1, \ldots, v_n) = \det[v_1 \ \cdots \ v_n]$, viewed as a function of the n column vectors $v_1, \ldots, v_n$ in $\mathbb{R}^n$, is n-linear.

Definition 3.10. A k-linear function $f \colon V^k \to \mathbb{R}$ is *symmetric* if

$$f\left(v_{\sigma(1)}, \ldots, v_{\sigma(k)}\right) = f(v_1, \ldots, v_k)$$

for all permutations $\sigma \in S_k$; it is *alternating* if

$$f\left(v_{\sigma(1)}, \ldots, v_{\sigma(k)}\right) = (\operatorname{sgn} \sigma) f(v_1, \ldots, v_k)$$

for all $\sigma \in S_k$.

Examples.
 (i) The dot product $f(v, w) = v \bullet w$ on $\mathbb{R}^n$ is symmetric.
 (ii) The determinant $f(v_1, \ldots, v_n) = \det[v_1 \ \cdots \ v_n]$ on $\mathbb{R}^n$ is alternating.
 (iii) The cross product $v \times w$ on $\mathbb{R}^3$ is alternating.
 (iv) For any two linear functions f, $g \colon V \to \mathbb{R}$ on a vector space V, the function $f \wedge g \colon V \times V \to \mathbb{R}$ defined by

$$(f \wedge g)(u, v) = f(u)g(v) - f(v)g(u)$$

is alternating. This is a special case of the *wedge product*, which we will soon define.

We are especially interested in the space $A_k(V)$ of all alternating k-linear functions on a vector space V for $k > 0$. These are also called *alternating k-tensors*, *k-covectors*, or *multicovectors of degree k* on V. For $k = 0$, we define a *0-covector* to be a constant, so that $A_0(V)$ is the vector space $\mathbb{R}$. A 1-covector is simply a covector.

3.4 The Permutation Action on Multilinear Functions

If f is a k-linear function on a vector space V and σ is a permutation in S_k, we define a new k-linear function σf by

$$(\sigma f)(v_1, \ldots, v_k) = f\left(v_{\sigma(1)}, \ldots, v_{\sigma(k)}\right).$$

Thus, f is symmetric if and only if $\sigma f = f$ for all $\sigma \in S_k$ and f is alternating if and only if $\sigma f = (\operatorname{sgn} \sigma) f$ for all $\sigma \in S_k$.

When there is only one argument, the permutation group S_1 is the identity group and a 1-linear function is both symmetric and alternating. In particular,

$$A_1(V) = L_1(V) = V^{\vee}.$$

Lemma 3.11. *If $\sigma, \tau \in S_k$ and f is a k-linear function on V, then $\tau(\sigma f) = (\tau\sigma)f$.*

Proof. For $v_1, \ldots, v_k \in V$,

$$
\begin{aligned}
\tau(\sigma f)(v_1, \ldots, v_k) &= (\sigma f)\left(v_{\tau(1)}, \ldots, v_{\tau(k)}\right) \\
&= (\sigma f)(w_1, \ldots, w_k) \qquad\qquad (\text{letting } w_i = v_{\tau(i)}) \\
&= f\left(w_{\sigma(1)}, \ldots, w_{\sigma(k)}\right) \\
&= f\left(v_{\tau(\sigma(1))}, \ldots, v_{\tau(\sigma(k))}\right) = f\left(v_{(\tau\sigma)(1)}, \ldots, v_{(\tau\sigma)(k)}\right) \\
&= (\tau\sigma)f(v_1, \ldots, v_k). \qquad\qquad\qquad\qquad\qquad \square
\end{aligned}
$$

In general, if G is a group and X is a set, a map

$$
\begin{aligned}
G \times X &\to X, \\
(\sigma, x) &\mapsto \sigma \cdot x
\end{aligned}
$$

is called a *left action* of G on X if

(i) $e \cdot x = x$, where e is the identity element in G and x is any element in X, and
(ii) $\tau \cdot (\sigma \cdot x) = (\tau\sigma) \cdot x$ for all $\tau, \sigma \in G$ and $x \in X$.

The *orbit* of an element $x \in X$ is defined to be the set $Gx := \{\sigma \cdot x \in X \mid \sigma \in G\}$. In this terminology, we have defined a left action of the permutation group S_k on the space $L_k(V)$ of k-linear functions on V. Note that each permutation acts as a linear function on the vector space $L_k(V)$ since σf is $\mathbb{R}$-linear in f.

A *right action* of G on X is defined similarly; it is a map $X \times G \to X$ such that

(i) $x \cdot e = x$, and
(ii) $(x \cdot \sigma) \cdot \tau = x \cdot (\sigma\tau)$

for all $\sigma, \tau \in G$ and $x \in X$.

Remark. In some books the notation for σf is f^σ. In that notation, $(f^\sigma)^\tau = f^{\tau\sigma}$, not $f^{\sigma\tau}$.

3.5 The Symmetrizing and Alternating Operators

Given any k-linear function f on a vector space V, there is a way to make a symmetric k-linear function Sf from it:

$$
(Sf)(v_1, \ldots, v_k) = \sum_{\sigma \in S_k} f\left(v_{\sigma(1)}, \ldots, v_{\sigma(k)}\right)
$$

or, in our new shorthand,

$$
Sf = \sum_{\sigma \in S_k} \sigma f.
$$

Similarly, there is a way to make an alternating k-linear function from f. Define

$$
Af = \sum_{\sigma \in S_k} (\operatorname{sgn}\sigma)\sigma f.
$$

Proposition 3.12. *If f is a k-linear function on a vector space V, then*

(i) *the k-linear function Sf is symmetric, and*
(ii) *the k-linear function Af is alternating.*

Proof. We prove (ii) only, leaving (i) as an exercise. For $\tau \in S_k$,

$$
\begin{aligned}
\tau(Af) &= \sum_{\sigma \in S_k} (\operatorname{sgn}\sigma)\tau(\sigma f) \\
&= \sum_{\sigma \in S_k} (\operatorname{sgn}\sigma)(\tau\sigma)f && \text{(by Lemma 3.11)} \\
&= (\operatorname{sgn}\tau) \sum_{\sigma \in S_k} (\operatorname{sgn}\tau\sigma)(\tau\sigma)f && \text{(by (3.2))} \\
&= (\operatorname{sgn}\tau)Af,
\end{aligned}
$$

since as σ runs through all permutations in S_k, so does $\tau\sigma$. □

Exercise 3.13 (Symmetrizing operator).* Show that the k-linear function Sf is symmetric.

Lemma 3.14. *If f is an alternating k-linear function on a vector space V, then $Af = (k!)f$.*

Proof. Since for alternating f we have $\sigma f = (\operatorname{sgn}\sigma)f$, and $\operatorname{sgn}\sigma$ is ± 1, we must have

$$
Af = \sum_{\sigma \in S_k} (\operatorname{sgn}\sigma)\sigma f = \sum_{\sigma \in S_k} (\operatorname{sgn}\sigma)(\operatorname{sgn}\sigma)f = (k!)f. \qquad \square
$$

Exercise 3.15 (Alternating operator).* If f is a 3-linear function on a vector space V and $v_1, v_2, v_3 \in V$, what is $(Af)(v_1, v_2, v_3)$?

3.6 The Tensor Product

Let f be a k-linear function and g an ℓ-linear function on a vector space V. Their *tensor product* is the $(k+\ell)$-linear function $f \otimes g$ defined by

$$
(f \otimes g)(v_1, \ldots, v_{k+\ell}) = f(v_1, \ldots, v_k)g(v_{k+1}, \ldots, v_{k+\ell}).
$$

Example 3.16 *(Bilinear maps).* Let $e_1, \ldots, e_n$ be a basis for a vector space V, $\alpha^1, \ldots, \alpha^n$ the dual basis in $V^\vee$, and $\langle\,,\,\rangle \colon V \times V \to \mathbb{R}$ a bilinear map on V. Set $g_{ij} = \langle e_i, e_j \rangle \in \mathbb{R}$. If $v = \sum v^i e_i$ and $w = \sum w^i e_i$, then as we observed in Example 3.3, $v^i = \alpha^i(v)$ and $w^j = \alpha^j(w)$. By bilinearity, we can express $\langle\,,\,\rangle$ in terms of the tensor product:

$$
\langle v, w \rangle = \sum v^i w^j \langle e_i, e_j \rangle = \sum \alpha^i(v)\alpha^j(w)g_{ij} = \sum g_{ij}(\alpha^i \otimes \alpha^j)(v, w).
$$

Hence, $\langle\,,\,\rangle = \sum g_{ij}\alpha^i \otimes \alpha^j$. This notation is often used in differential geometry to describe an inner product on a vector space.

Exercise 3.17 (Associativity of the tensor product). Check that the tensor product of multilinear functions is associative: if f, g, and h are multilinear functions on V, then

$$
(f \otimes g) \otimes h = f \otimes (g \otimes h).
$$

3.7 The Wedge Product

If two multilinear functions f and g on a vector space V are alternating, then we would like to have a product that is alternating as well. This motivates the definition of the *wedge product*, also called the *exterior product*: for $f \in A_k(V)$ and $g \in A_\ell(V)$,

$$f \wedge g = \frac{1}{k!\ell!} A(f \otimes g); \tag{3.4}$$

or explicitly,

$$(f \wedge g)(v_1, \ldots, v_{k+\ell})$$
$$= \frac{1}{k!\ell!} \sum_{\sigma \in S_{k+\ell}} (\operatorname{sgn} \sigma) f\left(v_{\sigma(1)}, \ldots, v_{\sigma(k)}\right) g\left(v_{\sigma(k+1)}, \ldots, v_{\sigma(k+\ell)}\right). \tag{3.5}$$

By Proposition 3.12, $f \wedge g$ is alternating.

When $k = 0$, the element $f \in A_0(V)$ is simply a constant c. In this case, the wedge product $c \wedge g$ is scalar multiplication, since the right-hand side of (3.5) is

$$\frac{1}{\ell!} \sum_{\sigma \in S_\ell} (\operatorname{sgn} \sigma) cg\left(v_{\sigma(1)}, \ldots, v_{\sigma(\ell)}\right) = cg(v_1, \ldots, v_\ell).$$

Thus $c \wedge g = cg$ for $c \in \mathbb{R}$ and $g \in A_\ell(V)$.

The coefficient $1/k!\ell!$ in the definition of the wedge product compensates for repetitions in the sum: for every permutation $\sigma \in S_{k+\ell}$, there are $k!$ permutations τ in S_k that permute the first k arguments $v_{\sigma(1)}, \ldots, v_{\sigma(k)}$ and leave the arguments of g alone; for all τ in S_k, the resulting permutations $\sigma\tau$ in $S_{k+\ell}$ contribute the same term to the sum, since

$$(\operatorname{sgn} \sigma\tau) f\left(v_{\sigma\tau(1)}, \ldots, v_{\sigma\tau(k)}\right) = (\operatorname{sgn} \sigma\tau)(\operatorname{sgn} \tau) f\left(v_{\sigma(1)}, \ldots, v_{\sigma(k)}\right)$$
$$= (\operatorname{sgn} \sigma) f\left(v_{\sigma(1)}, \ldots, v_{\sigma(k)}\right),$$

where the first equality follows from the fact that $(\tau(1), \ldots, \tau(k))$ is a permutation of $(1, \ldots, k)$. So we divide by $k!$ to get rid of the $k!$ repeating terms in the sum coming from permutations of the k arguments of f; similarly, we divide by $\ell!$ on account of the ℓ arguments of g.

Example 3.18. For $f \in A_2(V)$ and $g \in A_1(V)$,

$$A(f \otimes g)(v_1, v_2, v_3) = \quad f(v_1, v_2)g(v_3) - f(v_1, v_3)g(v_2) + f(v_2, v_3)g(v_1)$$
$$- f(v_2, v_1)g(v_3) + f(v_3, v_1)g(v_2) - f(v_3, v_2)g(v_1).$$

Among these six terms, there are three pairs of equal terms, which we have lined up vertically in the display above:

$$f(v_1, v_2)g(v_3) = -f(v_2, v_1)g(v_3), \quad \text{and so on.}$$

Therefore, after dividing by 2,

$$(f \wedge g)(v_1, v_2, v_3) = f(v_1, v_2)g(v_3) - f(v_1, v_3)g(v_2) + f(v_2, v_3)g(v_1).$$

One way to avoid redundancies in the definition of $f \wedge g$ is to stipulate that in the sum (3.5), $\sigma(1), \ldots, \sigma(k)$ be in ascending order and $\sigma(k+1), \ldots, \sigma(k+\ell)$ also be in ascending order. We call a permutation $\sigma \in S_{k+\ell}$ a (k, ℓ)-*shuffle* if

$$\sigma(1) < \cdots < \sigma(k) \quad \text{and} \quad \sigma(k+1) < \cdots < \sigma(k+\ell).$$

By the paragraph before Example 3.18, one may rewrite (3.5) as

$$
\begin{aligned}
(f \wedge g)&(v_1, \ldots, v_{k+\ell}) \\
&= \sum_{\substack{(k,\ell)\text{-shuffles} \\ \sigma}} (\operatorname{sgn} \sigma) f\left(v_{\sigma(1)}, \ldots, v_{\sigma(k)}\right) g\left(v_{\sigma(k+1)}, \ldots, v_{\sigma(k+\ell)}\right). \quad (3.6)
\end{aligned}
$$

Written this way, the definition of $(f \wedge g)(v_1, \ldots, v_{k+\ell})$ is a sum of $\binom{k+\ell}{k}$ terms, instead of $(k+\ell)!$ terms.

Example 3.19 (*Wedge product of two covectors*).* If f and g are covectors on a vector space V and $v_1, v_2 \in V$, then by (3.6),

$$(f \wedge g)(v_1, v_2) = f(v_1)g(v_2) - f(v_2)g(v_1).$$

Exercise 3.20 (Wedge product of two 2-covectors). For $f, g \in A_2(V)$, write out the definition of $f \wedge g$ using $(2,2)$-shuffles.

3.8 Anticommutativity of the Wedge Product

It follows directly from the definition of the wedge product (3.5) that $f \wedge g$ is bilinear in f and in g.

Proposition 3.21. *The wedge product is* anticommutative: *if $f \in A_k(V)$ and $g \in A_\ell(V)$, then*

$$f \wedge g = (-1)^{k\ell} g \wedge f.$$

Proof. Define $\tau \in S_{k+\ell}$ to be the permutation

$$
\tau = \begin{bmatrix} 1 & \cdots & \ell & \ell+1 & \cdots & \ell+k \\ k+1 & \cdots & k+\ell & 1 & \cdots & k \end{bmatrix}.
$$

This means that

$$\tau(1) = k+1, \ldots, \tau(\ell) = k+\ell, \tau(\ell+1) = 1, \ldots, \tau(\ell+k) = k.$$

Then

$$
\begin{aligned}
\sigma(1) &= \sigma\tau(\ell+1), \ldots, \sigma(k) = \sigma\tau(\ell+k), \\
\sigma(k+1) &= \sigma\tau(1), \ldots, \sigma(k+\ell) = \sigma\tau(\ell).
\end{aligned}
$$

For any $v_1, \ldots, v_{k+\ell} \in V$,

$$A(f \otimes g)(v_1, \ldots, v_{k+\ell})$$

$$= \sum_{\sigma \in S_{k+\ell}} (\operatorname{sgn} \sigma) f\left(v_{\sigma(1)}, \ldots, v_{\sigma(k)}\right) g\left(v_{\sigma(k+1)}, \ldots, v_{\sigma(k+\ell)}\right)$$

$$= \sum_{\sigma \in S_{k+\ell}} (\operatorname{sgn} \sigma) f\left(v_{\sigma\tau(\ell+1)}, \ldots, v_{\sigma\tau(\ell+k)}\right) g\left(v_{\sigma\tau(1)}, \ldots, v_{\sigma\tau(\ell)}\right)$$

$$= (\operatorname{sgn} \tau) \sum_{\sigma \in S_{k+\ell}} (\operatorname{sgn} \sigma\tau) g\left(v_{\sigma\tau(1)}, \ldots, v_{\sigma\tau(\ell)}\right) f\left(v_{\sigma\tau(\ell+1)}, \ldots, v_{\sigma\tau(\ell+k)}\right)$$

$$= (\operatorname{sgn} \tau) A(g \otimes f)(v_1, \ldots, v_{k+\ell}).$$

The last equality follows from the fact that as σ runs through all permutations in $S_{k+\ell}$, so does $\sigma\tau$.

We have proven

$$A(f \otimes g) = (\operatorname{sgn} \tau) A(g \otimes f).$$

Dividing by $k! \ell!$ gives

$$f \wedge g = (\operatorname{sgn} \tau) g \wedge f.$$

Exercise 3.22 (Sign of a permutation).* Show that $\operatorname{sgn} \tau = (-1)^{k\ell}$.   □

Corollary 3.23. *If f is a multicovector of odd degree on V, then $f \wedge f = 0$.*

Proof. Let k be the degree of f. By anticommutativity,

$$f \wedge f = (-1)^{k^2} f \wedge f = -f \wedge f,$$

since k is odd. Hence, $2f \wedge f = 0$. Dividing by 2 gives $f \wedge f = 0$.   □

3.9 Associativity of the Wedge Product

The wedge product of a k-covector f and an ℓ-covector g on a vector space V is by definition the $(k+\ell)$-covector

$$f \wedge g = \frac{1}{k!\ell!} A(f \otimes g).$$

To prove the associativity of the wedge product, we will follow Godbillon [14] by first proving a lemma on the alternating operator A.

Lemma 3.24. *Suppose f is a k-linear function and g an ℓ-linear function on a vector space V. Then*

(i) $A(A(f) \otimes g) = k! A(f \otimes g)$, *and*
(ii) $A(f \otimes A(g)) = \ell! A(f \otimes g)$.

Proof. (i) By definition,

$$A(A(f) \otimes g) = \sum_{\sigma \in S_{k+\ell}} (\operatorname{sgn} \sigma) \sigma \left(\sum_{\tau \in S_k} (\operatorname{sgn} \tau)(\tau f) \otimes g \right).$$

We can view $\tau \in S_k$ also as a permutation in $S_{k+\ell}$ fixing $k+1,\ldots,k+\ell$. Viewed this way, τ satisfies

$$(\tau f) \otimes g = \tau(f \otimes g).$$

Hence,

$$A(A(f) \otimes g) = \sum_{\sigma \in S_{k+\ell}} \sum_{\tau \in S_k} (\operatorname{sgn} \sigma)(\operatorname{sgn} \tau)(\sigma \tau)(f \otimes g). \qquad (3.7)$$

For each $\mu \in S_{k+\ell}$ and each $\tau \in S_k$, there is a unique element $\sigma = \mu \tau^{-1} \in S_{k+\ell}$ such that $\mu = \sigma \tau$, so each $\mu \in S_{k+\ell}$ appears once in the double sum (3.7) for each $\tau \in S_k$, and hence $k!$ times in total. So the double sum (3.7) can be rewritten as

$$A(A(f) \otimes g) = k! \sum_{\mu \in S_{k+\ell}} (\operatorname{sgn} \mu) \mu(f \otimes g) = k! A(f \otimes g).$$

The equality in (ii) is proved in the same way. $\qquad \square$

Proposition 3.25 (Associativity of the wedge product). *Let V be a real vector space and f,g,h alternating multilinear functions on V of degrees k,ℓ,m, respectively. Then*

$$(f \wedge g) \wedge h = f \wedge (g \wedge h).$$

Proof. By the definition of the wedge product,

$$
\begin{aligned}
(f \wedge g) \wedge h &= \frac{1}{(k+\ell)! m!} A((f \wedge g) \otimes h) \\
&= \frac{1}{(k+\ell)! m!} \frac{1}{k! \ell!} A(A(f \otimes g) \otimes h) \\
&= \frac{(k+\ell)!}{(k+\ell)! m! k! \ell!} A((f \otimes g) \otimes h) \quad \text{(by Lemma 3.24(i))} \\
&= \frac{1}{k! \ell! m!} A((f \otimes g) \otimes h).
\end{aligned}
$$

Similarly,

$$
\begin{aligned}
f \wedge (g \wedge h) &= \frac{1}{k!(\ell+m)!} A\left(f \otimes \frac{1}{\ell! m!} A(g \otimes h)\right) \\
&= \frac{1}{k! \ell! m!} A(f \otimes (g \otimes h)).
\end{aligned}
$$

Since the tensor product is associative, we conclude that

$$(f \wedge g) \wedge h = f \wedge (g \wedge h). \qquad \square$$

By associativity, we can omit the parentheses in a multiple wedge product such as $(f \wedge g) \wedge h$ and write simply $f \wedge g \wedge h$.

Corollary 3.26. *Under the hypotheses of the proposition,*

$$f \wedge g \wedge h = \frac{1}{k! \ell! m!} A(f \otimes g \otimes h).$$

This corollary easily generalizes to an arbitrary number of factors: if $f_i \in A_{d_i}(V)$, then

$$f_1 \wedge \cdots \wedge f_r = \frac{1}{(d_1)! \cdots (d_r)!} A(f_1 \otimes \cdots \otimes f_r). \tag{3.8}$$

In particular, we have the following proposition. We use the notation $[b^i_j]$ to denote the matrix whose (i, j)-entry is b^i_j.

Proposition 3.27 (Wedge product of 1-covectors). *If* $\alpha^1, \ldots, \alpha^k$ *are linear functions on a vector space V and $v_1, \ldots, v_k \in V$, then*

$$(\alpha^1 \wedge \cdots \wedge \alpha^k)(v_1, \ldots, v_k) = \det[\alpha^i(v_j)].$$

Proof. By (3.8),

$$
\begin{aligned}
(\alpha^1 \wedge \cdots \wedge \alpha^k)(v_1, \ldots, v_k) &= A(\alpha^1 \otimes \cdots \otimes \alpha^k)(v_1, \ldots, v_k) \\
&= \sum_{\sigma \in S_k} (\operatorname{sgn} \sigma) \alpha^1 \left(v_{\sigma(1)} \right) \cdots \alpha^k \left(v_{\sigma(k)} \right) \\
&= \det[\alpha^i(v_j)]. \qquad \square
\end{aligned}
$$

An algebra A over a field K is said to be *graded* if it can be written as a direct sum $A = \bigoplus_{k=0}^{\infty} A^k$ of vector spaces over K such that the multiplication map sends $A^k \times A^\ell$ to $A^{k+\ell}$. The notation $A = \bigoplus_{k=0}^{\infty} A^k$ means that each nonzero element of A is uniquely a *finite* sum

$$a = a_{i_1} + \cdots + a_{i_m},$$

where $a_{i_j} \neq 0 \in A^{i_j}$. A graded algebra $A = \bigoplus_{k=0}^{\infty} A^k$ is said to be *anticommutative* or *graded commutative* if for all $a \in A^k$ and $b \in A^\ell$,

$$ab = (-1)^{k\ell} ba.$$

A *homomorphism of graded algebras* is an algebra homomorphism that preserves the degree.

Example. The polynomial algebra $A = \mathbb{R}[x, y]$ is graded by degree; A^k consists of all homogeneous polynomials of total degree k in the variables x and y.

For a finite-dimensional vector space V, say of dimension n, define

$$A_*(V) = \bigoplus_{k=0}^{\infty} A_k(V) = \bigoplus_{k=0}^{n} A_k(V).$$

With the wedge product of multicovectors as multiplication, $A_*(V)$ becomes an anticommutative graded algebra, called the *exterior algebra* or the *Grassmann algebra* of multicovectors on the vector space V.

3.10 A Basis for k-Covectors

Let $e_1,\ldots,e_n$ be a basis for a real vector space V, and let $\alpha^1,\ldots,\alpha^n$ be the dual basis for $V^\vee$. Introduce the multi-index notation

$$I = (i_1,\ldots,i_k)$$

and write e_I for $(e_{i_1},\ldots,e_{i_k})$ and α^I for $\alpha^{i_1} \wedge \cdots \wedge \alpha^{i_k}$.

A k-linear function f on V is completely determined by its values on all k-tuples $(e_{i_1},\ldots,e_{i_k})$. If f is alternating, then it is completely determined by its values on $(e_{i_1},\ldots,e_{i_k})$ with $1 \le i_1 < \cdots < i_k \le n$; that is, it suffices to consider e_I with I in *strictly ascending* order.

Lemma 3.28. *Let $e_1,\ldots,e_n$ be a basis for a vector space V and let $\alpha^1,\ldots,\alpha^n$ be its dual basis in $V^\vee$. If $I = (1 \le i_1 < \cdots < i_k \le n)$ and $J = (1 \le j_1 < \cdots < j_k \le n)$ are strictly ascending multi-indices of length k, then*

$$\alpha^I(e_J) = \delta^I_J = \begin{cases} 1 & \text{for } I = J, \\ 0 & \text{for } I \ne J. \end{cases}$$

Proof. By Proposition 3.27,

$$\alpha^I(e_J) = \det[\alpha^i(e_j)]_{i \in I, j \in J}.$$

If $I = J$, then $[\alpha^i(e_j)]$ is the identity matrix and its determinant is 1.

If $I \ne J$, we compare them term by term until the terms differ:

$$i_1 = j_1, \quad \ldots, \quad i_{\ell-1} = j_{\ell-1}, \;\; i_\ell \ne j_\ell, \;\; \ldots \;.$$

Without loss of generality, we may assume $i_\ell < j_\ell$. Then i_ℓ will be different from $j_1,\ldots,j_{\ell-1}$ (because these are the same as $i_1,\ldots,i_\ell$, and I is strictly ascending), and i_ℓ will also be different from $j_\ell, j_{\ell+1},\ldots,j_k$ (because J is strictly ascending). Thus, i_ℓ will be different from $j_1,\ldots,j_k$, and the ℓth row of the matrix $[a^i(e_j)]$ will be all zero. Hence, $\det[a^i(e_j)] = 0$. $\qquad\square$

Proposition 3.29. *The alternating k-linear functions α^I, $I = (i_1 < \cdots < i_k)$, form a basis for the space $A_k(V)$ of alternating k-linear functions on V.*

Proof. First, we show linear independence. Suppose $\sum c_I \alpha^I = 0$, $c_I \in \mathbb{R}$, and I runs over all strictly ascending multi-indices of length k. Applying both sides to e_J, $J = (j_1 < \cdots < j_k)$, we get by Lemma 3.28,

$$0 = \sum_I c_I \alpha^I(e_J) = \sum_I c_I \delta^I_J = c_J,$$

since among all strictly ascending multi-indices I of length k, there is only one equal to J. This proves that the α^I are linearly independent.

To show that the α^I span $A_k(V)$, let $f \in A_k(V)$. We claim that

$$f = \sum f(e_I)\alpha^I,$$

where I runs over all strictly ascending multi-indices of length k. Let $g = \sum f(e_I)\alpha^I$. By k-linearity and the alternating property, if two k-covectors agree on all e_J, where $J = (j_1 < \cdots < j_k)$, then they are equal. But

$$g(e_J) = \sum f(e_I)\alpha^I(e_J) = \sum f(e_I)\delta_J^I = f(e_J).$$

Therefore, $f = g = \sum f(e_I)\alpha^I$. □

Corollary 3.30. *If the vector space V has dimension n, then the vector space $A_k(V)$ of k-covectors on V has dimension $\binom{n}{k}$.*

Proof. A strictly ascending multi-index $I = (i_1 < \cdots < i_k)$ is obtained by choosing a subset of k numbers from $1, \ldots, n$. This can be done in $\binom{n}{k}$ ways. □

Corollary 3.31. *If $k > \dim V$, then $A_k(V) = 0$.*

Proof. In $\alpha^{i_1} \wedge \cdots \wedge \alpha^{i_k}$, at least two of the factors must be the same, say $\alpha^j = \alpha^\ell = \alpha$. Because α is a 1-covector, $\alpha \wedge \alpha = 0$ by Corollary 3.23, so $\alpha^{i_1} \wedge \cdots \wedge \alpha^{i_k} = 0$. □

Problems

3.1. Tensor product of covectors
Let $e_1, \ldots, e_n$ be a basis for a vector space V and let $\alpha^1, \ldots, \alpha^n$ be its dual basis in $V^\vee$. Suppose $[g_{ij}] \in \mathbb{R}^{n \times n}$ is an $n \times n$ matrix. Define a bilinear function $f \colon V \times V \to \mathbb{R}$ by

$$f(v, w) = \sum_{1 \le i, j \le n} g_{ij} v^i w^j$$

for $v = \sum v^i e_i$ and $w = \sum w^j e_j$ in V. Describe f in terms of the tensor products of α^i and α^j, $1 \le i, j \le n$.

3.2. Hyperplanes
(a) Let V be a vector space of dimension n and $f \colon V \to \mathbb{R}$ a nonzero linear functional. Show that $\dim \ker f = n - 1$. A linear subspace of V of dimension $n - 1$ is called a *hyperplane* in V.

(b) Show that a nonzero linear functional on a vector space V is determined up to a multiplicative constant by its kernel, a hyperplane in V. In other words, if f and $g \colon V \to \mathbb{R}$ are nonzero linear functionals and $\ker f = \ker g$, then $g = cf$ for some constant $c \in \mathbb{R}$.

3.3. A basis for k-tensors
Let V be a vector space of dimension n with basis $e_1, \ldots, e_n$. Let $\alpha^1, \ldots, \alpha^n$ be the dual basis for $V^\vee$. Show that a basis for the space $L_k(V)$ of k-linear functions on V is $\{\alpha^{i_1} \otimes \cdots \otimes \alpha^{i_k}\}$ for all multi-indices $(i_1, \ldots, i_k)$ (not just the strictly ascending multi-indices as for $A_k(L)$). In particular, this shows that $\dim L_k(V) = n^k$. (This problem generalizes Problem 3.1.)

3.4. A characterization of alternating k-tensors

Let f be a k-tensor on a vector space V. Prove that f is alternating if and only if f changes sign whenever two successive arguments are interchanged:

$$f(\ldots,v_{i+1},v_i,\ldots) = -f(\ldots,v_i,v_{i+1},\ldots)$$

for $i = 1,\ldots,k-1$.

3.5. Another characterization of alternating k-tensors

Let f be a k-tensor on a vector space V. Prove that f is alternating if and only if $f(v_1,\ldots,v_k) = 0$ whenever two of the vectors $v_1,\ldots,v_k$ are equal.

3.6. Wedge product and scalars

Let V be a vector space. For $a,b \in \mathbb{R}$, $f \in A_k(V)$, and $g \in A_\ell(V)$, show that $af \wedge bg = (ab)\, f \wedge g$.

3.7. Transformation rule for a wedge product of covectors

Suppose two sets of covectors on a vector space V, $\beta^1,\ldots,\beta^k$ and $\gamma^1,\ldots,\gamma^k$, are related by

$$\beta^i = \sum_{j=1}^{k} a_j^i \gamma^j, \quad i = 1,\ldots,k,$$

for a $k \times k$ matrix $A = [a_j^i]$. Show that

$$\beta^1 \wedge \cdots \wedge \beta^k = (\det A)\, \gamma^1 \wedge \cdots \wedge \gamma^k.$$

3.8. Transformation rule for k-covectors

Let f be a k-covector on a vector space V. Suppose two sets of vectors $u_1,\ldots,u_k$ and $v_1,\ldots,v_k$ in V are related by

$$u_j = \sum_{i=1}^{k} a_j^i v_i, \quad j = 1,\ldots,k,$$

for a $k \times k$ matrix $A = [a_j^i]$. Show that

$$f(u_1,\ldots,u_k) = (\det A) f(v_1,\ldots,v_k).$$

3.9. Vanishing of a covector of top degree

Let V be a vector space of dimension n. Prove that if an n-covector ω vanishes on a basis $e_1,\ldots,e_n$ for V, then ω is the zero covector on V.

3.10.* Linear independence of covectors

Let $\alpha^1,\ldots,\alpha^k$ be 1-covectors on a vector space V. Show that $\alpha^1 \wedge \cdots \wedge \alpha^k \neq 0$ if and only if $\alpha^1,\ldots,\alpha^k$ are linearly independent in the dual space $V^\vee$.

3.11.* Exterior multiplication

Let α be a nonzero 1-covector and γ a k-covector on a finite-dimensional vector space V. Show that $\alpha \wedge \gamma = 0$ if and only if $\gamma = \alpha \wedge \beta$ for some $(k-1)$-covector β on V.

§4 Differential Forms on $\mathbb{R}^n$

Just as a vector field assigns a tangent vector to each point of an open subset U of $\mathbb{R}^n$, so dually a differential k-form assigns a k-covector on the tangent space to each point of U. The wedge product of differential forms is defined pointwise as the wedge product of multicovectors. Since differential forms exist on an open set, not just at a single point, there is a notion of differentiation for differential forms. In fact, there is a unique one, called the *exterior derivative*, characterized by three natural properties. Although we define it using the standard coordinates of $\mathbb{R}^n$, the exterior derivative turns out to be independent of coordinates, as we shall see later, and is therefore intrinsic to a manifold. It is the ultimate abstract extension to a manifold of the gradient, curl, and divergence of vector calculus in $\mathbb{R}^3$. Differential forms extend Grassmann's exterior algebra from the tangent space at a point globally to an entire manifold. Since its creation around the turn of the twentieth century, generally credited to É. Cartan [5] and H. Poincaré [34], the calculus of differential forms has had far-reaching consequences in geometry, topology, and physics. In fact, certain physical concepts such as electricity and magnetism are best formulated in terms of differential forms.

In this section we will study the simplest case, that of differential forms on an open subset of $\mathbb{R}^n$. Even in this setting, differential forms already provide a way to unify the main theorems of vector calculus in $\mathbb{R}^3$.

4.1 Differential 1-Forms and the Differential of a Function

The *cotangent space* to $\mathbb{R}^n$ at p, denoted by $T_p^*(\mathbb{R}^n)$ or $T_p^*\mathbb{R}^n$, is defined to be the dual space $(T_p\mathbb{R}^n)^\vee$ of the tangent space $T_p(\mathbb{R}^n)$. Thus, an element of the cotangent space $T_p^*(\mathbb{R}^n)$ is a covector or a linear functional on the tangent space $T_p(\mathbb{R}^n)$. In parallel with the definition of a vector field, a *covector field* or a *differential 1-form* on an open subset U of $\mathbb{R}^n$ is a function ω that assigns to each point p in U a covector $\omega_p \in T_p^*(\mathbb{R}^n)$,

$$\omega: U \to \bigcup_{p\in U} T_p^*(\mathbb{R}^n),$$
$$p \mapsto \omega_p \in T_p^*(\mathbb{R}^n).$$

Note that in the union $\bigcup_{p\in U} T_p^*(\mathbb{R}^n)$, the sets $T_p^*(\mathbb{R}^n)$ are all disjoint. We call a differential 1-form a 1-*form* for short.

From any C^∞ function $f: U \to \mathbb{R}$, we can construct a 1-form df, called the *differential* of f, as follows. For $p \in U$ and $X_p \in T_pU$, define

$$(df)_p(X_p) = X_p f.$$

A few words may be in order about the definition of the differential. The directional derivative of a function in the direction of a tangent vector at a point p sets up a bilinear pairing

$$T_p(\mathbb{R}^n) \times C_p^\infty(\mathbb{R}^n) \to \mathbb{R},$$
$$(X_p, f) \mapsto \langle X_p, f \rangle = X_p f.$$

One may think of a tangent vector as a function on the second argument of this pairing: $\langle X_p, \cdot \rangle$. The differential $(df)_p$ at p is a function on the first argument of the pairing:

$$(df)_p = \langle \cdot, f \rangle.$$

The value of the differential df at p is also written $df|_p$.

Let $x^1, \dots, x^n$ be the standard coordinates on $\mathbb{R}^n$. We saw in Subsection 2.3 that the set $\{\partial/\partial x^1|_p, \dots, \partial/\partial x^n|_p\}$ is a basis for the tangent space $T_p(\mathbb{R}^n)$.

Proposition 4.1. *If $x^1, \dots, x^n$ are the standard coordinates on $\mathbb{R}^n$, then at each point $p \in \mathbb{R}^n$, $\{(dx^1)_p, \dots, (dx^n)_p\}$ is the basis for the cotangent space $T_p^*(\mathbb{R}^n)$ dual to the basis $\{\partial/\partial x^1|_p, \dots, \partial/\partial x^n|_p\}$ for the tangent space $T_p(\mathbb{R}^n)$.*

Proof. By definition,

$$(dx^i)_p \left(\left. \frac{\partial}{\partial x^j} \right|_p \right) = \left. \frac{\partial}{\partial x^j} \right|_p x^i = \delta_j^i. \qquad \square$$

If ω is a 1-form on an open subset U of $\mathbb{R}^n$, then by Proposition 4.1, at each point p in U, ω can be written as a linear combination

$$\omega_p = \sum a_i(p)(dx^i)_p,$$

for some $a_i(p) \in \mathbb{R}$. As p varies over U, the coefficients a_i become functions on U, and we may write $\omega = \sum a_i \, dx^i$. The covector field ω is said to be C^∞ *on U* if the coefficient functions a_i are all C^∞ on U.

If x, y, and z are the coordinates on $\mathbb{R}^3$, then dx, dy, and dz are 1-forms on $\mathbb{R}^3$. In this way, we give meaning to what was merely a notation in elementary calculus.

Proposition 4.2 (The differential in terms of coordinates). *If $f : U \to \mathbb{R}$ is a C^∞ function on an open set U in $\mathbb{R}^n$, then*

$$df = \sum \frac{\partial f}{\partial x^i} dx^i. \tag{4.1}$$

Proof. By Proposition 4.1, at each point p in U,

$$(df)_p = \sum a_i(p)(dx^i)_p \tag{4.2}$$

for some real numbers $a_i(p)$ depending on p. Thus, $df = \sum a_i \, dx^i$ for some real functions a_i on U. To find a_j, apply both sides of (4.2) to the coordinate vector field $\partial/\partial x^j$:

$$df \left(\frac{\partial}{\partial x^j} \right) = \sum_i a_i \, dx^i \left(\frac{\partial}{\partial x^j} \right) = \sum_i a_i \delta_j^i = a_j.$$

On the other hand, by the definition of the differential,

$$df \left(\frac{\partial}{\partial x^j} \right) = \frac{\partial f}{\partial x^j}. \qquad \square$$

Equation (4.1) shows that if f is a C^∞ function, then the 1-form df is also C^∞.

Example. Differential 1-forms arise naturally even if one is interested only in tangent vectors. Every tangent vector $X_p \in T_p(\mathbb{R}^n)$ is a linear combination of the standard basis vectors:

$$X_p = \sum_i b^i(X_p) \left. \frac{\partial}{\partial x^i} \right|_p .$$

In Example 3.3 we saw that at each point $p \in \mathbb{R}^n$, we have $b^i(X_p) = (dx^i)_p(X_p)$. Hence, the coefficient b^i of a vector at p with respect to the standard basis $\partial/\partial x^1|_p$, ..., $\partial/\partial x^n|_p$ is none other than the dual covector $dx^i|_p$ on $\mathbb{R}^n$. As p varies, $b^i = dx^i$.

4.2 Differential k-Forms

More generally, a *differential form ω of degree k* or a *k-form* on an open subset U of $\mathbb{R}^n$ is a function that assigns to each point p in U an alternating k-linear function on the tangent space $T_p(\mathbb{R}^n)$, i.e., $\omega_p \in A_k(T_p\mathbb{R}^n)$. Since $A_1(T_p\mathbb{R}^n) = T_p^*(\mathbb{R}^n)$, the definition of a k-form generalizes that of a 1-form in Subsection 4.1.

By Proposition 3.29, a basis for $A_k(T_p\mathbb{R}^n)$ is

$$dx_p^I = dx_p^{i_1} \wedge \cdots \wedge dx_p^{i_k}, \quad 1 \le i_1 < \cdots < i_k \le n.$$

Therefore, at each point p in U, ω_p is a linear combination

$$\omega_p = \sum a_I(p)\, dx_p^I, \quad 1 \le i_1 < \cdots < i_k \le n,$$

and a k-form ω on U is a linear combination

$$\omega = \sum a_I\, dx^I,$$

with function coefficients $a_I : U \to \mathbb{R}$. We say that a k-form ω is C^∞ on U if all the coefficients a_I are C^∞ functions on U.

Denote by $\Omega^k(U)$ the vector space of C^∞ k-forms on U. A 0-form on U assigns to each point p in U an element of $A_0(T_p\mathbb{R}^n) = \mathbb{R}$. Thus, a 0-form on U is simply a function on U, and $\Omega^0(U) = C^\infty(U)$.

There are no nonzero differential forms of degree $> n$ on an open subset of $\mathbb{R}^n$. This is because if $\deg dx^I > n$, then in the expression dx^I at least two of the 1-forms dx^{i_α} must be the same, forcing $dx^I = 0$.

The *wedge product* of a k-form ω and an ℓ-form τ on an open set U is defined pointwise:

$$(\omega \wedge \tau)_p = \omega_p \wedge \tau_p, \quad p \in U.$$

In terms of coordinates, if $\omega = \sum_I a_I\, dx^I$ and $\tau = \sum_J b_J\, dx^J$, then

$$\omega \wedge \tau = \sum_{I,J} (a_I b_J)\, dx^I \wedge dx^J.$$

In this sum, if I and J are not disjoint on the right-hand side, then $dx^I \wedge dx^J = 0$. Hence, the sum is actually over disjoint multi-indices:

$$\omega \wedge \tau = \sum_{I,J \text{ disjoint}} (a_I b_J) \, dx^I \wedge dx^J,$$

which shows that the wedge product of two C^∞ forms is C^∞. So the wedge product is a bilinear map

$$\wedge: \Omega^k(U) \times \Omega^\ell(U) \to \Omega^{k+\ell}(U).$$

By Propositions 3.21 and 3.25, the wedge product of differential forms is anticommutative and associative.

In case one of the factors has degree 0, say $k = 0$, the wedge product

$$\wedge: \Omega^0(U) \times \Omega^\ell(U) \to \Omega^\ell(U)$$

is the pointwise multiplication of a C^∞ ℓ-form by a C^∞ function:

$$(f \wedge \omega)_p = f(p) \wedge \omega_p = f(p)\omega_p,$$

since as we noted in Subsection 3.7, the wedge product with a 0-covector is scalar multiplication. Thus, if $f \in C^\infty(U)$ and $\omega \in \Omega^\ell(U)$, then $f \wedge \omega = f\omega$.

Example. Let x, y, z be the coordinates on $\mathbb{R}^3$. The C^∞ 1-forms on $\mathbb{R}^3$ are

$$f \, dx + g \, dy + h \, dz,$$

where f, g, h range over all C^∞ functions on $\mathbb{R}^3$. The C^∞ 2-forms are

$$f \, dy \wedge dz + g \, dx \wedge dz + h \, dx \wedge dy$$

and the C^∞ 3-forms are

$$f \, dx \wedge dy \wedge dz.$$

Exercise 4.3 (A basis for 3-covectors).* Let x^1, x^2, x^3, x^4 be the coordinates on $\mathbb{R}^4$ and p a point in $\mathbb{R}^4$. Write down a basis for the vector space $A_3(T_p(\mathbb{R}^4))$.

With the wedge product as multiplication and the degree of a form as the grading, the direct sum $\Omega^*(U) = \bigoplus_{k=0}^n \Omega^k(U)$ becomes an anticommutative graded algebra over $\mathbb{R}$. Since one can multiply C^∞ k-forms by C^∞ functions, the set $\Omega^k(U)$ of C^∞ k-forms on U is both a vector space over $\mathbb{R}$ and a module over $C^\infty(U)$, and so the direct sum $\Omega^*(U) = \bigoplus_{k=0}^n \Omega^k(U)$ is also a module over the ring $C^\infty(U)$ of C^∞ functions.

4.3 Differential Forms as Multilinear Functions on Vector Fields

If ω is a C^∞ 1-form and X is a C^∞ vector field on an open set U in $\mathbb{R}^n$, we define a function $\omega(X)$ on U by the formula

$$\omega(X)_p = \omega_p(X_p), \quad p \in U.$$

Written out in coordinates,

$$\omega = \sum a_i \, dx^i, \qquad X = \sum b^j \frac{\partial}{\partial x^j} \qquad \text{for some } a_i, b^j \in C^\infty(U),$$

so

$$\omega(X) = \left(\sum a_i \, dx^i \right) \left(\sum b^j \frac{\partial}{\partial x^j} \right) = \sum a_i b^i,$$

which shows that $\omega(X)$ is C^∞ on U. Thus, a C^∞ 1-form on U gives rise to a map from $\mathfrak{X}(U)$ to $C^\infty(U)$.

This function is actually linear over the ring $C^\infty(U)$; i.e., if $f \in C^\infty(U)$, then $\omega(fX) = f\omega(X)$. To show this, it suffices to evaluate $\omega(fX)$ at an arbitrary point $p \in U$:

$$
\begin{aligned}
(\omega(fX))_p &= \omega_p(f(p)X_p) && \text{(definition of } \omega(fX)) \\
&= f(p)\omega_p(X_p) && (\omega_p \text{ is } \mathbb{R}\text{-linear}) \\
&= (f\omega(X))_p && \text{(definition of } f\omega(X)).
\end{aligned}
$$

Let $\mathcal{F}(U) = C^\infty(U)$. In this notation, a 1-form ω on U gives rise to an $\mathcal{F}(U)$-linear map $\mathfrak{X}(U) \to \mathcal{F}(U)$, $X \mapsto \omega(X)$. Similarly, a k-form ω on U gives rise to a k-linear map over $\mathcal{F}(U)$,

$$\underbrace{\mathfrak{X}(U) \times \cdots \times \mathfrak{X}(U)}_{k \text{ times}} \to \mathcal{F}(U),$$

$$(X_1, \ldots, X_k) \mapsto \omega(X_1, \ldots, X_k).$$

Exercise 4.4 (Wedge product of a 2-form with a 1-form).* Let ω be a 2-form and τ a 1-form on $\mathbb{R}^3$. If X, Y, Z are vector fields on M, find an explicit formula for $(\omega \wedge \tau)(X, Y, Z)$ in terms of the values of ω and τ on the vector fields X, Y, Z.

4.4 The Exterior Derivative

To define the *exterior derivative* of a C^∞ k-form on an open subset U of $\mathbb{R}^n$, we first define it on 0-forms: the exterior derivative of a C^∞ function $f \in C^\infty(U)$ is defined to be its differential $df \in \Omega^1(U)$; in terms of coordinates, Proposition 4.2 gives

$$df = \sum \frac{\partial f}{\partial x^i} \, dx^i.$$

Definition 4.5. For $k \geq 1$, if $\omega = \sum_I a_I \, dx^I \in \Omega^k(U)$, then

$$d\omega = \sum_I da_I \wedge dx^I = \sum_I \left(\sum_j \frac{\partial a_I}{\partial x^j} \, dx^j \right) \wedge dx^I \in \Omega^{k+1}(U).$$

Example. Let ω be the 1-form $f \, dx + g \, dy$ on $\mathbb{R}^2$, where f and g are C^∞ functions on $\mathbb{R}^2$. To simplify the notation, write $f_x = \partial f / \partial x$, $f_y = \partial f / \partial y$. Then

$$d\omega = df \wedge dx + dg \wedge dy$$
$$= (f_x dx + f_y dy) \wedge dx + (g_x dx + g_y dy) \wedge dy$$
$$= (g_x - f_y) dx \wedge dy.$$

In this computation $dy \wedge dx = -dx \wedge dy$ and $dx \wedge dx = dy \wedge dy = 0$ by the anticommutative property of the wedge product (Proposition 3.21 and Corollary 3.23).

Definition 4.6. Let $A = \bigoplus_{k=0}^{\infty} A^k$ be a graded algebra over a field K. An *antiderivation* of the graded algebra A is a K-linear map $D: A \to A$ such that for $a \in A^k$ and $b \in A^\ell$,

$$D(ab) = (Da)b + (-1)^k a Db. \tag{4.3}$$

If there is an integer m such that the antiderivation D sends A^k to A^{k+m} for all k, then we say that it is an antiderivation of *degree m*. By defining $A_k = 0$ for $k < 0$, we can extend the grading of a graded algebra A to negative integers. With this extension, the degree m of an antiderivation can be negative. (An example of an antiderivation of degree -1 is interior multiplication, to be discussed in Subsection 20.4.)

Proposition 4.7.
 (i) *The exterior differentiation* $d: \Omega^*(U) \to \Omega^*(U)$ *is an antiderivation of degree* 1:

$$d(\omega \wedge \tau) = (d\omega) \wedge \tau + (-1)^{\deg \omega} \omega \wedge d\tau.$$

 (ii) $d^2 = 0$.
 (iii) *If* $f \in C^\infty(U)$ *and* $X \in \mathfrak{X}(U)$, *then* $(df)(X) = Xf$.

Proof.
(i) Since both sides of (4.3) are linear in ω and in τ, it suffices to check the equality for $\omega = f\, dx^I$ and $\tau = g\, dx^J$. Then

$$d(\omega \wedge \tau) = d(fg\, dx^I \wedge dx^J)$$
$$= \sum \frac{\partial(fg)}{\partial x^i}\, dx^i \wedge dx^I \wedge dx^J$$
$$= \sum \frac{\partial f}{\partial x^i} g\, dx^i \wedge dx^I \wedge dx^J + \sum f \frac{\partial g}{\partial x^i}\, dx^i \wedge dx^I \wedge dx^J.$$

In the second sum, moving the 1-form $(\partial g / \partial x^i)\, dx^i$ across the k-form dx^I results in the sign $(-1)^k$ by anticommutativity. Hence,

$$d(\omega \wedge \tau) = \sum \frac{\partial f}{\partial x^i}\, dx^i \wedge dx^I \wedge g\, dx^J + (-1)^k \sum f\, dx^I \wedge \frac{\partial g}{\partial x^i}\, dx^i \wedge dx^J$$
$$= d\omega \wedge \tau + (-1)^k \omega \wedge d\tau.$$

(ii) Again by the $\mathbb{R}$-linearity of d, it suffices to show that $d^2\omega = 0$ for $\omega = f\, dx^I$. We compute:

$$d^2(f\,dx^I) = d\left(\sum \frac{\partial f}{\partial x^i}\,dx^i \wedge dx^I\right) = \sum \frac{\partial^2 f}{\partial x^j \partial x^i}\,dx^j \wedge dx^i \wedge dx^I.$$

In this sum if $i = j$, then $dx^j \wedge dx^i = 0$; if $i \neq j$, then $\partial^2 f/\partial x^i \partial x^j$ is symmetric in i and j, but $dx^j \wedge dx^i$ is alternating in i and j, so the terms with $i \neq j$ pair up and cancel each other. For example,

$$\frac{\partial^2 f}{\partial x^1 \partial x^2}\,dx^1 \wedge dx^2 + \frac{\partial^2 f}{\partial x^2 \partial x^1}\,dx^2 \wedge dx^1$$

$$= \frac{\partial^2 f}{\partial x^1 \partial x^2}\,dx^1 \wedge dx^2 + \frac{\partial^2 f}{\partial x^1 \partial x^2}(-dx^1 \wedge dx^2) = 0.$$

Therefore, $d^2(f\,dx^I) = 0$.

(iii) This is simply the definition of the exterior derivative of a function as the differential of the function. $\square$

Proposition 4.8 (Characterization of the exterior derivative). *The three properties of Proposition 4.7 uniquely characterize exterior differentiation on an open set U in $\mathbb{R}^n$; that is, if $D\colon \Omega^*(U) \to \Omega^*(U)$ is* (i) *an antiderivation of degree 1 such that* (ii) $D^2 = 0$ *and* (iii) $(Df)(X) = Xf$ *for $f \in C^\infty(U)$ and $X \in \mathfrak{X}(U)$, then $D = d$.*

Proof. Since every k-form on U is a sum of terms such as $f\,dx^{i_1} \wedge \cdots \wedge dx^{i_k}$, by linearity it suffices to show that $D = d$ on a k-form of this type. By (iii), $Df = df$ on C^∞ functions. It follows that $D\,dx^i = DDx^i = 0$ by (ii). A simple induction on k, using the antiderivation property of D, proves that for all k and all multi-indices I of length k,

$$D(dx^I) = D(dx^{i_1} \wedge \cdots \wedge dx^{i_k}) = 0. \qquad (4.4)$$

Finally, for every k-form $f\,dx^I$,

$$\begin{aligned} D(f\,dx^I) &= (Df) \wedge dx^I + f D(dx^I) &&\text{(by (i))}\\ &= (df) \wedge dx^I &&\text{(by (ii) and (4.4))}\\ &= d(f\,dx^I) &&\text{(definition of } d\text{).} \end{aligned}$$

Hence, $D = d$ on $\Omega^*(U)$. $\square$

4.5 Closed Forms and Exact Forms

A k-form ω on U is *closed* if $d\omega = 0$; it is *exact* if there is a $(k-1)$-form τ such that $\omega = d\tau$ on U. Since $d(d\tau) = 0$, every exact form is closed. In the next section we will discuss the meaning of closed and exact forms in the context of vector calculus on $\mathbb{R}^3$.

Exercise 4.9 (A closed 1-form on the punctured plane). Define a 1-form ω on $\mathbb{R}^2 - \{0\}$ by

$$\omega = \frac{1}{x^2 + y^2}(-y\,dx + x\,dy).$$

Show that ω is closed.

A collection of vector spaces $\{V^k\}_{k=0}^{\infty}$ with linear maps $d_k \colon V^k \to V^{k+1}$ such that $d_{k+1} \circ d_k = 0$ is called a *differential complex* or a *cochain complex*. For any open subset U of $\mathbb{R}^n$, the exterior derivative d makes the vector space $\Omega^*(U)$ of C^∞ forms on U into a cochain complex, called the *de Rham complex* of U:

$$0 \to \Omega^0(U) \overset{d}{\to} \Omega^1(U) \overset{d}{\to} \Omega^2(U) \to \cdots .$$

The closed forms are precisely the elements of the kernel of d, and the exact forms are the elements of the image of d.

4.6 Applications to Vector Calculus

The theory of differential forms unifies many theorems in vector calculus on $\mathbb{R}^3$. We summarize here some results from vector calculus and then show how they fit into the framework of differential forms.

By a *vector-valued function* on an open subset U of $\mathbb{R}^3$, we mean a function $\mathbf{F} = \langle P, Q, R \rangle \colon U \to \mathbb{R}^3$. Such a function assigns to each point $p \in U$ a vector $\mathbf{F}_p \in \mathbb{R}^3 \simeq T_p(\mathbb{R}^3)$. Hence, a vector-valued function on U is precisely a vector field on U. Recall the three operators gradient, curl, and divergence on scalar- and vector-valued functions on U:

$$\{\text{scalar func.}\} \overset{\text{grad}}{\longrightarrow} \{\text{vector func.}\} \overset{\text{curl}}{\longrightarrow} \{\text{vector func.}\} \overset{\text{div}}{\longrightarrow} \{\text{scalar func.}\},$$

$$\operatorname{grad} f = \begin{bmatrix} \partial/\partial x \\ \partial/\partial y \\ \partial/\partial z \end{bmatrix} f = \begin{bmatrix} f_x \\ f_y \\ f_z \end{bmatrix},$$

$$\operatorname{curl} \begin{bmatrix} P \\ Q \\ R \end{bmatrix} = \begin{bmatrix} \partial/\partial x \\ \partial/\partial y \\ \partial/\partial z \end{bmatrix} \times \begin{bmatrix} P \\ Q \\ R \end{bmatrix} = \begin{bmatrix} R_y - Q_z \\ -(R_x - P_z) \\ Q_x - P_y \end{bmatrix},$$

$$\operatorname{div} \begin{bmatrix} P \\ Q \\ R \end{bmatrix} = \begin{bmatrix} \partial/\partial x \\ \partial/\partial y \\ \partial/\partial z \end{bmatrix} \cdot \begin{bmatrix} P \\ Q \\ R \end{bmatrix} = P_x + Q_y + R_z.$$

Since every 1-form on U is a linear combination with function coefficients of dx, dy, and dz, we can identify 1-forms with vector fields on U via

$$P\,dx + Q\,dy + R\,dz \longleftrightarrow \begin{bmatrix} P \\ Q \\ R \end{bmatrix}.$$

Similarly, 2-forms on U can also be identified with vector fields on U:

$$P\,dy \wedge dz + Q\,dz \wedge dx + R\,dx \wedge dy \longleftrightarrow \begin{bmatrix} P \\ Q \\ R \end{bmatrix},$$

and 3-forms on U can be identified with functions on U:

$$f \, dx \wedge dy \wedge dz \longleftrightarrow f.$$

In terms of these identifications, the exterior derivative of a 0-form f is

$$df = \frac{\partial f}{\partial x} dx + \frac{\partial f}{\partial y} dy + \frac{\partial f}{\partial z} dz \longleftrightarrow \begin{bmatrix} \partial f/\partial x \\ \partial f/\partial y \\ \partial f/\partial x \end{bmatrix} = \operatorname{grad} f;$$

the exterior derivative of a 1-form is

$$\begin{aligned} d(P\,dx + Q\,dy + R\,dz) \\ = (R_y - Q_z)\,dy \wedge dz - (R_x - P_z)\,dz \wedge dx + (Q_x - P_y)\,dx \wedge dy, \end{aligned} \tag{4.5}$$

which corresponds to

$$\operatorname{curl} \begin{bmatrix} P \\ Q \\ R \end{bmatrix} = \begin{bmatrix} R_y - Q_z \\ -(R_x - P_z) \\ Q_x - P_y \end{bmatrix};$$

the exterior derivative of a 2-form is

$$\begin{aligned} d(P\,dy \wedge dz + Q\,dz \wedge dx + R\,dx \wedge dy) \\ = (P_x + Q_y + R_z)\,dx \wedge dy \wedge dz, \end{aligned} \tag{4.6}$$

which corresponds to

$$\operatorname{div} \begin{bmatrix} P \\ Q \\ R \end{bmatrix} = P_x + Q_y + R_z.$$

Thus, after appropriate identifications, the exterior derivatives d on 0-forms, 1-forms, and 2-forms are simply the three operators grad, curl, and div. In summary, on an open subset U of $\mathbb{R}^3$, there are identifications

$$\Omega^0(U) \xrightarrow{\ d\ } \Omega^1(U) \xrightarrow{\ d\ } \Omega^2(U) \xrightarrow{\ d\ } \Omega^3(U)$$
$$\simeq\downarrow \qquad\quad \simeq\downarrow \qquad\quad \simeq\downarrow \qquad\quad \simeq\downarrow$$
$$C^\infty(U) \xrightarrow[\text{grad}]{} \mathfrak{X}(U) \xrightarrow[\text{curl}]{} \mathfrak{X}(U) \xrightarrow[\text{div}]{} C^\infty(U).$$

Under these identifications, a vector field $\langle P, Q, R \rangle$ on $\mathbb{R}^3$ is the gradient of a C^∞ function f if and only if the corresponding 1-form $P\,dx + Q\,dy + R\,dz$ is df.

Next we recall three basic facts from calculus concerning grad, curl, and div.

Proposition A. $\operatorname{curl}(\operatorname{grad} f) = \begin{bmatrix} 0 \\ 0 \\ 0 \end{bmatrix}.$

Proposition B. $\mathrm{div}\left(\mathrm{curl}\begin{bmatrix}P\\Q\\R\end{bmatrix}\right)=0.$

Proposition C. *On* $\mathbb{R}^3$, *a vector field* $\mathbf{F}$ *is the gradient of some scalar function* f *if and only if* $\mathrm{curl}\,\mathbf{F}=0.$

Propositions A and B express the property $d^2=0$ of the exterior derivative on open subsets of $\mathbb{R}^3$; these are easy computations. Proposition C expresses the fact that a 1-form on $\mathbb{R}^3$ is exact if and only if it is closed. Proposition C need not be true on a region other than $\mathbb{R}^3$, as the following well-known example from calculus shows.

Example. If $U=\mathbb{R}^3-\{z\text{-axis}\}$, and $\mathbf{F}$ is the vector field

$$\mathbf{F}=\left\langle\frac{-y}{x^2+y^2},\frac{x}{x^2+y^2},0\right\rangle$$

on $\mathbb{R}^3$, then $\mathrm{curl}\,\mathbf{F}=\mathbf{0}$, but $\mathbf{F}$ is not the gradient of any C^∞ function on U. The reason is that if $\mathbf{F}$ were the gradient of a C^∞ function f on U, then by the fundamental theorem for line integrals, the line integral

$$\int_C-\frac{y}{x^2+y^2}\,dx+\frac{x}{x^2+y^2}\,dy$$

over any closed curve C would be zero. However, on the unit circle C in the (x,y)-plane, with $x=\cos t$ and $y=\sin t$ for $0\le t\le 2\pi$, this integral is

$$\int_C-y\,dx+x\,dy=\int_0^{2\pi}-(\sin t)\,d\cos t+(\cos t)\,d\sin t=2\pi.$$

In terms of differential forms, the 1-form

$$\omega=\frac{-y}{x^2+y^2}\,dx+\frac{x}{x^2+y^2}\,dy$$

is closed but not exact on U. (This 1-form is defined by the same formula as the 1-form ω in Exercise 4.9, but is defined on a different space.)

It turns out that whether Proposition C is true for a region U depends only on the topology of U. One measure of the failure of a closed k-form to be exact is the quotient vector space

$$H^k(U):=\frac{\{\text{closed }k\text{-forms on }U\}}{\{\text{exact }k\text{-forms on }U\}},$$

called the kth *de Rham cohomology* of U.

The generalization of Proposition C to any differential form on $\mathbb{R}^n$ is called the *Poincaré lemma*: for $k\ge 1$, every closed k-form on $\mathbb{R}^n$ is exact. This is of course equivalent to the vanishing of the kth de Rham cohomology $H^k(\mathbb{R}^n)$ for $k \geq 1$. We will prove it in Section 27.

The theory of differential forms allows us to generalize vector calculus from $\mathbb{R}^3$ to $\mathbb{R}^n$ and indeed to a manifold of any dimension. The general Stokes theorem for a manifold that we will prove in Subsection 23.5 subsumes and unifies the fundamental theorem for line integrals, Green's theorem in the plane, the classical Stokes theorem for a surface in $\mathbb{R}^3$, and the divergence theorem. As a first step in this program, we begin the next chapter with the definition of a manifold.

4.7 Convention on Subscripts and Superscripts

In differential geometry it is customary to index vector fields with subscripts $e_1, \ldots, e_n$, and differential forms with superscripts $\omega^1, \ldots, \omega^n$. Being 0-forms, coordinate functions take superscripts: $x^1, \ldots, x^n$. Their differentials, being 1-forms, should also have superscripts, and indeed they do: $dx^1, \ldots, dx^n$. Coordinate vector fields $\partial/\partial x^1, \ldots, \partial/\partial x^n$ are considered to have subscripts because the i in $\partial/\partial x^i$, although a superscript for x^i, is in the lower half of the fraction.

Coefficient functions can have superscripts or subscripts depending on whether they are the coefficient functions of a vector field or of a differential form. For a vector field $X = \sum a^i e_i$, the coefficient functions a^i have superscripts; the idea is that the superscript in a^i "cancels out" the subscript in e_i. For the same reason, the coefficient functions b_j in a differential form $\omega = \sum b_j dx^j$ have subscripts.

The beauty of this convention is that there is a "conservation of indices" on the two sides of an equality sign. For example, if $X = \sum a^i \partial/\partial x^i$, then

$$a^i = (dx^i)(X).$$

Here both sides have a net superscript i. As another example, if $\omega = \sum b_j dx^j$, then

$$\omega(X) = \left(\sum b_j dx^j \right) \left(\sum a^i \frac{\partial}{\partial x^i} \right) = \sum b_i a^i;$$

after cancellation of superscripts and subscripts, both sides of the equality sign have zero net index. This convention is a useful mnemonic aid in some of the transformation formulas of differential geometry.

Problems

4.1. A 1-form on $\mathbb{R}^3$

Let ω be the 1-form $z\,dx - dz$ and let X be the vector field $y\partial/\partial x + x\partial/\partial y$ on $\mathbb{R}^3$. Compute $\omega(X)$ and $d\omega$.

4.2. A 2-form on $\mathbb{R}^3$

At each point $p \in \mathbb{R}^3$, define a bilinear function ω_p on $T_p(\mathbb{R}^3)$ by

$$\omega_p(\mathbf{a},\mathbf{b}) = \omega_p\left(\begin{bmatrix} a^1 \\ a^2 \\ a^3 \end{bmatrix}, \begin{bmatrix} b^1 \\ b^2 \\ b^3 \end{bmatrix}\right) = p^3 \det \begin{bmatrix} a^1 & b^1 \\ a^2 & b^2 \end{bmatrix},$$

for tangent vectors $\mathbf{a},\mathbf{b} \in T_p(\mathbb{R}^3)$, where p^3 is the third component of $p = (p^1, p^2, p^3)$. Since ω_p is an alternating bilinear function on $T_p(\mathbb{R}^3)$, ω is a 2-form on $\mathbb{R}^3$. Write ω in terms of the standard basis $dx^i \wedge dx^j$ at each point.

4.3. Exterior calculus
Suppose the standard coordinates on $\mathbb{R}^2$ are called r and θ (this $\mathbb{R}^2$ is the (r,θ)-plane, not the (x,y)-plane). If $x = r\cos\theta$ and $y = r\sin\theta$, calculate dx, dy, and $dx \wedge dy$ in terms of dr and $d\theta$.

4.4. Exterior calculus
Suppose the standard coordinates on $\mathbb{R}^3$ are called ρ, ϕ, and θ. If $x = \rho\sin\phi\cos\theta$, $y = \rho\sin\phi\sin\theta$, and $z = \rho\cos\phi$, calculate dx, dy, dz, and $dx \wedge dy \wedge dz$ in terms of $d\rho$, $d\phi$, and $d\theta$.

4.5. Wedge product
Let α be a 1-form and β a 2-form on $\mathbb{R}^3$. Then

$$\alpha = a_1 \, dx^1 + a_2 \, dx^2 + a_3 \, dx^3,$$
$$\beta = b_1 \, dx^2 \wedge dx^3 + b_2 \, dx^3 \wedge dx^1 + b_3 \, dx^1 \wedge dx^2.$$

Simplify the expression $\alpha \wedge \beta$ as much as possible.

4.6. Wedge product and cross product
The correspondence between differential forms and vector fields on an open subset of $\mathbb{R}^3$ in Subsection 4.6 also makes sense pointwise. Let V be a vector space of dimension 3 with basis e_1, e_2, e_3, and dual basis $\alpha^1, \alpha^2, \alpha^3$. To a 1-covector $\alpha = a_1 \alpha^1 + a_2 \alpha^2 + a_3 \alpha^3$ on V, we associate the vector $\mathbf{v}_\alpha = \langle a_1, a_2, a_3 \rangle \in \mathbb{R}^3$. To the 2-covector

$$\gamma = c_1 \alpha^2 \wedge \alpha^3 + c_2 \alpha^3 \wedge \alpha^1 + c_3 \alpha^1 \wedge \alpha^2$$

on V, we associate the vector $\mathbf{v}_\gamma = \langle c_1, c_2, c_3 \rangle \in \mathbb{R}^3$. Show that under this correspondence, the wedge product of 1-covectors corresponds to the cross product of vectors in $\mathbb{R}^3$: if $\alpha = a_1 \alpha^1 + a_2 \alpha^2 + a_3 \alpha^3$ and $\beta = b_1 \alpha^1 + b_2 \alpha^2 + b_3 \alpha^3$, then $\mathbf{v}_{\alpha \wedge \beta} = \mathbf{v}_\alpha \times \mathbf{v}_\beta$.

4.7. Commutator of derivations and antiderivations
Let $A = \oplus_{k=-\infty}^{\infty} A^k$ be a graded algebra over a field K with $A^k = 0$ for $k < 0$. Let m be an integer. A *superderivation of A of degree m* is a K-linear map $D \colon A \to A$ such that for all k, $D(A^k) \subset A^{k+m}$ and for all $a \in A^k$ and $b \in A^\ell$,

$$D(ab) = (Da)b + (-1)^{km} a(Db).$$

If D_1 and D_2 are two superderivations of A of respective degrees m_1 and m_2, define their *commutator* to be

$$[D_1, D_2] = D_1 \circ D_2 - (-1)^{m_1 m_2} D_2 \circ D_1.$$

Show that $[D_1, D_2]$ is a superderivation of degree $m_1 + m_2$. (A superderivation is said to be *even* or *odd* depending on the parity of its degree. An even superderivation is a derivation; an odd superderivation is an antiderivation.)

Chapter 2

Manifolds

Intuitively, a manifold is a generalization of curves and surfaces to higher dimensions. It is locally Euclidean in that every point has a neighborhood, called a chart, homeomorphic to an open subset of $\mathbb{R}^n$. The coordinates on a chart allow one to carry out computations as though in a Euclidean space, so that many concepts from $\mathbb{R}^n$, such as differentiability, point-derivations, tangent spaces, and differential forms, carry over to a manifold.

Bernhard Riemann

(1826–1866)

Like most fundamental mathematical concepts, the idea of a manifold did not originate with a single person, but is rather the distillation of years of collective activity. In his masterpiece *Disquisitiones generales circa superficies curvas* ("General Investigations of Curved Surfaces") published in 1827, Carl Friedrich Gauss freely used local coordinates on a surface, and so he already had the idea of charts. Moreover, he appeared to be the first to consider a surface as an abstract space existing in its own right, independent of a particular embedding in a Euclidean space. Bernhard Riemann's inaugural lecture *Über die Hypothesen, welche der Geometrie zu Grunde liegen* ("On the hypotheses that underlie geometry") in Göttingen in 1854 laid the foundations of higher-dimensional differential geometry. Indeed, the word "manifold" is a direct translation of the German word "Mannigfaltigkeit," which Riemann used to describe the objects of his inquiry. This was followed by the work of Henri Poincaré in the late nineteenth century on homology, in which locally Euclidean spaces figured prominently. The late nineteenth and early twentieth centuries were also a period of feverish development in point-set topology. It was not until 1931 that one finds the modern definition of a manifold based on point-set topology and a group of transition functions [37].

L. W. Tu, *An Introduction to Manifolds*, Universitext, DOI 10.1007/978-1-4419-7400-6_2,
© Springer Science+Business Media, LLC 2011

In this chapter we give the basic definitions and properties of a smooth manifold and of smooth maps between manifolds. Initially, the only way we have to verify that a space is a manifold is to exhibit a collection of C^∞ compatible charts covering the space. In Section 7 we describe a set of sufficient conditions under which a quotient topological space becomes a manifold, giving us a second way to construct manifolds.

§5 Manifolds

While there are many kinds of manifolds—for example, topological manifolds, C^k-manifolds, analytic manifolds, and complex manifolds—in this book we are concerned mainly with smooth manifolds. Starting with topological manifolds, which are Hausdorff, second countable, locally Euclidean spaces, we introduce the concept of a maximal C^∞ atlas, which makes a topological manifold into a smooth manifold. This is illustrated with a few simple examples.

5.1 Topological Manifolds

We first recall a few definitions from point-set topology. For more details, see Appendix A. A topological space is *second countable* if it has a countable basis. A *neighborhood* of a point p in a topological space M is any open set containing p. An *open cover* of M is a collection $\{U_\alpha\}_{\alpha \in A}$ of open sets in M whose union $\bigcup_{\alpha \in A} U_\alpha$ is M.

Definition 5.1. A topological space M is *locally Euclidean of dimension n* if every point p in M has a neighborhood U such that there is a homeomorphism ϕ from U onto an open subset of $\mathbb{R}^n$. We call the pair $(U, \phi \colon U \to \mathbb{R}^n)$ a *chart*, U a *coordinate neighborhood* or a *coordinate open set*, and ϕ a *coordinate map* or a *coordinate system* on U. We say that a chart (U, ϕ) is *centered* at $p \in U$ if $\phi(p) = 0$.

Definition 5.2. A *topological manifold* is a Hausdorff, second countable, locally Euclidean space. It is said to be of *dimension n* if it is locally Euclidean of dimension n.

For the dimension of a topological manifold to be well defined, we need to know that for $n \neq m$ an open subset of $\mathbb{R}^n$ is not homeomorphic to an open subset of $\mathbb{R}^m$. This fact, called *invariance of dimension*, is indeed true, but is not easy to prove directly. We will not pursue this point, since we are mainly interested in *smooth* manifolds, for which the analogous result is easy to prove (Corollary 8.7). Of course, if a topological manifold has several connected components, it is possible for each component to have a different dimension.

Example. The Euclidean space $\mathbb{R}^n$ is covered by a single chart $(\mathbb{R}^n, \mathbb{1}_{\mathbb{R}^n})$, where $\mathbb{1}_{\mathbb{R}^n} \colon \mathbb{R}^n \to \mathbb{R}^n$ is the identity map. It is the prime example of a topological manifold. Every open subset of $\mathbb{R}^n$ is also a topological manifold, with chart $(U, \mathbb{1}_U)$.

Recall that the Hausdorff condition and second countability are "hereditary properties"; that is, they are inherited by subspaces: a subspace of a Hausdorff space is Hausdorff (Proposition A.19) and a subspace of a second-countable space is second countable (Proposition A.14). So any subspace of $\mathbb{R}^n$ is automatically Hausdorff and second countable.

Example 5.3 (*A cusp*). The graph of $y = x^{2/3}$ in $\mathbb{R}^2$ is a topological manifold (Figure 5.1(a)). By virtue of being a subspace of $\mathbb{R}^2$, it is Hausdorff and second countable. It is locally Euclidean, because it is homeomorphic to $\mathbb{R}$ via $(x, x^{2/3}) \mapsto x$.

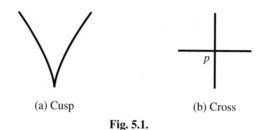

(a) Cusp   (b) Cross

Fig. 5.1.

Example 5.4 (*A cross*). Show that the cross in $\mathbb{R}^2$ in Figure 5.1 with the subspace topology is not locally Euclidean at the intersection p, and so cannot be a topological manifold.

Solution. Suppose the cross is locally Euclidean of dimension n at the point p. Then p has a neighborhood U homeomorphic to an open ball $B := B(0, \varepsilon) \subset \mathbb{R}^n$ with p mapping to 0. The homeomorphism $U \to B$ restricts to a homeomorphism $U - \{p\} \to B - \{0\}$. Now $B - \{0\}$ is either connected if $n \geq 2$ or has two connected components if $n = 1$. Since $U - \{p\}$ has four connected components, there can be no homeomorphism from $U - \{p\}$ to $B - \{0\}$. This contradiction proves that the cross is not locally Euclidean at p. $\qquad\square$

5.2 Compatible Charts

Suppose $(U, \phi \colon U \to \mathbb{R}^n)$ and $(V, \psi \colon V \to \mathbb{R}^n)$ are two charts of a topological manifold. Since $U \cap V$ is open in U and $\phi \colon U \to \mathbb{R}^n$ is a homeomorphism onto an open subset of $\mathbb{R}^n$, the image $\phi(U \cap V)$ will also be an open subset of $\mathbb{R}^n$. Similarly, $\psi(U \cap V)$ is an open subset of $\mathbb{R}^n$.

Definition 5.5.   Two charts $(U, \phi \colon U \to \mathbb{R}^n)$, $(V, \psi \colon V \to \mathbb{R}^n)$ of a topological manifold are C^∞-*compatible* if the two maps

$$\phi \circ \psi^{-1}: \psi(U \cap V) \to \phi(U \cap V), \quad \psi \circ \phi^{-1}: \phi(U \cap V) \to \psi(U \cap V)$$

are C^∞ (Figure 5.2). These two maps are called the *transition functions* between the charts. If $U \cap V$ is empty, then the two charts are automatically C^∞-compatible. To simplify the notation, we will sometimes write $U_{\alpha\beta}$ for $U_\alpha \cap U_\beta$ and $U_{\alpha\beta\gamma}$ for $U_\alpha \cap U_\beta \cap U_\gamma$.

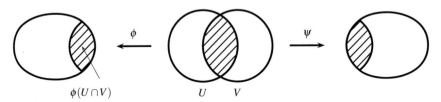

Fig. 5.2. The transition function $\psi \circ \phi^{-1}$ is defined on $\phi(U \cap V)$.

Since we are interested only in C^∞-compatible charts, we often omit mention of "C^∞" and speak simply of compatible charts.

Definition 5.6. A C^∞ *atlas* or simply an *atlas* on a locally Euclidean space M is a collection $\mathfrak{U} = \{(U_\alpha, \phi_\alpha)\}$ of pairwise C^∞-compatible charts that cover M, i.e., such that $M = \bigcup_\alpha U_\alpha$.

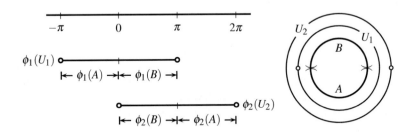

Fig. 5.3. A C^∞ atlas on a circle.

Example 5.7 (*A C^∞ atlas on a circle*). The unit circle S^1 in the complex plane $\mathbb{C}$ may be described as the set of points $\{e^{it} \in \mathbb{C} \mid 0 \le t \le 2\pi\}$. Let U_1 and U_2 be the two open subsets of S^1 (see Figure 5.3)

$$U_1 = \{e^{it} \in \mathbb{C} \mid -\pi < t < \pi\},$$
$$U_2 = \{e^{it} \in \mathbb{C} \mid 0 < t < 2\pi\},$$

and define $\phi_\alpha : U_\alpha \to \mathbb{R}$ for $\alpha = 1, 2$ by

$$\phi_1(e^{it}) = t, \quad -\pi < t < \pi,$$
$$\phi_2(e^{it}) = t, \quad 0 < t < 2\pi.$$

Both ϕ_1 and ϕ_2 are branches of the complex log function $(1/i) \log z$ and are homeomorphisms onto their respective images. Thus, (U_1, ϕ_1) and (U_2, ϕ_2) are charts on S^1. The intersection $U_1 \cap U_2$ consists of two connected components,

$$A = \{e^{it} \mid -\pi < t < 0\},$$
$$B = \{e^{it} \mid 0 < t < \pi\},$$

with

$$\phi_1(U_1 \cap U_2) = \phi_1(A \amalg B) = \phi_1(A) \amalg \phi_1(B) =]-\pi, 0[\amalg]0, \pi[,$$
$$\phi_2(U_1 \cap U_2) = \phi_2(A \amalg B) = \phi_2(A) \amalg \phi_2(B) =]\pi, 2\pi[\amalg]0, \pi[.$$

Here we use the notation $A \amalg B$ to indicate a union in which the two subsets A and B are disjoint. The transition function $\phi_2 \circ \phi_1^{-1} : \phi_1(A \amalg B) \to \phi_2(A \amalg B)$ is given by

$$(\phi_2 \circ \phi_1^{-1})(t) = \begin{cases} t + 2\pi & \text{for } t \in]-\pi, 0[, \\ t & \text{for } t \in]0, \pi[. \end{cases}$$

Similarly,

$$(\phi_1 \circ \phi_2^{-1})(t) = \begin{cases} t - 2\pi & \text{for } t \in]\pi, 2\pi[, \\ t & \text{for } t \in]0, \pi[. \end{cases}$$

Therefore, (U_1, ϕ_1) and (U_2, ϕ_2) are C^∞-compatible charts and form a C^∞ atlas on S^1.

Although the C^∞ compatibility of charts is clearly reflexive and symmetric, it is not transitive. The reason is as follows. Suppose (U_1, ϕ_1) is C^∞-compatible with (U_2, ϕ_2), and (U_2, ϕ_2) is C^∞-compatible with (U_3, ϕ_3). Note that the three coordinate functions are simultaneously defined only on the triple intersection U_{123}. Thus, the composite

$$\phi_3 \circ \phi_1^{-1} = (\phi_3 \circ \phi_2^{-1}) \circ (\phi_2 \circ \phi_1^{-1})$$

is C^∞, but only on $\phi_1(U_{123})$, not necessarily on $\phi_1(U_{13})$ (Figure 5.4). A priori we know nothing about $\phi_3 \circ \phi_1^{-1}$ on $\phi_1(U_{13} - U_{123})$ and so we cannot conclude that (U_1, ϕ_1) and (U_3, ϕ_3) are C^∞-compatible.

We say that a chart (V, ψ) is *compatible* with an atlas $\{(U_\alpha, \phi_\alpha)\}$ if it is compatible with all the charts (U_α, ϕ_α) of the atlas.

Lemma 5.8. Let $\{(U_\alpha, \phi_\alpha)\}$ be an atlas on a locally Euclidean space. If two charts (V, ψ) and (W, σ) are both compatible with the atlas $\{(U_\alpha, \phi_\alpha)\}$, then they are compatible with each other.

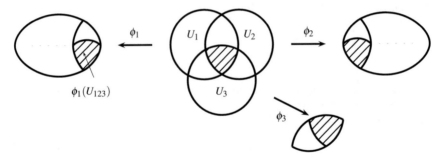

Fig. 5.4. The transition function $\phi_3 \circ \phi_1^{-1}$ is C^∞ on $\phi_1(U_{123})$.

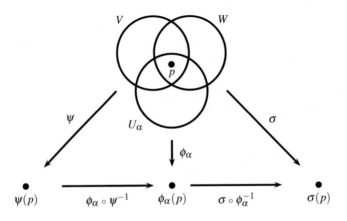

Fig. 5.5. Two charts (V, ψ), (W, σ) compatible with an atlas.

Proof. (See Figure 5.5.) Let $p \in V \cap W$. We need to show that $\sigma \circ \psi^{-1}$ is C^∞ at $\psi(p)$. Since $\{(U_\alpha, \phi_\alpha)\}$ is an atlas for M, $p \in U_\alpha$ for some α. Then p is in the triple intersection $V \cap W \cap U_\alpha$.

By the remark above, $\sigma \circ \psi^{-1} = (\sigma \circ \phi_\alpha^{-1}) \circ (\phi_\alpha \circ \psi^{-1})$ is C^∞ on $\psi(V \cap W \cap U_\alpha)$, hence at $\psi(p)$. Since p was an arbitrary point of $V \cap W$, this proves that $\sigma \circ \psi^{-1}$ is C^∞ on $\psi(V \cap W)$. Similarly, $\psi \circ \sigma^{-1}$ is C^∞ on $\sigma(V \cap W)$.   □

Note that in an equality such as $\sigma \circ \psi^{-1} = (\sigma \circ \phi_\alpha^{-1}) \circ (\phi_\alpha \circ \psi^{-1})$ in the proof above, the maps on the two sides of the equality sign have different domains. What the equality means is that the two maps are equal on their common domain.

5.3 Smooth Manifolds

An atlas $\mathfrak{M}$ on a locally Euclidean space is said to be *maximal* if it is not contained in a larger atlas; in other words, if $\mathfrak{U}$ is any other atlas containing $\mathfrak{M}$, then $\mathfrak{U} = \mathfrak{M}$.

Definition 5.9. A *smooth* or *C^∞ manifold* is a topological manifold M together with a maximal atlas. The maximal atlas is also called a *differentiable structure* on M. A manifold is said to have dimension n if all of its connected components have dimension n. A 1-dimensional manifold is also called a *curve*, a 2-dimensional manifold a *surface*, and an n-dimensional manifold an *n-manifold*.

In Corollary 8.7 we will prove that if an open set $U \subset \mathbb{R}^n$ is diffeomorphic to an open set $V \subset \mathbb{R}^m$, then $n = m$. As a consequence, the dimension of a manifold at a point is well defined.

In practice, to check that a topological manifold M is a smooth manifold, it is not necessary to exhibit a maximal atlas. The existence of *any* atlas on M will do, because of the following proposition.

Proposition 5.10. *Any atlas $\mathfrak{U} = \{(U_\alpha, \phi_\alpha)\}$ on a locally Euclidean space is contained in a unique maximal atlas.*

Proof. Adjoin to the atlas $\mathfrak{U}$ all charts (V_i, ψ_i) that are compatible with $\mathfrak{U}$. By Proposition 5.8 the charts (V_i, ψ_i) are compatible with one another. So the enlarged collection of charts is an atlas. Any chart compatible with the new atlas must be compatible with the original atlas $\mathfrak{U}$ and so by construction belongs to the new atlas. This proves that the new atlas is maximal.

Let $\mathfrak{M}$ be the maximal atlas containing $\mathfrak{U}$ that we have just constructed. If $\mathfrak{M}'$ is another maximal atlas containing $\mathfrak{U}$, then all the charts in $\mathfrak{M}'$ are compatible with $\mathfrak{U}$ and so by construction must belong to $\mathfrak{M}$. This proves that $\mathfrak{M}' \subset \mathfrak{M}$. Since both are maximal, $\mathfrak{M}' = \mathfrak{M}$. Therefore, the maximal atlas containing $\mathfrak{U}$ is unique. $\square$

In summary, to show that a topological space M is a C^∞ manifold, it suffices to check that

(i) M is Hausdorff and second countable,
(ii) M has a C^∞ atlas (not necessarily maximal).

From now on, a "manifold" will mean a C^∞ manifold. We use the terms "smooth" and "C^∞" interchangeably. In the context of manifolds, we denote the standard coordinates on $\mathbb{R}^n$ by $r^1, \ldots, r^n$. If $(U, \phi: U \to \mathbb{R}^n)$ is a chart of a manifold, we let $x^i = r^i \circ \phi$ be the ith component of ϕ and write $\phi = (x^1, \ldots, x^n)$ and $(U, \phi) = (U, x^1, \ldots, x^n)$. Thus, for $p \in U$, $(x^1(p), \ldots, x^n(p))$ is a point in $\mathbb{R}^n$. The functions $x^1, \ldots, x^n$ are called *coordinates* or *local coordinates* on U. By abuse of notation, we sometimes omit the p. So the notation $(x^1, \ldots, x^n)$ stands alternately for local coordinates on the open set U and for a point in $\mathbb{R}^n$. By a *chart (U, ϕ) about p* in a manifold M, we will mean a chart in the differentiable structure of M such that $p \in U$.

5.4 Examples of Smooth Manifolds

Example 5.11 (*Euclidean space*). The Euclidean space $\mathbb{R}^n$ is a smooth manifold with a single chart $(\mathbb{R}^n, r^1, \ldots, r^n)$, where $r^1, \ldots, r^n$ are the standard coordinates on $\mathbb{R}^n$.

Example 5.12 (*Open subset of a manifold*). Any open subset V of a manifold M is also a manifold. If $\{(U_\alpha, \phi_\alpha)\}$ is an atlas for M, then $\{(U_\alpha \cap V, \phi_\alpha|_{U_\alpha \cap V}\}$ is an atlas for V, where $\phi_\alpha|_{U_\alpha \cap V} : U_\alpha \cap V \to \mathbb{R}^n$ denotes the restriction of ϕ_α to the subset $U_\alpha \cap V$.

Example 5.13 (*Manifolds of dimension zero*). In a manifold of dimension zero, every singleton subset is homeomorphic to $\mathbb{R}^0$ and so is open. Thus, a zero-dimensional manifold is a discrete set. By second countability, this discrete set must be countable.

Example 5.14 (*Graph of a smooth function*). For a subset of $A \subset \mathbb{R}^n$ and a function $f : A \to \mathbb{R}^m$, the *graph* of f is defined to be the subset (Figure 5.6)

$$\Gamma(f) = \{(x, f(x)) \in A \times \mathbb{R}^m\}.$$

If U is an open subset of $\mathbb{R}^n$ and $f : U \to \mathbb{R}^n$ is C^∞, then the two maps

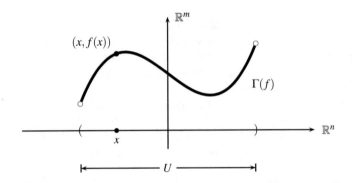

Fig. 5.6. The graph of a smooth function $f : \mathbb{R}^n \supset U \to \mathbb{R}^m$.

$$\phi : \Gamma(f) \to U, \qquad (x, f(x)) \mapsto x,$$

and

$$(1, f) : U \to \Gamma(f), \qquad x \mapsto (x, f(x)),$$

are continuous and inverse to each other, and so are homeomorphisms. The graph $\Gamma(f)$ of a C^∞ function $f : U \to \mathbb{R}^m$ has an atlas with a single chart $(\Gamma(f), \phi)$, and is therefore a C^∞ manifold. This shows that many of the familiar surfaces of calculus, for example an elliptic paraboloid or a hyperbolic paraboloid, are manifolds.

Example 5.15 (*General linear groups*). For any two positive integers m and n let $\mathbb{R}^{m \times n}$ be the vector space of all $m \times n$ matrices. Since $\mathbb{R}^{m \times n}$ is isomorphic to $\mathbb{R}^{mn}$, we give it the topology of $\mathbb{R}^{mn}$. The *general linear group* $\mathrm{GL}(n, \mathbb{R})$ is by definition

$$\mathrm{GL}(n, \mathbb{R}) := \{A \in \mathbb{R}^{n \times n} \mid \det A \neq 0\} = \det^{-1}(\mathbb{R} - \{0\}).$$

Since the determinant function

$$\det\colon \mathbb{R}^{n\times n} \to \mathbb{R}$$

is continuous, $\mathrm{GL}(n,\mathbb{R})$ is an open subset of $\mathbb{R}^{n\times n} \simeq \mathbb{R}^{n^2}$ and is therefore a manifold.

The *complex general linear group* $\mathrm{GL}(n,\mathbb{C})$ is defined to be the group of non-singular $n \times n$ complex matrices. Since an $n \times n$ matrix A is nonsingular if and only if $\det A \neq 0$, $\mathrm{GL}(n,\mathbb{C})$ is an open subset of $\mathbb{C}^{n\times n} \simeq \mathbb{R}^{2n^2}$, the vector space of $n \times n$ complex matrices. By the same reasoning as in the real case, $\mathrm{GL}(n,\mathbb{C})$ is a manifold of dimension $2n^2$.

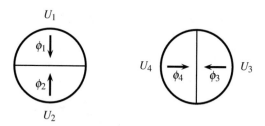

Fig. 5.7. Charts on the unit circle.

Example 5.16 (*Unit circle in the* (x,y)-*plane*). In Example 5.7 we found a C^∞ atlas with two charts on the unit circle S^1 in the complex plane $\mathbb{C}$. It follows that S^1 is a manifold. We now view S^1 as the unit circle in the real plane $\mathbb{R}^2$ with defining equation $x^2 + y^2 = 1$, and describe a C^∞ atlas with four charts on it.

We can cover S^1 with four open sets: the upper and lower semicircles U_1, U_2, and the right and left semicircles U_3, U_4 (Figure 5.7). On U_1 and U_2, the coordinate function x is a homeomorphism onto the open interval $]-1,1[$ on the x-axis. Thus, $\phi_i(x,y) = x$ for $i = 1,2$. Similarly, on U_3 and U_4, y is a homeomorphism onto the open interval $]-1,1[$ on the y-axis, and so $\phi_i(x,y) = y$ for $i = 3,4$.

It is easy to check that on every nonempty pairwise intersection $U_\alpha \cap U_\beta$, $\phi_\beta \circ \phi_\alpha^{-1}$ is C^∞. For example, on $U_1 \cap U_3$,

$$(\phi_3 \circ \phi_1^{-1})(x) = \phi_3\left(x, \sqrt{1-x^2}\right) = \sqrt{1-x^2},$$

which is C^∞. On $U_2 \cap U_4$,

$$(\phi_4 \circ \phi_2^{-1})(x) = \phi_4\left(x, -\sqrt{1-x^2}\right) = -\sqrt{1-x^2},$$

which is also C^∞. Thus, $\{(U_i, \phi_i)\}_{i=1}^4$ is a C^∞ atlas on S^1.

Example 5.17 (*Product manifold*). If M and N are C^∞ manifolds, then $M \times N$ with its product topology is Hausdorff and second countable (Corollary A.21 and Proposition A.22). To show that $M \times N$ is a manifold, it remains to exhibit an atlas on it. Recall that the product of two set maps $f\colon X \to X'$ and $g\colon Y \to Y'$ is

$$f \times g\colon X \times Y \to X' \times Y', \quad (f \times g)(x,y) = (f(x), g(y)).$$

Proposition 5.18 (An atlas for a product manifold). *If* $\{(U_\alpha, \phi_\alpha)\}$ *and* $\{(V_i, \psi_i)\}$
are C^∞ *atlases for the manifolds M and N of dimensions m and n, respectively, then
the collection*

$$\{(U_\alpha \times V_i, \phi_\alpha \times \psi_i : U_\alpha \times V_i \to \mathbb{R}^m \times \mathbb{R}^n)\}$$

of charts is a C^∞ *atlas on* $M \times N$. *Therefore,* $M \times N$ *is a* C^∞ *manifold of dimension*
$m + n$.

Proof. Problem 5.5.   □

Example. It follows from Proposition 5.18 that the infinite cylinder $S^1 \times \mathbb{R}$ and the
torus $S^1 \times S^1$ are manifolds (Figure 5.8).

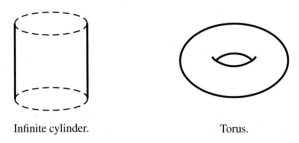

Infinite cylinder.   Torus.

Fig. 5.8.

Since $M \times N \times P = (M \times N) \times P$ is the successive product of pairs of spaces, if
M, N, and P are manifolds, then so is $M \times N \times P$. Thus, the n-dimensional torus
$S^1 \times \cdots \times S^1$ (n times) is a manifold.

Remark. Let S^n be the unit sphere

$$(x^1)^2 + (x^2)^2 + \cdots + (x^{n+1})^2 = 1$$

in $\mathbb{R}^{n+1}$. Using Problem 5.3 as a guide, it is easy to write down a C^∞ atlas on S^n,
showing that S^n has a differentiable structure. The manifold S^n with this differentiable structure is called the *standard n-sphere*.

One of the most surprising achievements in topology was John Milnor's dis-
covery [27] in 1956 of exotic 7-spheres, smooth manifolds homeomorphic but not
diffeomorphic to the standard 7-sphere. In 1963, Michel Kervaire and John Milnor
[24] determined that there are exactly 28 nondiffeomorphic differentiable structures
on S^7.

It is known that in dimensions < 4 every topological manifold has a unique differentiable structure and in dimensions > 4 every compact topological manifold has
a finite number of differentiable structures. Dimension 4 is a mystery. It is not known

whether S^4 has a finite or infinite number of differentiable structures. The statement that S^4 has a unique differentiable structure is called the *smooth Poincaré conjecture*. As of this writing in 2010, the conjecture is still open.

There are topological manifolds with no differentiable structure. Michel Kervaire was the first to construct an example [23].

Problems

5.1. The real line with two origins
Let A and B be two points not on the real line $\mathbb{R}$. Consider the set $S = (\mathbb{R} - \{0\}) \cup \{A, B\}$ (see Figure 5.9).

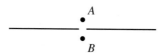

Fig. 5.9. Real line with two origins.

For any two positive real numbers c, d, define

$$I_A(-c, d) = \,]-c, 0[\,\cup\, \{A\} \,\cup\,]0, d[$$

and similarly for $I_B(-c, d)$, with B instead of A. Define a topology on S as follows: On $(\mathbb{R} - \{0\})$, use the subspace topology inherited from $\mathbb{R}$, with open intervals as a basis. A basis of neighborhoods at A is the set $\{I_A(-c, d) \mid c, d > 0\}$; similarly, a basis of neighborhoods at B is $\{I_B(-c, d) \mid c, d > 0\}$.

(a) Prove that the map $h \colon I_A(-c, d) \to \,]-c, d[$ defined by

$$h(x) = x \quad \text{for } x \in \,]-c, 0[\,\cup\,]0, d[,$$
$$h(A) = 0$$

is a homeomorphism.

(b) Show that S is locally Euclidean and second countable, but not Hausdorff.

5.2. A sphere with a hair
A fundamental theorem of topology, the theorem on invariance of dimension, states that if two nonempty open sets $U \subset \mathbb{R}^n$ and $V \subset \mathbb{R}^m$ are homeomorphic, then $n = m$ (for a proof, see [18, p. 126]). Use the idea of Example 5.4 as well as the theorem on invariance of dimension to prove that the sphere with a hair in $\mathbb{R}^3$ (Figure 5.10) is not locally Euclidean at q. Hence it cannot be a topological manifold.

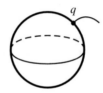

Fig. 5.10. A sphere with a hair.

5.3. Charts on a sphere

Let S^2 be the unit sphere

$$x^2 + y^2 + z^2 = 1$$

in $\mathbb{R}^3$. Define in S^2 the six charts corresponding to the six hemispheres—the front, rear, right, left, upper, and lower hemispheres (Figure 5.11):

$$
\begin{aligned}
U_1 &= \{(x,y,z) \in S^2 \mid x > 0\}, & \phi_1(x,y,z) &= (y,z), \\
U_2 &= \{(x,y,z) \in S^2 \mid x < 0\}, & \phi_2(x,y,z) &= (y,z), \\
U_3 &= \{(x,y,z) \in S^2 \mid y > 0\}, & \phi_3(x,y,z) &= (x,z), \\
U_4 &= \{(x,y,z) \in S^2 \mid y < 0\}, & \phi_4(x,y,z) &= (x,z), \\
U_5 &= \{(x,y,z) \in S^2 \mid z > 0\}, & \phi_5(x,y,z) &= (x,y), \\
U_6 &= \{(x,y,z) \in S^2 \mid z < 0\}, & \phi_6(x,y,z) &= (x,y).
\end{aligned}
$$

Describe the domain $\phi_4(U_{14})$ of $\phi_1 \circ \phi_4^{-1}$ and show that $\phi_1 \circ \phi_4^{-1}$ is C^∞ on $\phi_4(U_{14})$. Do the same for $\phi_6 \circ \phi_1^{-1}$.

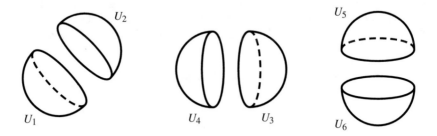

Fig. 5.11. Charts on the unit sphere.

5.4.* Existence of a coordinate neighborhood

Let $\{(U_\alpha, \phi_\alpha)\}$ be the maximal atlas on a manifold M. For any open set U in M and a point $p \in U$, prove the existence of a coordinate open set U_α such that $p \in U_\alpha \subset U$.

5.5. An atlas for a product manifold

Prove Proposition 5.18.

§6 Smooth Maps on a Manifold

Now that we have defined smooth manifolds, it is time to consider maps between them. Using coordinate charts, one can transfer the notion of smooth maps from Euclidean spaces to manifolds. By the C^∞ compatibility of charts in an atlas, the smoothness of a map turns out to be independent of the choice of charts and is therefore well defined. We give various criteria for the smoothness of a map as well as examples of smooth maps.

Next we transfer the notion of partial derivatives from Euclidean space to a coordinate chart on a manifold. Partial derivatives relative to coordinate charts allow us to generalize the inverse function theorem to manifolds. Using the inverse function theorem, we formulate a criterion for a set of smooth functions to serve as local coordinates near a point.

6.1 Smooth Functions on a Manifold

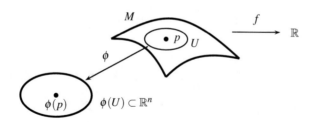

Fig. 6.1. Checking that a function f is C^∞ at p by pulling back to $\mathbb{R}^n$.

Definition 6.1. Let M be a smooth manifold of dimension n. A function $f\colon M \to \mathbb{R}$ is said to be C^∞ or *smooth at a point* p in M if there is a chart (U,ϕ) about p in M such that $f \circ \phi^{-1}$, a function defined on the open subset $\phi(U)$ of $\mathbb{R}^n$, is C^∞ at $\phi(p)$ (see Figure 6.1). The function f is said to be C^∞ *on M* if it is C^∞ at every point of M.

Remark 6.2. The definition of the smoothness of a function f at a point is independent of the chart (U,ϕ), for if $f \circ \phi^{-1}$ is C^∞ at $\phi(p)$ and (V,ψ) is any other chart about p in M, then on $\psi(U \cap V)$,

$$f \circ \psi^{-1} = (f \circ \phi^{-1}) \circ (\phi \circ \psi^{-1}),$$

which is C^∞ at $\psi(p)$ (see Figure 6.2).

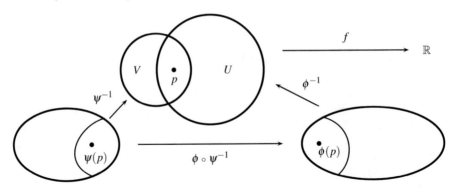

Fig. 6.2. Checking that a function f is C^∞ at p via two charts.

In Definition 6.1, $f: M \to \mathbb{R}$ is not assumed to be continuous. However, if f is C^∞ at $p \in M$, then $f \circ \phi^{-1}: \phi(U) \to \mathbb{R}$, being a C^∞ function at the point $\phi(p)$ in an open subset of $\mathbb{R}^n$, is continuous at $\phi(p)$. As a composite of continuous functions, $f = (f \circ \phi^{-1}) \circ \phi$ is continuous at p. Since we are interested only in functions that are smooth on an open set, there is no loss of generality in assuming at the outset that f is continuous.

Proposition 6.3 (Smoothness of a real-valued function). *Let M be a manifold of dimension n, and $f: M \to \mathbb{R}$ a real-valued function on M. The following are equivalent:*

(i) *The function $f: M \to \mathbb{R}$ is C^∞.*
(ii) *The manifold M has an atlas such that for every chart (U, ϕ) in the atlas, $f \circ \phi^{-1}: \mathbb{R}^n \supset \phi(U) \to \mathbb{R}$ is C^∞.*
(iii) *For every chart (V, ψ) on M, the function $f \circ \psi^{-1}: \mathbb{R}^n \supset \psi(V) \to \mathbb{R}$ is C^∞.*

Proof. We will prove the proposition as a cyclic chain of implications.
(ii) $\Rightarrow$ (i): This follows directly from the definition of a C^∞ function, since by (ii) every point $p \in M$ has a coordinate neighborhood (U, ϕ) such that $f \circ \phi^{-1}$ is C^∞ at $\phi(p)$.
(i) $\Rightarrow$ (iii): Let (V, ψ) be an arbitrary chart on M and let $p \in V$. By Remark 6.2, $f \circ \psi^{-1}$ is C^∞ at $\psi(p)$. Since p was an arbitrary point of V, $f \circ \psi^{-1}$ is C^∞ on $\psi(V)$.
(iii) $\Rightarrow$ (ii): Obvious. $\square$

The smoothness conditions of Proposition 6.3 will be a recurrent motif throughout the book: to prove the smoothness of an object, it is sufficient that a smoothness criterion hold on the charts of some atlas. Once the object is shown to be smooth, it then follows that the same smoothness criterion holds on *every* chart on the manifold.

Definition 6.4. Let $F: N \to M$ be a map and h a function on M. The *pullback* of h by F, denoted by F^*h, is the composite function $h \circ F$.

In this terminology, a function f on M is C^∞ on a chart (U, ϕ) if and only if its pullback $(\phi^{-1})^*f$ by ϕ^{-1} is C^∞ on the subset $\phi(U)$ of Euclidean space.

6.2 Smooth Maps Between Manifolds

We emphasize again that unless otherwise specified, by a manifold we always mean a C^∞ manifold. We use the terms "C^∞" and "smooth" interchangeably. An atlas or a chart on a smooth manifold means an atlas or a chart contained in the differentiable structure of the smooth manifold. We generally denote a manifold by M and its dimension by n. However, when speaking of two manifolds simultaneously, as in a map $f\colon N \to M$, we will let the dimension of N be n and that of M be m.

Definition 6.5. Let N and M be manifolds of dimension n and m, respectively. A continuous map $F\colon N \to M$ is C^∞ *at a point* p in N if there are charts (V, ψ) about $F(p)$ in M and (U, ϕ) about p in N such that the composition $\psi \circ F \circ \phi^{-1}$, a map from the open subset $\phi(F^{-1}(V) \cap U)$ of $\mathbb{R}^n$ to $\mathbb{R}^m$, is C^∞ at $\phi(p)$ (see Figure 6.3). The continuous map $F\colon N \to M$ is said to be C^∞ if it is C^∞ at every point of N.

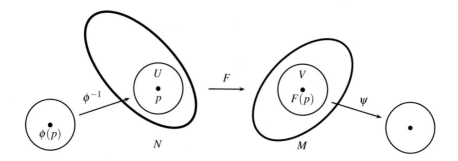

Fig. 6.3. Checking that a map $F\colon N \to M$ is C^∞ at p.

In Definition 6.5, we assume $F\colon N \to M$ continuous to ensure that $F^{-1}(V)$ is an open set in N. Thus, C^∞ maps between manifolds are by definition continuous.

Remark 6.6 (*Smooth maps into* $\mathbb{R}^m$). In case $M = \mathbb{R}^m$, we can take $(\mathbb{R}^m, \mathbb{1}_{\mathbb{R}^m})$ as a chart about $F(p)$ in $\mathbb{R}^m$. According to Definition 6.5, $F\colon N \to \mathbb{R}^m$ is C^∞ at $p \in N$ if and only if there is a chart (U, ϕ) about p in N such that $F \circ \phi^{-1}\colon \phi(U) \to \mathbb{R}^m$ is C^∞ at $\phi(p)$. Letting $m = 1$, we recover the definition of a function being C^∞ at a point.

We show now that the definition of the smoothness of a map $F\colon N \to M$ at a point is independent of the choice of charts. This is analogous to how the smoothness of a function $N \to \mathbb{R}$ at $p \in N$ is independent of the choice of a chart on N about p.

Proposition 6.7. *Suppose* $F\colon N \to M$ *is* C^∞ *at* $p \in N$. *If* (U, ϕ) *is any chart about* p *in* N *and* (V, ψ) *is any chart about* $F(p)$ *in* M, *then* $\psi \circ F \circ \phi^{-1}$ *is* C^∞ *at* $\phi(p)$.

Proof. Since F is C^∞ at $p \in N$, there are charts (U_α, ϕ_α) about p in N and (V_β, ψ_β) about $F(p)$ in M such that $\psi_\beta \circ F \circ \phi_\alpha^{-1}$ is C^∞ at $\phi_\alpha(p)$. By the C^∞ compatibility of charts in a differentiable structure, both $\phi_\alpha \circ \phi^{-1}$ and $\psi \circ \psi_\beta^{-1}$ are C^∞ on open subsets of Euclidean spaces. Hence, the composite

$$\psi \circ F \circ \phi^{-1} = (\psi \circ \psi_\beta^{-1}) \circ (\psi_\beta \circ F \circ \phi_\alpha^{-1}) \circ (\phi_\alpha \circ \phi^{-1})$$

is C^∞ at $\phi(p)$. $\square$

The next proposition gives a way to check smoothness of a map without specifying a point in the domain.

Proposition 6.8 (Smoothness of a map in terms of charts). *Let N and M be smooth manifolds, and $F: N \to M$ a continuous map. The following are equivalent:*

(i) *The map $F: N \to M$ is C^∞.*
(ii) *There are atlases $\mathfrak{U}$ for N and $\mathfrak{V}$ for M such that for every chart (U, ϕ) in $\mathfrak{U}$ and (V, ψ) in $\mathfrak{V}$, the map*

$$\psi \circ F \circ \phi^{-1}: \phi(U \cap F^{-1}(V)) \to \mathbb{R}^m$$

is C^∞.
(iii) *For every chart (U, ϕ) on N and (V, ψ) on M, the map*

$$\psi \circ F \circ \phi^{-1}: \phi(U \cap F^{-1}(V)) \to \mathbb{R}^m$$

is C^∞.

Proof. (ii) $\Rightarrow$ (i): Let $p \in N$. Suppose (U, ϕ) is a chart about p in $\mathfrak{U}$ and (V, ψ) is a chart about $F(p)$ in $\mathfrak{V}$. By (ii), $\psi \circ F \circ \phi^{-1}$ is C^∞ at $\phi(p)$. By the definition of a C^∞ map, $F: N \to M$ is C^∞ at p. Since p was an arbitrary point of N, the map $F: N \to M$ is C^∞.
(i) $\Rightarrow$ (iii): Suppose (U, ϕ) and (V, ψ) are charts on N and M respectively such that $U \cap F^{-1}(V) \neq \varnothing$. Let $p \in U \cap F^{-1}(V)$. Then (U, ϕ) is a chart about p and (V, ψ) is a chart about $F(p)$. By Proposition 6.7, $\psi \circ F \circ \phi^{-1}$ is C^∞ at $\phi(p)$. Since $\phi(p)$ was an arbitrary point of $\phi(U \cap F^{-1}(V))$, the map $\psi \circ F \circ \phi^{-1}: \phi(U \cap F^{-1}(V)) \to \mathbb{R}^m$ is C^∞.
(iii) $\Rightarrow$ (ii): Clear. $\square$

Proposition 6.9 (Composition of C^∞ maps). *If $F: N \to M$ and $G: M \to P$ are C^∞ maps of manifolds, then the composite $G \circ F: N \to P$ is C^∞.*

Proof. Let (U, ϕ), (V, ψ), and (W, σ) be charts on N, M, and P respectively. Then

$$\sigma \circ (G \circ F) \circ \phi^{-1} = (\sigma \circ G \circ \psi^{-1}) \circ (\psi \circ F \circ \phi^{-1}).$$

Since F and G are C^∞, by Proposition 6.8(i)$\Rightarrow$(iii), $\sigma \circ G \circ \psi^{-1}$ and $\psi \circ F \circ \phi^{-1}$ are C^∞. As a composite of C^∞ maps of open subsets of Euclidean spaces, $\sigma \circ (G \circ F) \circ \phi^{-1}$ is C^∞. By Proposition 6.8(iii)$\Rightarrow$(i), $G \circ F$ is C^∞. $\square$

6.3 Diffeomorphisms

A *diffeomorphism* of manifolds is a bijective C^∞ map $F\colon N \to M$ whose inverse F^{-1} is also C^∞. According to the next two propositions, coordinate maps are diffeomorphisms, and conversely, every diffeomorphism of an open subset of a manifold with an open subset of a Euclidean space can serve as a coordinate map.

Proposition 6.10. *If (U,ϕ) is a chart on a manifold M of dimension n, then the coordinate map $\phi\colon U \to \phi(U) \subset \mathbb{R}^n$ is a diffeomorphism.*

Proof. By definition, ϕ is a homeomorphism, so it suffices to check that both ϕ and ϕ^{-1} are smooth. To test the smoothness of $\phi\colon U \to \phi(U)$, we use the atlas $\{(U,\phi)\}$ with a single chart on U and the atlas $\{(\phi(U), \mathbb{1}_{\phi(U)})\}$ with a single chart on $\phi(U)$. Since $\mathbb{1}_{\phi(U)} \circ \phi \circ \phi^{-1}\colon \phi(U) \to \phi(U)$ is the identity map, it is C^∞. By Proposition 6.8(ii)$\Rightarrow$(i), ϕ is C^∞.

To test the smoothness of $\phi^{-1}\colon \phi(U) \to U$, we use the same atlases as above. Since $\phi \circ \phi^{-1} \circ \mathbb{1}_{\phi(U)} = \mathbb{1}_{\phi(U)}\colon \phi(U) \to \phi(U)$, the map ϕ^{-1} is also C^∞. $\square$

Proposition 6.11. *Let U be an open subset of a manifold M of dimension n. If $F\colon U \to F(U) \subset \mathbb{R}^n$ is a diffeomorphism onto an open subset of $\mathbb{R}^n$, then (U,F) is a chart in the differentiable structure of M.*

Proof. For any chart (U_α, ϕ_α) in the maximal atlas of M, both ϕ_α and ϕ_α^{-1} are C^∞ by Proposition 6.10. As composites of C^∞ maps, both $F \circ \phi_\alpha^{-1}$ and $\phi_\alpha \circ F^{-1}$ are C^∞. Hence, (U,F) is compatible with the maximal atlas. By the maximality of the atlas, the chart (U,F) is in the atlas. $\square$

6.4 Smoothness in Terms of Components

In this subsection we derive a criterion that reduces the smoothness of a map to the smoothness of real-valued functions on open sets.

Proposition 6.12 (Smoothness of a vector-valued function). *Let N be a manifold and $F\colon N \to \mathbb{R}^m$ a continuous map. The following are equivalent:*

 (i) *The map $F\colon N \to \mathbb{R}^m$ is C^∞.*
 (ii) *The manifold N has an atlas such that for every chart (U,ϕ) in the atlas, the map $F \circ \phi^{-1}\colon \phi(U) \to \mathbb{R}^m$ is C^∞.*
 (iii) *For every chart (U,ϕ) on N, the map $F \circ \phi^{-1}\colon \phi(U) \to \mathbb{R}^m$ is C^∞.*

Proof. (ii) $\Rightarrow$ (i): In Proposition 6.8(ii), take $\mathfrak{V}$ to be the atlas with the single chart $(\mathbb{R}^m, \mathbb{1}_{\mathbb{R}^m})$ on $M = \mathbb{R}^m$.
(i) $\Rightarrow$ (iii): In Proposition 6.8(iii), let (V, ψ) be the chart $(\mathbb{R}^m, \mathbb{1}_{\mathbb{R}^m})$ on $M = \mathbb{R}^m$.
(iii) $\Rightarrow$ (ii): Obvious. $\square$

Proposition 6.13 (Smoothness in terms of components). *Let N be a manifold. A vector-valued function $F\colon N \to \mathbb{R}^m$ is C^∞ if and only if its component functions $F^1, \ldots, F^m\colon N \to \mathbb{R}$ are all C^∞.*

Proof.

The map $F \colon N \to \mathbb{R}^m$ is C^∞

$\Longleftrightarrow$ for every chart (U, ϕ) on N, the map $F \circ \phi^{-1} \colon \phi(U) \to \mathbb{R}^m$ is C^∞ (by Proposition 6.12)

$\Longleftrightarrow$ for every chart (U, ϕ) on N, the functions $F^i \circ \phi^{-1} \colon \phi(U) \to \mathbb{R}$ are all C^∞ (definition of smoothness for maps of Euclidean spaces)

$\Longleftrightarrow$ the functions $F^i \colon N \to \mathbb{R}$ are all C^∞ (by Proposition 6.3). $\qquad \square$

Exercise 6.14 (Smoothness of a map to a circle).* Prove that the map $F \colon \mathbb{R} \to S^1$, $F(t) = (\cos t, \sin t)$ is C^∞.

Proposition 6.15 (Smoothness of a map in terms of vector-valued functions). *Let $F \colon N \to M$ be a continuous map between two manifolds of dimensions n and m respectively. The following are equivalent:*

(i) *The map $F \colon N \to M$ is C^∞.*
(ii) *The manifold M has an atlas such that for every chart $(V, \psi) = (V, y^1, \ldots, y^m)$ in the atlas, the vector-valued function $\psi \circ F \colon F^{-1}(V) \to \mathbb{R}^m$ is C^∞.*
(iii) *For every chart $(V, \psi) = (V, y^1, \ldots, y^m)$ on M, the vector-valued function $\psi \circ F \colon F^{-1}(V) \to \mathbb{R}^m$ is C^∞.*

Proof. (ii) $\Rightarrow$ (i): Let $\mathfrak{V}$ be the atlas for M in (ii), and let $\mathfrak{U} = \{(U, \phi)\}$ be an arbitrary atlas for N. For each chart (V, ψ) in the atlas $\mathfrak{V}$, the collection $\{(U \cap F^{-1}(V), \phi|_{U \cap F^{-1}(V)})\}$ is an atlas for $F^{-1}(V)$. Since $\psi \circ F \colon F^{-1}(V) \to \mathbb{R}^m$ is C^∞, by Proposition 6.12(i)$\Rightarrow$(iii),

$$\psi \circ F \circ \phi^{-1} \colon \phi(U \cap F^{-1}(V)) \to \mathbb{R}^m$$

is C^∞. It then follows from Proposition 6.8(ii)$\Rightarrow$(i) that $F \colon N \to M$ is C^∞.

(i) $\Rightarrow$ (iii): Being a coordinate map, ψ is C^∞ (Proposition 6.10). As the composite of two C^∞ maps, $\psi \circ F$ is C^∞.

(iii) $\Rightarrow$ (ii): Obvious. $\qquad \square$

By Proposition 6.13, this smoothness criterion for a map translates into a smoothness criterion in terms of the components of the map.

Proposition 6.16 (Smoothness of a map in terms of components). *Let $F \colon N \to M$ be a continuous map between two manifolds of dimensions n and m respectively. The following are equivalent:*

(i) *The map $F \colon N \to M$ is C^∞.*
(ii) *The manifold M has an atlas such that for every chart $(V, \psi) = (V, y^1, \ldots, y^m)$ in the atlas, the components $y^i \circ F \colon F^{-1}(V) \to \mathbb{R}$ of F relative to the chart are all C^∞.*
(iii) *For every chart $(V, \psi) = (V, y^1, \ldots, y^m)$ on M, the components $y^i \circ F \colon F^{-1}(V) \to \mathbb{R}$ of F relative to the chart are all C^∞.*

6.5 Examples of Smooth Maps

We have seen that coordinate maps are smooth. In this subsection we look at a few more examples of smooth maps.

Example 6.17 (*Smoothness of a projection map*). Let M and N be manifolds and $\pi \colon M \times N \to M$, $\pi(p,q) = p$ the projection to the first factor. Prove that π is a C^∞ map.

Solution. Let (p,q) be an arbitrary point of $M \times N$. Suppose $(U,\phi) = (U, x^1, \ldots, x^m)$ and $(V, \psi) = (V, y^1, \ldots, y^n)$ are coordinate neighborhoods of p and q in M and N respectively. By Proposition 5.18, $(U \times V, \phi \times \psi) = (U \times V, x^1, \ldots, x^m, y^1, \ldots, y^n)$ is a coordinate neighborhood of (p,q). Then

$$\left(\phi \circ \pi \circ (\phi \times \psi)^{-1} \right) (a^1, \ldots, a^m, b^1, \ldots, b^n) = (a^1, \ldots, a^m),$$

which is a C^∞ map from $(\phi \times \psi)(U \times V)$ in $\mathbb{R}^{m+n}$ to $\phi(U)$ in $\mathbb{R}^m$, so π is C^∞ at (p,q). Since (p,q) was an arbitrary point in $M \times N$, π is C^∞ on $M \times N$.

Exercise 6.18 (Smoothness of a map to a Cartesian product).* Let M_1, M_2, and N be manifolds of dimensions m_1, m_2, and n respectively. Prove that a map $(f_1, f_2) \colon N \to M_1 \times M_2$ is C^∞ if and only if $f_i \colon N \to M_i$, $i = 1, 2$, are both C^∞.

Example 6.19. In Examples 5.7 and 5.16 we showed that the unit circle S^1 defined by $x^2 + y^2 = 1$ in $\mathbb{R}^2$ is a C^∞ manifold. Prove that a C^∞ function $f(x,y)$ on $\mathbb{R}^2$ restricts to a C^∞ function on S^1.

Solution. To avoid confusing functions with points, we will denote a point on S^1 as $p = (a,b)$ and use x, y to mean the standard coordinate functions on $\mathbb{R}^2$. Thus, $x(a,b) = a$ and $y(a,b) = b$. Suppose we can show that x and y restrict to C^∞ functions on S^1. By Exercise 6.18, the inclusion map $i \colon S^1 \to \mathbb{R}^2$, $i(p) = (x(p), y(p))$ is then C^∞ on S^1. As the composition of C^∞ maps, $f|_{S^1} = f \circ i$ will be C^∞ on S^1 (Proposition 6.9).

Consider first the function x. We use the atlas (U_i, ϕ_i) from Example 5.16. Since x is a coordinate function on U_1 and on U_2, by Proposition 6.10 it is C^∞ on $U_1 \cup U_2 = S^1 - \{(\pm 1, 0)\}$. To show that x is C^∞ on U_3, it suffices to check the smoothness of $x \circ \phi_3^{-1} \colon \phi_3(U_3) \to \mathbb{R}$:

$$\left(x \circ \phi_3^{-1} \right)(b) = x \left(\sqrt{1 - b^2}, b \right) = \sqrt{1 - b^2}.$$

On U_3, we have $b \neq \pm 1$, so that $\sqrt{1 - b^2}$ is a C^∞ function of b. Hence, x is C^∞ on U_3.
    On U_4,

$$\left(x \circ \phi_4^{-1} \right)(b) = x \left(-\sqrt{1 - b^2}, b \right) = -\sqrt{1 - b^2},$$

which is C^∞ because b is not equal to ± 1. Since x is C^∞ on the four open sets U_1, U_2, U_3, and U_4, which cover S^1, x is C^∞ on S^1.
    The proof that y is C^∞ on S^1 is similar.

Armed with the definition of a smooth map between manifolds, we can define a Lie group.

Definition 6.20. A *Lie group*[1] is a C^∞ manifold G having a group structure such that the multiplication map

$$\mu : G \times G \to G$$

and the inverse map

$$\iota : G \to G, \quad \iota(x) = x^{-1},$$

are both C^∞.

Similarly, a *topological group* is a topological space having a group structure such that the multiplication and inverse maps are both continuous. Note that a topological group is required to be a topological space, but not a topological manifold.

Examples.
(i) The Euclidean space $\mathbb{R}^n$ is a Lie group under addition.
(ii) The set $\mathbb{C}^\times$ of nonzero complex numbers is a Lie group under multiplication.
(iii) The unit circle S^1 in $\mathbb{C}^\times$ is a Lie group under multiplication.
(iv) The Cartesian product $G_1 \times G_2$ of two Lie groups (G_1, μ_1) and (G_2, μ_2) is a Lie group under coordinatewise multiplication $\mu_1 \times \mu_2$.

Example 6.21 (*General linear group*). In Example 5.15 we defined the general linear group

$$\mathrm{GL}(n, \mathbb{R}) = \{A = [a_{ij}] \in \mathbb{R}^{n \times n} \mid \det A \neq 0\}.$$

As an open subset of $\mathbb{R}^{n \times n}$, it is a manifold. Since the (i, j)-entry of the product of two matrices A and B in $\mathrm{GL}(n, \mathbb{R})$,

$$(AB)_{ij} = \sum_{k=1}^{n} a_{ik} b_{kj},$$

is a polynomial in the coordinates of A and B, matrix multiplication

$$\mu : \mathrm{GL}(n, \mathbb{R}) \times \mathrm{GL}(n, \mathbb{R}) \to \mathrm{GL}(n, \mathbb{R})$$

is a C^∞ map.

Recall that the (i, j)-*minor* of a matrix A is the determinant of the submatrix of A obtained by deleting the ith row and the jth column of A. By Cramer's rule from linear algebra, the (i, j)-entry of A^{-1} is

$$(A^{-1})_{ij} = \frac{1}{\det A} \cdot (-1)^{i+j}((j, i)\text{-minor of } A),$$

which is a C^∞ function of the a_{ij}'s provided $\det A \neq 0$. Therefore, the inverse map $\iota : \mathrm{GL}(n, \mathbb{R}) \to \mathrm{GL}(n, \mathbb{R})$ is also C^∞. This proves that $\mathrm{GL}(n, \mathbb{R})$ is a Lie group.

---

[1]Lie groups and Lie algebras are named after the Norwegian mathematician Sophus Lie (1842–1899). In this context, "Lie" is pronounced "lee," not "lye."

In Section 15 we will study less obvious examples of Lie groups.

NOTATION. The notation for matrices presents a special challenge. An $n \times n$ matrix A can represent a linear transformation $y = Ax$, with $x, y \in \mathbb{R}^n$. In this case, $y^i = \sum_j a^i_j x^j$, so $A = [a^i_j]$. An $n \times n$ matrix can also represent a bilinear form $\langle x, y \rangle = x^T A y$ with $x, y \in \mathbb{R}^n$. In this case, $\langle x, y \rangle = \sum_{i,j} x^i a_{ij} y^j$, so $A = [a_{ij}]$. In the absence of any context, we will write a matrix as $A = [a_{ij}]$, using a lowercase letter a to denote an entry of a matrix A and using a double subscript $(\ )_{ij}$ to denote the (i, j)-entry.

6.6 Partial Derivatives

On a manifold M of dimension n, let (U, ϕ) be a chart and f a C^∞ function As a function into $\mathbb{R}^n$, ϕ has n components $x^1, \ldots, x^n$. This means that if $r^1, \ldots, r^n$ are the standard coordinates on $\mathbb{R}^n$, then $x^i = r^i \circ \phi$. For $p \in U$, we define the *partial derivative $\partial f / \partial x^i$ of f with respect to x^i at p* to be

$$\left. \frac{\partial}{\partial x^i} \right|_p f := \frac{\partial f}{\partial x^i}(p) := \frac{\partial \left(f \circ \phi^{-1} \right)}{\partial r^i}(\phi(p)) := \left. \frac{\partial}{\partial r^i} \right|_{\phi(p)} \left(f \circ \phi^{-1} \right).$$

Since $p = \phi^{-1}(\phi(p))$, this equation may be rewritten in the form

$$\frac{\partial f}{\partial x^i} \left(\phi^{-1}(\phi(p)) \right) = \frac{\partial \left(f \circ \phi^{-1} \right)}{\partial r^i}(\phi(p)).$$

Thus, as functions on $\phi(U)$,

$$\frac{\partial f}{\partial x^i} \circ \phi^{-1} = \frac{\partial \left(f \circ \phi^{-1} \right)}{\partial r^i}.$$

The partial derivative $\partial f / \partial x^i$ is C^∞ on U because its pullback $(\partial f / \partial x^i) \circ \phi^{-1}$ is C^∞ on $\phi(U)$.

In the next proposition we see that partial derivatives on a manifold satisfy the same duality property $\partial r^i / \partial r^j = \delta^i_j$ as the coordinate functions r^i on $\mathbb{R}^n$.

Proposition 6.22. *Suppose $(U, x^1, \ldots, x^n)$ is a chart on a manifold. Then $\partial x^i / \partial x^j = \delta^i_j$.*

Proof. At a point $p \in U$, by the definition of $\partial / \partial x^j |_p$,

$$\frac{\partial x^i}{\partial x^j}(p) = \frac{\partial \left(x^i \circ \phi^{-1} \right)}{\partial r^j}(\phi(p)) = \frac{\partial \left(r^i \circ \phi \circ \phi^{-1} \right)}{\partial r^j}(\phi(p)) = \frac{\partial r^i}{\partial r^j}(\phi(p)) = \delta^i_j. \quad \square$$

Definition 6.23. Let $F: N \to M$ be a smooth map, and let $(U, \phi) = (U, x^1, \ldots, x^n)$ and $(V, \psi) = (V, y^1, \ldots, y^m)$ be charts on N and M respectively such that $F(U) \subset V$. Denote by

$$F^i := y^i \circ F = r^i \circ \psi \circ F: U \to \mathbb{R}$$

the ith component of F in the chart (V, ψ). Then the matrix $[\partial F^i/\partial x^j]$ is called the *Jacobian matrix* of F relative to the charts (U, ϕ) and (V, ψ). In case N and M have the same dimension, the determinant $\det[\partial F^i/\partial x^j]$ is called the *Jacobian determinant* of F relative to the two charts. The Jacobian determinant is also written as $\partial(F^1, \ldots, F^n)/\partial(x^1, \ldots, x^n)$.

When M and N are open subsets of Euclidean spaces and the charts are $(U, r^1, \ldots, r^n)$ and $(V, r^1, \ldots, r^m)$, the Jacobian matrix $[\partial F^i/\partial r^j]$, where $F^i = r^i \circ F$, is the usual Jacobian matrix from calculus.

Example 6.24 (Jacobian matrix of a transition map). Let $(U, \phi) = (U, x^1, \ldots, x^n)$ and $(V, \psi) = (V, y^1, \ldots, y^n)$ be overlapping charts on a manifold M. The transition map $\psi \circ \phi^{-1} \colon \phi(U \cap V) \to \psi(U \cap V)$ is a diffeomorphism of open subsets of $\mathbb{R}^n$. Show that its Jacobian matrix $J(\psi \circ \phi^{-1})$ at $\phi(p)$ is the matrix $[\partial y^i/\partial x^j]$ of partial derivatives at p.

Solution. By definition, $J(\psi \circ \phi^{-1}) = [\partial(\psi \circ \phi^{-1})^i/\partial r^j]$, where

$$\frac{\partial \left(\psi \circ \phi^{-1} \right)^i}{\partial r^j}(\phi(p)) = \frac{\partial \left(r^i \circ \psi \circ \phi^{-1} \right)}{\partial r^j}(\phi(p)) = \frac{\partial \left(y^i \circ \phi^{-1} \right)}{\partial r^j}(\phi(p)) = \frac{\partial y^i}{\partial x^j}(p).$$

6.7 The Inverse Function Theorem

By Proposition 6.11, any diffeomorphism $F \colon U \to F(U) \subset \mathbb{R}^n$ of an open subset U of a manifold may be thought of as a coordinate system on U. We say that a C^∞ map $F \colon N \to M$ is *locally invertible* or a *local diffeomorphism* at $p \in N$ if p has a neighborhood U on which $F|_U \colon U \to F(U)$ is a diffeomorphism.

Given n smooth functions $F^1, \ldots, F^n$ in a neighborhood of a point p in a manifold N of dimension n, one would like to know whether they form a coordinate system, possibly on a smaller neighborhood of p. This is equivalent to whether $F = (F^1, \ldots, F^n) \colon N \to \mathbb{R}^n$ is a local diffeomorphism at p. The inverse function theorem provides an answer.

Theorem 6.25 (Inverse function theorem for $\mathbb{R}^n$). *Let $F \colon W \to \mathbb{R}^n$ be a C^∞ map defined on an open subset W of $\mathbb{R}^n$. For any point p in W, the map F is locally invertible at p if and only if the Jacobian determinant $\det[\partial F^i/\partial r^j(p)]$ is not zero.*

This theorem is usually proved in an undergraduate course on real analysis. See Appendix B for a discussion of this and related theorems. Because the inverse function theorem for $\mathbb{R}^n$ is a local result, it easily translates to manifolds.

Theorem 6.26 (Inverse function theorem for manifolds). *Let $F \colon N \to M$ be a C^∞ map between two manifolds of the same dimension, and $p \in N$. Suppose for some charts $(U, \phi) = (U, x^1, \ldots, x^n)$ about p in N and $(V, \psi) = (V, y^1, \ldots, y^n)$ about $F(p)$ in M, $F(U) \subset V$. Set $F^i = y^i \circ F$. Then F is locally invertible at p if and only if its Jacobian determinant $\det[\partial F^i/\partial x^j(p)]$ is nonzero.*

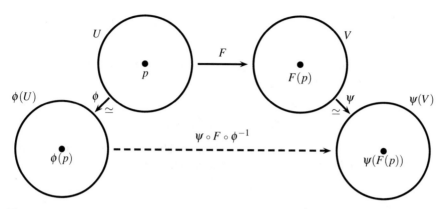

Fig. 6.4. The map F is locally invertible at p because $\psi \circ F \circ \phi^{-1}$ is locally invertible at $\phi(p)$.

Proof. Since $F^i = y^i \circ F = r^i \circ \psi \circ F$, the Jacobian matrix of F relative to the charts (U, ϕ) and (V, ψ) is

$$\left[\frac{\partial F^i}{\partial x^j}(p)\right] = \left[\frac{\partial (r^i \circ \psi \circ F)}{\partial x^j}(p)\right] = \left[\frac{\partial (r^i \circ \psi \circ F \circ \phi^{-1})}{\partial r^j}(\phi(p))\right],$$

which is precisely the Jacobian matrix at $\phi(p)$ of the map

$$\psi \circ F \circ \phi^{-1} \colon \mathbb{R}^n \supset \phi(U) \to \psi(V) \subset \mathbb{R}^n$$

between two open subsets of $\mathbb{R}^n$. By the inverse function theorem for $\mathbb{R}^n$,

$$\det\left[\frac{\partial F^i}{\partial x^j}(p)\right] = \det\left[\frac{\partial r^i \circ (\psi \circ F \circ \phi^{-1})}{\partial r^j}(\phi(p))\right] \neq 0$$

if and only if $\psi \circ F \circ \phi^{-1}$ is locally invertible at $\phi(p)$. Since ψ and ϕ are diffeomorphisms (Proposition 6.10), this last statement is equivalent to the local invertibility of F at p (see Figure 6.4). $\square$

We usually apply the inverse function theorem in the following form.

Corollary 6.27. *Let N be a manifold of dimension n. A set of n smooth functions $F^1, \ldots, F^n$ defined on a coordinate neighborhood $(U, x^1, \ldots, x^n)$ of a point $p \in N$ forms a coordinate system about p if and only if the Jacobian determinant $\det[\partial F^i / \partial x^j(p)]$ is nonzero.*

Proof. Let $F = (F^1, \ldots, F^n) \colon U \to \mathbb{R}^n$. Then

    $\det[\partial F^i / \partial x^j(p)] \neq 0$

$\Longleftrightarrow$ $F \colon U \to \mathbb{R}^n$ is locally invertible at p (by the inverse function theorem)

$\Longleftrightarrow$ there is a neighborhood W of p in N such that $F \colon W \to F(W)$ is a diffeomorphism (by the definition of local invertibility)

$\Longleftrightarrow$ $(W, F^1, \ldots, F^n)$ is a coordinate chart about p in the differentiable structure of N (by Proposition 6.11). $\square$

Example. Find all points in $\mathbb{R}^2$ in a neighborhood of which the functions $x^2 + y^2 - 1, y$ can serve as a local coordinate system.

Solution. Define $F: \mathbb{R}^2 \to \mathbb{R}^2$ by

$$F(x, y) = \left(x^2 + y^2 - 1, y\right).$$

The map F can serve as a coordinate map in a neighborhood of p if and only if it is a local diffeomorphism at p. The Jacobian determinant of F is

$$\frac{\partial \left(F^1, F^2\right)}{\partial (x, y)} = \det \begin{bmatrix} 2x & 2y \\ 0 & 1 \end{bmatrix} = 2x.$$

By the inverse function theorem, F is a local diffeomorphism at $p = (x, y)$ if and only if $x \neq 0$. Thus, F can serve as a coordinate system at any point p not on the y-axis.

Problems

6.1. Differentiable structures on $\mathbb{R}$
Let $\mathbb{R}$ be the real line with the differentiable structure given by the maximal atlas of the chart $(\mathbb{R}, \phi = \mathbb{1}: \mathbb{R} \to \mathbb{R})$, and let $\mathbb{R}'$ be the real line with the differentiable structure given by the maximal atlas of the chart $(\mathbb{R}, \psi: \mathbb{R} \to \mathbb{R})$, where $\psi(x) = x^{1/3}$.

(a) Show that these two differentiable structures are distinct.
(b) Show that there is a diffeomorphism between $\mathbb{R}$ and $\mathbb{R}'$. (*Hint*: The identity map $\mathbb{R} \to \mathbb{R}$ is not the desired diffeomorphism; in fact, this map is not smooth.)

6.2. The smoothness of an inclusion map
Let M and N be manifolds and let q_0 be a point in N. Prove that the inclusion map $i_{q_0}: M \to M \times N$, $i_{q_0}(p) = (p, q_0)$, is C^∞.

6.3.* Group of automorphisms of a vector space
Let V be a finite-dimensional vector space over $\mathbb{R}$, and $\mathrm{GL}(V)$ the group of all linear automorphisms of V. Relative to an ordered basis $e = (e_1, \ldots, e_n)$ for V, a linear automorphism $L \in \mathrm{GL}(V)$ is represented by a matrix $[a^i_j]$ defined by

$$L(e_j) = \sum_i a^i_j e_i.$$

The map

$$\phi_e: \mathrm{GL}(V) \to \mathrm{GL}(n, \mathbb{R}),$$

$$L \mapsto [a^i_j],$$

is a bijection with an open subset of $\mathbb{R}^{n \times n}$ that makes $\mathrm{GL}(V)$ into a C^∞ manifold, which we denote temporarily by $\mathrm{GL}(V)_e$. If $\mathrm{GL}(V)_u$ is the manifold structure induced from another ordered basis $u = (u_1, \ldots, u_n)$ for V, show that $\mathrm{GL}(V)_e$ is the same as $\mathrm{GL}(V)_u$.

6.4. Local coordinate systems
Find all points in $\mathbb{R}^3$ in a neighborhood of which the functions $x, x^2 + y^2 + z^2 - 1, z$ can serve as a local coordinate system.

§7 Quotients

Gluing the edges of a malleable square is one way to create new surfaces. For example, gluing together the top and bottom edges of a square gives a cylinder; gluing together the boundaries of the cylinder with matching orientations gives a torus (Figure 7.1). This gluing process is called an *identification* or a *quotient construction*.

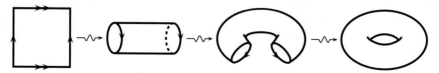

Fig. 7.1. Gluing the edges of a malleable square.

The quotient construction is a process of simplification. Starting with an equivalence relation on a set, we identify each equivalence class to a point. Mathematics abounds in quotient constructions, for example, the quotient group, quotient ring, or quotient vector space in algebra. If the original set is a topological space, it is always possible to give the quotient set a topology so that the natural projection map becomes continuous. However, even if the original space is a manifold, a quotient space is often not a manifold. The main results of this section give conditions under which a quotient space remains second countable and Hausdorff. We then study real projective space as an example of a quotient manifold.

Real projective space can be interpreted as a quotient of a sphere with antipodal points identified, or as the set of lines through the origin in a vector space. These two interpretations give rise to two distinct generalizations—covering maps on the one hand and Grassmannians of k-dimensional subspaces of a vector space on the other. In one of the exercises, we carry out an extensive investigation of $G(2,4)$, the Grassmannian of 2-dimensional subspaces of $\mathbb{R}^4$.

7.1 The Quotient Topology

Recall that an equivalence relation on a set S is a reflexive, symmetric, and transitive relation. The *equivalence class* $[x]$ of $x \in S$ is the set of all elements in S equivalent to x. An equivalence relation on S partitions S into disjoint equivalence classes. We denote the set of equivalence classes by $S/\sim$ and call this set the *quotient* of S by the equivalence relation $\sim$. There is a natural *projection map* $\pi\colon S \to S/\sim$ that sends $x \in S$ to its equivalence class $[x]$.

Assume now that S is a topological space. We define a topology on $S/\sim$ by declaring a set U in $S/\sim$ to be *open* if and only if $\pi^{-1}(U)$ is open in S. Clearly, both the empty set $\varnothing$ and the entire quotient $S/\sim$ are open. Further, since

$$\pi^{-1}\left(\bigcup_\alpha U_\alpha\right) = \bigcup_\alpha \pi^{-1}(U_\alpha)$$

and

$$\pi^{-1}\left(\bigcap_i U_i\right) = \bigcap_i \pi^{-1}(U_i),$$

the collection of open sets in $S/\sim$ is closed under arbitrary unions and finite intersections, and is therefore a topology. It is called the *quotient topology* on $S/\sim$. With this topology, $S/\sim$ is called the *quotient space* of S by the equivalence relation $\sim$. With the quotient topology on $S/\sim$, the projection map $\pi\colon S \to S/\sim$ is automatically continuous, because the inverse image of an open set in $S/\sim$ is by definition open in S.

7.2 Continuity of a Map on a Quotient

Let $\sim$ be an equivalence relation on the topological space S and give $S/\sim$ the quotient topology. Suppose a function $f\colon S \to Y$ from S to another topological space Y is constant on each equivalence class. Then it induces a map $\bar{f}\colon S/\sim \to Y$ by

$$\bar{f}([p]) = f(p) \quad \text{for } p \in S.$$

In other words, there is a commutative diagram

Proposition 7.1. *The induced map $\bar{f}\colon S/\sim \to Y$ is continuous if and only if the map $f\colon S \to Y$ is continuous.*

Proof.
($\Rightarrow$) If $\bar{f}$ is continuous, then as the composite $\bar{f} \circ \pi$ of continuous functions, f is also continuous.

($\Leftarrow$) Suppose f is continuous. Let V be open in Y. Then $f^{-1}(V) = \pi^{-1}(\bar{f}^{-1}(V))$ is open in S. By the definition of quotient topology, $\bar{f}^{-1}(V)$ is open in $S/\sim$. Since V was arbitrary, $\bar{f}\colon S/\sim \to Y$ is continuous. $\qquad\square$

This proposition gives a useful criterion for checking whether a function $\bar{f}$ on a quotient space $S/\sim$ is continuous: simply lift the function $\bar{f}$ to $f := f \circ \pi$ on S and check the continuity of the lifted map f on S. For examples of this, see Example 7.2 and Proposition 7.3.

7.3 Identification of a Subset to a Point

If A is a subspace of a topological space S, we can define a relation $\sim$ on S by declaring

$$x \sim x \quad \text{for all } x \in S$$

(so the relation is reflexive) and

$$x \sim y \quad \text{for all } x, y \in A.$$

This is an equivalence relation on S. We say that the quotient space $S/\sim$ is obtained from S by *identifying A to a point*.

Example 7.2. Let I be the unit interval $[0,1]$ and $I/\sim$ the quotient space obtained from I by identifying the two points $\{0,1\}$ to a point. Denote by S^1 the unit circle in the complex plane. The function $f \colon I \to S^1$, $f(x) = \exp(2\pi i x)$, assumes the same value at 0 and 1 (Figure 7.2), and so induces a function $\bar{f} \colon I/\sim \to S^1$.

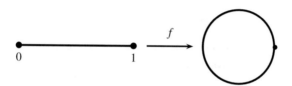

Fig. 7.2. The unit circle as a quotient space of the unit interval.

Proposition 7.3. *The function $\bar{f} \colon I/\sim \to S^1$ is a homeomorphism.*

Proof. Since f is continuous, $\bar{f}$ is also continuous by Proposition 7.1. Clearly, $\bar{f}$ is a bijection. As the continuous image of the compact set I, the quotient $I/\sim$ is compact. Thus, $\bar{f}$ is a continuous bijection from the compact space $I/\sim$ to the Hausdorff space S^1. By Corollary A.36, $\bar{f}$ is a homeomorphism.   □

7.4 A Necessary Condition for a Hausdorff Quotient

The quotient construction does not in general preserve the Hausdorff property or second countability. Indeed, since every singleton set in a Hausdorff space is closed, if $\pi \colon S \to S/\sim$ is the projection and the quotient $S/\sim$ is Hausdorff, then for any $p \in S$, its image $\{\pi(p)\}$ is closed in $S/\sim$. By the continuity of π, the inverse image $\pi^{-1}(\{\pi(p)\}) = [p]$ is closed in S. This gives a necessary condition for a quotient space to be Hausdorff.

Proposition 7.4. *If the quotient space $S/\sim$ is Hausdorff, then the equivalence class $[p]$ of any point p in S is closed in S.*

Example. Define an equivalence relation $\sim$ on $\mathbb{R}$ by identifying the open interval $]0,\infty[$ to a point. Then the quotient space $\mathbb{R}/\sim$ is not Hausdorff because the equivalence class $]0,\infty[$ of $\sim$ in $\mathbb{R}$ corresponding to the point $]0,\infty[$ in $\mathbb{R}/\sim$ is not a closed subset of $\mathbb{R}$.

7.5 Open Equivalence Relations

In this section we follow the treatment of Boothby [3] and derive conditions under which a quotient space is Hausdorff or second countable. Recall that a map $f\colon X \to Y$ of topological spaces is *open* if the image of any open set under f is open.

Definition 7.5.   An equivalence relation $\sim$ on a topological space S is said to be *open* if the projection map $\pi\colon S \to S/\sim$ is open.

In other words, the equivalence relation $\sim$ on S is open if and only if for every open set U in S, the set

$$\pi^{-1}(\pi(U)) = \bigcup_{x\in U} [x]$$

of all points equivalent to some point of U is open.

Example 7.6. The projection map to a quotient space is in general not open. For example, let $\sim$ be the equivalence relation on the real line $\mathbb{R}$ that identifies the two points 1 and -1, and $\pi\colon \mathbb{R} \to \mathbb{R}/\sim$ the projection map.

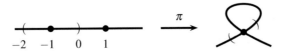

Fig. 7.3. A projection map that is not open.

The map π is open if and only if for every open set V in $\mathbb{R}$, its image $\pi(V)$ is open in $\mathbb{R}/\sim$, which by the definition of the quotient topology means that $\pi^{-1}(\pi(V))$ is open in $\mathbb{R}$. Now let V be the open interval $]-2,0[$ in $\mathbb{R}$. Then

$$\pi^{-1}(\pi(V)) = \,]-2,0[\,\cup\{1\},$$

which is not open in $\mathbb{R}$ (Figure 7.3). Therefore, the projection map $\pi\colon \mathbb{R} \to \mathbb{R}/\sim$ is not an open map.

Given an equivalence relation $\sim$ on S, let R be the subset of $S \times S$ that defines the relation

$$R = \{(x,y) \in S \times S \mid x \sim y\}.$$

We call R the *graph* of the equivalence relation $\sim$.

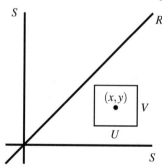

Fig. 7.4. The graph R of an equivalence relation and an open set $U \times V$ disjoint from R.

Theorem 7.7. *Suppose $\sim$ is an open equivalence relation on a topological space S. Then the quotient space $S/\!\sim$ is Hausdorff if and only if the graph R of $\sim$ is closed in $S \times S$.*

Proof. There is a sequence of equivalent statements:

R is closed in $S \times S$

$\Longleftrightarrow$ $(S \times S) - R$ is open in $S \times S$

$\Longleftrightarrow$ for every $(x,y) \in S \times S - R$, there is a basic open set $U \times V$ containing (x,y) such that $(U \times V) \cap R = \varnothing$ (Figure 7.4)

$\Longleftrightarrow$ for every pair $x \nsim y$ in S, there exist neighborhoods U of x and V of y in S such that no element of U is equivalent to an element of V

$\Longleftrightarrow$ for any two points $[x] \neq [y]$ in $S/\!\sim$, there exist neighborhoods U of x and V of y in S such that $\pi(U) \cap \pi(V) = \varnothing$ in $S/\!\sim$.   (∗)

We now show that this last statement (∗) is equivalent to $S/\!\sim$ being Hausdorff. First assume (∗). Since $\sim$ is an open equivalence relation, $\pi(U)$ and $\pi(V)$ are disjoint open sets in $S/\!\sim$ containing $[x]$ and $[y]$ respectively. Therefore, $S/\!\sim$ is Hausdorff.

Conversely, suppose $S/\!\sim$ is Hausdorff. Let $[x] \neq [y]$ in $S/\!\sim$. Then there exist disjoint open sets A and B in $S/\!\sim$ such that $[x] \in A$ and $[y] \in B$. By the surjectivity of π, we have $A = \pi(\pi^{-1}A)$ and $B = \pi(\pi^{-1}B)$ (see Problem 7.1). Let $U = \pi^{-1}A$ and $V = \pi^{-1}B$. Then $x \in U$, $y \in V$, and $A = \pi(U)$ and $B = \pi(V)$ are disjoint open sets in $S/\!\sim$. $\square$

If the equivalence relation $\sim$ is equality, then the quotient space $S/\!\sim$ is S itself and the graph R of $\sim$ is simply the diagonal

$$\Delta = \{(x,x) \in S \times S\}.$$

In this case, Theorem 7.7 becomes the following well-known characterization of a Hausdorff space by its diagonal (cf. Problem A.6).

Corollary 7.8. *A topological space S is Hausdorff if and only if the diagonal Δ in $S \times S$ is closed.*

Theorem 7.9. *Let ∼ be an open equivalence relation on a topological space S with projection* $\pi \colon S \to S/\sim$. *If* $\mathcal{B} = \{B_\alpha\}$ *is a basis for S, then its image* $\{\pi(B_\alpha)\}$ *under* π *is a basis for* $S/\sim$.

Proof. Since π is an open map, $\{\pi(B_\alpha)\}$ is a collection of open sets in $S/\sim$. Let W be an open set in $S/\sim$ and $[x] \in W$, $x \in S$. Then $x \in \pi^{-1}(W)$. Since $\pi^{-1}(W)$ is open, there is a basic open set $B \in \mathcal{B}$ such that

$$x \in B \subset \pi^{-1}(W).$$

Then

$$[x] = \pi(x) \in \pi(B) \subset W,$$

which proves that $\{\pi(B_\alpha)\}$ is a basis for $S/\sim$. □

Corollary 7.10. *If ∼ is an open equivalence relation on a second-countable space S, then the quotient space* $S/\sim$ *is second countable.*

7.6 Real Projective Space

Define an equivalence relation on $\mathbb{R}^{n+1} - \{0\}$ by

$$x \sim y \quad \Longleftrightarrow \quad y = tx \text{ for some nonzero real number } t,$$

where $x, y \in \mathbb{R}^{n+1} - \{0\}$. The *real projective space* $\mathbb{R}P^n$ is the quotient space of $\mathbb{R}^{n+1} - \{0\}$ by this equivalence relation. We denote the equivalence class of a point $(a^0, \dots, a^n) \in \mathbb{R}^{n+1} - \{0\}$ by $[a^0, \dots, a^n]$ and let $\pi \colon \mathbb{R}^{n+1} - \{0\} \to \mathbb{R}P^n$ be the projection. We call $[a^0, \dots, a^n]$ *homogeneous coordinates* on $\mathbb{R}P^n$.

Geometrically, two nonzero points in $\mathbb{R}^{n+1}$ are equivalent if and only if they lie on the same line through the origin, so $\mathbb{R}P^n$ can be interpreted as the set of all lines through the origin in $\mathbb{R}^{n+1}$. Each line through the origin in $\mathbb{R}^{n+1}$ meets the unit

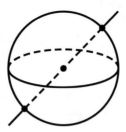

Fig. 7.5. A line through 0 in $\mathbb{R}^3$ corresponds to a pair of antipodal points on S^2.

sphere S^n in a pair of antipodal points, and conversely, a pair of antipodal points on S^n determines a unique line through the origin (Figure 7.5). This suggests that we define an equivalence relation ∼ on S^n by identifying antipodal points:

$$x \sim y \iff x = \pm y, \quad x, y \in S^n.$$

We then have a bijection $\mathbb{R}P^n \leftrightarrow S^n/\sim$.

Exercise 7.11 (Real projective space as a quotient of a sphere).* For $x = (x^1, \ldots, x^n) \in \mathbb{R}^n$, let $\|x\| = \sqrt{\sum_i (x^i)^2}$ be the modulus of x. Prove that the map $f : \mathbb{R}^{n+1} - \{0\} \to S^n$ given by

$$f(x) = \frac{x}{\|x\|}$$

induces a homeomorphism $\bar{f} : \mathbb{R}P^n \to S^n/\sim$. (*Hint*: Find an inverse map

$$\bar{g} : S^n/\sim \to \mathbb{R}P^n$$

and show that both $\bar{f}$ and $\bar{g}$ are continuous.)

Example 7.12 (*The real projective line* $\mathbb{R}P^1$).

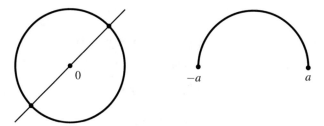

Fig. 7.6. The real projective line $\mathbb{R}P^1$ as the set of lines through 0 in $\mathbb{R}^2$.

Each line through the origin in $\mathbb{R}^2$ meets the unit circle in a pair of antipodal points. By Exercise 7.11, $\mathbb{R}P^1$ is homeomorphic to the quotient $S^1/\sim$, which is in turn homeomorphic to the closed upper semicircle with the two endpoints identified (Figure 7.6). Thus, $\mathbb{R}P^1$ is homeomorphic to S^1.

Example 7.13 (*The real projective plane* $\mathbb{R}P^2$). By Exercise 7.11, there is a homeomorphism

$$\mathbb{R}P^2 \simeq S^2/\{\text{antipodal points}\} = S^2/\sim.$$

For points not on the equator, each pair of antipodal points contains a unique point in the upper hemisphere. Thus, there is a bijection between $S^2/\sim$ and the quotient of the closed upper hemisphere in which each pair of antipodal points on the equator is identified. It is not difficult to show that this bijection is a homeomorphism (see Problem 7.2).

Let H^2 be the closed upper hemisphere

$$H^2 = \{(x, y, z) \in \mathbb{R}^3 \mid x^2 + y^2 + z^2 = 1, \; z \geq 0\}$$

and let D^2 be the closed unit disk

$$D^2 = \{(x,y) \in \mathbb{R}^2 \mid x^2 + y^2 \leq 1\}.$$

These two spaces are homeomorphic to each other via the continuous map

$$\varphi \colon H^2 \to D^2,$$
$$\varphi(x,y,z) = (x,y),$$

and its inverse

$$\psi \colon D^2 \to H^2,$$
$$\psi(x,y) = \left(x, y, \sqrt{1 - x^2 - y^2}\right).$$

On H^2, define an equivalence relation $\sim$ by identifying the antipodal points on the equator:

$$(x,y,0) \sim (-x,-y,0), \quad x^2 + y^2 = 1.$$

On D^2, define an equivalence relation $\sim$ by identifying the antipodal points on the boundary circle:

$$(x,y) \sim (-x,-y), \quad x^2 + y^2 = 1.$$

Then φ and ψ induce homeomorphisms

$$\bar{\varphi} \colon H^2/\!\sim \; \to D^2/\!\sim, \quad \bar{\psi} \colon D^2/\!\sim \; \to H^2/\!\sim .$$

In summary, there is a sequence of homeomorphisms

$$\mathbb{R}P^2 \;\overset{\simeq}{\to}\; S^2/\!\sim \;\overset{\simeq}{\to}\; H^2/\!\sim \;\overset{\simeq}{\to}\; D^2/\!\sim$$

that identifies the real projective plane as the quotient of the closed disk D^2 with the antipodal points on its boundary identified. This may be the best way to picture $\mathbb{R}P^2$ (Figure 7.7).

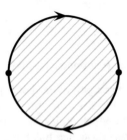

Fig. 7.7. The real projective plane as the quotient of a disk.

The real projective plane $\mathbb{R}P^2$ cannot be embedded as a submanifold of $\mathbb{R}^3$. However, if we allow self-intersection, then we can map $\mathbb{R}P^2$ into $\mathbb{R}^3$ as a cross-cap (Figure 7.8). This map is not one-to-one.

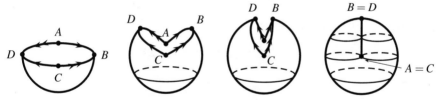

Fig. 7.8. The real projective plane immersed as a cross-cap in $\mathbb{R}^3$.

Proposition 7.14. *The equivalence relation $\sim$ on $\mathbb{R}^{n+1} - \{0\}$ in the definition of $\mathbb{R}P^n$ is an open equivalence relation.*

Proof. For an open set $U \subset \mathbb{R}^{n+1} - \{0\}$, the image $\pi(U)$ is open in $\mathbb{R}P^n$ if and only if $\pi^{-1}(\pi(U))$ is open in $\mathbb{R}^{n+1} - \{0\}$. But $\pi^{-1}(\pi(U))$ consists of all nonzero scalar multiples of points of U; that is,

$$\pi^{-1}(\pi(U)) = \bigcup_{t \in \mathbb{R}^\times} tU = \bigcup_{t \in \mathbb{R}^\times} \{tp \mid p \in U\}.$$

Since multiplication by $t \in \mathbb{R}^\times$ is a homeomorphism of $\mathbb{R}^{n+1} - \{0\}$, the set tU is open for any t. Therefore, their union $\bigcup_{t \in \mathbb{R}^\times} tU = \pi^{-1}(\pi(U))$ is also open. $\square$

Corollary 7.15. *The real projective space $\mathbb{R}P^n$ is second countable.*

Proof. Apply Corollary 7.10. $\square$

Proposition 7.16. *The real projective space $\mathbb{R}P^n$ is Hausdorff.*

Proof. Let $S = \mathbb{R}^{n+1} - \{0\}$ and consider the set

$$R = \{(x,y) \in S \times S \mid y = tx \text{ for some } t \in \mathbb{R}^\times\}.$$

If we write x and y as column vectors, then $[x\ y]$ is an $(n+1) \times 2$ matrix, and R may be characterized as the set of matrices $[x\ y]$ in $S \times S$ of rank ≤ 1. By a standard fact from linear algebra, $\mathrm{rk}[x\ y] \leq 1$ is equivalent to the vanishing of all 2×2 minors of $[x\ y]$ (see Problem B.1). As the zero set of finitely many polynomials, R is a closed subset of $S \times S$. Since $\sim$ is an open equivalence relation on S, and R is closed in $S \times S$, by Theorem 7.7 the quotient $S/\sim\ \simeq \mathbb{R}P^n$ is Hausdorff. $\square$

7.7 The Standard C^∞ Atlas on a Real Projective Space

Let $[a^0, \ldots, a^n]$ be homogeneous coordinates on the projective space $\mathbb{R}P^n$. Although a^0 is not a well-defined function on $\mathbb{R}P^n$, the condition $a^0 \neq 0$ is independent of the choice of a representative for $[a^0, \ldots, a^n]$. Hence, the condition $a^0 \neq 0$ makes sense on $\mathbb{R}P^n$, and we may define

$$U_0 := \{[a^0, \ldots, a^n] \in \mathbb{R}P^n \mid a^0 \neq 0\}.$$

Similarly, for each $i = 1, \ldots, n$, let

$$U_i := \{[a^0, \ldots, a^n] \in \mathbb{R}P^n \mid a^i \neq 0\}.$$

Define

$$\phi_0 \colon U_0 \to \mathbb{R}^n$$

by

$$[a^0, \ldots, a^n] \mapsto \left(\frac{a^1}{a^0}, \ldots, \frac{a^n}{a^0} \right).$$

This map has a continuous inverse

$$(b^1, \ldots, b^n) \mapsto [1, b^1, \ldots, b^n]$$

and is therefore a homeomorphism. Similarly, there are homeomorphisms for each $i = 1, \ldots, n$:

$$\phi_i \colon U_i \to \mathbb{R}^n,$$

$$[a^0, \ldots, a^n] \mapsto \left(\frac{a^0}{a^i}, \ldots, \frac{\widehat{a^i}}{a^i}, \ldots, \frac{a^n}{a^i} \right),$$

where the caret sign $\widehat{}$ over a^i/a^i means that that entry is to be omitted. This proves that $\mathbb{R}P^n$ is locally Euclidean with the (U_i, ϕ_i) as charts.

On the intersection $U_0 \cap U_1$, we have $a^0 \neq 0$ and $a^1 \neq 0$, and there are two coordinate systems

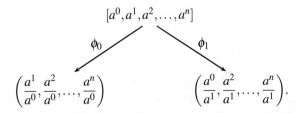

We will refer to the coordinate functions on U_0 as $x^1, \ldots, x^n$, and the coordinate functions on U_1 as $y^1, \ldots, y^n$. On U_0,

$$x^i = \frac{a^i}{a^0}, \quad i = 1, \ldots, n,$$

and on U_1,

$$y^1 = \frac{a^0}{a^1}, \quad y^2 = \frac{a^2}{a^1}, \quad \ldots, \quad y^n = \frac{a^n}{a^1}.$$

Then on $U_0 \cap U_1$,

$$y^1 = \frac{1}{x^1}, \quad y^2 = \frac{x^2}{x^1}, \quad y^3 = \frac{x^3}{x^1}, \quad \ldots, \quad y^n = \frac{x^n}{x^1},$$

so

$$(\phi_1 \circ \phi_0^{-1})(x) = \left(\frac{1}{x^1}, \frac{x^2}{x^1}, \frac{x^3}{x^1}, \ldots, \frac{x^n}{x^1} \right).$$

This is a C^∞ function because $x^1 \neq 0$ on $\phi_0(U_0 \cap U_1)$. On any other $U_i \cap U_j$ an analogous formula holds. Therefore, the collection $\{(U_i, \phi_i)\}_{i=0,\ldots,n}$ is a C^∞ atlas for $\mathbb{R}P^n$, called the *standard atlas*. This concludes the proof that $\mathbb{R}P^n$ is a C^∞ manifold.

Problems

7.1. Image of the inverse image of a map
Let $f: X \to Y$ be a map of sets, and let $B \subset Y$. Prove that $f(f^{-1}(B)) = B \cap f(X)$. Therefore, if f is surjective, then $f(f^{-1}(B)) = B$.

7.2. Real projective plane
Let H^2 be the closed upper hemisphere in the unit sphere S^2, and let $i: H^2 \to S^2$ be the inclusion map. In the notation of Example 7.13, prove that the induced map $f: H^2/\!\!\sim\; \to S^2/\!\!\sim$ is a homeomorphism. (*Hint*: Imitate Proposition 7.3.)

7.3. Closedness of the diagonal of a Hausdorff space
Deduce Theorem 7.7 from Corollary 7.8. (*Hint*: To prove that if $S/\!\!\sim$ is Hausdorff, then the graph R of $\sim$ is closed in $S \times S$, use the continuity of the projection map $\pi: S \to S/\!\!\sim$. To prove the reverse implication, use the openness of π.)

7.4.* Quotient of a sphere with antipodal points identified
Let S^n be the unit sphere centered at the origin in $\mathbb{R}^{n+1}$. Define an equivalence relation $\sim$ on S^n by identifying antipodal points:

$$x \sim y \iff x = \pm y, \quad x, y \in S^n.$$

(a) Show that $\sim$ is an open equivalence relation.
(b) Apply Theorem 7.7 and Corollary 7.8 to prove that the quotient space $S^n/\!\!\sim$ is Hausdorff, without making use of the homeomorphism $\mathbb{R}P^n \simeq S^n/\!\!\sim$.

7.5.* Orbit space of a continuous group action
Suppose a right action of a topological group G on a topological space S is continuous; this simply means that the map $S \times G \to S$ describing the action is continuous. Define two points x, y of S to be equivalent if they are in the same orbit; i.e., there is an element $g \in G$ such that $y = xg$. Let S/G be the quotient space; it is called the *orbit space* of the action. Prove that the projection map $\pi: S \to S/G$ is an open map. (This problem generalizes Proposition 7.14, in which $G = \mathbb{R}^\times = \mathbb{R} - \{0\}$ and $S = \mathbb{R}^{n+1} - \{0\}$. Because $\mathbb{R}^\times$ is commutative, a left $\mathbb{R}^\times$-action becomes a right $\mathbb{R}^\times$-action if scalar multiplication is written on the right.)

7.6. Quotient of $\mathbb{R}$ by $2\pi\mathbb{Z}$
Let the additive group $2\pi\mathbb{Z}$ act on $\mathbb{R}$ on the right by $x \cdot 2\pi n = x + 2\pi n$, where n is an integer. Show that the orbit space $\mathbb{R}/2\pi\mathbb{Z}$ is a smooth manifold.

7.7. The circle as a quotient space

(a) Let $\{(U_\alpha, \phi_\alpha)\}_{\alpha=1}^2$ be the atlas of the circle S^1 in Example 5.7, and let $\bar{\phi}_\alpha$ be the map ϕ_α followed by the projection $\mathbb{R} \to \mathbb{R}/2\pi\mathbb{Z}$. On $U_1 \cap U_2 = A \amalg B$, since ϕ_1 and ϕ_2 differ by an integer multiple of 2π, $\bar{\phi}_1 = \bar{\phi}_2$. Therefore, $\bar{\phi}_1$ and $\bar{\phi}_2$ piece together to give a well-defined map $\bar{\phi}: S^1 \to \mathbb{R}/2\pi\mathbb{Z}$. Prove that $\bar{\phi}$ is C^∞.

(b) The complex exponential $\mathbb{R} \to S^1$, $t \mapsto e^{it}$, is constant on each orbit of the action of $2\pi\mathbb{Z}$ on $\mathbb{R}$. Therefore, there is an induced map $F: \mathbb{R}/2\pi\mathbb{Z} \to S^1$, $F([t]) = e^{it}$. Prove that F is C^∞.

(c) Prove that $F: \mathbb{R}/2\pi\mathbb{Z} \to S^1$ is a diffeomorphism.

7.8. The Grassmannian $G(k,n)$

The Grassmannian $G(k,n)$ is the set of all k-planes through the origin in $\mathbb{R}^n$. Such a k-plane is a linear subspace of dimension k of $\mathbb{R}^n$ and has a basis consisting of k linearly independent vectors $a_1, \ldots, a_k$ in $\mathbb{R}^n$. It is therefore completely specified by an $n \times k$ matrix $A = [a_1 \cdots a_k]$ of rank k, where the *rank* of a matrix A, denoted by $\mathrm{rk}\, A$, is defined to be the number of linearly independent columns of A. This matrix is called a *matrix representative* of the k-plane. (For properties of the rank, see the problems in Appendix B.)

Two bases $a_1, \ldots, a_k$ and $b_1, \ldots, b_k$ determine the same k-plane if there is a change-of-basis matrix $g = [g_{ij}] \in GL(k, \mathbb{R})$ such that

$$b_j = \sum_i a_i g_{ij}, \quad 1 \le i, j \le k.$$

In matrix notation, $B = Ag$.

Let $F(k,n)$ be the set of all $n \times k$ matrices of rank k, topologized as a subspace of $\mathbb{R}^{n \times k}$, and $\sim$ the equivalence relation

$$A \sim B \quad \text{iff} \quad \text{there is a matrix } g \in GL(k, \mathbb{R}) \text{ such that } B = Ag.$$

In the notation of Problem B.3, $F(k,n)$ is the set $D_{\max}$ in $\mathbb{R}^{n \times k}$ and is therefore an open subset. There is a bijection between $G(k,n)$ and the quotient space $F(k,n)/\sim$. We give the Grassmannian $G(k,n)$ the quotient topology on $F(k,n)/\sim$.

(a) Show that $\sim$ is an open equivalence relation. (*Hint*: Either mimic the proof of Proposition 7.14 or apply Problem 7.5.)

(b) Prove that the Grassmannian $G(k,n)$ is second countable. (*Hint*: Apply Corollary 7.10.)

(c) Let $S = F(k,n)$. Prove that the graph R in $S \times S$ of the equivalence relation $\sim$ is closed. (*Hint*: Two matrices $A = [a_1 \cdots a_k]$ and $B = [b_1 \cdots b_k]$ in $F(k,n)$ are equivalent if and only if every column of B is a linear combination of the columns of A if and only if $\mathrm{rk}[A\ B] \le k$ if and only if all $(k+1) \times (k+1)$ minors of $[A\ B]$ are zero.)

(d) Prove that the Grassmannian $G(k,n)$ is Hausdorff. (*Hint*: Mimic the proof of Proposition 7.16.)

Next we want to find a C^∞ atlas on the Grassmannian $G(k,n)$. For simplicity, we specialize to $G(2,4)$. For any 4×2 matrix A, let A_{ij} be the 2×2 submatrix consisting of its ith row and jth row. Define

$$V_{ij} = \{A \in F(2,4) \mid A_{ij} \text{ is nonsingular}\}.$$

Because the complement of V_{ij} in $F(2,4)$ is defined by the vanishing of $\det A_{ij}$, we conclude that V_{ij} is an open subset of $F(2,4)$.

(e) Prove that if $A \in V_{ij}$, then $Ag \in V_{ij}$ for any nonsingular matrix $g \in GL(2, \mathbb{R})$.

Define $U_{ij} = V_{ij}/\sim$. Since $\sim$ is an open equivalence relation, $U_{ij} = V_{ij}/\sim$ is an open subset of $G(2,4)$.

For $A \in V_{12}$,

$$A \sim AA_{12}^{-1} = \begin{bmatrix} 1 & 0 \\ 0 & 1 \\ * & * \\ * & * \end{bmatrix} = \begin{bmatrix} I \\ A_{34}A_{12}^{-1} \end{bmatrix}.$$

This shows that the matrix representatives of a 2-plane in U_{12} have a canonical form B in which B_{12} is the identity matrix.

(f) Show that the map $\tilde{\phi}_{12} \colon V_{12} \to \mathbb{R}^{2\times 2}$,

$$\tilde{\phi}_{12}(A) = A_{34}A_{12}^{-1},$$

induces a homeomorphism $\phi_{12} \colon U_{12} \to \mathbb{R}^{2\times 2}$.

(g) Define similarly homeomorphisms $\phi_{ij} \colon U_{ij} \to \mathbb{R}^{2\times 2}$. Compute $\phi_{12} \circ \phi_{23}^{-1}$, and show that it is C^∞.

(h) Show that $\{U_{ij} \mid 1 \le i < j \le 4\}$ is an open cover of $G(2,4)$ and that $G(2,4)$ is a smooth manifold.

Similar consideration shows that $F(k,n)$ has an open cover $\{V_I\}$, where I is a strictly ascending multi-index $1 \le i_1 < \cdots < i_k \le n$. For $A \in F(k,n)$, let A_I be the $k \times k$ submatrix of A consisting of i_1th, ..., i_kth rows of A. Define

$$V_I = \{A \in G(k,n) \mid \det A_I \ne 0\}.$$

Next define $\tilde{\phi}_I \colon V_I \to \mathbb{R}^{(n-k)\times k}$ by

$$\tilde{\phi}_I(A) = (AA_I^{-1})_{I'},$$

where $(\ )_{I'}$ denotes the $(n-k) \times k$ submatrix obtained from the complement I' of the multi-index I. Let $U_I = V_I/\sim$. Then $\tilde{\phi}$ induces a homeomorphism $\phi \colon U_I \to \mathbb{R}^{(n-k)\times k}$. It is not difficult to show that $\{(U_I, \phi_I)\}$ is a C^∞ atlas for $G(k,n)$. Therefore the Grassmannian $G(k,n)$ is a C^∞ manifold of dimension $k(n-k)$.

7.9.* Compactness of real projective space

Show that the real projective space $\mathbb{R}P^n$ is compact. (*Hint*: Use Exercise 7.11.)

Chapter 3

The Tangent Space

By definition, the tangent space to a manifold at a point is the vector space of derivations at the point. A smooth map of manifolds induces a linear map, called its *differential*, of tangent spaces at corresponding points. In local coordinates, the differential is represented by the Jacobian matrix of partial derivatives of the map. In this sense, the differential of a map between manifolds is a generalization of the derivative of a map between Euclidean spaces.

A basic principle in manifold theory is the linearization principle, according to which a manifold can be approximated near a point by its tangent space at the point, and a smooth map can be approximated by the differential of the map. In this way, one turns a topological problem into a linear problem. A good example of the linearization principle is the inverse function theorem, which reduces the local invertibility of a smooth map to the invertibility of its differential at a point.

Using the differential, we classify maps having maximal rank at a point into immersions and submersions at the point, depending on whether the differential is injective or surjective there. A point where the differential is surjective is a *regular point* of the map. The regular level set theorem states that a level set all of whose points are regular is a regular submanifold, i.e., a subset that locally looks like a coordinate k-plane in $\mathbb{R}^n$. This theorem gives a powerful tool for proving that a topological space is a manifold.

We then introduce categories and functors, a framework for comparing structural similarities. After this interlude, we return to the study of maps via their differentials. From the rank of the differential, one obtains three local normal forms for smooth maps—the constant rank theorem, the immersion theorem, and the submersion theorem, corresponding to constant-rank differentials, injective differentials, and surjective differentials respectively. We give three proofs of the regular level set theorem, a first proof (Theorem 9.9), using the inverse function theorem, that actually produces explicit local coordinates, and two more proofs (p. 119) that are corollaries of the constant rank theorem and the submersion theorem.

The collection of tangent spaces to a manifold can be given the structure of a *vector bundle*; it is then called the *tangent bundle* of the manifold. Intuitively, a vector bundle over a manifold is a locally trivial family of vector spaces parametrized

L. W. Tu, *An Introduction to Manifolds,* Universitext, DOI 10.1007/978-1-4419-7400-6_3,

by points of the manifold. A smooth map of manifolds induces, via its differential at each point, a bundle map of the corresponding tangent bundles. In this way we obtain a covariant functor from the category of smooth manifolds and smooth maps to the category of vector bundles and bundle maps. Vector fields, which manifest themselves in the physical world as velocity, force, electricity, magnetism, and so on, may be viewed as sections of the tangent bundle over a manifold.

Smooth C^∞ bump functions and partitions of unity are an indispensable technical tool in the theory of smooth manifolds. Using C^∞ bump functions, we give several criteria for a vector field to be smooth. The chapter ends with integral curves, flows, and the Lie bracket of smooth vector fields.

§8 The Tangent Space

In Section 2 we saw that for any point p in an open set U in $\mathbb{R}^n$ there are two equivalent ways to define a tangent vector at p:

(i) as an arrow (Figure 8.1), represented by a column vector;

Fig. 8.1. A tangent vector in $\mathbb{R}^n$ as an arrow and as a column vector.

(ii) as a point-derivation of C_p^∞, the algebra of germs of C^∞ functions at p.

Both definitions generalize to a manifold. In the arrow approach, one defines a tangent vector at p in a manifold M by first choosing a chart (U, ϕ) at p and then decreeing a tangent vector at p to be an arrow at $\phi(p)$ in $\phi(U)$. This approach, while more visual, is complicated to work with, since a different chart (V, ψ) at p would give rise to a different set of tangent vectors at p and one would have to decide how to identify the arrows at $\phi(p)$ in U with the arrows at $\psi(p)$ in $\psi(V)$.

The cleanest and most intrinsic definition of a tangent vector at p in M is as a point-derivation, and this is the approach we adopt.

8.1 The Tangent Space at a Point

Just as for $\mathbb{R}^n$, we define a *germ* of a C^∞ function at p in M to be an equivalence class of C^∞ functions defined in a neighborhood of p in M, two such functions being equivalent if they agree on some, possibly smaller, neighborhood of p. The set of germs of C^∞ real-valued functions at p in M is denoted by $C_p^\infty(M)$. The addition and multiplication of functions make $C_p^\infty(M)$ into a ring; with scalar multiplication by real numbers, $C_p^\infty(M)$ becomes an algebra over $\mathbb{R}$.

Generalizing a derivation at a point in $\mathbb{R}^n$, we define a *derivation at a point* in a manifold M, or a *point-derivation* of $C_p^\infty(M)$, to be a linear map $D: C_p^\infty(M) \to \mathbb{R}$ such that

$$D(fg) = (Df)g(p) + f(p)Dg.$$

Definition 8.1. A *tangent vector* at a point p in a manifold M is a derivation at p.

Just as for $\mathbb{R}^n$, the tangent vectors at p form a vector space $T_p(M)$, called the *tangent space of M at p*. We also write T_pM instead of $T_p(M)$.

Remark 8.2 (*Tangent space to an open subset*). If U is an open set containing p in M, then the algebra $C_p^\infty(U)$ of germs of C^∞ functions in U at p is the same as $C_p^\infty(M)$. Hence, $T_pU = T_pM$.

Given a coordinate neighborhood $(U,\phi) = (U,x^1,\dots,x^n)$ about a point p in a manifold M, we recall the definition of the partial derivatives $\partial/\partial x^i$ first introduced in Section 6. Let $r^1,\dots,r^n$ be the standard coordinates on $\mathbb{R}^n$. Then

$$x^i = r^i \circ \phi : U \to \mathbb{R}.$$

If f is a smooth function in a neighborhood of p, we set

$$\left.\frac{\partial}{\partial x^i}\right|_p f = \left.\frac{\partial}{\partial r^i}\right|_{\phi(p)} \left(f \circ \phi^{-1}\right) \in \mathbb{R}.$$

It is easily checked that $\partial/\partial x^i|_p$ satisfies the derivation property and so is a tangent vector at p.

When M is one-dimensional and t is a local coordinate, it is customary to write $d/dt|_p$ instead of $\partial/\partial t|_p$ for the coordinate vector at the point p. To simplify the notation, we will sometimes write $\partial/\partial x^i$ instead of $\partial/\partial x^i|_p$ if it is understood at which point the tangent vector is located.

8.2 The Differential of a Map

Let $F: N \to M$ be a C^∞ map between two manifolds. At each point $p \in N$, the map F induces a linear map of tangent spaces, called its *differential at p*,

$$F_* : T_pN \to T_{F(p)}M$$

as follows. If $X_p \in T_pN$, then $F_*(X_p)$ is the tangent vector in $T_{F(p)}M$ defined by

$$(F_*(X_p))f = X_p(f \circ F) \in \mathbb{R} \quad \text{for } f \in C_{F(p)}^\infty(M). \tag{8.1}$$

Here f is a germ at $F(p)$, represented by a C^∞ function in a neighborhood of $F(p)$. Since (8.1) is independent of the representative of the germ, in practice we can be cavalier about the distinction between a germ and a representative function for the germ.

Exercise 8.3 (The differential of a map). Check that $F_*(X_p)$ is a derivation at $F(p)$ and that $F_*: T_pN \to T_{F(p)}M$ is a linear map.

To make the dependence on p explicit we sometimes write $F_{*,p}$ instead of F_*.

Example 8.4 (Differential of a map between Euclidean spaces). Suppose $F: \mathbb{R}^n \to \mathbb{R}^m$ is smooth and p is a point in $\mathbb{R}^n$. Let $x^1, \ldots, x^n$ be the coordinates on $\mathbb{R}^n$ and $y^1, \ldots, y^m$ the coordinates on $\mathbb{R}^m$. Then the tangent vectors $\partial/\partial x^1|_p, \ldots, \partial/\partial x^n|_p$ form a basis for the tangent space $T_p(\mathbb{R}^n)$ and $\partial/\partial y^1|_{F(p)}, \ldots, \partial/\partial y^m|_{F(p)}$ form a basis for the tangent space $T_{F(p)}(\mathbb{R}^m)$. The linear map $F_*: T_p(\mathbb{R}^n) \to T_{F(p)}(\mathbb{R}^m)$ is described by a matrix $[a_j^i]$ relative to these two bases:

$$F_*\left(\frac{\partial}{\partial x^j}\bigg|_p\right) = \sum_k a_j^k \frac{\partial}{\partial y^k}\bigg|_{F(p)}, \quad a_j^k \in \mathbb{R}. \tag{8.2}$$

Let $F^i = y^i \circ F$ be the ith component of F. We can find a_j^i by evaluating the right-hand side (RHS) and left-hand side (LHS) of (8.2) on y^i:

$$\text{RHS} = \sum_k a_j^k \frac{\partial}{\partial y^k}\bigg|_{F(p)} y^i = \sum_k a_j^k \delta_k^i = a_j^i,$$

$$\text{LHS} = F_*\left(\frac{\partial}{\partial x^j}\bigg|_p\right) y^i = \frac{\partial}{\partial x^j}\bigg|_p (y^i \circ F) = \frac{\partial F^i}{\partial x^j}(p).$$

So the matrix of F_* relative to the bases $\{\partial/\partial x^j|_p\}$ and $\{\partial/\partial y^i|_{F(p)}\}$ is $[\partial F^i/\partial x^j(p)]$. This is precisely the Jacobian matrix of the derivative of F at p. Thus, the differential of a map between manifolds generalizes the derivative of a map between Euclidean spaces.

8.3 The Chain Rule

Let $F: N \to M$ and $G: M \to P$ be smooth maps of manifolds, and $p \in N$. The differentials of F at p and G at $F(p)$ are linear maps

$$T_pN \xrightarrow{F_{*,p}} T_{F(p)}M \xrightarrow{G_{*,F(p)}} T_{G(F(p))}P.$$

Theorem 8.5 (The chain rule). *If $F: N \to M$ and $G: M \to P$ are smooth maps of manifolds and $p \in N$, then*

$$(G \circ F)_{*,p} = G_{*,F(p)} \circ F_{*,p}.$$

Proof. Let $X_p \in T_pN$ and let f be a smooth function at $G(F(p))$ in P. Then

$$((G \circ F)_*X_p)f = X_p(f \circ G \circ F)$$

and

$$((G_* \circ F_*)X_p)f = (G_*(F_*X_p))f = (F_*X_p)(f \circ G) = X_p(f \circ G \circ F). \qquad \square$$

Example 8.13 shows that when written out in terms of matrices, the chain rule of Theorem 8.5 assumes a more familiar form as a sum of products of partial derivatives.

Remark. The differential of the identity map $1_M \colon M \to M$ at any point p in M is the identity map

$$1_{T_pM} \colon T_pM \to T_pM,$$

because

$$((1_M)_* X_p)f = X_p(f \circ 1_M) = X_p f,$$

for any $X_p \in T_pM$ and $f \in C_p^\infty(M)$.

Corollary 8.6. *If $F \colon N \to M$ is a diffeomorphism of manifolds and $p \in N$, then $F_* \colon T_pN \to T_{F(p)}M$ is an isomorphism of vector spaces.*

Proof. To say that F is a diffeomorphism means that it has a differentiable inverse $G \colon M \to N$ such that $G \circ F = 1_N$ and $F \circ G = 1_M$. By the chain rule,

$$(G \circ F)_* = G_* \circ F_* = (1_N)_* = 1_{T_pN},$$
$$(F \circ G)_* = F_* \circ G_* = (1_M)_* = 1_{T_{F(p)}M}.$$

Hence, F_* and G_* are isomorphisms. $\square$

Corollary 8.7 (Invariance of dimension). *If an open set $U \subset \mathbb{R}^n$ is diffeomorphic to an open set $V \subset \mathbb{R}^m$, then $n = m$.*

Proof. Let $F \colon U \to V$ be a diffeomorphism and let $p \in U$. By Corollary 8.6, $F_{*,p} \colon T_pU \to T_{F(p)}V$ is an isomorphism of vector spaces. Since there are vector space isomorphisms $T_pU \simeq \mathbb{R}^n$ and $T_{F(p)} \simeq \mathbb{R}^m$, we must have that $n = m$. $\square$

8.4 Bases for the Tangent Space at a Point

As usual, we denote by $r^1, \ldots, r^n$ the standard coordinates on $\mathbb{R}^n$, and if (U, ϕ) is a chart about a point p in a manifold M of dimension n, we set $x^i = r^i \circ \phi$. Since $\phi \colon U \to \mathbb{R}^n$ is a diffeomorphism onto its image (Proposition 6.10), by Corollary 8.6 the differential

$$\phi_* \colon T_pM \to T_{\phi(p)}\mathbb{R}^n$$

is a vector space isomorphism. In particular, the tangent space T_pM has the same dimension n as the manifold M.

Proposition 8.8. *Let $(U, \phi) = (U, x^1, \ldots, x^n)$ be a chart about a point p in a manifold M. Then*

$$\phi_* \left(\frac{\partial}{\partial x^i} \bigg|_p \right) = \frac{\partial}{\partial r^i} \bigg|_{\phi(p)}.$$

Proof. For any $f \in C^\infty_{\phi(p)}(\mathbb{R}^n)$,

$$\phi_* \left(\frac{\partial}{\partial x^i} \bigg|_p \right) f = \frac{\partial}{\partial x^i} \bigg|_p (f \circ \phi) \qquad \text{(definition of } \phi_*)$$

$$= \frac{\partial}{\partial r^i} \bigg|_{\phi(p)} (f \circ \phi \circ \phi^{-1}) \quad \text{(definition of } \partial/\partial x^i|_p)$$

$$= \frac{\partial}{\partial r^i} \bigg|_{\phi(p)} f. \qquad \qquad \square$$

Proposition 8.9. *If* $(U, \phi) = (U, x^1, \dots, x^n)$ *is a chart containing p, then the tangent space T_pM has basis*

$$\frac{\partial}{\partial x^1} \bigg|_p, \dots, \frac{\partial}{\partial x^n} \bigg|_p.$$

Proof. An isomorphism of vector spaces carries a basis to a basis. By Proposition 8.8 the isomorphism $\phi_* : T_pM \to T_{\phi(p)}(\mathbb{R}^n)$ maps $\partial/\partial x^1|_p, \dots, \partial/\partial x^n|_p$ to $\partial/\partial r^1|_{\phi(p)}, \dots, \partial/\partial r^n|_{\phi(p)}$, which is a basis for the tangent space $T_{\phi(p)}(\mathbb{R}^n)$. Therefore, $\partial/\partial x^1|_p, \dots, \partial/\partial x^n|_p$ is a basis for T_pM. $\qquad \square$

Proposition 8.10 (Transition matrix for coordinate vectors). *Suppose* $(U, x^1, \dots, x^n)$ *and* $(V, y^1, \dots, y^n)$ *are two coordinate charts on a manifold M. Then*

$$\frac{\partial}{\partial x^j} = \sum_i \frac{\partial y^i}{\partial x^j} \frac{\partial}{\partial y^i}$$

on $U \cap V$.

Proof. At each point $p \in U \cap V$, the sets $\{\partial/\partial x^j|_p\}$ and $\{\partial/\partial y^i|_p\}$ are both bases for the tangent space T_pM, so there is a matrix $[a^i_j(p)]$ of real numbers such that on $U \cap V$,

$$\frac{\partial}{\partial x^j} = \sum_k a^k_j \frac{\partial}{\partial y^k}.$$

Applying both sides of the equation to y^i, we get

$$\frac{\partial y^i}{\partial x^j} = \sum_k a^k_j \frac{\partial y^i}{\partial y^k}$$

$$= \sum_k a^k_j \delta^i_k \quad \text{(by Proposition 6.22)}$$

$$= a^i_j. \qquad \qquad \square$$

8.5 A Local Expression for the Differential

Given a smooth map $F : N \to M$ of manifolds and $p \in N$, let $(U, x^1, \ldots, x^n)$ be a chart about p in N and let $(V, y^1, \ldots, y^m)$ be a chart about $F(p)$ in M. We will find a local expression for the differential $F_{*,p} : T_p N \to T_{F(p)} M$ relative to the two charts.

By Proposition 8.9, $\{\partial/\partial x^j|_p\}_{j=1}^n$ is a basis for $T_p N$ and $\{\partial/\partial y^i|_{F(p)}\}_{i=1}^m$ is a basis for $T_{F(p)} M$. Therefore, the differential $F_* = F_{*,p}$ is completely determined by the numbers a_j^i such that

$$F_* \left(\frac{\partial}{\partial x^j} \bigg|_p \right) = \sum_{k=1}^m a_j^k \frac{\partial}{\partial y^k} \bigg|_{F(p)}, \quad j = 1, \ldots, n.$$

Applying both sides to y^i, we find that

$$a_j^i = \left(\sum_{k=1}^m a_j^k \frac{\partial}{\partial y^k} \bigg|_{F(p)} \right) y^i = F_* \left(\frac{\partial}{\partial x^j} \bigg|_p \right) y^i = \frac{\partial}{\partial x^j} \bigg|_p (y^i \circ F) = \frac{\partial F^i}{\partial x^j}(p).$$

We state this result as a proposition.

Proposition 8.11. *Given a smooth map $F : N \to M$ of manifolds and a point $p \in N$, let $(U, x^1, \ldots, x^n)$ and $(V, y^1, \ldots, y^m)$ be coordinate charts about p in N and $F(p)$ in M respectively. Relative to the bases $\{\partial/\partial x^j|_p\}$ for $T_p N$ and $\{\partial/\partial y^i|_{F(p)}\}$ for $T_{F(p)} M$, the differential $F_{*,p} : T_p N \to T_{F(p)} M$ is represented by the matrix $[\partial F^i/\partial x^j(p)]$, where $F^i = y^i \circ F$ is the ith component of F.*

This proposition is in the spirit of the "arrow" approach to tangent vectors. Here each tangent vector in $T_p N$ is represented by a column vector relative to the basis $\{\partial/\partial x^j|_p\}$, and the differential $F_{*,p}$ is represented by a matrix.

Remark 8.12 (*Inverse function theorem*). In terms of the differential, the inverse function theorem for manifolds (Theorem 6.26) has a coordinate-free description: a C^∞ map $F : N \to M$ between two manifolds of the same dimension is locally invertible at a point $p \in N$ if and only if its differential $F_{*,p} : T_p N \to T_{f(p)} M$ at p is an isomorphism.

Example 8.13 (*The chain rule in calculus notation*). Suppose $w = G(x, y, z)$ is a C^∞ function: $\mathbb{R}^3 \to \mathbb{R}$ and $(x, y, z) = F(t)$ is a C^∞ function: $\mathbb{R} \to \mathbb{R}^3$. Under composition,

$$w = (G \circ F)(t) = G(x(t), y(t), z(t))$$

becomes a C^∞ function of $t \in \mathbb{R}$. The differentials F_*, G_*, and $(G \circ F)_*$ are represented by the matrices

$$\begin{bmatrix} dx/dt \\ dy/dt \\ dz/dt \end{bmatrix}, \quad \begin{bmatrix} \dfrac{\partial w}{\partial x} & \dfrac{\partial w}{\partial y} & \dfrac{\partial w}{\partial z} \end{bmatrix}, \quad \text{and} \quad \frac{dw}{dt},$$

respectively. Since composition of linear maps is represented by matrix multiplication, in terms of matrices the chain rule $(G \circ F)_* = G_* \circ F_*$ is equivalent to

$$\frac{dw}{dt} = \begin{bmatrix} \dfrac{\partial w}{\partial x} & \dfrac{\partial w}{\partial y} & \dfrac{\partial w}{\partial z} \end{bmatrix} \begin{bmatrix} dx/dt \\ dy/dt \\ dz/dt \end{bmatrix} = \frac{\partial w}{\partial x}\frac{dx}{dt} + \frac{\partial w}{\partial y}\frac{dy}{dt} + \frac{\partial w}{\partial z}\frac{dz}{dt}.$$

This is the usual form of the chain rule taught in calculus.

8.6 Curves in a Manifold

A *smooth curve* in a manifold M is by definition a smooth map $c\colon]a,b[\to M$ from some open interval $]a,b[$ into M. Usually we assume $0 \in]a,b[$ and say that c is a *curve starting at p* if $c(0) = p$. The *velocity vector* $c'(t_0)$ of the curve c at time $t_0 \in]a,b[$ is defined to be

$$c'(t_0) := c_*\left(\left.\frac{d}{dt}\right|_{t_0}\right) \in T_{c(t_0)}M.$$

We also say that $c'(t_0)$ is the velocity of c at the point $c(t_0)$. Alternative notations for $c'(t_0)$ are

$$\frac{dc}{dt}(t_0) \quad \text{and} \quad \left.\frac{d}{dt}\right|_{t_0} c.$$

NOTATION. When $c\colon]a,b[\to \mathbb{R}$ is a curve with target space $\mathbb{R}$, the notation $c'(t)$ can be a source of confusion. Here t is the standard coordinate on the domain $]a,b[$. Let x be the standard coordinate on the target space $\mathbb{R}$. By our definition, $c'(t)$ is a tangent vector at $c(t)$, hence a multiple of $d/dx|_{c(t)}$. On the other hand, in calculus notation $c'(t)$ is the derivative of a real-valued function and is therefore a scalar. If it is necessary to distinguish between these two meanings of $c'(t)$ when c maps into $\mathbb{R}$, we will write $\dot{c}(t)$ for the calculus derivative.

Exercise 8.14 (Velocity vector versus the calculus derivative).* Let $c\colon]a,b[\to \mathbb{R}$ be a curve with target space $\mathbb{R}$. Verify that $c'(t) = \dot{c}(t)\,d/dx|_{c(t)}$.

Example. Define $c\colon \mathbb{R} \to \mathbb{R}^2$ by

$$c(t) = (t^2, t^3).$$

(See Figure 8.2.)
   Then $c'(t)$ is a linear combination of $\partial/\partial x$ and $\partial/\partial y$ at $c(t)$:

$$c'(t) = a\frac{\partial}{\partial x} + b\frac{\partial}{\partial y}.$$

To compute a, we evaluate both sides on x:

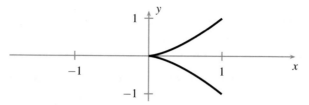

Fig. 8.2. A cuspidal cubic.

$$a = \left(a\frac{\partial}{\partial x} + b\frac{\partial}{\partial y}\right)x = c'(t)x = c_*\left(\frac{d}{dt}\right)x = \frac{d}{dt}(x \circ c) = \frac{d}{dt}t^2 = 2t.$$

Similarly,

$$b = \left(a\frac{\partial}{\partial x} + b\frac{\partial}{\partial y}\right)y = c'(t)y = c_*\left(\frac{d}{dt}\right)y = \frac{d}{dt}(y \circ c) = \frac{d}{dt}t^3 = 3t^2.$$

Thus,

$$c'(t) = 2t\frac{\partial}{\partial x} + 3t^2\frac{\partial}{\partial y}.$$

In terms of the basis $\partial/\partial x|_{c(t)}, \partial/\partial y|_{c(t)}$ for $T_{c(t)}(\mathbb{R}^2)$,

$$c'(t) = \begin{bmatrix} 2t \\ 3t^2 \end{bmatrix}.$$

More generally, as in this example, to compute the velocity vector of a smooth curve c in $\mathbb{R}^n$, one can simply differentiate the components of c. This shows that our definition of the velocity vector of a curve agrees with the usual definition in vector calculus.

Proposition 8.15 (Velocity of a curve in local coordinates). *Let $c\colon]a,b[\to M$ be a smooth curve, and let $(U,x^1,\ldots,x^n)$ be a coordinate chart about $c(t)$. Write $c^i = x^i \circ c$ for the ith component of c in the chart. Then $c'(t)$ is given by*

$$c'(t) = \sum_{i=1}^n \dot{c}^i(t) \left.\frac{\partial}{\partial x^i}\right|_{c(t)}.$$

Thus, relative to the basis $\{\partial/\partial x^i|_p\}$ for $T_{c(t)}M$, the velocity $c'(t)$ is represented by the column vector

$$\begin{bmatrix} \dot{c}^1(t) \\ \vdots \\ \dot{c}^n(t) \end{bmatrix}.$$

Proof. Problem 8.5.   □

Every smooth curve c at p in a manifold M gives rise to a tangent vector $c'(0)$ in T_pM. Conversely, one can show that every tangent vector $X_p \in T_pM$ is the velocity vector of some curve at p, as follows.

Proposition 8.16 (Existence of a curve with a given initial vector). *For any point p in a manifold M and any tangent vector $X_p \in T_pM$, there are $\varepsilon > 0$ and a smooth curve $c\colon\ ]-\varepsilon, \varepsilon[\to M$ such that $c(0) = p$ and $c'(0) = X_p$.*

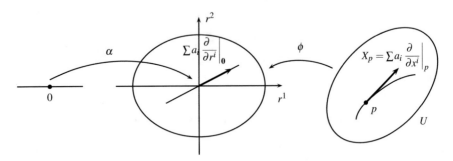

Fig. 8.3. Existence of a curve through a point with a given initial vector.

Proof. Let $(U,\phi) = (U,x^1,\ldots,x^n)$ be a chart centered at p; i.e., $\phi(p) = \mathbf{0} \in \mathbb{R}^n$. Suppose $X_p = \sum a^i \partial/\partial x^i|_p$ at p. Let $r^1,\ldots,r^n$ be the standard coordinates on $\mathbb{R}^n$. Then $x^i = r^i \circ \phi$. To find a curve c at p with $c'(0) = X_p$, start with a curve α in $\mathbb{R}^n$ with $\alpha(0) = \mathbf{0}$ and $\alpha'(0) = \sum a^i \partial/\partial r^i|_0$. We then map α to M via ϕ^{-1} (Figure 8.3). By Proposition 8.15, the simplest such α is

$$\alpha(t) = (a^1 t, \ldots, a^n t), \quad t \in\ ]-\varepsilon, \varepsilon[,$$

where ε is sufficiently small that $\alpha(t)$ lies in $\phi(U)$. Define $c = \phi^{-1} \circ \alpha\colon\ ]-\varepsilon, \varepsilon[\to M$. Then

$$c(0) = \phi^{-1}(\alpha(0)) = \phi^{-1}(\mathbf{0}) = p,$$

and by Proposition 8.8,

$$c'(0) = (\phi^{-1})_* \alpha_* \left(\left. \frac{d}{dt} \right|_{t=0} \right) = (\phi^{-1})_* \left(\sum a^i \left. \frac{\partial}{\partial r^i} \right|_0 \right) = \sum a^i \left. \frac{\partial}{\partial x^i} \right|_p = X_p. \quad \square$$

In Definition 8.1 we defined a tangent vector at a point p of a manifold abstractly as a derivation at p. Using curves, we can now interpret a tangent vector geometrically as a directional derivative.

Proposition 8.17. *Suppose X_p is a tangent vector at a point p of a manifold M and $f \in C_p^\infty(M)$. If $c\colon\ ]-\varepsilon, \varepsilon[\to M$ is a smooth curve starting at p with $c'(0) = X_p$, then*

$$X_p f = \left. \frac{d}{dt} \right|_0 (f \circ c).$$

Proof. By the definitions of $c'(0)$ and c_*,

$$X_p f = c'(0)f = c_* \left(\frac{d}{dt}\bigg|_0 \right) f = \frac{d}{dt}\bigg|_0 (f \circ c).$$ □

8.7 Computing the Differential Using Curves

We have introduced two ways of computing the differential of a smooth map, in terms of derivations at a point (equation (8.1)) and in terms of local coordinates (Proposition 8.11). The next proposition gives still another way of computing the differential $F_{*,p}$, this time using curves.

Proposition 8.18. *Let $F: N \to M$ be a smooth map of manifolds, $p \in N$, and $X_p \in T_pN$. If c is a smooth curve starting at p in N with velocity X_p at p, then*

$$F_{*,p}(X_p) = \frac{d}{dt}\bigg|_0 (F \circ c)(t).$$

In other words, $F_{*,p}(X_p)$ is the velocity vector of the image curve $F \circ c$ at $F(p)$.

Proof. By hypothesis, $c(0) = p$ and $c'(0) = X_p$. Then

$$F_{*,p}(X_p) = F_{*,p}(c'(0))$$

$$= (F_{*,p} \circ c_{*,0}) \left(\frac{d}{dt}\bigg|_0 \right)$$

$$= (F \circ c)_{*,0} \left(\frac{d}{dt}\bigg|_0 \right) \quad \text{(by the chain rule, Theorem 8.5)}$$

$$= \frac{d}{dt}\bigg|_0 (F \circ c)(t).$$ □

Example 8.19 (*Differential of left multiplication*). If g is a matrix in the general linear group $\mathrm{GL}(n, \mathbb{R})$, let $\ell_g: \mathrm{GL}(n, \mathbb{R}) \to \mathrm{GL}(n, \mathbb{R})$ be left multiplication by g; thus, $\ell_g(B) = gB$ for any $B \in \mathrm{GL}(n, \mathbb{R})$. Since $\mathrm{GL}(n, \mathbb{R})$ is an open subset of the vector space $\mathbb{R}^{n \times n}$, the tangent space $T_g(\mathrm{GL}(n, \mathbb{R}))$ can be identified with $\mathbb{R}^{n \times n}$. Show that with this identification the differential $(\ell_g)_{*,I}: T_I(\mathrm{GL}(n, \mathbb{R})) \to T_g(\mathrm{GL}(n, \mathbb{R}))$ is also left multiplication by g.

Solution. Let $X \in T_I(\mathrm{GL}(n, \mathbb{R})) = \mathbb{R}^{n \times n}$. To compute $(\ell_g)_{*,I}(X)$, choose a curve $c(t)$ in $\mathrm{GL}(n, \mathbb{R})$ with $c(0) = I$ and $c'(0) = X$. Then $\ell_g(c(t)) = gc(t)$ is simply matrix multiplication. By Proposition 8.18,

$$(\ell_g)_{*,I}(X) = \frac{d}{dt}\bigg|_{t=0} \ell_g(c(t)) = \frac{d}{dt}\bigg|_{t=0} gc(t) = gc'(0) = gX.$$

In this computation, $d/dt|_{t=0} \, gc(t) = gc'(0)$ by $\mathbb{R}$-linearity and Proposition 8.15. □

8.8 Immersions and Submersions

Just as the derivative of a map between Euclidean spaces is a linear map that best approximates the given map at a point, so the differential at a point serves the same purpose for a C^∞ map between manifolds. Two cases are especially important. A C^∞ map $F\colon N \to M$ is said to be an *immersion at* $p \in N$ if its differential $F_{*,p}\colon T_pN \to T_{F(p)}M$ is injective, and a *submersion at* p if $F_{*,p}$ is surjective. We call F an *immersion* if it is an immersion at every $p \in N$ and a *submersion* if it is a submersion at every $p \in N$.

Remark 8.20. Suppose N and M are manifolds of dimensions n and m respectively. Then $\dim T_pN = n$ and $\dim T_{F(p)}M = m$. The injectivity of the differential $F_{*,p}\colon T_pN \to T_{F(p)}M$ implies immediately that $n \le m$. Similarly, the surjectivity of the differential $F_{*,p}$ implies that $n \ge m$. Thus, if $F\colon N \to M$ is an immersion at a point of N, then $n \le m$ and if F is a submersion a point of N, then $n \ge m$.

Example 8.21. The prototype of an immersion is the inclusion of $\mathbb{R}^n$ in a higher-dimensional $\mathbb{R}^m$:

$$i(x^1,\ldots,x^n) = (x^1,\ldots,x^n,0,\ldots,0).$$

The prototype of a submersion is the projection of $\mathbb{R}^n$ onto a lower-dimensional $\mathbb{R}^m$:

$$\pi(x^1,\ldots,x^m,x^{m+1},\ldots,x^n) = (x^1,\ldots,x^m).$$

Example. If U is an open subset of a manifold M, then the inclusion $i\colon U \to M$ is both an immersion and a submersion. This example shows in particular that a submersion need not be onto.

In Section 11, we will undertake a more in-depth analysis of immersions and submersions. According to the immersion and submersion theorems to be proven there, every immersion is locally an inclusion and every submersion is locally a projection.

8.9 Rank, and Critical and Regular Points

The *rank* of a linear transformation $L\colon V \to W$ between finite-dimensional vector spaces is the dimension of the image $L(V)$ as a subspace of W, while the *rank* of a matrix A is the dimension of its column space. If L is represented by a matrix A relative to a basis for V and a basis for W, then the rank of L is the same as the rank of A, because the image $L(V)$ is simply the column space of A.

Now consider a smooth map $F\colon N \to M$ of manifolds. Its *rank* at a point p in N, denoted by $\operatorname{rk} F(p)$, is defined as the rank of the differential $F_{*,p}\colon T_pN \to T_{F(p)}M$. Relative to the coordinate neighborhoods $(U,x^1,\ldots,x^n)$ at p and $(V,y^1,\ldots,y^m)$ at $F(p)$, the differential is represented by the Jacobian matrix $[\partial F^i/\partial x^j(p)]$ (Proposition 8.11), so

$$\operatorname{rk} F(p) = \operatorname{rk} \left[\frac{\partial F^i}{\partial x^j}(p) \right].$$

Since the differential of a map is independent of coordinate charts, so is the rank of a Jacobian matrix.

Definition 8.22. A point p in N is a *critical point* of F if the differential

$$F_{*,p} : T_p N \to T_{F(p)} M$$

fails to be surjective. It is a *regular point* of F if the differential $F_{*,p}$ is surjective. In other words, p is a regular point of the map F if and only if F is a submersion at p. A point in M is a *critical value* if it is the image of a critical point; otherwise it is a *regular value* (Figure 8.4).

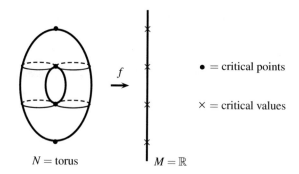

$N = \text{torus}$ $M = \mathbb{R}$

Fig. 8.4. Critical points and critical values of the function $f(x,y,z) = z$.

Two aspects of this definition merit elaboration:

(i) We do *not* define a regular value to be the image of a regular point. In fact, a regular value need not be in the image of F at all. Any point of M not in the image of F is automatically a regular value because it is not the image of a critical point.

(ii) A point c in M is a critical value if and only if *some* point in the preimage $F^{-1}(\{c\})$ is a critical point. A point c in the image of F is a regular value if and only if *every* point in the preimage $F^{-1}(\{c\})$ is a regular point.

Proposition 8.23. *For a real-valued function $f : M \to \mathbb{R}$, a point p in M is a critical point if and only if relative to some chart $(U, x^1, \ldots, x^n)$ containing p, all the partial derivatives satisfy*

$$\frac{\partial f}{\partial x^j}(p) = 0, \quad j = 1, \ldots, n.$$

Proof. By Proposition 8.11 the differential $f_{*,p} : T_p M \to T_{f(p)} \mathbb{R} \simeq \mathbb{R}$ is represented by the matrix

$$\left[\frac{\partial f}{\partial x^1}(p) \quad \cdots \quad \frac{\partial f}{\partial x^n}(p)\right].$$

Since the image of $f_{*,p}$ is a linear subspace of $\mathbb{R}$, it is either zero-dimensional or one-dimensional. In other words, $f_{*,p}$ is either the zero map or a surjective map. Therefore, $f_{*,p}$ fails to be surjective if and only if all the partial derivatives $\partial f/\partial x^i(p)$ are zero. $\qquad\square$

Problems

8.1.* Differential of a map
Let $F\colon \mathbb{R}^2 \to \mathbb{R}^3$ be the map

$$(u,v,w) = F(x,y) = (x,y,xy).$$

Let $p = (x,y) \in \mathbb{R}^2$. Compute $F_*(\partial/\partial x|_p)$ as a linear combination of $\partial/\partial u$, $\partial/\partial v$, and $\partial/\partial w$ at $F(p)$.

8.2. Differential of a linear map
Let $L\colon \mathbb{R}^n \to \mathbb{R}^m$ be a linear map. For any $p \in \mathbb{R}^n$, there is a canonical identification $T_p(\mathbb{R}^n) \overset{\sim}{\to} \mathbb{R}^n$ given by

$$\sum a^i \frac{\partial}{\partial x^i}\bigg|_p \mapsto \mathbf{a} = \langle a^1,\dots,a^n\rangle.$$

Show that the differential $L_{*,p}\colon T_p(\mathbb{R}^n) \to T_{L(p)}(\mathbb{R}^m)$ is the map $L\colon \mathbb{R}^n \to \mathbb{R}^m$ itself, with the identification of the tangent spaces as above.

8.3. Differential of a map
Fix a real number α and define $F\colon \mathbb{R}^2 \to \mathbb{R}^2$ by

$$\begin{bmatrix} u \\ v \end{bmatrix} = (u,v) = F(x,y) = \begin{bmatrix} \cos\alpha & -\sin\alpha \\ \sin\alpha & \cos\alpha \end{bmatrix}\begin{bmatrix} x \\ y \end{bmatrix}.$$

Let $X = -y\partial/\partial x + x\partial/\partial y$ be a vector field on $\mathbb{R}^2$. If $p = (x,y) \in \mathbb{R}^2$ and $F_*(X_p) = (a\partial/\partial u + b\partial/\partial v)|_{F(p)}$, find a and b in terms of x, y, and α.

8.4. Transition matrix for coordinate vectors
Let x,y be the standard coordinates on $\mathbb{R}^2$, and let U be the open set

$$U = \mathbb{R}^2 - \{(x,0) \mid x \geq 0\}.$$

On U the polar coordinates r,θ are uniquely defined by

$$x = r\cos\theta,$$
$$y = r\sin\theta, \; r > 0, \; 0 < \theta < 2\pi.$$

Find $\partial/\partial r$ and $\partial/\partial\theta$ in terms of $\partial/\partial x$ and $\partial/\partial y$.

8.5.* Velocity of a curve in local coordinates
Prove Proposition 8.15.

8.6. Velocity vector
Let $p = (x, y)$ be a point in $\mathbb{R}^2$. Then

$$c_p(t) = \begin{bmatrix} \cos 2t & -\sin 2t \\ \sin 2t & \cos 2t \end{bmatrix} \begin{bmatrix} x \\ y \end{bmatrix}, \quad t \in \mathbb{R},$$

is a curve with initial point p in $\mathbb{R}^2$. Compute the velocity vector $c_p'(0)$.

8.7.* Tangent space to a product
If M and N are manifolds, let $\pi_1 : M \times N \to M$ and $\pi_2 : M \times N \to N$ be the two projections. Prove that for $(p, q) \in M \times N$,

$$(\pi_{1*}, \pi_{2*}) : T_{(p,q)}(M \times N) \to T_p M \times T_q N$$

is an isomorphism.

8.8. Differentials of multiplication and inverse
Let G be a Lie group with multiplication map $\mu : G \times G \to G$, inverse map $\iota : G \to G$, and identity element e.

(a) Show that the differential at the identity of the multiplication map μ is addition:

$$\mu_{*,(e,e)} : T_e G \times T_e G \to T_e G,$$
$$\mu_{*,(e,e)}(X_e, Y_e) = X_e + Y_e.$$

(Hint: First, compute $\mu_{*,(e,e)}(X_e, 0)$ and $\mu_{*,(e,e)}(0, Y_e)$ using Proposition 8.18.)
(b) Show that the differential at the identity of ι is the negative:

$$\iota_{*,e} : T_e G \to T_e G,$$
$$\iota_{*,e}(X_e) = -X_e.$$

(Hint: Take the differential of $\mu(c(t), (\iota \circ c)(t)) = e$.)

8.9.* Transforming vectors to coordinate vectors
Let $X_1, \ldots, X_n$ be n vector fields on an open subset U of a manifold of dimension n. Suppose that at $p \in U$, the vectors $(X_1)_p, \ldots, (X_n)_p$ are linearly independent. Show that there is a chart $(V, x^1, \ldots, x^n)$ about p such that $(X_i)_p = (\partial/\partial x^i)_p$ for $i = 1, \ldots, n$.

8.10. Local maxima
A real-valued function $f : M \to \mathbb{R}$ on a manifold is said to have a *local maximum* at $p \in M$ if there is a neighborhood U of p such that $f(p) \geq f(q)$ for all $q \in U$.

(a)* Prove that if a differentiable function $f : I \to \mathbb{R}$ defined on an open interval I has a local maximum at $p \in I$, then $f'(p) = 0$.
(b) Prove that a local maximum of a C^∞ function $f : M \to \mathbb{R}$ is a critical point of f. (Hint: Let X_p be a tangent vector in $T_p M$ and let $c(t)$ be a curve in M starting at p with initial vector X_p. Then $f \circ c$ is a real-valued function with a local maximum at 0. Apply (a).)

§9 Submanifolds

We now have two ways of showing that a given topological space is a manifold:

(a) by checking directly that the space is Hausdorff, second countable, and has a C^∞ atlas;
(b) by exhibiting it as an appropriate quotient space. Section 7 lists some conditions under which a quotient space is a manifold.

In this section we introduce the concept of a *regular submanifold* of a manifold, a subset that is locally defined by the vanishing of some of the coordinate functions. Using the inverse function theorem, we derive a criterion, called the *regular level set theorem*, that can often be used to show that a level set of a C^∞ map of manifolds is a regular submanifold and therefore a manifold.

Although the regular level set theorem is a simple consequence of the constant rank theorem and the submersion theorem to be discussed in Section 11, deducing it directly from the inverse function theorem has the advantage of producing explicit coordinate functions on the submanifold.

9.1 Submanifolds

The xy-plane in $\mathbb{R}^3$ is the prototype of a *regular submanifold* of a manifold. It is defined by the vanishing of the coordinate function z.

Definition 9.1.   A subset S of a manifold N of dimension n is a *regular submanifold* of dimension k if for every $p \in S$ there is a coordinate neighborhood $(U, \phi) = (U, x^1, \ldots, x^n)$ of p in the maximal atlas of N such that $U \cap S$ is defined by the vanishing of $n - k$ of the coordinate functions. By renumbering the coordinates, we may assume that these $n - k$ coordinate functions are $x^{k+1}, \ldots, x^n$.

We call such a chart (U, ϕ) in N an *adapted chart* relative to S. On $U \cap S$, $\phi = (x^1, \ldots, x^k, 0, \ldots, 0)$. Let

$$\phi_S \colon U \cap S \to \mathbb{R}^k$$

be the restriction of the first k components of ϕ to $U \cap S$, that is, $\phi_S = (x^1, \ldots, x^k)$. Note that $(U \cap S, \phi_S)$ is a chart for S in the subspace topology.

Definition 9.2.   If S is a regular submanifold of dimension k in a manifold N of dimension n, then $n - k$ is said to be the *codimension* of S in N.

Remark. As a topological space, a regular submanifold of N is required to have the subspace topology.

Example. In the definition of a regular submanifold, the dimension k of the submanifold may be equal to n, the dimension of the manifold. In this case, $U \cap S$ is defined by the vanishing of none of the coordinate functions and so $U \cap S = U$. Therefore, an open subset of a manifold is a regular submanifold of the same dimension.

Remark. There are other types of submanifolds, but unless otherwise specified, by a "submanifold" we will always mean a "regular submanifold."

Example. The interval $S := \,]-1,1[$ on the x-axis is a regular submanifold of the xy-plane (Figure 9.1). As an adapted chart, we can take the open square $U = \,]-1,1[\times \,]-1,1[$ with coordinates x,y. Then $U \cap S$ is precisely the zero set of y on U.

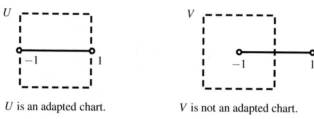

U is an adapted chart. V is not an adapted chart.

Fig. 9.1.

Note that if $V = \,]-2,0[\times \,]-1,1[$, then (V,x,y) is not an adapted chart relative to S, since $V \cap S$ is the open interval $]-1,0[$ on the x-axis, while the zero set of y on V is the open interval $]-2,0[$ on the x-axis.

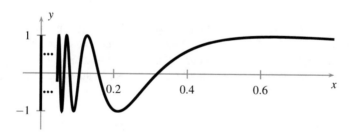

Fig. 9.2. The topologist's sine curve.

Example 9.3. Let Γ be the graph of the function $f(x) = \sin(1/x)$ on the interval $]0,1[$, and let S be the union of Γ and the open interval

$$I = \{(0,y) \in \mathbb{R}^2 \mid -1 < y < 1\}.$$

The subset S of $\mathbb{R}^2$ is not a regular submanifold for the following reason: if p is in the interval I, then there is no adapted chart containing p, since any sufficiently small neighborhood U of p in $\mathbb{R}^2$ intersects S in infinitely many components. (The closure of Γ in $\mathbb{R}^2$ is called the *topologist's sine curve* (Figure 9.2). It differs from S in including the endpoints $(1,\sin 1)$, $(0,1)$, and $(0,-1)$.)

Proposition 9.4. *Let S be a regular submanifold of N and $\mathfrak{U} = \{(U,\phi)\}$ a collection of compatible adapted charts of N that covers S. Then $\{(U \cap S, \phi_S)\}$ is an atlas for S. Therefore, a regular submanifold is itself a manifold. If N has dimension n and S is locally defined by the vanishing of $n-k$ coordinates, then $\dim S = k$.*

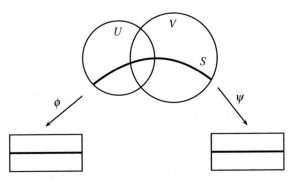

Fig. 9.3. Overlapping adapted charts relative to a regular submanifold S.

Proof. Let $(U,\phi) = (U, x^1, \ldots, x^n)$ and $(V, \psi) = (V, y^1, \ldots, y^n)$ be two adapted charts in the given collection (Figure 9.3). Assume that they intersect. As we remarked in Definition 9.1, in any adapted chart relative to a submanifold S it is possible to renumber the coordinates so that the last $n-k$ coordinates vanish on points of S. Then for $p \in U \cap V \cap S$,

$$\phi(p) = (x^1, \ldots, x^k, 0, \ldots, 0) \quad \text{and} \quad \psi(p) = (y^1, \ldots, y^k, 0, \ldots, 0),$$

so

$$\phi_S(p) = (x^1, \ldots, x^k) \quad \text{and} \quad \psi_S(p) = (y^1, \ldots, y^k).$$

Therefore,

$$\left(\psi_S \circ \phi_S^{-1} \right)(x^1, \ldots, x^k) = (y^1, \ldots, y^k).$$

Since $y^1, \ldots, y^k$ are C^∞ functions of $x^1, \ldots, x^k$ (because $\psi \circ \phi^{-1}(x^1, \ldots, x^k, 0, \ldots, 0)$ is C^∞), the transition function $\psi_S \circ \phi_S^{-1}$ is C^∞. Similarly, since $x^1, \ldots, x^k$ are C^∞ functions of $y^1, \ldots, y^k$, $\phi_S \circ \psi_S^{-1}$ is also C^∞. Hence, any two charts in $\{(U \cap S, \phi_S)\}$ are C^∞ compatible. Since $\{U \cap S\}_{U \in \mathfrak{U}}$ covers S, the collection $\{(U \cap S, \phi_S)\}$ is a C^∞ atlas on S. $\qquad\square$

9.2 Level Sets of a Function

A *level set* of a map $F: N \to M$ is a subset

$$F^{-1}(\{c\}) = \{p \in N \mid F(p) = c\}$$

for some $c \in M$. The usual notation for a level set is $F^{-1}(c)$, rather than the more correct $F^{-1}(\{c\})$. The value $c \in M$ is called the *level* of the level set $F^{-1}(c)$. If $F: N \to \mathbb{R}^m$, then $Z(F) := F^{-1}(0)$ is the *zero set* of F. Recall that c is a regular value of F if and only if either c is not in the image of F or at every point $p \in F^{-1}(c)$, the differential $F_{*,p}: T_pN \to T_{F(p)}M$ is surjective. The inverse image $F^{-1}(c)$ of a regular value c is called a *regular level set*. If the zero set $F^{-1}(0)$ is a regular level set of $F: N \to \mathbb{R}^m$, it is called a *regular zero set*.

Remark 9.5. If a regular level set $F^{-1}(c)$ is nonempty, say $p \in F^{-1}(c)$, then the map $F: N \to M$ is a submersion at p. By Remark 8.20, $\dim N \geq \dim M$.

Example 9.6 (The 2-sphere in $\mathbb{R}^3$). The unit 2-sphere

$$S^2 = \{(x,y,z) \in \mathbb{R}^3 \mid x^2 + y^2 + z^2 = 1\}$$

is the level set $g^{-1}(1)$ of level 1 of the function $g(x,y,z) = x^2 + y^2 + z^2$. We will use the inverse function theorem to find adapted charts of $\mathbb{R}^3$ that cover S^2. As the proof will show, the process is easier for a zero set, mainly because a regular submanifold is defined locally as the zero set of coordinate functions. To express S^2 as a zero set, we rewrite its defining equation as

$$f(x,y,z) = x^2 + y^2 + z^2 - 1 = 0.$$

Then $S^2 = f^{-1}(0)$.

Since

$$\frac{\partial f}{\partial x} = 2x, \qquad \frac{\partial f}{\partial y} = 2y, \qquad \frac{\partial f}{\partial z} = 2z,$$

the only critical point of f is $(0,0,0)$, which does not lie on the sphere S^2. Thus, all points on the sphere are regular points of f and 0 is a regular value of f.

Let p be a point of S^2 at which $(\partial f/\partial x)(p) = 2x(p) \neq 0$. Then the Jacobian matrix of the map $(f,y,z): \mathbb{R}^3 \to \mathbb{R}^3$ is

$$\begin{bmatrix} \dfrac{\partial f}{\partial x} & \dfrac{\partial f}{\partial y} & \dfrac{\partial f}{\partial z} \\[2mm] \dfrac{\partial y}{\partial x} & \dfrac{\partial y}{\partial y} & \dfrac{\partial y}{\partial z} \\[2mm] \dfrac{\partial z}{\partial x} & \dfrac{\partial z}{\partial y} & \dfrac{\partial z}{\partial z} \end{bmatrix} = \begin{bmatrix} \dfrac{\partial f}{\partial x} & \dfrac{\partial f}{\partial y} & \dfrac{\partial f}{\partial z} \\[2mm] 0 & 1 & 0 \\[2mm] 0 & 0 & 1 \end{bmatrix},$$

and the Jacobian determinant $\partial f/\partial x(p)$ is nonzero. By Corollary 6.27 of the inverse function theorem (Theorem 6.26), there is a neighborhood U_p of p in $\mathbb{R}^3$ such that

(U_p, f, y, z) is a chart in the atlas of $\mathbb{R}^3$. In this chart, the set $U_p \cap S^2$ is defined by the vanishing of the first coordinate f. Thus, (U_p, f, y, z) is an adapted chart relative to S^2, and $(U_p \cap S^2, y, z)$ is a chart for S^2.

Similarly, if $(\partial f / \partial y)(p) \neq 0$, then there is an adapted chart (V_p, x, f, z) containing p in which the set $V_p \cap S^2$ is the zero set of the second coordinate f. If $(\partial f / \partial z)(p) \neq 0$, then there is an adapted chart (W_p, x, y, f) containing p. Since for every $p \in S^2$, at least one of the partial derivatives $\partial f / \partial x(p)$, $\partial f / \partial y(p)$, $\partial f / \partial z(p)$ is nonzero, as p varies over all points of the sphere we obtain a collection of adapted charts of $\mathbb{R}^3$ covering S^2. Therefore, S^2 is a regular submanifold of $\mathbb{R}^3$. By Proposition 9.4, S^2 is a manifold of dimension 2.

This is an important example because one can generalize its proof almost verbatim to prove that if the zero set of a function $f : N \to \mathbb{R}$ is a regular level set, then it is a regular submanifold of N. The idea is that in a coordinate neighborhood $(U, x^1, \dots, x^n)$ if a partial derivative $\partial f / \partial x^i(p)$ is nonzero, then we can replace the coordinate x^i by f.

First we show that any regular level set $g^{-1}(c)$ of a C^∞ real function g on a manifold can be expressed as a regular zero set.

Lemma 9.7. *Let $g : N \to \mathbb{R}$ be a C^∞ function. A regular level set $g^{-1}(c)$ of level c of the function g is the regular zero set $f^{-1}(0)$ of the function $f = g - c$.*

Proof. For any $p \in N$,

$$g(p) = c \iff f(p) = g(p) - c = 0.$$

Hence, $g^{-1}(c) = f^{-1}(0)$. Call this set S. Because the differential $f_{*,p}$ equals $g_{*,p}$ at every point $p \in N$, the functions f and g have exactly the same critical points. Since g has no critical points in S, neither does f. $\square$

Theorem 9.8. *Let $g : N \to \mathbb{R}$ be a C^∞ function on the manifold N. Then a nonempty regular level set $S = g^{-1}(c)$ is a regular submanifold of N of codimension 1.*

Proof. Let $f = g - c$. By the preceding lemma, S equals $f^{-1}(0)$ and is a regular level set of f. Let $p \in S$. Since p is a regular point of f, relative to any chart $(U, x^1, \dots, x^n)$ about p, $(\partial f / \partial x^i)(p) \neq 0$ for some i. By renumbering $x^1, \dots, x^n$, we may assume that $(\partial f / \partial x^1)(p) \neq 0$.

The Jacobian matrix of the C^∞ map $(f, x^2, \dots, x^n) : U \to \mathbb{R}^n$ is

$$
\begin{bmatrix}
\dfrac{\partial f}{\partial x^1} & \dfrac{\partial f}{\partial x^2} & \cdots & \dfrac{\partial f}{\partial x^n} \\[2mm]
\dfrac{\partial x^2}{\partial x^1} & \dfrac{\partial x^2}{\partial x^2} & \cdots & \dfrac{\partial x^2}{\partial x^n} \\[2mm]
\vdots & \vdots & \ddots & \vdots \\[2mm]
\dfrac{\partial x^n}{\partial x^1} & \dfrac{\partial x^n}{\partial x^2} & \cdots & \dfrac{\partial x^n}{\partial x^n}
\end{bmatrix}
=
\begin{bmatrix}
\dfrac{\partial f}{\partial x^1} & * & \cdots & * \\[2mm]
0 & 1 & \cdots & 0 \\[2mm]
\vdots & \vdots & \ddots & \vdots \\[2mm]
0 & 0 & \cdots & 1
\end{bmatrix}.
$$

So the Jacobian determinant $\partial(f, x^2, \ldots, x^n)/\partial(x^1, x^2, \ldots, x^n)$ at p is $\partial f/\partial x^1(p) \neq 0$. By the inverse function theorem (Corollary 6.27), there is a neighborhood U_p of p on which $f, x^2, \ldots, x^n$ form a coordinate system. Relative to the chart $(U_p, f, x^2, \ldots, x^n)$ the level set $U_p \cap S$ is defined by setting the first coordinate f equal to 0, so $(U_p, f, x^2, \ldots, x^n)$ is an adapted chart relative to S. Since p was arbitrary, S is a regular submanifold of dimension $n - 1$ in N.   □

9.3 The Regular Level Set Theorem

The next step is to extend Theorem 9.8 to a regular level set of a map between smooth manifolds. This very useful theorem does not seem to have an agreed-upon name in the literature. It is known variously as the implicit function theorem, the preimage theorem [17], and the regular level set theorem [25], among other nomenclatures. We will follow [25] and call it the regular level set theorem.

Theorem 9.9 (Regular level set theorem). *Let $F \colon N \to M$ be a C^∞ map of manifolds, with $\dim N = n$ and $\dim M = m$. Then a nonempty regular level set $F^{-1}(c)$, where $c \in M$, is a regular submanifold of N of dimension equal to $n - m$.*

Proof. Choose a chart $(V, \psi) = (V, y^1, \ldots, y^m)$ of M centered at c, i.e., such that $\psi(c) = \mathbf{0}$ in $\mathbb{R}^m$. Then $F^{-1}(V)$ is an open set in N that contains $F^{-1}(c)$. Moreover, in $F^{-1}(V)$, $F^{-1}(c) = (\psi \circ F)^{-1}(\mathbf{0})$. So the level set $F^{-1}(c)$ is the zero set of $\psi \circ F$. If $F^i = y^i \circ F = r^i \circ (\psi \circ F)$, then $F^{-1}(c)$ is also the common zero set of the functions $F^1, \ldots, F^m$ on $F^{-1}(V)$.

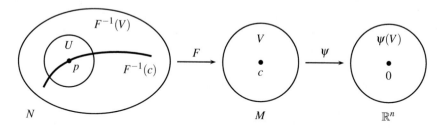

Fig. 9.4. The level set $F^{-1}(c)$ of F is the zero set of $\psi \circ F$.

Because the regular level set is assumed nonempty, $n \geq m$ (Remark 9.5). Fix a point $p \in F^{-1}(c)$ and let $(U, \phi) = (U, x^1, \ldots, x^n)$ be a coordinate neighborhood of p in N contained in $F^{-1}(V)$ (Figure 9.4). Since $F^{-1}(c)$ is a regular level set, $p \in F^{-1}(c)$ is a regular point of F. Therefore, the $m \times n$ Jacobian matrix $[\partial F^i/\partial x^j(p)]$ has rank m. By renumbering the F^i and x^j's, we may assume that the first $m \times m$ block $[\partial F^i/\partial x^j(p)]_{1 \leq i,j \leq m}$ is nonsingular.

Replace the first m coordinates $x^1, \ldots, x^m$ of the chart (U, ϕ) by $F^1, \ldots, F^m$. We claim that there is a neighborhood U_p of p such that $(U_p, F^1, \ldots, F^m, x^{m+1}, \ldots, x^n)$ is a chart in the atlas of N. It suffices to compute its Jacobian matrix at p:

$$
\begin{bmatrix}
\dfrac{\partial F^i}{\partial x^j} & \dfrac{\partial F^i}{\partial x^\beta} \\[2mm]
\dfrac{\partial x^\alpha}{\partial x^j} & \dfrac{\partial x^\alpha}{\partial x^\beta}
\end{bmatrix}
=
\begin{bmatrix}
\dfrac{\partial F^i}{\partial x^j} & * \\[2mm]
0 & I
\end{bmatrix},
$$

where $1 \le i, j \le m$ and $m+1 \le \alpha, \beta \le n$. Since this matrix has determinant

$$
\det \left[\frac{\partial F^i}{\partial x^j}(p) \right]_{1 \le i, j \le m} \ne 0,
$$

the inverse function theorem in the form of Corollary 6.27 implies the claim.

In the chart $(U_p, F^1, \ldots, F^m, x^{m+1}, \ldots, x^n)$, the set $S := f^{-1}(c)$ is obtained by setting the first m coordinate functions $F^1, \ldots, F^m$ equal to 0. So $(U_p, F^1, \ldots, F^m, x^{m+1}, \ldots, x^n)$ is an adapted chart for N relative to S. Since this is true about every point $p \in S$, S is a regular submanifold of N of dimension $n - m$. □

The proof of the regular level set theorem gives the following useful lemma.

Lemma 9.10. *Let $F \colon N \to \mathbb{R}^m$ be a C^∞ map on a manifold N of dimension n and let S be the level set $F^{-1}(0)$. If relative to some coordinate chart $(U, x^1, \ldots, x^n)$ about $p \in S$, the Jacobian determinant $\partial(F^1, \ldots, F^m)/\partial(x^{j_1}, \ldots, x^{j_m})(p)$ is nonzero, then in some neighborhood of p one may replace $x^{j_1}, \ldots, x^{j_m}$ by $F^1, \ldots, F^m$ to obtain an adapted chart for N relative to S.*

Remark. The regular level set theorem gives a sufficient but not necessary condition for a level set to be a regular submanifold. For example, if $f \colon \mathbb{R}^2 \to \mathbb{R}$ is the map $f(x, y) = y^2$, then the zero set $Z(f) = Z(y^2)$ is the x-axis, a regular submanifold of $\mathbb{R}^2$. However, since $\partial f/\partial x = 0$ and $\partial f/\partial y = 2y = 0$ on the x-axis, every point in $Z(f)$ is a critical point of f. Thus, although $Z(f)$ is a regular submanifold of $\mathbb{R}^2$, it is not a regular level set of f.

9.4 Examples of Regular Submanifolds

Example 9.11 (*Hypersurface*). Show that the solution set S of $x^3 + y^3 + z^3 = 1$ in $\mathbb{R}^3$ is a manifold of dimension 2.

Solution. Let $f(x, y, z) = x^3 + y^3 + z^3$. Then $S = f^{-1}(1)$. Since $\partial f/\partial x = 3x^2$, $\partial f/\partial y = 3y^2$, and $\partial f/\partial z = 3z^2$, the only critical point of f is $(0, 0, 0)$, which is not in S. Thus, 1 is a regular value of $f \colon \mathbb{R}^3 \to \mathbb{R}$. By the regular level set theorem (Theorem 9.9), S is a regular submanifold of $\mathbb{R}^3$ of dimension 2. So S is a manifold (Proposition 9.4). □

Example 9.12 (*Solution set of two polynomial equations*). Decide whether the subset S of $\mathbb{R}^3$ defined by the two equations

$$
x^3 + y^3 + z^3 = 1,
$$
$$
x + y + z = 0
$$

is a regular submanifold of $\mathbb{R}^3$.

Solution. Define $F\colon \mathbb{R}^3 \to \mathbb{R}^2$ by

$$(u,v) = F(x,y,z) = (x^3 + y^3 + z^3, x + y + z).$$

Then S is the level set $F^{-1}(1,0)$. The Jacobian matrix of F is

$$J(F) = \begin{bmatrix} u_x & u_y & u_z \\ v_x & v_y & v_z \end{bmatrix} = \begin{bmatrix} 3x^2 & 3y^2 & 3z^2 \\ 1 & 1 & 1 \end{bmatrix},$$

where $u_x = \partial u/\partial x$ and so forth. The critical points of F are the points (x,y,z) where the matrix $J(F)$ has rank < 2. That is precisely where all 2×2 minors of $J(F)$ are zero:

$$\begin{vmatrix} 3x^2 & 3y^2 \\ 1 & 1 \end{vmatrix} = 0, \qquad \begin{vmatrix} 3x^2 & 3z^2 \\ 1 & 1 \end{vmatrix} = 0. \tag{9.1}$$

(The third condition

$$\begin{vmatrix} 3y^2 & 3z^2 \\ 1 & 1 \end{vmatrix} = 0$$

is a consequence of these two.) Solving (9.1), we get $y = \pm x$, $z = \pm x$. Since $x + y + z = 0$ on S, this implies that $(x,y,z) = (0,0,0)$. Since $(0,0,0)$ does not satisfy the first equation $x^3 + y^3 + z^3 = 1$, there are no critical points of F on S. Therefore, S is a regular level set. By the regular level set theorem, S is a regular submanifold of $\mathbb{R}^3$ of dimension 1. $\qquad \square$

Example 9.13 (*Special linear group*). As a set, the *special linear group* $\mathrm{SL}(n,\mathbb{R})$ is the subset of $\mathrm{GL}(n,\mathbb{R})$ consisting of matrices of determinant 1. Since

$$\det(AB) = (\det A)(\det B) \quad \text{and} \quad \det(A^{-1}) = \frac{1}{\det A},$$

$\mathrm{SL}(n,\mathbb{R})$ is a subgroup of $\mathrm{GL}(n,\mathbb{R})$. To show that it is a regular submanifold, we let $f\colon \mathrm{GL}(n,\mathbb{R}) \to \mathbb{R}$ be the determinant map $f(A) = \det A$, and apply the regular level set theorem to $f^{-1}(1) = \mathrm{SL}(n,\mathbb{R})$. We need to check that 1 is a regular value of f.

Let $a_{ij}, 1 \le i \le n, 1 \le j \le n$, be the standard coordinates on $\mathbb{R}^{n \times n}$, and let S_{ij} denote the submatrix of $A = [a_{ij}] \in \mathbb{R}^{n \times n}$ obtained by deleting its ith row and jth column. Then $m_{ij} := \det S_{ij}$ is the (i,j)-*minor* of A. From linear algebra we have a formula for computing the determinant by expanding along any row or any column: if we expand along the ith row, we obtain

$$f(A) = \det A = (-1)^{i+1} a_{i1} m_{i1} + (-1)^{i+2} a_{i2} m_{i2} + \cdots + (-1)^{i+n} a_{in} m_{in}. \tag{9.2}$$

Therefore

$$\frac{\partial f}{\partial a_{ij}} = (-1)^{i+j} m_{ij}.$$

Hence, a matrix $A \in \mathrm{GL}(n,\mathbb{R})$ is a critical point of f if and only if all the $(n-1) \times (n-1)$ minors m_{ij} of A are 0. By (9.2) such a matrix A has determinant 0. Since every matrix in $\mathrm{SL}(n,\mathbb{R})$ has determinant 1, all the matrices in $\mathrm{SL}(n,\mathbb{R})$ are regular points of the determinant function. By the regular level set theorem (Theorem 9.9), $\mathrm{SL}(n,\mathbb{R})$ is a regular submanifold of $\mathrm{GL}(n,\mathbb{R})$ of codimension 1; i.e.,

$$\dim \mathrm{SL}(n,\mathbb{R}) = \dim \mathrm{GL}(n,\mathbb{R}) - 1 = n^2 - 1.$$

Problems

9.1. Regular values
Define $f\colon \mathbb{R}^2 \to \mathbb{R}$ by
$$f(x,y) = x^3 - 6xy + y^2.$$
Find all values $c \in \mathbb{R}$ for which the level set $f^{-1}(c)$ is a regular submanifold of $\mathbb{R}^2$.

9.2. Solution set of one equation
Let x, y, z, w be the standard coordinates on $\mathbb{R}^4$. Is the solution set of $x^5 + y^5 + z^5 + w^5 = 1$ in $\mathbb{R}^4$ a smooth manifold? Explain why or why not. (Assume that the subset is given the subspace topology.)

9.3. Solution set of two equations
Is the solution set of the system of equations
$$x^3 + y^3 + z^3 = 1, \quad z = xy,$$
in $\mathbb{R}^3$ a smooth manifold? Prove your answer.

9.4.* Regular submanifolds
Suppose that a subset S of $\mathbb{R}^2$ has the property that locally on S one of the coordinates is a C^∞ function of the other coordinate. Show that S is a regular submanifold of $\mathbb{R}^2$. (Note that the unit circle defined by $x^2 + y^2 = 1$ has this property. At every point of the circle, there is a neighborhood in which y is a C^∞ function of x or x is a C^∞ function of y.)

9.5. Graph of a smooth function
Show that the graph $\Gamma(f)$ of a smooth function $f\colon \mathbb{R}^2 \to \mathbb{R}$,
$$\Gamma(f) = \{(x,y,f(x,y)) \in \mathbb{R}^3\},$$
is a regular submanifold of $\mathbb{R}^3$.

9.6. Euler's formula
A polynomial $F(x_0,\ldots,x_n) \in \mathbb{R}[x_0,\ldots,x_n]$ is *homogeneous of degree* k if it is a linear combination of monomials $x_0^{i_0} \cdots x_n^{i_n}$ of degree $\sum_{j=0}^n i_j = k$. Let $F(x_0,\ldots,x_n)$ be a homogeneous polynomial of degree k. Clearly, for any $t \in \mathbb{R}$,
$$F(tx_0,\ldots,tx_n) = t^k F(x_0,\ldots,x_n). \tag{9.3}$$

Show that
$$\sum_{i=0}^n x_i \frac{\partial F}{\partial x_i} = kF.$$

9.7. Smooth projective hypersurface
On the projective space $\mathbb{R}P^n$ a homogeneous polynomial $F(x_0,\ldots,x_n)$ of degree k is not a function, since its value at a point $[a_0,\ldots,a_n]$ is not unique. However, the zero set in $\mathbb{R}P^n$ of a homogeneous polynomial $F(x_0,\ldots,x_n)$ is well defined, since $F(a_0,\ldots,a_n) = 0$ if and only if
$$F(ta_0,\ldots,ta_n) = t^k F(a_0,\ldots,a_n) = 0 \quad \text{for all } t \in \mathbb{R}^\times := \mathbb{R} - \{0\}.$$

The zero set of finitely many homogeneous polynomials in $\mathbb{R}P^n$ is called a *real projective variety*. A projective variety defined by a single homogeneous polynomial of degree k is called a *hypersurface* of degree k. Show that the hypersurface $Z(F)$ defined by $F(x_0, x_1, x_2) = 0$ is smooth if $\partial F / \partial x_0$, $\partial F / \partial x_1$, and $\partial F / \partial x_2$ are not simultaneously zero on $Z(F)$. (*Hint*: The standard coordinates on U_0, which is homeomorphic to $\mathbb{R}^2$, are $x = x_1/x_0$, $y = x_2/x_0$ (see Subsection 7.7). In U_0, $F(x_0, x_1, x_2) = x_0^k F(1, x_1/x_0, x_2/x_0) = x_0^k F(1, x, y)$. Define $f(x, y) = F(1, x, y)$. Then f and F have the same zero set in U_0.)

9.8. Product of regular submanifolds

If S_i is a regular submanifold of the manifold M_i for $i = 1, 2$, prove that $S_1 \times S_2$ is a regular submanifold of $M_1 \times M_2$.

9.9. Complex special linear group

The *complex special linear group* $\mathrm{SL}(n, \mathbb{C})$ is the subgroup of $\mathrm{GL}(n, \mathbb{C})$ consisting of $n \times n$ complex matrices of determinant 1. Show that $\mathrm{SL}(n, \mathbb{C})$ is a regular submanifold of $\mathrm{GL}(n, \mathbb{C})$ and determine its dimension. (This problem requires a rudimentary knowledge of complex analysis.)

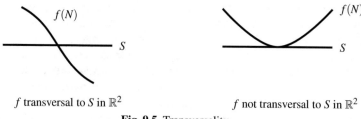

f transversal to S in $\mathbb{R}^2$ f not transversal to S in $\mathbb{R}^2$

Fig. 9.5. Transversality.

9.10. The transversality theorem

A C^∞ map $f \colon N \to M$ is said to be *transversal* to a submanifold $S \subset M$ (Figure 9.5) if for every $p \in f^{-1}(S)$,

$$f_*(T_p N) + T_{f(p)} S = T_{f(p)} M. \tag{9.4}$$

(If A and B are subspaces of a vector space, their sum $A + B$ is the subspace consisting of all $a + b$ with $a \in A$ and $b \in B$. The sum need not be a direct sum.) The goal of this exercise is to prove the *transversality theorem*: if a C^∞ map $f \colon N \to M$ is transversal to a regular submanifold S of codimension k in M, then $f^{-1}(S)$ is a regular submanifold of codimension k in N.

When S consists of a single point c, transversality of f to S simply means that $f^{-1}(c)$ is a regular level set. Thus the transversality theorem is a generalization of the regular level set theorem. It is especially useful in giving conditions under which the intersection of two submanifolds is a submanifold.

Let $p \in f^{-1}(S)$ and $(U, x^1, \ldots, x^m)$ be an adapted chart centered at $f(p)$ for M relative to S such that $U \cap S = Z(x^{m-k+1}, \ldots, x^m)$, the zero set of the functions $x^{m-k+1}, \ldots, x^m$. Define $g \colon U \to \mathbb{R}^k$ to be the map

$$g = (x^{m-k+1}, \ldots, x^m).$$

(a) Show that $f^{-1}(U) \cap f^{-1}(S) = (g \circ f)^{-1}(0)$.
(b) Show that $f^{-1}(U) \cap f^{-1}(S)$ is a regular level set of the function $g \circ f \colon f^{-1}(U) \to \mathbb{R}^k$.
(c) Prove the transversality theorem.

§10 Categories and Functors

Many of the problems in mathematics share common features. For example, in topology one is interested in knowing whether two topological spaces are homeomorphic and in group theory one is interested in knowing whether two groups are isomorphic. This has given rise to the theory of categories and functors, which tries to clarify the structural similarities among different areas of mathematics.

A category is essentially a collection of objects and arrows between objects. These arrows, called morphisms, satisfy the abstract properties of maps and are often structure-preserving maps. Smooth manifolds and smooth maps form a category, and so do vector spaces and linear maps. A functor from one category to another preserves the identity morphism and the composition of morphisms. It provides a way to simplify problems in the first category, for the target category of a functor is usually simpler than the original category. The tangent space construction with the differential of a smooth map is a functor from the category of smooth manifolds with a distinguished point to the category of vector spaces. The existence of the tangent space functor shows that if two manifolds are diffeomorphic, then their tangent spaces at corresponding points must be isomorphic, thereby proving the smooth invariance of dimension. Invariance of dimension in the continuous category of topological spaces and continuous maps is more difficult to prove, precisely because there is no tangent space functor in the continuous category.

Much of algebraic topology is the study of functors, for example, the homology, cohomology, and homotopy functors. For a functor to be truly useful, it should be simple enough to be computable, yet complex enough to preserve essential features of the original category. For smooth manifolds, this delicate balance is achieved in the de Rham cohomology functor. In the rest of the book, we will be introducing various functors of smooth manifolds, such as the tangent bundle and differential forms, culminating in de Rham cohomology.

In this section, after defining categories and functors, we study the dual construction on vector spaces as a nontrivial example of a functor.

10.1 Categories

A *category* consists of a collection of elements, called *objects*, and for any two objects A and B, a set $\operatorname{Mor}(A,B)$ of elements, called *morphisms* from A to B, such that given any morphism $f \in \operatorname{Mor}(A,B)$ and any morphism $g \in \operatorname{Mor}(B,C)$, the *composite* $g \circ f \in \operatorname{Mor}(A,C)$ is defined. Furthermore, the composition of morphisms is required to satisfy two properties:

(i) the identity axiom: for each object A, there is an identity morphism $\mathbb{1}_A \in \operatorname{Mor}(A,A)$ such that for any $f \in \operatorname{Mor}(A,B)$ and $g \in \operatorname{Mor}(B,A)$,

$$f \circ \mathbb{1}_A = f \quad \text{and} \quad \mathbb{1}_A \circ g = g;$$

(ii) the associative axiom: for $f \in \mathrm{Mor}(A,B)$, $g \in \mathrm{Mor}(B,C)$, and $h \in \mathrm{Mor}(C,D)$,

$$h \circ (g \circ f) = (h \circ g) \circ f.$$

If $f \in \mathrm{Mor}(A,B)$, we often write $f \colon A \to B$.

Example. The collection of groups and group homomorphisms forms a category in which the objects are groups and for any two groups A and B, $\mathrm{Mor}(A,B)$ is the set of group homomorphisms from A to B.

Example. The collection of all vector spaces over $\mathbb{R}$ and $\mathbb{R}$-linear maps forms a category in which the objects are real vector spaces and for any two real vector spaces V and W, $\mathrm{Mor}(V,W)$ is the set $\mathrm{Hom}(V,W)$ of linear maps from V to W.

Example. The collection of all topological spaces together with continuous maps between them is called the *continuous category*.

Example. The collection of smooth manifolds together with smooth maps between them is called the *smooth category*.

Example. We call a pair (M,q), where M is a manifold and q a point in M, a *pointed manifold*. Given any two such pairs (N,p) and (M,q), let $\mathrm{Mor}((N,p),(M,q))$ be the set of all smooth maps $F \colon N \to M$ such that $F(p) = q$. This gives rise to the *category of pointed manifolds*.

Definition 10.1.   Two objects A and B in a category are said to be *isomorphic* if there are morphisms $f \colon A \to B$ and $g \colon B \to A$ such that

$$g \circ f = \mathbb{1}_A \quad \text{and} \quad f \circ g = \mathbb{1}_B.$$

In this case both f and g are called *isomorphisms*.

The usual notation for an isomorphism is "$\simeq$". Thus, $A \simeq B$ can mean, for example, a group isomorphism, a vector space isomorphism, a homeomorphism, or a diffeomorphism, depending on the category and the context.

10.2 Functors

Definition 10.2.   A *(covariant) functor* $\mathcal{F}$ from one category $\mathcal{C}$ to another category $\mathcal{D}$ is a map that associates to each object A in $\mathcal{C}$ an object $\mathcal{F}(A)$ in $\mathcal{D}$ and to each morphism $f \colon A \to B$ a morphism $\mathcal{F}(f) \colon \mathcal{F}(A) \to \mathcal{F}(B)$ such that

(i) $\mathcal{F}(\mathbb{1}_A) = \mathbb{1}_{\mathcal{F}(A)}$,
(ii) $\mathcal{F}(f \circ g) = \mathcal{F}(f) \circ \mathcal{F}(g)$.

Example. The tangent space construction is a functor from the category of pointed manifolds to the category of vector spaces. To each pointed manifold (N, p) we associate the tangent space $T_p N$ and to each smooth map $f : (N, p) \to (M, f(p))$ we associate the differential $f_{*, p} : T_p N \to T_{f(p)} M$.

The functorial property (i) holds because if $1 : N \to N$ is the identity map, then its differential $\mathbb{1}_{*, p} : T_p N \to T_p N$ is also the identity map.

The functorial property (ii) holds because in this context it is the chain rule

$$(g \circ f)_{*, p} = g_{*, f(p)} \circ f_{*, p}.$$

Proposition 10.3. *Let* $\mathcal{F} : \mathcal{C} \to \mathcal{D}$ *be a functor from a category* $\mathcal{C}$ *to a category* $\mathcal{D}$. *If* $f : A \to B$ *is an isomorphism in* $\mathcal{C}$, *then* $\mathcal{F}(f) : \mathcal{F}(A) \to \mathcal{F}(B)$ *is an isomorphism in* $\mathcal{D}$.

Proof. Problem 10.2. $\qquad\qquad\qquad\qquad\qquad\qquad\qquad\qquad\qquad\qquad\qquad\qquad$ □

Note that we can recast Corollaries 8.6 and 8.7 in a more functorial form. Suppose $f : N \to M$ is a diffeomorphism. Then (N, p) and $(M, f(p))$ are isomorphic objects in the category of pointed manifolds. By Proposition 10.3, the tangent spaces $T_p N$ and $T_{f(p)} M$ must be isomorphic as vector spaces and therefore have the same dimension. It follows that the dimension of a manifold is invariant under a diffeomorphism.

If in the definition of a covariant functor we reverse the direction of the arrow for the morphism $\mathcal{F}(f)$, then we obtain a *contravariant functor*. More precisely, the definition is as follows.

Definition 10.4. A *contravariant functor* $\mathcal{F}$ from one category $\mathcal{C}$ to another category $\mathcal{D}$ is a map that associates to each object A in $\mathcal{C}$ an object $\mathcal{F}(A)$ in $\mathcal{D}$ and to each morphism $f : A \to B$ a morphism $\mathcal{F}(f) : \mathcal{F}(B) \to \mathcal{F}(A)$ such that

(i) $\mathcal{F}(\mathbb{1}_A) = \mathbb{1}_{\mathcal{F}(A)}$;
(ii) $\mathcal{F}(f \circ g) = \mathcal{F}(g) \circ \mathcal{F}(f)$. (Note the reversal of order.)

Example. Smooth functions on a manifold give rise to a contravariant functor that associates to each manifold M the algebra $\mathcal{F}(M) = C^\infty(M)$ of C^∞ functions on M and to each smooth map $F : N \to M$ of manifolds the pullback map $\mathcal{F}(F) = F^* : C^\infty(M) \to C^\infty(N)$, $F^*(h) = h \circ F$ for $h \in C^\infty(M)$. It is easy to verify that the pullback satisfies the two functorial properties:

(i) $(\mathbb{1}_M)^* = \mathbb{1}_{C^\infty(M)}$,
(ii) if $F : N \to M$ and $G : M \to P$ are C^∞ maps, then $(G \circ F)^* = F^* \circ G^* : C^\infty(P) \to C^\infty(N)$.

Another example of a contravariant functor is the dual of a vector space, which we review in the next section.

10.3 The Dual Functor and the Multicovector Functor

Let V be a real vector space. Recall that its dual space $V^\vee$ is the vector space of all *linear functionals* on V, i.e., linear functions $\alpha\colon V \to \mathbb{R}$. We also write

$$V^\vee = \operatorname{Hom}(V,\mathbb{R}).$$

If V is a finite-dimensional vector space with basis $\{e_1,\ldots,e_n\}$, then by Proposition 3.1 its dual space $V^\vee$ has as a basis the collection of linear functionals $\{\alpha^1,\ldots,\alpha^n\}$ defined by

$$\alpha^i(e_j) = \delta^i_j, \quad 1 \le i,j \le n.$$

Since a linear function on V is determined by what it does on a basis of V, this set of equations defines α^i uniquely.

A linear map $L\colon V \to W$ of vector spaces induces a linear map $L^\vee$, called the *dual* of L, as follows. To every linear functional $\alpha\colon W \to \mathbb{R}$, the dual map $L^\vee$ associates the linear functional

$$V \xrightarrow{L} W \xrightarrow{\alpha} \mathbb{R}.$$

Thus, the dual map $L^\vee\colon W^\vee \to V^\vee$ is given by

$$L^\vee(\alpha) = \alpha \circ L \quad \text{for } \alpha \in W^\vee.$$

Note that the dual of L reverses the direction of the arrow.

Proposition 10.5 (Functorial properties of the dual). *Suppose V, W, and S are real vector spaces.*

(i) *If $\mathbb{1}_V\colon V \to V$ is the identity map on V, then $\mathbb{1}_V^\vee\colon V^\vee \to V^\vee$ is the identity map on $V^\vee$.*

(ii) *If $f\colon V \to W$ and $g\colon W \to S$ are linear maps, then $(g \circ f)^\vee = f^\vee \circ g^\vee$.*

Proof. Problem 10.3. $\square$

According to this proposition, the dual construction $\mathcal{F}\colon (\ ) \mapsto (\ )^\vee$ is a contravariant functor from the category of vector spaces to itself: for V a real vector space, $\mathcal{F}(V) = V^\vee$ and for $f \in \operatorname{Hom}(V,W)$, $\mathcal{F}(f) = f^\vee \in \operatorname{Hom}(W^\vee,V^\vee)$. Consequently, if $f\colon V \to W$ is an isomorphism, then so is its dual $f^\vee\colon W^\vee \to V^\vee$ (cf. Proposition 10.3).

Fix a positive integer k. For any linear map $L\colon V \to W$ of vector spaces, define the *pullback map* $L^*\colon A_k(W) \to A_k(V)$ to be

$$(L^*f)(v_1,\ldots,v_k) = f(L(v_1),\ldots,L(v_k))$$

for $f \in A_k(W)$ and $v_1,\ldots,v_k \in V$. From the definition, it is easy to see that L^* is a linear map: $L^*(af+bg) = aL^*f + bL^*g$ for $a,b \in \mathbb{R}$ and $f,g \in A_k(W)$.

Proposition 10.6. *The pullback of covectors by a linear map satisfies the two functorial properties:*

(i) *If $\mathbb{1}_V\colon V \to V$ is the identity map on V, then $\mathbb{1}_V^* = \mathbb{1}_{A_k(V)}$, the identity map on* $A_k(V)$.

(ii) *If $K\colon U \to V$ and $L\colon V \to W$ are linear maps of vector spaces, then*

$$(L \circ K)^* = K^* \circ L^* \colon A_k(W) \to A_k(U).$$

Proof. Problem 10.6.   □

To each vector space V, we associate the vector space $A_k(V)$ of all k-covectors on V, and to each linear map $L\colon V \to W$ of vector spaces, we associate the pullback $A_k(L) = L^* \colon A_k(W) \to A_k(V)$. Then $A_k(\ )$ is a contravariant functor from the category of vector spaces and linear maps to itself.

When $k = 1$, for any vector space V, the space $A_1(V)$ is the dual space, and for any linear map $L\colon V \to W$, the pullback map $A_1(L) = L^*$ is the dual map $L^\vee \colon W^\vee \to V^\vee$. Thus, the multicovector functor $A_k(\ )$ generalizes the dual functor $(\ )^\vee$.

Problems

10.1. Differential of the inverse map
If $F\colon N \to M$ is a diffeomorphism of manifolds and $p \in N$, prove that $(F^{-1})_{*,F(p)} = (F_{*,p})^{-1}$.

10.2. Isomorphism under a functor
Prove Proposition 10.3.

10.3. Functorial properties of the dual
Prove Proposition 10.5.

10.4. Matrix of the dual map
Suppose a linear transformation $L\colon V \to \bar{V}$ is represented by the matrix $A = [a^i_j]$ relative to bases $e_1, \ldots, e_n$ for V and $\bar{e}_1, \ldots, \bar{e}_m$ for $\bar{V}$:

$$L(e_j) = \sum_i a^i_j \bar{e}_i.$$

Let $\alpha^1, \ldots, \alpha^n$ and $\bar{\alpha}^1, \ldots, \bar{\alpha}^m$ be the dual bases for $V^\vee$ and $\bar{V}^\vee$, respectively. Prove that if $L^\vee(\bar{\alpha}^i) = \sum_j b^i_j \alpha^j$, then $b^i_j = a^i_j$.

10.5. Injectivity of the dual map

(a) Suppose V and W are vector spaces of possibly infinite dimension over a field K. Show that if a linear map $L\colon V \to W$ is surjective, then its dual $L^\vee \colon W^\vee \to V^\vee$ is injective.
(b) Suppose V and W are finite-dimensional vector spaces over a field K. Prove the converse of the implication in (a).

10.6. Functorial properties of the pullback
Prove Proposition 10.6.

10.7. Pullback in the top dimension
Show that if $L\colon V \to V$ is a linear operator on a vector space V of dimension n, then the pullback $L^* \colon A_n(V) \to A_n(V)$ is multiplication by the determinant of L.

§11 The Rank of a Smooth Map

In this section we analyze the local structure of a smooth map through its rank. Recall that the rank of a smooth map $f: N \to M$ at a point $p \in N$ is the rank of its differential at p. Two cases are of special interest: that in which the map f has maximal rank at a point and that in which it has constant rank in a neighborhood. Let $n = \dim N$ and $m = \dim M$. In case $f: N \to M$ has maximal rank at p, there are three not mutually exclusive possibilities:

(i) If $n = m$, then by the inverse function theorem, f is a local diffeomorphism at p.
(ii) If $n \leq m$, then the maximal rank is n and f is an *immersion* at p.
(iii) If $n \geq m$, then the maximal rank is m and f is a *submersion* at p.

Because manifolds are locally Euclidean, theorems on the rank of a smooth map between Euclidean spaces (Appendix B) translate easily to theorems about manifolds. This leads to the constant rank theorem for manifolds, which gives a simple normal form for a smooth map having constant rank on an open set (Theorem 11.1). As an immediate consequence, we obtain a criterion for a level set to be a regular submanifold, which, following [25], we call the constant-rank level set theorem. As we explain in Subsection 11.2, maximal rank at a point implies constant rank in a neighborhood, so immersions and submersions are maps of constant rank. The constant rank theorem specializes to the immersion theorem and the submersion theorem, giving simple normal forms for an immersion and a submersion. The regular level set theorem, which we encountered in Subsection 9.3, is now seen to be a consequence of the submersion theorem and a special case of the constant-rank level set theorem.

By the regular level set theorem, the *preimage* of a regular value of a smooth map is a manifold. The *image* of a smooth map, on the other hand, does not generally have a nice structure. Using the immersion theorem we derive conditions under which the image of a smooth map is a manifold.

11.1 Constant Rank Theorem

Suppose $f: N \to M$ is a C^∞ map of manifolds and we want to show that the level set $f^{-1}(c)$ is a manifold for some c in M. In order to apply the regular level set theorem, we need the differential f_* to have maximal rank at every point of $f^{-1}(c)$. Sometimes this is not true; even if true, it may be difficult to show. In such cases, the constant-rank level set theorem can be helpful. It has one cardinal virtue: it is not necessary to know precisely the rank of f; it suffices that the rank be constant.

The constant rank theorem for Euclidean spaces (Theorem B.4) has an immediate analogue for manifolds.

Theorem 11.1 (Constant rank theorem). *Let N and M be manifolds of dimensions n and m respectively. Suppose $f: N \to M$ has constant rank k in a neighborhood of* *a point p in N. Then there are charts* (U, ϕ) *centered at p in N and* (V, ψ) *centered at* $f(p)$ *in M such that for* $(r^1, \ldots, r^n)$ *in* $\phi(U)$,

$$(\psi \circ f \circ \phi^{-1})(r^1, \ldots, r^n) = (r^1, \ldots, r^k, 0, \ldots, 0). \tag{11.1}$$

Proof. Choose a chart $(\bar{U}, \bar{\phi})$ about p in N and $(\bar{V}, \bar{\psi})$ about $f(p)$ in M. Then $\bar{\psi} \circ f \circ \bar{\phi}^{-1}$ is a map between open subsets of Euclidean spaces. Because $\bar{\phi}$ and $\bar{\psi}$ are diffeomorphisms, $\bar{\psi} \circ f \circ \bar{\phi}^{-1}$ has the same constant rank k as f in a neighborhood of $\bar{\phi}(p)$ in $\mathbb{R}^n$. By the constant rank theorem for Euclidean spaces (Theorem B.4) there are a diffeomorphism G of a neighborhood of $\bar{\phi}(p)$ in $\mathbb{R}^n$ and a diffeomorphism F of a neighborhood of $(\bar{\psi} \circ f)(p)$ in $\mathbb{R}^m$ such that

$$(F \circ \bar{\psi} \circ f \circ \bar{\phi}^{-1} \circ G^{-1})(r^1, \ldots, r^n) = (r^1, \ldots, r^k, 0, \ldots, 0).$$

Set $\phi = G \circ \bar{\phi}$ and $\psi = F \circ \bar{\psi}$.   □

In the constant rank theorem, it is possible that the normal form (11.1) for the function f has no zeros at all: if the rank k equals m, then

$$(\psi \circ f \circ \phi^{-1})(r^1, \ldots, r^n) = (r^1, \ldots, r^m).$$

From this theorem, the constant-rank level set theorem follows easily. By a *neighborhood* of a subset A of a manifold M we mean an open set containing A.

Theorem 11.2 (Constant-rank level set theorem). *Let* $f : N \to M$ *be a* C^∞ *map of manifolds and* $c \in M$. *If* f *has constant rank* k *in a neighborhood of the level set* $f^{-1}(c)$ *in N, then* $f^{-1}(c)$ *is a regular submanifold of N of codimension k.*

Proof. Let p be an arbitrary point in $f^{-1}(c)$. By the constant rank theorem there are a coordinate chart $(U, \phi) = (U, x^1, \ldots, x^n)$ centered at $p \in N$ and a coordinate chart $(V, \psi) = (V, y^1, \ldots, y^m)$ centered at $f(p) = c \in M$ such that

$$(\psi \circ f \circ \phi^{-1})(r^1, \ldots, r^n) = (r^1, \ldots, r^k, 0, \ldots, 0) \in \mathbb{R}^m.$$

This shows that the level set $(\psi \circ f \circ \phi^{-1})^{-1}(0)$ is defined by the vanishing of the coordinates $r^1, \ldots, r^k$.

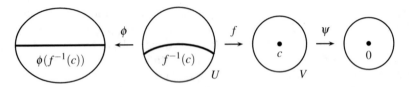

Fig. 11.1. Constant-rank level set.

The image of the level set $f^{-1}(c)$ under ϕ is the level set $(\psi \circ f \circ \phi^{-1})^{-1}(0)$ (Figure 11.1), since

$$\phi(f^{-1}(c)) = \phi(f^{-1}(\psi^{-1}(0))) = (\psi \circ f \circ \phi^{-1})^{-1}(0).$$

Thus, the level set $f^{-1}(c)$ in U is defined by the vanishing of the coordinate functions $x^1, \ldots, x^k$, where $x^i = r^i \circ \phi$. This proves that $f^{-1}(c)$ is a regular submanifold of N of codimension k. $\qquad\square$

Example 11.3 (*Orthogonal group*). The *orthogonal group* $O(n)$ is defined to be the subgroup of $GL(n, \mathbb{R})$ consisting of matrices A such that $A^T A = I$, the $n \times n$ identity matrix. Using the constant rank theorem, prove that $O(n)$ is a regular submanifold of $GL(n, \mathbb{R})$.

Solution. Define $f : GL(n, \mathbb{R}) \to GL(n, \mathbb{R})$ by $f(A) = A^T A$. Then $O(n)$ is the level set $f^{-1}(I)$. For any two matrices $A, B \in GL(n, \mathbb{R})$, there is a unique matrix $C \in GL(n, \mathbb{R})$ such that $B = AC$. Denote by ℓ_C and $r_C : GL(n, \mathbb{R}) \to GL(n, \mathbb{R})$ the left and right multiplication by C, respectively. Since

$$f(AC) = (AC)^T AC = C^T A^T AC = C^T f(A)C,$$

we have

$$(f \circ r_C)(A) = (\ell_{C^T} \circ r_C \circ f)(A).$$

Since this is true for all $A \in GL(n, \mathbb{R})$,

$$f \circ r_C = \ell_{C^T} \circ r_C \circ f.$$

By the chain rule,

$$f_{*,AC} \circ (r_C)_{*,A} = (\ell_{C^T})_{*,A^T AC} \circ (r_C)_{*,A^T A} \circ f_{*,A}. \tag{11.2}$$

Since left and right multiplications are diffeomorphisms, their differentials are isomorphisms. Composition with an isomorphism does not change the rank of a linear map. Hence, in (11.2),

$$\operatorname{rk} f_{*,AC} = \operatorname{rk} f_{*,A}.$$

Since AC and A are two arbitrary points of $GL(n, \mathbb{R})$, this proves that the differential of f has constant rank on $GL(n, \mathbb{R})$. By the constant-rank level set theorem, the orthogonal group $O(n) = f^{-1}(I)$ is a regular submanifold of $GL(n, \mathbb{R})$.

NOTATION. If $f : N \to M$ is a map with constant rank k in a neighborhood of a point $p \in N$, its local normal form (11.1) relative to the charts $(U, \phi) = (U, x^1, \ldots, x^n)$ and $(V, \psi) = (V, y^1, \ldots, y^m)$ in the constant rank theorem (Theorem 11.1) can be expressed in terms of the local coordinates $x^1, \ldots, x^n$ and $y^1, \ldots, y^m$ as follows.

First note that for any $q \in U$,

$$\phi(q) = (x^1(q), \ldots, x^n(q)) \text{ and } \psi(f(q)) = (y^1(f(q)), \ldots, y^n(f(q))).$$

Thus,

$$(y^1(f(q)),\ldots,y^m(f(q))) = \psi(f(q)) = (\psi \circ f \circ \phi^{-1})(\phi(q))$$
$$= (\psi \circ f \circ \phi^{-1})(x^1(q),\ldots,x^n(q))$$
$$= (x^1(q),\ldots,x^k(q)),0,\ldots,0) \quad \text{(by (11.1))}.$$

As functions on U,

$$(y^1 \circ f,\ldots,y^m \circ f) = (x^1,\ldots,x^k,0,\ldots,0). \tag{11.3}$$

We can rewrite (11.3) in the following form: relative to the charts $(U,x^1,\ldots,x^n)$ and $(V,y^1,\ldots,y^m)$, the map f is given by

$$(x^1,\ldots,x^n) \mapsto (x^1,\ldots,x^k,0,\ldots,0).$$

11.2 The Immersion and Submersion Theorems

In this subsection we explain why immersions and submersions have constant rank. The constant rank theorem gives local normal forms for immersions and submersions, called the immersion theorem and the submersion theorem respectively. From the submersion theorem and the constant-rank level set theorem, we get two more proofs of the regular level set theorem.

Consider a C^∞ map $f: N \to M$. Let $(U,\phi) = (U,x^1,\ldots,x^n)$ be a chart about p in N and $(V,\psi) = (V,y^1,\ldots,y^m)$ a chart about $f(p)$ in M. Write $f^i = y^i \circ f$ for the ith component of f in the chart $(V,y^1,\ldots,y^m)$. Relative to the charts (U,ϕ) and (V,ψ), the linear map $f_{*,p}$ is represented by the matrix $[\partial f^i/\partial x^j(p)]$ (Proposition 8.11). Hence,

$$\begin{aligned} f_{*,p} \text{ is injective} &\iff n \le m \text{ and } \mathrm{rk}[\partial f^i/\partial x^j(p)] = n, \\ f_{*,p} \text{ is surjective} &\iff n \ge m \text{ and } \mathrm{rk}[\partial f^i/\partial x^j(p)] = m. \end{aligned} \tag{11.4}$$

The rank of a matrix is the number of linearly independent rows of the matrix; it is also the number of linearly independent columns. Thus, the maximum possible rank of an $m \times n$ matrix is the minimum of m and n. It follows from (11.4) that being an immersion or a submersion at p is equivalent to the maximality of $\mathrm{rk}[\partial f^i/\partial x^j(p)]$.

Having maximal rank at a point is an *open condition* in the sense that the set

$$D_{\max}(f) = \{p \in U \mid f_{*,p} \text{ has maximal rank at } p\}$$

is an open subset of U. To see this, suppose k is the maximal rank of f. Then

$$\begin{aligned} \mathrm{rk}\, f_{*,p} = k &\iff \mathrm{rk}[\partial f^i/\partial x^j(p)] = k \\ &\iff \mathrm{rk}[\partial f^i/\partial x^j(p)] \ge k \quad \text{(since k is maximal)}. \end{aligned}$$

So the complement $U - D_{\max}(f)$ is defined by

$$\mathrm{rk}[\partial f^i/\partial x^j(p)] < k,$$

which is equivalent to the vanishing of all $k \times k$ minors of the matrix $[\partial f^i/\partial x^j(p)]$. As the zero set of finitely many continuous functions, $U - D_{\max}(f)$ is closed and so $D_{\max}(f)$ is open. In particular, if f has maximal rank at p, then it has maximal rank at all points in some neighborhood of p. We have proven the following proposition.

Proposition 11.4. *Let N and M be manifolds of dimensions n and m respectively. If a C^∞ map $f: N \to M$ is an immersion at a point $p \in N$, then it has constant rank n in a neighborhood of p. If a C^∞ map $f: N \to M$ is a submersion at a point $p \in N$, then it has constant rank m in a neighborhood of p.*

Example. While maximal rank at a point implies constant rank in a neighborhood, the converse is not true. The map $f: \mathbb{R}^2 \to \mathbb{R}^3$, $f(x,y) = (x,0,0)$, has constant rank 1, but it does not have maximal rank at any point.

By Proposition 11.4, the following theorems are simply special cases of the constant rank theorem.

Theorem 11.5. *Let N and M be manifolds of dimensions n and m respectively.*

(i) **(Immersion theorem)** *Suppose $f: N \to M$ is an immersion at $p \in N$. Then there are charts (U, ϕ) centered at p in N and (V, ψ) centered at $f(p)$ in M such that in a neighborhood of $\phi(p)$,*

$$(\psi \circ f \circ \phi^{-1})(r^1, \ldots, r^n) = (r^1, \ldots, r^n, 0, \ldots, 0).$$

(ii) **(Submersion theorem)** *Suppose $f: N \to M$ is a submersion at p in N. Then there are charts (U, ϕ) centered at p in N and (V, ψ) centered at $f(p)$ in M such that in a neighborhood of $\phi(p)$,*

$$(\psi \circ f \circ \phi^{-1})(r^1, \ldots, r^m, r^{m+1}, \ldots, r^n) = (r^1, \ldots, r^m).$$

Corollary 11.6. *A submersion $f: N \to M$ of manifolds is an open map.*

Proof. Let W be an open subset of N. We need to show that its image $f(W)$ is open in M. Choose a point $f(p)$ in $f(W)$, with $p \in W$. By the submersion theorem, f is locally a projection. Since a projection is an open map (Problem A.7), there is an open neighborhood U of p in W such that $f(U)$ is open in M. Clearly,

$$f(p) \in f(U) \subset f(W).$$

Since $f(p) \in f(W)$ was arbitrary, $f(W)$ is open in M. □

The regular level set theorem (Theorem 9.9) is an easy corollary of the submersion theorem. Indeed, for a C^∞ map $f: N \to M$ of manifolds, a level set $f^{-1}(c)$ is regular if and only if f is a submersion at every point $p \in f^{-1}(c)$. Fix one such point $p \in f^{-1}(c)$ and let (U, ϕ) and (V, ψ) be the charts in the submersion theorem. Then $\psi \circ f \circ \phi^{-1} = \pi: \mathbb{R}^n \supset \phi(U) \to \mathbb{R}^m$ is the projection to the first m coordinates, $\pi(r^1, \ldots, r^n) = (r^1, \ldots, r^m)$. It follows that on U,

$$\psi \circ f = \pi \circ \phi = (r^1, \ldots, r^m) \circ \phi = (x^1, \ldots, x^m).$$

Therefore,

$$f^{-1}(c) = f^{-1}(\psi^{-1}(0)) = (\psi \circ f)^{-1}(0) = Z(\psi \circ f) = Z(x^1, \ldots, x^m),$$

showing that in the chart $(U, x^1, \ldots, x^n)$, the level set $f^{-1}(c)$ is defined by the vanishing of the m coordinate functions $x^1, \ldots, x^m$. Therefore, $(U, x^1, \ldots, x^n)$ is an adapted chart for N relative to $f^{-1}(c)$. This gives a second proof that the regular level set $f^{-1}(c)$ is a regular submanifold of N.

Since the submersion theorem is a special case of the constant rank theorem, it is not surprising that the regular level set theorem is also a special case of the constant-rank level set theorem. On a regular level set $f^{-1}(c)$, the map $f \colon N \to M$ has maximal rank m at every point. Since the maximality of the rank of f is an open condition, a regular level set $f^{-1}(c)$ has a neighborhood on which f has constant rank m. By the constant-rank level set theorem (Theorem 11.2), $f^{-1}(c)$ is a regular submanifold of N, giving us a third proof of the regular level set theorem.

11.3 Images of Smooth Maps

The following are all examples of C^∞ maps $f \colon N \to M$, with $N = \mathbb{R}$ and $M = \mathbb{R}^2$.

Example 11.7. $f(t) = (t^2, t^3)$.

This f is one-to-one, because $t \mapsto t^3$ is one-to-one. Since $f'(0) = (0,0)$, the differential $f_{*,0} \colon T_0 \mathbb{R} \to T_{(0,0)} \mathbb{R}^2$ is the zero map and hence not injective; so f is not an immersion at 0. Its image is the cuspidal cubic $y^2 = x^3$ (Figure 11.2).

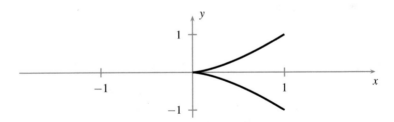

Fig. 11.2. A cuspidal cubic, not an immersion.

Example 11.8. $f(t) = (t^2 - 1, t^3 - t)$.

Since the equation $f'(t) = (2t, 3t^2 - 1) = (0,0)$ has no solution in t, this map f is an immersion. It is not one-to-one, because it maps both $t = 1$ and $t = -1$ to the origin. To find an equation for the image $f(N)$, let $x = t^2 - 1$ and $y = t^3 - t$. Then $y = t(t^2 - 1) = tx$; so

$$y^2 = t^2 x^2 = (x+1)x^2.$$

Thus the image of f is the nodal cubic $y^2 = x^2(x+1)$ (Figure 11.3).

Example 11.9. The map f in Figure 11.4 is a one-to-one immersion but its image, with the subspace topology induced from $\mathbb{R}^2$, is not homeomorphic to the domain $\mathbb{R}$, because there are points near $f(p)$ in the image that correspond to points in $\mathbb{R}$ far away from p. More precisely, if U is an interval about p as shown, there is no neighborhood V of $f(p)$ in $f(N)$ such that $f^{-1}(V) \subset U$; hence, f^{-1} is not continuous.

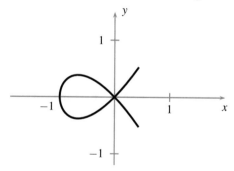

Fig. 11.3. A nodal cubic, an immersion but not one-to-one.

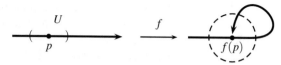

Fig. 11.4. A one-to-one immersion that is not an embedding.

Example 11.10. The manifold M in Figure 11.5 is the union of the graph of $y = \sin(1/x)$ on the interval $]0,1[$, the open line segment from $y = 0$ to $y = 1$ on the y-axis, and a smooth curve joining $(0,0)$ and $(1,\sin 1)$. The map f is a one-to-one immersion whose image with the subspace topology is not homeomorphic to $\mathbb{R}$.

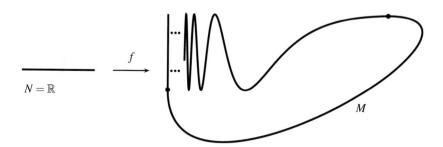

Fig. 11.5. A one-to-one immersion that is not an embedding.

Notice that in these examples the image $f(N)$ is not a regular submanifold of $M = \mathbb{R}^2$. We would like conditions on the map f so that its image $f(N)$ would be a regular submanifold of M.

Definition 11.11. A C^∞ map $f \colon N \to M$ is called an *embedding* if

(i) it is a one-to-one immersion and

(ii) the image $f(N)$ with the subspace topology is homeomorphic to N under f.
(The phrase "one-to-one" in this definition is redundant, since a homeomorphism is necessarily one-to-one.)

Remark. Unfortunately, there is quite a bit of terminological confusion in the literature concerning the use of the word "submanifold." Many authors give the image $f(N)$ of a one-to-one immersion $f: N \rightarrow M$ not the subspace topology, but the topology inherited from f; i.e., a subset $f(U)$ of $f(N)$ is said to be open if and only if U is open in N. With this topology, $f(N)$ is by definition homeomorphic to N. These authors define a submanifold to be the image of any one-to-one immersion with the topology and differentiable structure inherited from f. Such a set is sometimes called an *immersed submanifold* of M. Figures 11.4 and 11.5 show two examples of immersed submanifolds. If the underlying set of an immersed submanifold is given the subspace topology, then the resulting space need not be a manifold at all!

For us, a submanifold without any qualifying adjective is always a *regular submanifold*. To recapitulate, a regular submanifold of a manifold M is a subset S of M with the subspace topology such that every point of S has a neighborhood $U \cap S$ defined by the vanishing of coordinate functions on U, where U is a chart in M.

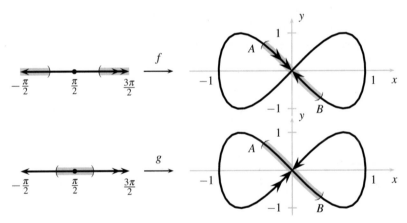

Fig. 11.6. The figure-eight as two distinct immersed submanifolds of $\mathbb{R}^2$.

Example 11.12 (*The figure-eight*). The figure-eight is the image of a one-to-one immersion

$$f(t) = (\cos t, \sin 2t), \quad -\pi/2 < t < 3\pi/2$$

(Figure 11.6). As such, it is an immersed submanifold of $\mathbb{R}^2$, with a topology and manifold structure induced from the open interval $]-\pi/2, 3\pi/2[$ by f. Because of the presence of a cross at the origin, it cannot be a regular submanifold of $\mathbb{R}^2$. In fact, with the subspace topology of $\mathbb{R}^2$, the figure-eight is not even a manifold.

The figure-eight is also the image of the one-to-one immersion

$$g(t) = (\cos t, -\sin 2t), \quad -\pi/2 < t < 3\pi/2$$

(Figure 11.6). The maps f and g induce distinct immersed submanifold structures on the figure-eight. For example, the open interval from A to B in Figure 11.6 is an open set in the topology induced from g, but it is not an open set in the topology induced from f, since its inverse image under f contains an isolated point $\pi/2$.

We will use the phrase "near p" to mean "in a neighborhood of p."

Theorem 11.13. *If $f: N \to M$ is an embedding, then its image $f(N)$ is a regular submanifold of M.*

Proof. Let $p \in N$. By the immersion theorem (Theorem 11.5), there are local coordinates $(U, x^1, \ldots, x^n)$ near p and $(V, y^1, \ldots, y^m)$ near $f(p)$ such that $f: U \to V$ has the form

$$(x^1, \ldots, x^n) \mapsto (x^1, \ldots, x^n, 0, \ldots, 0).$$

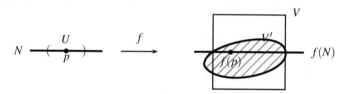

Fig. 11.7. The image of an embedding is a regular submanifold.

Thus, $f(U)$ is defined in V by the vanishing of the coordinates $y^{n+1}, \ldots, y^m$. This alone does not prove that $f(N)$ is a regular submanifold, since $V \cap f(N)$ may be larger than $f(U)$. (Think about Examples 11.9 and 11.10.) We need to show that in some neighborhood of $f(p)$ in V, the set $f(N)$ is defined by the vanishing of $m - n$ coordinates.

Since $f(N)$ with the subspace topology is homeomorphic to N, the image $f(U)$ is open in $f(N)$. By the definition of the subspace topology, there is an open set V' in M such that $V' \cap f(N) = f(U)$ (Figure 11.7). In $V \cap V'$,

$$V \cap V' \cap f(N) = V \cap f(U) = f(U),$$

and $f(U)$ is defined by the vanishing of $y^{n+1}, \ldots, y^m$. Thus, $(V \cap V', y^1, \ldots, y^m)$ is an adapted chart containing $f(p)$ for $f(N)$. Since $f(p)$ is an arbitrary point of $f(N)$, this proves that $f(N)$ is a regular submanifold of M. $\qquad\square$

Theorem 11.14. *If N is a regular submanifold of M, then the inclusion $i: N \to M$, $i(p) = p$, is an embedding.*

Proof. Since a regular submanifold has the subspace topology and $i(N)$ also has the subspace topology, $i: N \to i(N)$ is a homeomorphism. It remains to show that $i: N \to M$ is an immersion.

Let $p \in N$. Choose an adapted chart $(V, y^1, \ldots, y^n, y^{n+1}, \ldots, y^m)$ for M about p such that $V \cap N$ is the zero set of $y^{n+1}, \ldots, y^m$. Relative to the charts $(V \cap N, y^1, \ldots, y^n)$ for N and $(V, y^1, \ldots, y^m)$ for M, the inclusion i is given by

$$(y^1, \ldots, y^n) \mapsto (y^1, \ldots, y^n, 0, \ldots, 0),$$

which shows that i is an immersion.   □

In the literature the image of an embedding is often called an *embedded submanifold*. Theorems 11.13 and 11.14 show that an embedded submanifold and a regular submanifold are one and the same thing.

11.4 Smooth Maps into a Submanifold

Suppose $f: N \to M$ is a C^∞ map whose image $f(N)$ lies in a subset $S \subset M$. If S is a manifold, is the induced map $\tilde{f}: N \to S$ also C^∞? This question is more subtle than it looks, because the answer depends on whether S is a regular submanifold or an immersed submanifold of M.

Example. Consider the one-to-one immersions f and $g: I \to \mathbb{R}^2$ in Example 11.12, where I is the open interval $] - \pi/2, 3\pi/2 [$ in $\mathbb{R}$. Let S be the figure-eight in $\mathbb{R}^2$ with the immersed submanifold structure induced from g. Because the image of $f: I \to \mathbb{R}^2$ lies in S, the C^∞ map f induces a map $\tilde{f}: I \to S$.

The open interval from A to B in Figure 11.6 is an open neighborhood of the origin 0 in S. Its inverse image under $\tilde{f}$ contains the point $\pi/2$ as an isolated point and is therefore not open. This shows that although $f: I \to \mathbb{R}^2$ is C^∞, the induced map $\tilde{f}: I \to S$ is not continuous and therefore not C^∞.

Theorem 11.15. *Suppose $f: N \to M$ is C^∞ and the image of f lies in a subset S of M. If S is a regular submanifold of M, then the induced map $\tilde{f}: N \to S$ is C^∞.*

Proof. Let $p \in N$. Denote the dimensions of N, M, and S by n, m, and s, respectively. By hypothesis, $f(p) \in S \subset M$. Since S is a regular submanifold of M, there is an adapted coordinate chart $(V, \psi) = (V, y^1, \ldots, y^m)$ for M about $f(p)$ such that $S \cap V$ is the zero set of $y^{s+1}, \ldots, y^m$, with coordinate map $\psi_S = (y^1, \ldots, y^s)$. By the continuity of f, it is possible to choose a neighborhood of p with $f(U) \subset V$. Then $f(U) \subset V \cap S$, so that for $q \in U$,

$$(\psi \circ f)(q) = (y^1(f(q)), \ldots, y^s(f(q)), 0, \ldots, 0).$$

It follows that on U,

$$\psi_S \circ \tilde{f} = (y^1 \circ f, \ldots, y^s \circ f).$$

Since $y^1 \circ f, \ldots, y^s \circ f$ are C^∞ on U, by Proposition 6.16, $\tilde{f}$ is C^∞ on U and hence at p. Since p was an arbitrary point of N, the map $\tilde{f}: N \to S$ is C^∞.   □

Example 11.16 (*Multiplication map of* SL$(n,\mathbb{R})$). The multiplication map

$$\mu\colon \mathrm{GL}(n,\mathbb{R}) \times \mathrm{GL}(n,\mathbb{R}) \to \mathrm{GL}(n,\mathbb{R}),$$
$$(A,B) \mapsto AB,$$

is clearly C^∞ because

$$(AB)_{ij} = \sum_{k=1}^{n} a_{ik}b_{kj}$$

is a polynomial and hence a C^∞ function of the coordinates a_{ik} and b_{kj}. However, one cannot conclude in the same way that the multiplication map

$$\bar{\mu}\colon \mathrm{SL}(n,\mathbb{R}) \times \mathrm{SL}(n,\mathbb{R}) \to \mathrm{SL}(n,\mathbb{R})$$

is C^∞. This is because $\{a_{ij}\}_{1\le i,j\le n}$ is not a coordinate system on SL$(n,\mathbb{R})$; there is one coordinate too many (See Problem 11.6).

Since SL$(n,\mathbb{R}) \times$ SL$(n,\mathbb{R})$ is a regular submanifold of GL$(n,\mathbb{R}) \times$ GL$(n,\mathbb{R})$, the inclusion map

$$i\colon \mathrm{SL}(n,\mathbb{R}) \times \mathrm{SL}(n,\mathbb{R}) \to \mathrm{GL}(n,\mathbb{R}) \times \mathrm{GL}(n,\mathbb{R})$$

is C^∞ by Theorem 11.14; therefore, the composition

$$\mu \circ i\colon \mathrm{SL}(n,\mathbb{R}) \times \mathrm{SL}(n,\mathbb{R}) \to \mathrm{GL}(n,\mathbb{R})$$

is also C^∞. Because the image of $\mu \circ i$ lies in SL$(n,\mathbb{R})$, and SL$(n,\mathbb{R})$ is a regular submanifold of GL$(n,\mathbb{R})$ (see Example 9.13), by Theorem 11.15 the induced map

$$\bar{\mu}\colon \mathrm{SL}(n,\mathbb{R}) \times \mathrm{SL}(n,\mathbb{R}) \to \mathrm{SL}(n,\mathbb{R})$$

is C^∞.

11.5 The Tangent Plane to a Surface in $\mathbb{R}^3$

Suppose $f(x^1,x^2,x^3)$ is a real-valued function on $\mathbb{R}^3$ with no critical points on its zero set $N = f^{-1}(0)$. By the regular level set theorem, N is a regular submanifold of $\mathbb{R}^3$. By Theorem 11.14 the inclusion $i\colon N \to \mathbb{R}^3$ is an embedding, so at any point p in N, $i_{*,p}\colon T_pN \to T_p\mathbb{R}^3$ is injective. We may therefore think of the tangent plane T_pN as a plane in $T_p\mathbb{R}^3 \simeq \mathbb{R}^3$ (Figure 11.8). We would like to find the equation of this plane.

Suppose $v = \sum v^i \partial/\partial x^i|_p$ is a vector in T_pN. Under the linear isomorphism $T_p\mathbb{R}^3 \simeq \mathbb{R}^3$, we identify v with the vector $\langle v^1, v^2, v^3 \rangle$ in $\mathbb{R}^3$. Let $c(t)$ be a curve lying in N with $c(0) = p$ and $c'(0) = \langle v^1, v^2, v^3 \rangle$. Since $c(t)$ lies in N, $f(c(t)) = 0$ for all t. By the chain rule,

$$0 = \frac{d}{dt}f(c(t)) = \sum_{i=1}^{3} \frac{\partial f}{\partial x^i}(c(t))(c^i)'(t).$$

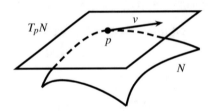

Fig. 11.8. Tangent plane to a surface N at p.

At $t = 0$,

$$0 = \sum_{i=1}^{3} \frac{\partial f}{\partial x^i}(c(0))(c^i)'(0) = \sum_{i=1}^{3} \frac{\partial f}{\partial x^i}(p)v^i.$$

Since the vector $v = \langle v^1, v^2, v^3 \rangle$ represents the arrow from the point $p = (p^1, p^2, p^3)$ to $x = (x^1, x^2, x^3)$ in the tangent plane, one usually makes the substitution $v^i = x^i - p^i$. This amounts to translating the tangent plane from the origin to p. Thus the tangent plane to N at p is defined by the equation

$$\sum_{i=1}^{3} \frac{\partial f}{\partial x^i}(p)(x^i - p^i) = 0. \tag{11.5}$$

One interpretation of this equation is that the gradient vector $\langle \partial f/\partial x^1(p), \partial f/\partial x^2(p), \partial f/\partial x^3(p) \rangle$ of f at p is normal to any vector in the tangent plane.

Example 11.17 (*Tangent plane to a sphere*). Let $f(x, y, z) = x^2 + y^2 + z^2 - 1$. To get the equation of the tangent plane to the unit sphere $S^2 = f^{-1}(0)$ in $\mathbb{R}^3$ at $(a, b, c) \in S^2$, we compute

$$\frac{\partial f}{\partial x} = 2x, \quad \frac{\partial f}{\partial y} = 2y, \quad \frac{\partial f}{\partial z} = 2z.$$

At $p = (a, b, c)$,

$$\frac{\partial f}{\partial x}(p) = 2a, \quad \frac{\partial f}{\partial y}(p) = 2b, \quad \frac{\partial f}{\partial z}(p) = 2c.$$

By (11.5) the equation of the tangent plane to the sphere at (a, b, c) is

$$2a(x - a) + 2b(y - b) + 2c(z - c) = 0,$$

or

$$ax + by + cz = 1,$$

since $a^2 + b^2 + c^2 = 1$.

Problems

11.1. Tangent vectors to a sphere

The unit sphere S^n in $\mathbb{R}^{n+1}$ is defined by the equation $\sum_{i=1}^{n+1}(x^i)^2 = 1$. For $p = (p^1, \ldots, p^{n+1}) \in S^n$, show that a necessary and sufficient condition for

$$X_p = \sum a^i \, \partial/\partial x^i|_p \in T_p(\mathbb{R}^{n+1})$$

to be tangent to S^n at p is $\sum a^i p^i = 0$.

11.2. Tangent vectors to a plane curve

(a) Let $i\colon S^1 \hookrightarrow \mathbb{R}^2$ be the inclusion map of the unit circle. In this problem, we denote by x, y the standard coordinates on $\mathbb{R}^2$ and by $\bar{x}, \bar{y}$ their restrictions to S^1. Thus, $\bar{x} = i^*x$ and $\bar{y} = i^*y$. On the upper semicircle $U = \{(a, b) \in S^1 \mid b > 0\}$, $\bar{x}$ is a local coordinate, so that $\partial/\partial\bar{x}$ is defined. Prove that for $p \in U$,

$$i_*\left(\left.\frac{\partial}{\partial\bar{x}}\right|_p\right) = \left.\left(\frac{\partial}{\partial x} + \frac{\partial\bar{y}}{\partial\bar{x}}\frac{\partial}{\partial y}\right)\right|_p.$$

Thus, although $i_*\colon T_p S^1 \to T_p \mathbb{R}^2$ is injective, $\partial/\partial\bar{x}|_p$ cannot be identified with $\partial/\partial x|_p$ (Figure 11.9).

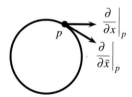

Fig. 11.9. Tangent vector $\partial/\partial\bar{x}|_p$ to a circle.

(b) Generalize (a) to a smooth curve C in $\mathbb{R}^2$, letting U be a chart in C on which $\bar{x}$, the restriction of x to C, is a local coordinate.

11.3.* Critical points of a smooth map on a compact manifold

Show that a smooth map f from a compact manifold N to $\mathbb{R}^m$ has a critical point. (*Hint:* Let $\pi\colon \mathbb{R}^m \to \mathbb{R}$ be the projection to the first factor. Consider the composite map $\pi \circ f\colon N \to \mathbb{R}$. A second proof uses Corollary 11.6 and the connectedness of $\mathbb{R}^m$.)

11.4. Differential of an inclusion map

On the upper hemisphere of the unit sphere S^2, we have the coordinate map $\phi = (u, v)$, where

$$u(a, b, c) = a \quad \text{and} \quad v(a, b, c) = b.$$

So the derivations $\partial/\partial u|_p, \partial/\partial v|_p$ are tangent vectors of S^2 at any point $p = (a, b, c)$ on the upper hemisphere. Let $i\colon S^2 \to \mathbb{R}^3$ be the inclusion and x, y, z the standard coordinates on $\mathbb{R}^3$. The differential $i_*\colon T_p S^2 \to T_p \mathbb{R}^3$ maps $\partial/\partial u|_p, \partial/\partial v|_p$ into $T_p \mathbb{R}^3$. Thus,

$$i_* \left(\left. \frac{\partial}{\partial u} \right|_p \right) = \alpha^1 \left. \frac{\partial}{\partial x} \right|_p + \beta^1 \left. \frac{\partial}{\partial y} \right|_p + \gamma^1 \left. \frac{\partial}{\partial z} \right|_p ,$$

$$i_* \left(\left. \frac{\partial}{\partial v} \right|_p \right) = \alpha^2 \left. \frac{\partial}{\partial x} \right|_p + \beta^2 \left. \frac{\partial}{\partial y} \right|_p + \gamma^2 \left. \frac{\partial}{\partial z} \right|_p ,$$

for some constants $\alpha^i, \beta^i, \gamma^i$. Find $(\alpha^i, \beta^i, \gamma^i)$ for $i = 1, 2$.

11.5. One-to-one immersion of a compact manifold

Prove that if N is a compact manifold, then a one-to-one immersion $f \colon N \to M$ is an embedding.

11.6. Multiplication map in $\mathrm{SL}(n, \mathbb{R})$

Let $f \colon \mathrm{GL}(n, \mathbb{R}) \to \mathbb{R}$ be the determinant map $f(A) = \det A = \det[a_{ij}]$. For $A \in \mathrm{SL}(n, \mathbb{R})$, there is at least one (k, ℓ) such that the partial derivative $\partial f / \partial a_{k\ell}(A)$ is nonzero (Example 9.13). Use Lemma 9.10 and the implicit function theorem to prove that

(a) there is a neighborhood of A in $\mathrm{SL}(n, \mathbb{R})$ in which a_{ij}, $(i, j) \neq (k, \ell)$, form a coordinate system, and $a_{k\ell}$ is a C^∞ function of the other entries a_{ij}, $(i, j) \neq (k, \ell)$;

(b) the multiplication map

$$\bar{\mu} \colon \mathrm{SL}(n, \mathbb{R}) \times \mathrm{SL}(n, \mathbb{R}) \to \mathrm{SL}(n, \mathbb{R})$$

is C^∞.

§12 The Tangent Bundle

A smooth vector bundle over a smooth manifold M is a smoothly varying family of vector spaces, parametrized by M, that locally looks like a product. Vector bundles and bundle maps form a category, and have played a fundamental role in geometry and topology since their appearance in the 1930s [39].

The collection of tangent spaces to a manifold has the structure of a vector bundle over the manifold, called the *tangent bundle*. A smooth map between two manifolds induces, via its differential at each point, a bundle map of the corresponding tangent bundles. Thus, the tangent bundle construction is a functor from the category of smooth manifolds to the category of vector bundles.

At first glance it might appear that the tangent bundle functor is not a simplification, since a vector bundle is a manifold plus an additional structure. However, because the tangent bundle is canonically associated to a manifold, invariants of the tangent bundle will give rise to invariants for the manifold. For example, the Chern–Weil theory of characteristic classes, which we treat in another volume, uses differential geometry to construct invariants for vector bundles. Applied to the tangent bundle, characteristic classes lead to numerical diffeomorphism invariants for a manifold called *characteristic numbers*. Characteristic numbers generalize, for instance, the classical Euler characteristic.

For us in this book the importance of the vector bundle point of view comes from its role in unifying concepts. A *section* of a vector bundle $\pi\colon E \to M$ is a map from M to E that maps each point of M into the fiber of the bundle over the point. As we shall see, both vector fields and differential forms on a manifold are sections of vector bundles over the manifold.

In the following pages we construct the tangent bundle of a manifold and show that it is a smooth vector bundle. We then discuss criteria for a section of a smooth vector bundle to be smooth.

12.1 The Topology of the Tangent Bundle

Let M be a smooth manifold. Recall that at each point $p \in M$, the tangent space T_pM is the vector space of all point-derivations of $C_p^\infty(M)$, the algebra of germs of C^∞ functions at p. The *tangent bundle* of M is the union of all the tangent spaces of M:

$$TM = \bigcup_{p \in M} T_pM.$$

In general, if $\{A_i\}_{i \in I}$ is a collection of subsets of a set S, then their *disjoint union* is defined to be the set

$$\coprod_{i \in I} A_i := \bigcup_{i \in I} (\{i\} \times A_i).$$

The subsets A_i may overlap, but in the disjoint union they are replaced by nonoverlapping copies.

In the definition of the tangent bundle, the union $\bigcup_{p\in M} T_p M$ is (up to notation) the same as the disjoint union $\coprod_{p\in M} T_p M$, since for distinct points p and q in M, the tangent spaces $T_p M$ and $T_q M$ are already disjoint.

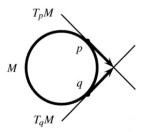

Fig. 12.1. Tangent spaces to a circle.

In a pictorial representation of tangent spaces such as Figure 12.1, where M is the unit circle, it may look as though the two tangent spaces $T_p M$ and $T_q M$ intersect. In fact, the intersection point of the two lines in Figure 12.1 represents distinct tangent vectors in $T_p M$ and $T_q M$, so that $T_p M$ and $T_q M$ are disjoint even in the figure.

There is a natural map $\pi\colon TM \to M$ given by $\pi(v) = p$ if $v \in T_p M$. (We use the word "natural" to mean that the map does not depend on any choice, for example, the choice of an atlas or of local coordinates for M.) As a matter of notation, we sometimes write a tangent vector $v \in T_p M$ as a pair (p, v), to make explicit the point $p \in M$ at which v is a tangent vector.

As defined, TM is a set, with no topology or manifold structure. We will make it into a smooth manifold and show that it is a C^∞ vector bundle over M. The first step is to give it a topology.

If $(U, \phi) = (U, x^1, \ldots, x^n)$ is a coordinate chart on M, let

$$TU = \bigcup_{p\in U} T_p U = \bigcup_{p\in U} T_p M.$$

(We saw in Remark 8.2 that $T_p U = T_p M$.) At a point $p \in U$, a basis for $T_p M$ is the set of coordinate vectors $\partial/\partial x^1|_p, \ldots, \partial/\partial x^n|_p$, so a tangent vector $v \in T_p M$ is uniquely a linear combination

$$v = \sum_{i}^{n} c^i \frac{\partial}{\partial x^i}\bigg|_p.$$

In this expression, the coefficients $c^i = c^i(v)$ depend on v and so are functions on TU. Let $\bar{x}^i = x^i \circ \pi$ and define the map $\tilde{\phi}\colon TU \to \phi(U) \times \mathbb{R}^n$ by

$$v \mapsto (x^1(p), \ldots, x^n(p), c^1(v), \ldots, c^n(v)) = (\bar{x}^1, \ldots, \bar{x}^n, c^1, \ldots, c^n)(v). \qquad (12.1)$$

Then $\tilde{\phi}$ has inverse

$$(\phi(p), c^1, \ldots, c^n) \mapsto \sum c^i \left. \frac{\partial}{\partial x^i} \right|_p$$

and is therefore a bijection. This means we can use $\tilde{\phi}$ to transfer the topology of $\phi(U) \times \mathbb{R}^n$ to TU: a set A in TU is open if and only if $\tilde{\phi}(A)$ is open in $\phi(U) \times \mathbb{R}^n$, where $\phi(U) \times \mathbb{R}^n$ is given its standard topology as an open subset of $\mathbb{R}^{2n}$. By definition, TU, with the topology induced by $\tilde{\phi}$, is homeomorphic to $\phi(U) \times \mathbb{R}^n$. If V is an open subset of U, then $\phi(V) \times \mathbb{R}^n$ is an open subset of $\phi(U) \times \mathbb{R}^n$. Hence, the relative topology on TV as a subset of TU is the same as the topology induced from the bijection $\tilde{\phi}|_{TV} : TV \to \phi(V) \times \mathbb{R}^n$.

Let $\phi_* : T_p U \to T_{\phi(p)}(\mathbb{R}^n)$ be the differential of the coordinate map ϕ at p. Since $\phi_*(v) = \sum c^i \partial/\partial r^i|_{\phi(p)} \in T_{\phi(p)}(\mathbb{R}^n) \simeq \mathbb{R}^n$ by Proposition 8.8, we may identify $\phi_*(v)$ with the column vector $\langle c^1, \ldots, c^n \rangle$ in $\mathbb{R}^n$. So another way to describe $\tilde{\phi}$ is $\tilde{\phi} = (\phi \circ \pi, \phi_*)$.

Let $\mathcal{B}$ be the collection of all open subsets of $T(U_\alpha)$ as U_α runs over all coordinate open sets in M:

$$\mathcal{B} = \bigcup_\alpha \{A \mid A \text{ open in } T(U_\alpha), U_\alpha \text{ a coordinate open set in } M\}.$$

Lemma 12.1. (i) *For any manifold M, the set TM is the union of all $A \in \mathcal{B}$.*
(ii) *Let U and V be coordinate open sets in a manifold M. If A is open in TU and B is open in TV, then $A \cap B$ is open in $T(U \cap V)$.*

Proof. (i) Let $\{(U_\alpha, \phi_\alpha)\}$ be the maximal atlas for M. Then

$$TM = \bigcup_\alpha T(U_\alpha) \subset \bigcup_{A \in \mathcal{B}} A \subset TM,$$

so equality holds everywhere.
(ii) Since $T(U \cap V)$ is a subspace of TU, by the definition of relative topology, $A \cap T(U \cap V)$ is open in $T(U \cap V)$. Similarly, $B \cap T(U \cap V)$ is open in $T(U \cap V)$. But

$$A \cap B \subset TU \cap TV = T(U \cap V).$$

Hence,

$$A \cap B = A \cap B \cap T(U \cap V) = (A \cap T(U \cap V)) \cap (B \cap T(U \cap V))$$

is open in $T(U \cap V)$. □

It follows from this lemma that the collection $\mathcal{B}$ satisfies the conditions (i) and (ii) of Proposition A.8 for a collection of subsets to be a basis for some topology on TM. We give the tangent bundle TM the topology generated by the basis $\mathcal{B}$.

Lemma 12.2. *A manifold M has a countable basis consisting of coordinate open sets.*

Proof. Let $\{(U_\alpha, \phi_\alpha)\}$ be the maximal atlas on M and $\mathcal{B} = \{B_i\}$ a countable basis for M. For each coordinate open set U_α and point $p \in U_\alpha$, choose a basic open set $B_{p,\alpha} \in \mathcal{B}$ such that

$$p \in B_{p,\alpha} \subset U_\alpha.$$

The collection $\{B_{p,\alpha}\}$, without duplicate elements, is a subcollection of $\mathcal{B}$ and is therefore countable.

For any open set U in M and a point $p \in U$, there is a coordinate open set U_α such that

$$p \in U_\alpha \subset U.$$

Hence,

$$p \in B_{p,\alpha} \subset U,$$

which shows that $\{B_{p,\alpha}\}$ is a basis for M. □

Proposition 12.3. *The tangent bundle TM of a manifold M is second countable.*

Proof. Let $\{U_i\}_{i=1}^\infty$ be a countable basis for M consisting of coordinate open sets. Let ϕ_i be the coordinate map on U_i. Since TU_i is homeomorphic to the open subset $\phi_i(U_i) \times \mathbb{R}^n$ of $\mathbb{R}^{2n}$ and any subset of a Euclidean space is second countable (Example A.13 and Proposition A.14), TU_i is second countable. For each i, choose a countable basis $\{B_{i,j}\}_{j=1}^\infty$ for TU_i. Then $\{B_{i,j}\}_{i,j=1}^\infty$ is a countable basis for the tangent bundle. □

Proposition 12.4. *The tangent bundle TM of a manifold M is Hausdorff.*

Proof. Problem 12.1. □

12.2 The Manifold Structure on the Tangent Bundle

Next we show that if $\{(U_\alpha, \phi_\alpha)\}$ is a C^∞ atlas for M, then $\{(TU_\alpha, \tilde{\phi}_\alpha)\}$ is a C^∞ atlas for the tangent bundle TM, where $\tilde{\phi}_\alpha$ is the map on TU_α induced by ϕ_α as in (12.1). It is clear that $TM = \bigcup_\alpha TU_\alpha$. It remains to check that on $(TU_\alpha) \cap (TU_\beta)$, $\tilde{\phi}_\alpha$ and $\tilde{\phi}_\beta$ are C^∞ compatible.

Recall that if $(U, x^1, \ldots, x^n)$, $(V, y^1, \ldots, y^n)$ are two charts on M, then for any $p \in U \cap V$ there are two bases singled out for the tangent space T_pM: $\{\partial/\partial x^j|_p\}_{j=1}^n$ and $\{\partial/\partial y^i|_p\}_{i=1}^n$. So any tangent vector $v \in T_pM$ has two descriptions:

$$v = \sum_j a^j \frac{\partial}{\partial x^j}\bigg|_p = \sum_i b^i \frac{\partial}{\partial y^i}\bigg|_p. \tag{12.2}$$

It is easy to compare them. By applying both sides to x^k, we find that

$$a^k = \left(\sum_j a^j \frac{\partial}{\partial x^j}\right) x^k = \left(\sum_i b^i \frac{\partial}{\partial y^i}\right) x^k = \sum_i b^i \frac{\partial x^k}{\partial y^i}.$$

Similarly, applying both sides of (12.2) to y^k gives

$$b^k = \sum_j a^j \frac{\partial y^k}{\partial x^j}. \tag{12.3}$$

Returning to the atlas $\{(U_\alpha, \phi_\alpha)\}$, we write $U_{\alpha\beta} = U_\alpha \cap U_\beta$, $\phi_\alpha = (x^1, \ldots, x^n)$ and $\phi_\beta = (y^1, \ldots, y^n)$. Then

$$\tilde{\phi}_\beta \circ \tilde{\phi}_\alpha^{-1} \colon \phi_\alpha(U_{\alpha\beta}) \times \mathbb{R}^n \to \phi_\beta(U_{\alpha\beta}) \times \mathbb{R}^n$$

is given by

$$(\phi_\alpha(p), a^1, \ldots, a^n) \mapsto \left(p, \sum_j a^j \left. \frac{\partial}{\partial x^j} \right|_p \right) \mapsto ((\phi_\beta \circ \phi_\alpha^{-1})(\phi_\alpha(p)), b^1, \ldots, b^n),$$

where by (12.3) and Example 6.24,

$$b^i = \sum_j a^j \frac{\partial y^i}{\partial x^j}(p) = \sum_j a^j \frac{\partial(\phi_\beta \circ \phi_\alpha^{-1})^i}{\partial r^j}(\phi_\alpha(p)).$$

By the definition of an atlas, $\phi_\beta \circ \phi_\alpha^{-1}$ is C^∞. Therefore, $\tilde{\phi}_\beta \circ \tilde{\phi}_\alpha^{-1}$ is C^∞. This completes the proof that the tangent bundle TM is a C^∞ manifold, with $\{(TU_\alpha, \tilde{\phi}_\alpha)\}$ as a C^∞ atlas.

12.3 Vector Bundles

On the tangent bundle TM of a smooth manifold M, the natural projection map $\pi \colon TM \to M$, $\pi(p, v) = p$ makes TM into a C^∞ *vector bundle* over M, which we now define.

Given any map $\pi \colon E \to M$, we call the inverse image $\pi^{-1}(p) := \pi^{-1}(\{p\})$ of a point $p \in M$ the *fiber at p*. The fiber at p is often written E_p. For any two maps $\pi \colon E \to M$ and $\pi' \colon E' \to M$ with the same target space M, a map $\phi \colon E \to E'$ is said to be *fiber-preserving* if $\phi(E_p) \subset E'_p$ for all $p \in M$.

Exercise 12.5 (Fiber-preserving maps). Given two maps $\pi \colon E \to M$ and $\pi' \colon E' \to M$, check that a map $\phi \colon E \to E'$ is fiber-preserving if and only if the diagram

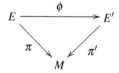

commutes.

A surjective smooth map $\pi \colon E \to M$ of manifolds is said to be *locally trivial of rank r* if

(i) each fiber $\pi^{-1}(p)$ has the structure of a vector space of dimension r;
(ii) for each $p \in M$, there are an open neighborhood U of p and a fiber-preserving
     diffeomorphism $\phi \colon \pi^{-1}(U) \to U \times \mathbb{R}^r$ such that for every $q \in U$ the restriction

$$\phi|_{\pi^{-1}(q)} \colon \pi^{-1}(q) \to \{q\} \times \mathbb{R}^r$$

is a vector space isomorphism. Such an open set U is called a *trivializing open set* for E, and ϕ is called a *trivialization* of E over U.

The collection $\{(U,\phi)\}$, with $\{U\}$ an open cover of M, is called a *local trivialization* for E, and $\{U\}$ is called a *trivializing open cover* of M for E.

A C^∞ *vector bundle of rank r* is a triple (E, M, π) consisting of manifolds E and M and a surjective smooth map $\pi \colon E \to M$ that is locally trivial of rank r. The manifold E is called the *total space* of the vector bundle and M the *base space*. By abuse of language, we say that E is a *vector bundle over M*. For any regular submanifold $S \subset M$, the triple $(\pi^{-1}S, S, \pi|_{\pi^{-1}S})$ is a C^∞ vector bundle over S, called the *restriction* of E to S. We will often write the restriction as $E|_S$ instead of $\pi^{-1}S$.

Properly speaking, the tangent bundle of a manifold M is a triple (TM, M, π), and TM is the total space of the tangent bundle. In common usage, TM is often referred to as the tangent bundle.

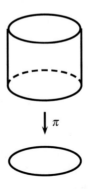

Fig. 12.2. A circular cylinder is a product bundle over a circle.

Example 12.6 (*Product bundle*).   Given a manifold M, let $\pi \colon M \times \mathbb{R}^r \to M$ be the projection to the first factor. Then $M \times \mathbb{R}^r \to M$ is a vector bundle of rank r, called the *product bundle* of rank r over M. The vector space structure on the fiber $\pi^{-1}(p) = \{(p,v) \mid v \in \mathbb{R}^r\}$ is the obvious one:

$$(p,u) + (p,v) = (p, u+v), \quad b \cdot (p,v) = (p, bv) \text{ for } b \in \mathbb{R}.$$

A local trivialization on $M \times R$ is given by the identity map $1_{M \times \mathbb{R}} \colon M \times \mathbb{R} \to M \times \mathbb{R}$. The infinite cylinder $S^1 \times \mathbb{R}$ is the product bundle of rank 1 over the circle (Figure 12.2).

Let $\pi\colon E \to M$ be a C^∞ vector bundle. Suppose $(U, \psi) = (U, x^1, \ldots, x^n)$ is a chart on M and

$$\phi\colon E|_U \overset{\sim}{\to} U \times \mathbb{R}^r, \quad \phi(e) = \big(\pi(e), c^1(e), \ldots, c^r(e)\big),$$

is a trivialization of E over U. Then

$$(\psi \times \mathbb{1}) \circ \phi = (x^1, \ldots, x^n, c^1, \ldots, c^r)\colon E|_U \overset{\sim}{\to} U \times \mathbb{R}^n \overset{\sim}{\to} \psi(U) \times \mathbb{R}^r \subset \mathbb{R}^n \times \mathbb{R}^r$$

is a diffeomorphism of $E|_U$ onto its image and so is a chart on E. We call $x^1, \ldots, x^n$ the *base coordinates* and $c^1, \ldots, c^r$ the *fiber coordinates* of the chart $(E|_U, (\psi \times \mathbb{1}) \circ \phi)$ on E. Note that the fiber coordinates c^i depend only on the trivialization ϕ of the bundle $E|_U$ and not on the trivialization ψ of the base U.

Let $\pi_E\colon E \to M$, $\pi_F\colon F \to N$ be two vector bundles, possibly of different ranks. A *bundle map* from E to F is a pair of maps $(f, \tilde{f})$, $f\colon M \to N$ and $\tilde{f}\colon E \to F$, such that

(i) the diagram

$$
\begin{array}{ccc}
E & \overset{\tilde{f}}{\longrightarrow} & F \\
{\scriptstyle \pi_E}\big\downarrow & & \big\downarrow{\scriptstyle \pi_F} \\
M & \underset{f}{\longrightarrow} & N
\end{array}
$$

is commutative, meaning $\pi_F \circ \tilde{f} = f \circ \pi_E$;

(ii) $\tilde{f}$ is linear on each fiber; i.e., for each $p \in M$, $\tilde{f}\colon E_p \to F_{f(p)}$ is a linear map of vector spaces.

The collection of all vector bundles together with bundle maps between them forms a category.

Example. A smooth map $f\colon N \to M$ of manifolds induces a bundle map $(f, \tilde{f})$, where $\tilde{f}\colon TN \to TM$ is given by

$$\tilde{f}(p, v) = (f(p), f_*(v)) \in \{f(p)\} \times T_{f(p)}M \subset TM$$

for all $v \in T_pN$. This gives rise to a covariant functor T from the category of smooth manifolds and smooth maps to the category of vector bundles and bundle maps: to each manifold M, we associate its tangent bundle $T(M)$, and to each C^∞ map $f\colon N \to M$ of manifolds, we associate the bundle map $T(f) = \big(f\colon N \to M, \tilde{f}\colon T(N) \to T(M)\big)$.

If E and F are two vector bundles over the same manifold M, then a bundle map from E to F *over M* is a bundle map in which the base map is the identity $\mathbb{1}_M$. For a fixed manifold M, we can also consider the category of all C^∞ vector bundles over M and C^∞ bundle maps over M. In this category it makes sense to speak of an isomorphism of vector bundles *over M*. Any vector bundle over M isomorphic over M to the product bundle $M \times \mathbb{R}^r$ is called a *trivial bundle*.

12.4 Smooth Sections

A *section* of a vector bundle $\pi\colon E \to M$ is a map $s\colon M \to E$ such that $\pi \circ s = \mathbb{1}_M$, the identity map on M. This condition means precisely that for each p in M, s maps p into the fiber E_p above p. Pictorially we visualize a section as a cross-section of the bundle (Figure 12.3). We say that a section is *smooth* if it is smooth as a map from M to E.

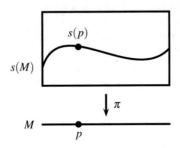

Fig. 12.3. A section of a vector bundle.

Definition 12.7. A *vector field* X on a manifold M is a function that assigns a tangent vector $X_p \in T_pM$ to each point $p \in M$. In terms of the tangent bundle, a vector field on M is simply a section of the tangent bundle $\pi\colon TM \to M$ and the vector field is *smooth* if it is smooth as a map from M to TM.

Example 12.8. The formula

$$X_{(x,y)} = -y\frac{\partial}{\partial x} + x\frac{\partial}{\partial y} = \begin{bmatrix} -y \\ x \end{bmatrix}$$

defines a smooth vector field on $\mathbb{R}^2$ (Figure 12.4, cf. Example 2.3).

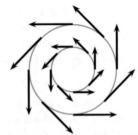

Fig. 12.4. The vector field $(-y,x)$ in $\mathbb{R}^2$.

Proposition 12.9. *Let s and t be C^∞ sections of a C^∞ vector bundle $\pi\colon E \to M$ and let f be a C^∞ real-valued function on M. Then*

(i) *the sum $s+t\colon M \to E$ defined by*

$$(s+t)(p) = s(p) + t(p) \in E_p, \quad p \in M,$$

is a C^∞ section of E.

(ii) *the product $fs\colon M \to E$ defined by*

$$(fs)(p) = f(p)s(p) \in E_p, \quad p \in M,$$

is a C^∞ section of E.

Proof.
(i) It is clear that $s+t$ is a section of E. To show that it is C^∞, fix a point $p \in M$ and let V be a trivializing open set for E containing p, with C^∞ trivialization

$$\phi\colon \pi^{-1}(V) \to V \times \mathbb{R}^r.$$

Suppose

$$(\phi \circ s)(q) = (q, a^1(q), \ldots, a^r(q))$$

and

$$(\phi \circ t)(q) = (q, b^1(q), \ldots, b^r(q))$$

for $q \in V$. Because s and t are C^∞ maps, a^i and b^i are C^∞ functions on V (Proposition 6.16). Since ϕ is linear on each fiber,

$$(\phi \circ (s+t))(q) = (q, a^1(q) + b^1(q), \ldots, a^r(q) + b^r(q)), \quad q \in V.$$

This proves that $s+t$ is a C^∞ map on V and hence at p. Since p is an arbitrary point of M, the section $s+t$ is C^∞ on M.

(ii) We omit the proof, since it is similar to that of (i). □

Denote the set of all C^∞ sections of E by $\Gamma(E)$. The proposition shows that $\Gamma(E)$ is not only a vector space over $\mathbb{R}$, but also a module over the ring $C^\infty(M)$ of C^∞ functions on M. For any open subset $U \subset M$, one can also consider the vector space $\Gamma(U,E)$ of C^∞ sections of E over U. Then $\Gamma(U,E)$ is both a vector space over $\mathbb{R}$ and a $C^\infty(U)$-module. Note that $\Gamma(M,E) = \Gamma(E)$. To contrast with sections over a proper subset U, a section over the entire manifold M is called a *global section*.

12.5 Smooth Frames

A *frame* for a vector bundle $\pi\colon E \to M$ over an open set U is a collection of sections $s_1, \ldots, s_r$ of E over U such that at each point $p \in U$, the elements $s_1(p), \ldots, s_r(p)$ form a basis for the fiber $E_p := \pi^{-1}(p)$. A frame $s_1, \ldots, s_r$ is said to be *smooth* or C^∞

if $s_1, \ldots, s_r$ are C^∞ as sections of E over U. A frame for the tangent bundle $TM \to M$ over an open set U is simply called a *frame on U*.

Example. The collection of vector fields $\partial/\partial x, \partial/\partial y, \partial/\partial z$ is a smooth frame on $\mathbb{R}^3$.

Example. Let M be a manifold and $e_1, \ldots, e_r$ the standard basis for $\mathbb{R}^n$. Define $\bar{e}_i \colon M \to M \times \mathbb{R}^r$ by $\bar{e}_i(p) = (p, e_i)$. Then $\bar{e}_1, \ldots, \bar{e}_r$ is a C^∞ frame for the product bundle $M \times \mathbb{R}^r \to M$.

Example 12.10 (*The frame of a trivialization*). Let $\pi \colon E \to M$ be a smooth vector bundle of rank r. If $\phi \colon E|_U \xrightarrow{\sim} U \times \mathbb{R}^r$ is a trivialization of E over an open set U, then ϕ^{-1} carries the C^∞ frame $\bar{e}_1, \ldots, \bar{e}_r$ of the product bundle $U \times \mathbb{R}^r$ to a C^∞ frame $t_1, \ldots, t_r$ for E over U:

$$t_i(p) = \phi^{-1}(\bar{e}_i(p)) = \phi^{-1}(p, e_i), \quad p \in U.$$

We call $t_1, \ldots, t_r$ the C^∞ *frame over U of the trivialization ϕ*.

Lemma 12.11. *Let $\phi \colon E|_U \to U \times \mathbb{R}^r$ be a trivialization over an open set U of a C^∞ vector bundle $E \to M$, and $t_1, \ldots, t_r$ the C^∞ frame over U of the trivialization. Then a section $s = \sum b^i t_i$ of E over U is C^∞ if and only if its coefficients b^i relative to the frame $t_1, \ldots, t_r$ are C^∞.*

Proof.
($\Leftarrow$) This direction is an immediate consequence of Proposition 12.9.

($\Rightarrow$) Suppose the section $s = \sum b^i t_i$ of E over U is C^∞. Then $\phi \circ s$ is C^∞. Note that

$$(\phi \circ s)(p) = \sum b^i(p)\phi(t_i(p)) = \sum b^i(p)(p, e_i) = \left(p, \sum b^i(p)e_i\right).$$

Thus, the $b^i(p)$ are simply the fiber coordinates of $s(p)$ relative to the trivialization ϕ. Since $\phi \circ s$ is C^∞, all the b^i are C^∞. □

Proposition 12.12 (Characterization of C^∞ sections). *Let $\pi \colon E \to M$ be a C^∞ vector bundle and U an open subset of M. Suppose $s_1, \ldots, s_r$ is a C^∞ frame for E over U. Then a section $s = \sum c^j s_j$ of E over U is C^∞ if and only if the coefficients c^j are C^∞ functions on U.*

Proof. If $s_1, \ldots, s_r$ is the frame of a trivialization of E over U, then the proposition is Lemma 12.11. We prove the proposition in general by reducing it to this case. One direction is quite easy. If the c^j's are C^∞ functions on U, then $s = \sum c^j s_j$ is a C^∞ section on U by Proposition 12.9.

Conversely, suppose $s = \sum c^j s_j$ is a C^∞ section of E over U. Fix a point $p \in U$ and choose a trivializing open set $V \subset U$ for E containing p, with C^∞ trivialization $\phi \colon \pi^{-1}(V) \to V \times \mathbb{R}^r$. Let $t_1, \ldots, t_r$ be the C^∞ frame of the trivialization ϕ (Example 12.10). If we write s and s_j in terms of the frame $t_1, \ldots, t_r$, say $s = \sum b^i t_i$ and $s_j = \sum a^i_j t_i$, the coefficients b^i, a^i_j will all be C^∞ functions on V by Lemma 12.11. Next, express $s = \sum c^j s_j$ in terms of the t_i's:

$$\sum b^i t_i = s = \sum c^j s_j = \sum_{i,j} c^j a^i_j t_i.$$

Comparing the coefficients of t_i gives $b^i = \sum_j c^j a^i_j$. In matrix notation,

$$b = \begin{bmatrix} b^1 \\ \vdots \\ b^r \end{bmatrix} = A \begin{bmatrix} c^1 \\ \vdots \\ c^r \end{bmatrix} = Ac.$$

At each point of V, being the transition matrix between two bases, the matrix A is invertible. By Cramer's rule, A^{-1} is a matrix of C^∞ functions on V (see Example 6.21). Hence, $c = A^{-1}b$ is a column vector of C^∞ functions on V. This proves that $c^1, \ldots, c^r$ are C^∞ functions at $p \in U$. Since p is an arbitrary point of U, the coefficients c^j are C^∞ functions on U. $\qquad\square$

Remark 12.13. If one replaces "smooth" by "continuous" throughout, the discussion in this subsection remains valid in the continuous category.

Problems

12.1.* Hausdorff condition on the tangent bundle
Prove Proposition 12.4.

12.2. Transition functions for the total space of the tangent bundle
Let $(U, \phi) = (U, x^1, \ldots, x^n)$ and $(V, \psi) = (V, y^1, \ldots, y^n)$ be overlapping coordinate charts on a manifold M. They induce coordinate charts $(TU, \tilde{\phi})$ and $(TV, \tilde{\psi})$ on the total space TM of the tangent bundle (see equation (12.1)), with transition function $\tilde{\psi} \circ \tilde{\phi}^{-1}$:

$$(x^1, \ldots, x^n, a^1, \ldots, a^n) \mapsto (y^1, \ldots, y^n, b^1, \ldots, b^n).$$

(a) Compute the Jacobian matrix of the transition function $\tilde{\psi} \circ \tilde{\phi}^{-1}$ at $\phi(p)$.
(b) Show that the Jacobian determinant of the transition function $\tilde{\psi} \circ \tilde{\phi}^{-1}$ at $\phi(p)$ is $(\det[\partial y^i / \partial x^j])^2$.

12.3. Smoothness of scalar multiplication
Prove Proposition 12.9(ii).

12.4. Coefficients relative to a smooth frame
Let $\pi \colon E \to M$ be a C^∞ vector bundle and $s_1, \ldots, s_r$ a C^∞ frame for E over an open set U in M. Then every $e \in \pi^{-1}(U)$ can be written uniquely as a linear combination

$$e = \sum_{j=1}^r c^j(e) s_j(p), \quad p = \pi(e) \in U.$$

Prove that $c^j \colon \pi^{-1}U \to \mathbb{R}$ is C^∞ for $j = 1, \ldots, r$. (*Hint:* First show that the coefficients of e relative to the frame $t_1, \ldots, t_r$ of a trivialization are C^∞.)

§13 Bump Functions and Partitions of Unity

A partition of unity on a manifold is a collection of nonnegative functions that sum
to 1. Usually one demands in addition that the partition of unity be *subordinate* to
an open cover $\{U_\alpha\}_{\alpha \in A}$. What this means is that the partition of unity $\{\rho_\alpha\}_{\alpha \in A}$ is
indexed by the same set as the open over $\{U_\alpha\}_{\alpha \in A}$ and for each α in the index A,
the support of ρ_α is contained in U_α. In particular, ρ_α vanishes outside U_α.

The existence of a C^∞ partition of unity is one of the most important technical
tools in the theory of C^∞ manifolds. It is the single feature that makes the behavior
of C^∞ manifolds so different from that of real-analytic or complex manifolds. In this
section we construct C^∞ bump functions on any manifold and prove the existence
of a C^∞ partition of unity on a compact manifold. The proof of the existence of a
C^∞ partition of unity on a general manifold is more technical and is postponed to
Appendix C.

A partition of unity is used in two ways: (1) to decompose a global object on a
manifold into a locally finite sum of local objects on the open sets U_α of an open
cover, and (2) to patch together local objects on the open sets U_α into a global object
on the manifold. Thus, a partition of unity serves as a bridge between global and
local analysis on a manifold. This is useful because while there are always local coordinates on a manifold, there may be no global coordinates. In subsequent sections
we will see examples of both uses of a C^∞ partition of unity.

13.1 C^∞ Bump Functions

Recall that $\mathbb{R}^\times$ denotes the set of nonzero real numbers. The *support* of a real-valued
function f on a manifold M is defined to be the closure in M of the subset on which
$f \neq 0$:

$$\operatorname{supp} f = \operatorname{cl}_M(f^{-1}(\mathbb{R}^\times)) = \text{closure of } \{q \in M \mid f(q) \neq 0\} \text{ in } M.^1$$

Let q be a point in M, and U a neighborhood of q. By a *bump function at q
supported in U* we mean any continuous nonnegative function ρ on M that is 1 in a
neighborhood of q with $\operatorname{supp} \rho \subset U$.

For example, Figure 13.1 is the graph of a bump function at 0 supported in the
open interval $]-2,2[$. The function is nonzero on the open interval $]-1,1[$ and is
zero otherwise. Its support is the closed interval $[-1,1]$.

Example. The support of the function $f\colon \ ]-1,1[\ \to \mathbb{R}$, $f(x) = \tan(\pi x/2)$, is the
open interval $]-1,1[$, not the closed interval $[-1,1]$, because the closure of $f^{-1}(\mathbb{R}^\times)$
is taken in the domain $]-1,1[$, not in $\mathbb{R}$.

---

[1]In this section a general point is often denoted by q, instead of p, because p resembles
too much ρ, the notation for a bump function.

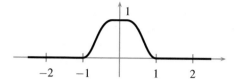

Fig. 13.1. A bump function at 0 on $\mathbb{R}$.

The only bump functions of interest to us are C^∞ bump functions. While the continuity of a function can often be seen by inspection, the smoothness of a function always requires a formula. Our goal in this subsection is to find a formula for a C^∞ bump function as in Figure 13.1.

Example. The graph of $y = x^{5/3}$ looks perfectly smooth (Figure 13.2), but it is in fact not smooth at $x = 0$, since its second derivative $y'' = (10/9)x^{-1/3}$ is not defined there.

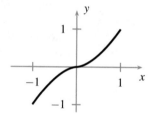

Fig. 13.2. The graph of $y = x^{5/3}$.

In Example 1.3 we introduced the C^∞ function

$$f(t) = \begin{cases} e^{-1/t} & \text{for } t > 0, \\ 0 & \text{for } t \le 0, \end{cases}$$

with graph as in Figure 13.3.

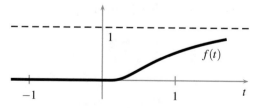

Fig. 13.3. The graph of $f(t)$.

The main challenge in building a smooth bump function from f is to construct a smooth version of a step function, that is, a C^∞ function $g\colon \mathbb{R} \to \mathbb{R}$ with graph as in Figure 13.4. Once we have such a C^∞ step function g, it is then simply a matter of

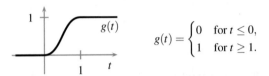

$$g(t) = \begin{cases} 0 & \text{for } t \le 0, \\ 1 & \text{for } t \ge 1. \end{cases}$$

Fig. 13.4. The graph of $g(t)$.

translating, reflecting, and scaling the function in order to make its graph look like Figure 13.1.

We seek $g(t)$ by dividing $f(t)$ by a positive function $\ell(t)$, for the quotient $f(t)/\ell(t)$ will then be zero for $t \le 0$. The denominator $\ell(t)$ should be a positive function that agrees with $f(t)$ for $t \ge 1$, for then $f(t)/\ell(t)$ will be identically 1 for $t \ge 1$. The simplest way to construct such an $\ell(t)$ is to add to $f(t)$ a nonnegative function that vanishes for $t \ge 1$. One such nonnegative function is $f(1-t)$. This suggests that we take $\ell(t) = f(t) + f(1-t)$ and consider

$$g(t) = \frac{f(t)}{f(t) + f(1-t)}. \tag{13.1}$$

Let us verify that the denominator $f(t) + f(1-t)$ is never zero. For $t > 0$, $f(t) > 0$ and therefore

$$f(t) + f(1-t) \ge f(t) > 0.$$

For $t \le 0$, $1 - t \ge 1$ and therefore

$$f(t) + f(1-t) \ge f(1-t) > 0.$$

In either case, $f(t) + f(1-t) \ne 0$. This proves that $g(t)$ is defined for all t. As the quotient of two C^∞ functions with denominator never zero, $g(t)$ is C^∞ for all t.

As noted above, for $t \le 0$, the numerator $f(t)$ equals 0, so $g(t)$ is identically zero for $t \le 0$. For $t \ge 1$, we have $1 - t \le 0$ and $f(1-t) = 0$, so $g(t) = f(t)/f(t)$ is identically 1 for $t \ge 1$. Thus, g is a C^∞ step function with the desired properties.

Given two positive real numbers $a < b$, we make a linear change of variables to map $[a^2, b^2]$ to $[0,1]$:

$$x \mapsto \frac{x - a^2}{b^2 - a^2}.$$

Let

$$h(x) = g\left(\frac{x - a^2}{b^2 - a^2}\right).$$

Then $h\colon \mathbb{R} \to [0,1]$ is a C^∞ step function such that

$$h(x) = \begin{cases} 0 & \text{for } x \le a^2, \\ 1 & \text{for } x \ge b^2. \end{cases}$$

(See Figure 13.5.)

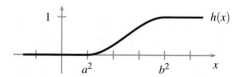

Fig. 13.5. The graph of $h(x)$.

Replace x by x^2 to make the function symmetric in x: $k(x) = h(x^2)$ (Figure 13.6).

Fig. 13.6. The graph of $k(x)$.

Finally, set

$$\rho(x) = 1 - k(x) = 1 - g\left(\frac{x^2 - a^2}{b^2 - a^2}\right).$$

This $\rho(x)$ is a C^∞ bump function at 0 in $\mathbb{R}$ that is identically 1 on $[-a, a]$ and has support in $[-b, b]$ (Figure 13.7). For any $q \in \mathbb{R}$, $\rho(x - q)$ is a C^∞ bump function at q.

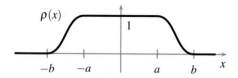

Fig. 13.7. A bump function at 0 on $\mathbb{R}$.

It is easy to extend the construction of a bump function from $\mathbb{R}$ to $\mathbb{R}^n$. To get a C^∞ bump function at $\mathbf{0}$ in $\mathbb{R}^n$ that is 1 on the closed ball $\bar{B}(\mathbf{0}, a)$ and has support in the closed ball $\bar{B}(\mathbf{0}, b)$, set

$$\sigma(x) = \rho(\|x\|) = 1 - g\left(\frac{\|x\| r^2 - a^2}{b^2 - a^2}\right). \tag{13.2}$$

As a composition of C^∞ functions, σ is C^∞. To get a C^∞ bump function at q in $\mathbb{R}^n$, take $\sigma(x - q)$.

Exercise 13.1 (Bump function supported in an open set).* Let q be a point and U any neighborhood of q in a manifold. Construct a C^∞ bump function at q supported in U.

In general, a C^∞ function on an open subset U of a manifold M cannot be extended to a C^∞ function on M; an example is the function $\sec(x)$ on the open interval $]-\pi/2, \pi/2[$ in $\mathbb{R}$. However, if we require that the global function on M agree with the given function only on some neighborhood of a point in U, then a C^∞ extension is possible.

Proposition 13.2 (C^∞ extension of a function). *Suppose f is a C^∞ function defined on a neighborhood U of a point p in a manifold M. Then there is a C^∞ function $\tilde{f}$ on M that agrees with f in some possibly smaller neighborhood of p.*

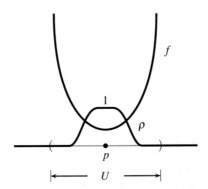

Fig. 13.8. Extending the domain of a function by multiplying by a bump function.

Proof. Choose a C^∞ bump function $\rho : M \to \mathbb{R}$ supported in U that is identically 1 in a neighborhood V of p (Figure 13.8). Define

$$\tilde{f}(q) = \begin{cases} \rho(q)f(q) & \text{for } q \text{ in } U, \\ 0 & \text{for } q \text{ not in } U. \end{cases}$$

As the product of two C^∞ functions on U, $\tilde{f}$ is C^∞ on U. If $q \notin U$, then $q \notin \operatorname{supp}\rho$, and so there is an open set containing q on which $\tilde{f}$ is 0, since $\operatorname{supp}\rho$ is closed. Therefore, $\tilde{f}$ is also C^∞ at every point $q \notin U$.

Finally, since $\rho \equiv 1$ on V, the function $\tilde{f}$ agrees with f on V.   □

13.2 Partitions of Unity

If $\{U_i\}_{i\in I}$ is a finite open cover of M, a C^∞ *partition of unity subordinate to* $\{U_i\}_{i\in I}$ is a collection of nonnegative C^∞ functions $\{\rho_i\colon M \to \mathbb{R}\}_{i\in I}$ such that $\operatorname{supp}\rho_i \subset U_i$ and

$$\sum \rho_i = 1. \tag{13.3}$$

When I is an infinite set, for the sum in (13.3) to make sense, we will impose a *local finiteness* condition. A collection $\{A_\alpha\}$ of subsets of a topological space S is said to be *locally finite* if every point q in S has a neighborhood that meets only finitely many of the sets A_α. In particular, every q in S is contained in only finitely many of the A_α's.

Example 13.3 (*An open cover that is not locally finite*). Let $U_{r,n}$ be the open interval $\left]r-\frac{1}{n},r+\frac{1}{n}\right[$ on the real line $\mathbb{R}$. The open cover $\{U_{r,n} \mid r \in \mathbb{Q}, n \in \mathbb{Z}^+\}$ of $\mathbb{R}$ is not locally finite.

Definition 13.4. A C^∞ *partition of unity* on a manifold is a collection of nonnegative C^∞ functions $\{\rho_\alpha\colon M \to \mathbb{R}\}_{\alpha\in A}$ such that

(i) the collection of supports, $\{\operatorname{supp}\rho_\alpha\}_{\alpha\in A}$, is locally finite,
(ii) $\sum \rho_\alpha = 1$.

Given an open cover $\{U_\alpha\}_{\alpha\in A}$ of M, we say that a partition of unity $\{\rho_\alpha\}_{\alpha\in A}$ is *subordinate to the open cover* $\{U_\alpha\}$ if $\operatorname{supp}\rho_\alpha \subset U_\alpha$ for every $\alpha \in A$.

Since the collection of supports, $\{\operatorname{supp}\rho_\alpha\}_{\alpha\in A}$, is locally finite (condition (i)), every point q lies in only finitely many of the sets $\operatorname{supp}\rho_\alpha$. Hence $\rho_\alpha(q) \neq 0$ for only finitely many α. It follows that the sum in (ii) is a finite sum at every point.

Example. Let U and V be the open intervals $]-\infty,2[$ and $]-1,\infty[$ in $\mathbb{R}$ respectively, and let ρ_V be a C^∞ function with graph as in Figure 13.9, for example the function $g(t)$ in (13.1). Define $\rho_U = 1 - \rho_V$. Then $\operatorname{supp}\rho_V \subset V$ and $\operatorname{supp}\rho_U \subset U$. Thus, $\{\rho_U, \rho_V\}$ is a partition of unity subordinate to the open cover $\{U,V\}$.

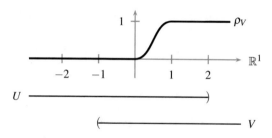

Fig. 13.9. A partition of unity $\{\rho_U, \rho_V\}$ subordinate to an open cover $\{U,V\}$.

Remark. Suppose $\{f_\alpha\}_{\alpha \in A}$ is a collection of C^∞ functions on a manifold M such that the collection of its supports, $\{\operatorname{supp} f_\alpha\}_{\alpha \in A}$, is locally finite. Then every point q in M has a neighborhood W_q that intersects $\operatorname{supp} f_\alpha$ for only finitely many α. Thus, on W_q the sum $\sum_{\alpha \in A} f_\alpha$ is actually a finite sum. This shows that the function $f = \sum f_\alpha$ is well defined and C^∞ on the manifold M. We call such a sum a *locally finite* sum.

13.3 Existence of a Partition of Unity

In this subsection we begin a proof of the existence of a C^∞ partition of unity on a manifold. Because the case of a compact manifold is somewhat easier and already has some of the features of the general case, for pedagogical reasons we give a separate proof for the compact case.

Lemma 13.5. *If $\rho_1, \ldots, \rho_m$ are real-valued functions on a manifold M, then*

$$\operatorname{supp} \left(\sum \rho_i \right) \subset \bigcup \operatorname{supp} \rho_i.$$

Proof. Problem 13.1. □

Proposition 13.6. *Let M be a compact manifold and $\{U_\alpha\}_{\alpha \in A}$ an open cover of M. There exists a C^∞ partition of unity $\{\rho_\alpha\}_{\alpha \in A}$ subordinate to $\{U_\alpha\}_{\alpha \in A}$.*

Proof. For each $q \in M$, find an open set U_α containing q from the given cover and let ψ_q be a C^∞ bump function at q supported in U_α (Exercise 13.1, p. 144). Because $\psi_q(q) > 0$, there is a neighborhood W_q of q on which $\psi_q > 0$. By the compactness of M, the open cover $\{W_q \mid q \in M\}$ has a finite subcover, say $\{W_{q_1}, \ldots, W_{q_m}\}$. Let $\psi_{q_1}, \ldots, \psi_{q_m}$ be the corresponding bump functions. Then $\psi := \sum \psi_{q_i}$ is positive at every point q in M because $q \in W_{q_i}$ for some i. Define

$$\varphi_i = \frac{\psi_{q_i}}{\psi}, \quad i = 1, \ldots, m.$$

Clearly, $\sum \varphi_i = 1$. Moreover, since $\psi > 0$, $\varphi_i(q) \neq 0$ if and only if $\psi_{q_i}(q) \neq 0$, so

$$\operatorname{supp} \varphi_i = \operatorname{supp} \psi_{q_i} \subset U_\alpha$$

for some $\alpha \in A$. This shows that $\{\varphi_i\}$ is a partition of unity such that for every i, $\operatorname{supp} \varphi_i \subset U_\alpha$ for some $\alpha \in A$.

The next step is to make the index set of the partition of unity the same as that of the open cover. For each $i = 1, \ldots, m$, choose $\tau(i) \in A$ to be an index such that

$$\operatorname{supp} \varphi_i \subset U_{\tau(i)}.$$

We group the collection of functions $\{\varphi_i\}$ into subcollections according to $\tau(i)$ and define for each $\alpha \in A$,

$$\rho_\alpha = \sum_{\tau(i) = \alpha} \varphi_i;$$

if there is no i for which $\tau(i) = \alpha$, the sum above is empty and we define $\rho_\alpha = 0$. Then

$$\sum_{\alpha \in A} \rho_\alpha = \sum_{\alpha \in A} \sum_{\tau(i)=\alpha} \varphi_i = \sum_{i=1}^{m} \varphi_i = 1.$$

Moreover, by Lemma 13.5,

$$\operatorname{supp} \rho_\alpha \subset \bigcup_{\tau(i)=\alpha} \operatorname{supp} \varphi_i \subset U_\alpha.$$

So $\{\rho_\alpha\}$ is a partition of unity subordinate to $\{U_\alpha\}$. □

To generalize the proof of Proposition 13.6 to an arbitrary manifold, it will be necessary to find an appropriate substitute for compactness. Since the proof is rather technical and is not necessary for the rest of the book, we put it in Appendix C. The statement is as follows.

Theorem 13.7 (Existence of a C^∞ partition of unity). *Let $\{U_\alpha\}_{\alpha \in A}$ be an open cover of a manifold M.*

(i) *There is a C^∞ partition of unity $\{\varphi_k\}_{k=1}^{\infty}$ with every φ_k having compact support such that for each k, $\operatorname{supp} \varphi_k \subset U_\alpha$ for some $\alpha \in A$.*

(ii) *If we do not require compact support, then there is a C^∞ partition of unity $\{\rho_\alpha\}$ subordinate to $\{U_\alpha\}$.*

Problems

13.1.* Support of a finite sum
Prove Lemma 13.5.

13.2.* Locally finite family and compact set
Let $\{A_\alpha\}$ be a locally finite family of subsets of a topological space S. Show that every compact set K in S has a neighborhood W that intersects only finitely many of the A_α.

13.3. Smooth Urysohn lemma

(a) Let A and B be two disjoint closed sets in a manifold M. Find a C^∞ function f on M such that f is identically 1 on A and identically 0 on B. (*Hint*: Consider a C^∞ partition of unity $\{\rho_{M-A}, \rho_{M-B}\}$ subordinate to the open cover $\{M - A, M - B\}$. This lemma is needed in Subsection 29.3.)

(b) Let A be a closed subset and U an open subset of a manifold M. Show that there is a C^∞ function f on M such that f is identically 1 on A and $\operatorname{supp} f \subset U$.

13.4. Support of the pullback of a function
Let $F : N \to M$ be a C^∞ map of manifolds and $h : M \to \mathbb{R}$ a C^∞ real-valued function. Prove that $\operatorname{supp} F^* h \subset F^{-1}(\operatorname{supp} h)$. (*Hint*: First show that $(F^* h)^{-1}(\mathbb{R}^\times) \subset F^{-1}(\operatorname{supp} h)$.)

13.5.* Support of the pullback by a projection

Let $f\colon M \to \mathbb{R}$ be a C^∞ function on a manifold M. If N is another manifold and $\pi\colon M \times N \to M$ is the projection onto the first factor, prove that

$$\operatorname{supp}(\pi^* f) = (\operatorname{supp} f) \times N.$$

13.6. Pullback of a partition of unity

Suppose $\{\rho_\alpha\}$ is a partition of unity on a manifold M subordinate to an open cover $\{U_\alpha\}$ of M and $F\colon N \to M$ is a C^∞ map. Prove that

(a) the collection of supports $\{\operatorname{supp} F^* \rho_\alpha\}$ is locally finite;
(b) the collection of functions $\{F^* \rho_\alpha\}$ is a partition of unity on N subordinate to the open cover $\{F^{-1}(U_\alpha)\}$ of N.

13.7.* Closure of a locally finite union

If $\{A_\alpha\}$ is a locally finite collection of subsets in a topological space, then

$$\overline{\bigcup A_\alpha} = \bigcup \overline{A_\alpha}, \tag{13.4}$$

where $\overline{A}$ denotes the closure of the subset A.

Remark. For any collection of subsets A_α, one always has

$$\bigcup \overline{A_\alpha} \subset \overline{\bigcup A_\alpha}.$$

However, the reverse inclusion is in general not true. For example, suppose A_n is the closed interval $[0, 1 - (1/n)]$ in $\mathbb{R}$. Then

$$\overline{\bigcup_{n=1}^{\infty} A_n} = \overline{[0,1)} = [0,1],$$

but

$$\bigcup_{n=1}^{\infty} \overline{A_n} = \bigcup_{n=1}^{\infty} \left[0, 1 - \frac{1}{n}\right] = [0,1).$$

If $\{A_\alpha\}$ is a finite collection, the equality (13.4) is easily shown to be true.

§14 Vector Fields

A vector field X on a manifold M is the assignment of a tangent vector $X_p \in T_pM$ to each point $p \in M$. More formally, a vector field on M is a section of the tangent bundle TM of M. It is natural to define a vector field as smooth if it is smooth as a section of the tangent bundle. In the first subsection we give two other characterizations of smooth vector fields, in terms of the coefficients relative to coordinate vector fields and in terms of smooth functions on the manifold.

Vector fields abound in nature, for example the velocity vector field of a fluid flow, the electric field of a charge, the gravitational field of a mass, and so on. The fluid flow model is in fact quite general, for as we will see shortly, every smooth vector field may be viewed locally as the velocity vector field of a fluid flow. The path traced out by a point under this flow is called an *integral curve* of the vector field. Integral curves are curves whose velocity vector field is the restriction of the given vector field to the curve. Finding the equation of an integral curve is equivalent to solving a system of first-order ordinary differential equations (ODE). Thus, the theory of ODE guarantees the existence of integral curves.

The set $\mathfrak{X}(M)$ of all C^∞ vector fields on a manifold M clearly has the structure of a vector space. We introduce a bracket operation $[\,,\,]$ that makes it into a Lie algebra. Because vector fields do not push forward under smooth maps, the Lie algebra $\mathfrak{X}(M)$ does not give rise to a functor on the category of smooth manifolds. Nonetheless, there is a notion of *related vector fields* that allows us to compare vector fields on two manifolds under a smooth map.

14.1 Smoothness of a Vector Field

In Definition 12.7 we defined a vector field X on a manifold M to be *smooth* if the map $X \colon M \to TM$ is smooth as a section of the tangent bundle $\pi \colon TM \to M$. In a coordinate chart $(U, \phi) = (U, x^1, \ldots, x^n)$ on M, the value of the vector field X at $p \in U$ is a linear combination

$$X_p = \sum a^i(p) \left. \frac{\partial}{\partial x^i} \right|_p .$$

As p varies in U, the coefficients a^i become functions on U.

As we learned in Subsections 12.1 and 12.2, the chart $(U, \phi) = (U, x^1, \ldots, x^n)$ on the manifold M induces a chart

$$(TU, \tilde{\phi}) = (TU, \bar{x}^1, \ldots, \bar{x}^n, c^1, \ldots, c^n)$$

on the tangent bundle TM, where $\bar{x}^i = \pi^* x^i = x^i \circ \pi$ and the c^i are defined by

$$v = \sum c^i(v) \left. \frac{\partial}{\partial x^i} \right|_p , \quad v \in T_pM.$$

Comparing coefficients in

$$X_p = \sum a^i(p) \left.\frac{\partial}{\partial x^i}\right|_p = \sum c^i(X_p) \left.\frac{\partial}{\partial x^i}\right|_p, \quad p \in U,$$

we get $a^i = c^i \circ X$ as functions on U. Being coordinates, the c^i are smooth functions on TU. Thus, if X is smooth and $(U, x^1, \ldots, x^n)$ is any chart on M, then the coefficients a^i of $X = \sum a^i \partial/\partial x^i$ relative to the frame $\partial/\partial x^i$ are smooth on U.

The converse is also true, as indicated in the following lemma.

Lemma 14.1 (Smoothness of a vector field on a chart). *Let* $(U, \phi) = (U, x^1, \ldots, x^n)$ *be a chart on a manifold M. A vector field* $X = \sum a^i \partial/\partial x^i$ *on U is smooth if and only if the coefficient functions* a^i *are all smooth on U.*

Proof. This lemma is a special case of Proposition 12.12, with E the tangent bundle of M and s_i the coordinate vector field $\partial/\partial x^i$.

Because we have an explicit description of the manifold structure on the tangent bundle TM, a direct proof of the lemma is also possible. Since $\tilde{\phi} \colon TU \overset{\sim}{\to} U \times \mathbb{R}^n$ is a diffeomorphism, $X \colon U \to TU$ is smooth if and only if $\tilde{\phi} \circ X \colon U \to U \times \mathbb{R}^n$ is smooth. For $p \in U$,

$$(\tilde{\phi} \circ X)(p) = \tilde{\phi}(X_p) = \left(x^1(p), \ldots, x^n(p), c^1(X_p), \ldots, c^n(X_p)\right)$$
$$= \left(x^1(p), \ldots, x^n(p), a^1(p), \ldots, a^n(p)\right).$$

As coordinate functions, $x^1, \ldots, x^n$ are C^∞ on U. Therefore, by Proposition 6.13, $\tilde{\phi} \circ X$ is smooth if and only if all the a^i are smooth on U.   □

This lemma leads to a characterization of the smoothness of a vector field on a manifold in terms of the coefficients of the vector field relative to coordinate frames.

Proposition 14.2 (Smoothness of a vector field in terms of coefficients). *Let X be a vector field on a manifold M. The following are equivalent:*

(i) *The vector field X is smooth on M.*
(ii) *The manifold M has an atlas such that on any chart* $(U, \phi) = (U, x^1, \ldots, x^n)$ *of the atlas, the coefficients* a^i *of* $X = \sum a^i \partial/\partial x^i$ *relative to the frame* $\partial/\partial x^i$ *are all smooth.*
(iii) *On any chart* $(U, \phi) = (U, x^1, \ldots, x^n)$ *on the manifold M, the coefficients* a^i *of* $X = \sum a^i \partial/\partial x^i$ *relative to the frame* $\partial/\partial x^i$ *are all smooth.*

Proof. (ii) ⇒ (i): Assume (ii). By the preceding lemma, X is smooth on every chart (U, ϕ) of an atlas of M. Thus, X is smooth on M.

(i) ⇒ (iii): A smooth vector field on M is smooth on every chart (U, ϕ) on M. The preceding lemma then implies (iii).

(iii) ⇒ (ii): Obvious.   □

Just as in Subsection 2.5, a vector field X on a manifold M induces a linear map on the algebra $C^\infty(M)$ of C^∞ functions on M: for $f \in C^\infty(M)$, define Xf to be the function

$$(Xf)(p) = X_p f, \quad p \in M.$$

In terms of its action as an operator on C^∞ functions, there is still another characterization of a smooth vector field.

Proposition 14.3 (Smoothness of a vector field in terms of functions). *A vector field X on M is smooth if and only if for every smooth function f on M, the function Xf is smooth on M.*

Proof.
($\Rightarrow$) Suppose X is smooth and $f \in C^\infty(M)$. By Proposition 14.2, on any chart $(U, x^1, \ldots, x^n)$ on M, the coefficients a^i of the vector field $X = \sum a^i\, \partial/\partial x^i$ are C^∞. It follows that $Xf = \sum a^i\, \partial f/\partial x^i$ is C^∞ on U. Since M can be covered by charts, Xf is C^∞ on M.

($\Leftarrow$) Let $(U, x^1, \ldots, x^n)$ be any chart on M. Suppose $X = \sum a^i\, \partial/\partial x^i$ on U and $p \in U$. By Proposition 13.2, for $k = 1, \ldots, n$, each x^k can be extended to a C^∞ function $\tilde{x}^k$ on M that agrees with x^k in a neighborhood V of p in U. Therefore, on V,

$$X\tilde{x}^k = \left(\sum a^i \frac{\partial}{\partial x^i}\right) \tilde{x}^k = \left(\sum a^i \frac{\partial}{\partial x^i}\right) x^k = a^k.$$

This proves that a^k is C^∞ at p. Since p is an arbitrary point in U, the function a^k is C^∞ on U. By the smoothness criterion of Proposition 14.2, X is smooth.

In this proof it is necessary to extend x^k to a C^∞ global function $\tilde{x}^k$ on M, for while it is true that $Xx^k = a^k$, the coordinate function x^k is defined only on U, not on M, and so the smoothness hypothesis on Xf does not apply to Xx^k. $\square$

By Proposition 14.3, we may view a C^∞ vector field X as a linear operator $X : C^\infty(M) \to C^\infty(M)$ on the algebra of C^∞ functions on M. As in Proposition 2.6, this linear operator $X : C^\infty(M) \to C^\infty(M)$ is a derivation: for all $f, g \in C^\infty(M)$,

$$X(fg) = (Xf)g + f(Xg).$$

In the following we think of C^∞ vector fields on M alternately as C^∞ sections of the tangent bundle TM and as derivations on the algebra $C^\infty(M)$ of C^∞ functions. In fact, it can be shown that these two descriptions of C^∞ vector fields are equivalent (Problem 19.12).

Proposition 13.2 on C^∞ extensions of functions has an analogue for vector fields.

Proposition 14.4 (C^∞ extension of a vector field). *Suppose X is a C^∞ vector field defined on a neighborhood U of a point p in a manifold M. Then there is a C^∞ vector field $\tilde{X}$ on M that agrees with X on some possibly smaller neighborhood of p.*

Proof. Choose a C^∞ bump function $\rho : M \to \mathbb{R}$ supported in U that is identically 1 in a neighborhood V of p (Figure 13.8). Define

$$\tilde{X}(q) = \begin{cases} \rho(q)X_q & \text{for } q \text{ in } U, \\ 0 & \text{for } q \text{ not in } U. \end{cases}$$

The rest of the proof is the same as in Proposition 13.2.   □

14.2 Integral Curves

In Example 12.8, it appears that through each point in the plane one can draw a circle whose velocity at any point is the given vector field at that point. Such a curve is an example of an *integral curve* of the vector field, which we now define.

Definition 14.5. Let X be a C^∞ vector field on a manifold M, and $p \in M$. An *integral curve* of X is a smooth curve $c : \,]a,b[\to M$ such that $c'(t) = X_{c(t)}$ for all $t \in \,]a,b[$. Usually we assume that the open interval $]a,b[$ contains 0. In this case, if $c(0) = p$, then we say that c is an integral curve *starting at p* and call p the *initial point* of c. To show the dependence of such an integral curve on the initial point p, we also write $c_t(p)$ instead of $c(t)$.

Definition 14.6. An integral curve is *maximal* if its domain cannot be extended to a larger interval.

Example. Recall the vector field $X_{(x,y)} = \langle -y, x \rangle$ on $\mathbb{R}^2$ (Figure 12.4). We will find an integral curve $c(t)$ of X starting at the point $(1,0) \in \mathbb{R}^2$. The condition for $c(t) = (x(t), y(t))$ to be an integral curve is $c'(t) = X_{c(t)}$, or

$$\begin{bmatrix} \dot{x}(t) \\ \dot{y}(t) \end{bmatrix} = \begin{bmatrix} -y(t) \\ x(t) \end{bmatrix},$$

so we need to solve the system of first-order ordinary differential equations

$$\dot{x} = -y, \tag{14.1}$$
$$\dot{y} = x, \tag{14.2}$$

with initial condition $(x(0), y(0)) = (1,0)$. From (14.1), $y = -\dot{x}$, so $\dot{y} = -\ddot{x}$. Substituting into (14.2) gives

$$\ddot{x} = -x.$$

It is well known that the general solution to this equation is

$$x = A\cos t + B\sin t. \tag{14.3}$$

Hence,

$$y = -\dot{x} = A\sin t - B\cos t. \tag{14.4}$$

The initial condition forces $A = 1$, $B = 0$, so the integral curve starting at $(1,0)$ is $c(t) = (\cos t, \sin t)$, which parametrizes the unit circle.

More generally, if the initial point of the integral curve, corresponding to $t = 0$, is $p = (x_0, y_0)$, then (14.3) and (14.4) give

$$A = x_0, \quad B = -y_0,$$

and the general solution to (14.1) and (14.2) is

$$x = x_0 \cos t - y_0 \sin t,$$
$$y = x_0 \sin t + y_0 \cos t, \quad t \in \mathbb{R}.$$

This can be written in matrix notation as

$$c(t) = \begin{bmatrix} x(t) \\ y(t) \end{bmatrix} = \begin{bmatrix} \cos t & -\sin t \\ \sin t & \cos t \end{bmatrix} \begin{bmatrix} x_0 \\ y_0 \end{bmatrix} = \begin{bmatrix} \cos t & -\sin t \\ \sin t & \cos t \end{bmatrix} p,$$

which shows that the integral curve of X starting at p can be obtained by rotating the point p counterclockwise about the origin through an angle t. Notice that

$$c_s(c_t(p)) = c_{s+t}(p),$$

since a rotation through an angle t followed by a rotation through an angle s is the same as a rotation through the angle $s+t$. For each $t \in \mathbb{R}$, $c_t \colon \mathbb{R}^2 \to \mathbb{R}^2$ is a diffeomorphism with inverse c_{-t}.

Let $\mathrm{Diff}(M)$ be the group of diffeomorphisms of a manifold M with itself, the group operation being composition. A homomorphism $c \colon \mathbb{R} \to \mathrm{Diff}(M)$ is called a *one-parameter group of diffeomorphisms* of M. In this example the integral curves of the vector field $X_{(x,y)} = \langle -y, x \rangle$ on $\mathbb{R}^2$ give rise to a one-parameter group of diffeomorphisms of $\mathbb{R}^2$.

Example. Let X be the vector field $x^2 \, d/dx$ on the real line $\mathbb{R}$. Find the maximal integral curve of X starting at $x = 2$.

Solution. Denote the integral curve by $x(t)$. Then

$$x'(t) = X_{x(t)} \quad \Longleftrightarrow \quad \dot{x}(t) \frac{d}{dx} = x^2 \frac{d}{dx},$$

where $x'(t)$ is the velocity vector of the curve $x(t)$, and $\dot{x}(t)$ is the calculus derivative of the real-valued function $x(t)$. Thus, $x(t)$ satisfies the differential equation

$$\frac{dx}{dt} = x^2, \quad x(0) = 2. \tag{14.5}$$

On can solve (14.5) by separation of variables:

$$\frac{dx}{x^2} = dt. \tag{14.6}$$

Integrating both sides of (14.6) gives

$$-\frac{1}{x} = t + C, \quad \text{or} \quad x = -\frac{1}{t+C},$$

for some constant C. The initial condition $x(0) = 2$ forces $C = -1/2$. Hence, $x(t) = 2/(1-2t)$. The maximal interval containing 0 on which $x(t)$ is defined is $]-\infty, 1/2[$.

From this example we see that it may not be possible to extend the domain of definition of an integral curve to the entire real line.

14.3 Local Flows

The two examples in the preceding section illustrate the fact that locally, finding an integral curve of a vector field amounts to solving a system of first-order ordinary differential equations with initial conditions. In general, if X is a smooth vector field on a manifold M, to find an integral curve $c(t)$ of X starting at p, we first choose a coordinate chart $(U, \phi) = (U, x^1, \ldots, x^n)$ about p. In terms of the local coordinates,

$$X_{c(t)} = \sum a^i(c(t)) \left.\frac{\partial}{\partial x^i}\right|_{c(t)},$$

and by Proposition 8.15,

$$c'(t) = \sum \dot{c}^i(t) \left.\frac{\partial}{\partial x^i}\right|_{c(t)},$$

where $c^i(t) = x^i \circ c(t)$ is the ith component of $c(t)$ in the chart (U, ϕ). The condition $c'(t) = X_{c(t)}$ is thus equivalent to

$$\dot{c}^i(t) = a^i(c(t)) \quad \text{for } i = 1, \ldots, n. \tag{14.7}$$

This is a system of ordinary differential equations (ODE); the initial condition $c(0) = p$ translates to $(c^1(0), \ldots, c^n(0)) = (p^1, \ldots, p^n)$. By an existence and uniqueness theorem from the theory of ODE, such a system always has a unique solution in the following sense.

Theorem 14.7. *Let V be an open subset of $\mathbb{R}^n$, p_0 a point in V, and $f \colon V \to \mathbb{R}^n$ a C^∞ function. Then the differential equation*

$$dy/dt = f(y), \quad y(0) = p_0,$$

has a unique C^∞ solution $y \colon]a(p_0), b(p_0)[\to V$, where $]a(p_0), b(p_0)[$ is the maximal open interval containing 0 on which y is defined.

The uniqueness of the solution means that if $z \colon]\delta, \varepsilon[\to V$ satisfies the same differential equation

$$dz/dt = f(z), \quad z(0) = p_0,$$

then the domain of definition $]\delta,\varepsilon[$ of z is a subset of $]a(p_0),b(p_0)[$ and $z(t) = y(t)$ on the interval $]\delta,\varepsilon[$.

For a vector field X on a chart U of a manifold and a point $p \in U$, this theorem guarantees the existence and uniqueness of a maximal integral curve starting at p.

Next we would like to study the dependence of an integral curve on its initial point. Again we study the problem locally on $\mathbb{R}^n$. The function y will now be a function of two arguments t and q, and the condition for y to be an integral curve starting at the point q is

$$\frac{\partial y}{\partial t}(t,q) = f(y(t,q)), \quad y(0,q) = q. \tag{14.8}$$

The following theorem from the theory of ODE guarantees the smooth dependence of the solution on the initial point.

Theorem 14.8. *Let V be an open subset of $\mathbb{R}^n$ and $f: V \to \mathbb{R}^n$ a C^∞ function on V. For each point $p_0 \in V$, there are a neighborhood W of p_0 in V, a number $\varepsilon > 0$, and a C^∞ function*

$$y:]-\varepsilon,\varepsilon[\times W \to V$$

such that

$$\frac{\partial y}{\partial t}(t,q) = f(y(t,q)), \quad y(0,q) = q$$

for all $(t,q) \in]-\varepsilon,\varepsilon[\times W$.

For a proof of these two theorems, see [7, Appendix C, pp. 359–366].

It follows from Theorem 14.8 and (14.8) that if X is any C^∞ vector field on a chart U and $p \in U$, then there are a neighborhood W of p in U, an $\varepsilon > 0$, and a C^∞ map

$$F:]-\varepsilon,\varepsilon[\times W \to U \tag{14.9}$$

such that for each $q \in W$, the function $F(t,q)$ is an integral curve of X starting at q. In particular, $F(0,q) = q$. We usually write $F_t(q)$ for $F(t,q)$.

Fig. 14.1. The flow line through q of a local flow.

Suppose s,t in the interval $]-\varepsilon,\varepsilon[$ are such that both $F_t(F_s(q))$ and $F_{t+s}(q)$ are defined. Then both $F_t(F_s(q))$ and $F_{t+s}(q)$ as functions of t are integral curves of X with initial point $F_s(q)$, which is the point corresponding to $t = 0$. By the uniqueness of the integral curve starting at a point,

$$F_t(F_s(q)) = F_{t+s}(q). \tag{14.10}$$

The map F in (14.9) is called a *local flow generated by the vector field X*. For each $q \in U$, the function $F_t(q)$ of t is called a *flow line* of the local flow. Each flow line is an integral curve of X. If a local flow F is defined on $\mathbb{R} \times M$, then it is called a *global flow*. Every smooth vector field has a local flow about any point, but not necessarily a global flow. A vector field having a global flow is called a *complete vector field*. If F is a global flow, then for every $t \in \mathbb{R}$,

$$F_t \circ F_{-t} = F_{-t} \circ F_t = F_0 = \mathbb{1}_M,$$

so $F_t : M \to M$ is a diffeomorphism. Thus, a global flow on M gives rise to a one-parameter group of diffeomorphisms of M.

This discussion suggests the following definition.

Definition 14.9. A *local flow* about a point p in an open set U of a manifold is a C^∞ function

$$F :]-\varepsilon, \varepsilon[\times W \to U,$$

where ε is a positive real number and W is a neighborhood of p in U, such that writing $F_t(q) = F(t,q)$, we have

(i) $F_0(q) = q$ for all $q \in W$,
(ii) $F_t(F_s(q)) = F_{t+s}(q)$ whenever both sides are defined.

If $F(t,q)$ is a local flow of the vector field X on U, then

$$F(0,q) = q \quad \text{and} \quad \frac{\partial F}{\partial t}(0,q) = X_{F(0,q)} = X_q.$$

Thus, one can recover the vector field from its flow.

Example. The function $F : \mathbb{R} \times \mathbb{R}^2 \to \mathbb{R}^2$,

$$F\left(t, \begin{bmatrix} x \\ y \end{bmatrix}\right) = \begin{bmatrix} \cos t & -\sin t \\ \sin t & \cos t \end{bmatrix} \begin{bmatrix} x \\ y \end{bmatrix},$$

is the global flow on $\mathbb{R}^2$ generated by the vector field

$$X_{(x,y)} = \frac{\partial F}{\partial t}(t,(x,y))\Big|_{t=0} = \begin{bmatrix} -\sin t & -\cos t \\ \cos t & -\sin t \end{bmatrix} \begin{bmatrix} x \\ y \end{bmatrix}\Big|_{t=0}$$

$$= \begin{bmatrix} 0 & -1 \\ 1 & 0 \end{bmatrix} \begin{bmatrix} x \\ y \end{bmatrix} = \begin{bmatrix} -y \\ x \end{bmatrix} = -y\frac{\partial}{\partial x} + x\frac{\partial}{\partial y}.$$

This is Example 12.8 again.

14.4 The Lie Bracket

Suppose X and Y are smooth vector fields on an open subset U of a manifold M. We view X and Y as derivations on $C^\infty(U)$. For a C^∞ function f on U, by Proposition 14.3 the function Yf is C^∞ on U, and the function $(XY)f := X(Yf)$ is also C^∞ on U. Moreover, because X and Y are both $\mathbb{R}$-linear maps from $C^\infty(U)$ to $C^\infty(U)$, the map $XY \colon C^\infty(U) \to C^\infty(U)$ is $\mathbb{R}$-linear. However, XY does not satisfy the derivation property: if $f, g \in C^\infty(U)$, then

$$XY(fg) = X((Yf)g + fYg)$$
$$= (XYf)g + (Yf)(Xg) + (Xf)(Yg) + f(XYg).$$

Looking more closely at this formula, we see that the two extra terms $(Yf)(Xg)$ and $(Xf)(Yg)$ that make XY not a derivation are symmetric in X and Y. Thus, if we compute $YX(fg)$ as well and subtract it from $XY(fg)$, the extra terms will disappear, and $XY - YX$ will be a derivation of $C^\infty(U)$.

Given two smooth vector fields X and Y on U and $p \in U$, we define their *Lie bracket* $[X, Y]$ at p to be

$$[X, Y]_p f = (X_p Y - Y_p X)f$$

for any germ f of a C^∞ function at p. By the same calculation as above, but now evaluated at p, it is easy to check that $[X, Y]_p$ is a derivation of $C_p^\infty(U)$ and is therefore a tangent vector at p (Definition 8.1). As p varies over U, $[X, Y]$ becomes a vector field on U.

Proposition 14.10. *If X and Y are smooth vector fields on M, then the vector field $[X, Y]$ is also smooth on M.*

Proof. By Proposition 14.3 it suffices to check that if f is a C^∞ function on M, then so is $[X, Y]f$. But

$$[X, Y]f = (XY - YX)f,$$

which is clearly C^∞ on M, since both X and Y are. $\qquad\square$

From this proposition, we see that the Lie bracket provides a product operation on the vector space $\mathfrak{X}(M)$ of all smooth vector fields on M. Clearly,

$$[Y, X] = -[X, Y].$$

Exercise 14.11 (Jacobi identity). Check the *Jacobi identity*:

$$\sum_{\text{cyclic}} [X, [Y, Z]] = 0.$$

This notation means that one permutes X, Y, Z cyclically and one takes the sum of the resulting terms. Written out,

$$\sum_{\text{cyclic}} [X, [Y, Z]] = [X, [Y, Z]] + [Y, [Z, X]] + [Z, [X, Y]].$$

Definition 14.12. Let K be a field. A *Lie algebra* over K is a vector space V over K together with a product $[\,,\,]: V \times V \to V$, called the *bracket*, satisfying the following properties: for all $a, b \in K$ and $X, Y, Z \in V$,

(i) (bilinearity) $[aX + bY, Z] = a[X, Z] + b[Y, Z]$,
$$[Z, aX + bY] = a[Z, X] + b[Z, Y],$$

(ii) (anticommutativity) $[Y, X] = -[X, Y]$,

(iii) (Jacobi identity) $\sum_{\text{cyclic}} [X, [Y, Z]] = 0$.

In practice, we will be concerned only with *real Lie algebras*, i.e., Lie algebras over $\mathbb{R}$. Unless otherwise specified, a Lie algebra in this book means a real Lie algebra.

Example. On any vector space V, define $[X, Y] = 0$ for all $X, Y \in V$. With this bracket, V becomes a Lie algebra, called an *abelian Lie algebra*.

Our definition of an algebra in Subsection 2.2 requires that the product be associative. An abelian Lie algebra is trivially associative, but in general the bracket of a Lie algebra need not be associative. So despite its name, a Lie algebra is in general not an algebra.

Example. If M is a manifold, then the vector space $\mathfrak{X}(M)$ of C^{∞} vector fields on M is a real Lie algebra with the Lie bracket $[\,,\,]$ as the bracket.

Example. Let $K^{n \times n}$ be the vector space of all $n \times n$ matrices over a field K. Define for $X, Y \in K^{n \times n}$,
$$[X, Y] = XY - YX,$$

where XY is the matrix product of X and Y. With this bracket, $K^{n \times n}$ becomes a Lie algebra. The bilinearity and anticommutativity of $[\,,\,]$ are immediate, while the Jacobi identity follows from the same computation as in Exercise 14.11.

More generally, if A is any algebra over a field K, then the product
$$[x, y] = xy - yx, \quad x, y \in A,$$

makes A into a Lie algebra over K.

Definition 14.13. A *derivation* of a Lie algebra V over a field K is a K-linear map $D: V \to V$ satisfying the product rule
$$D[Y, Z] = [DY, Z] + [Y, DZ] \quad \text{for } Y, Z \in V.$$

Example. Let V be a Lie algebra over a field K. For each X in V, define $\text{ad}_X: V \to V$ by
$$\text{ad}_X(Y) = [X, Y].$$

We may rewrite the Jacobi identity in the form

$$[X,[Y,Z]] = [[X,Y],Z] + [Y,[X,Z]]$$

or

$$\mathrm{ad}_X[Y,Z] = [\mathrm{ad}_X Y, Z] + [Y, \mathrm{ad}_X Z],$$

which shows that $\mathrm{ad}_X : V \to V$ is a derivation of V.

14.5 The Pushforward of Vector Fields

Let $F : N \to M$ be a smooth map of manifolds and let $F_* : T_pN \to T_{F(p)}M$ be its differential at a point p in N. If $X_p \in T_pN$, we call $F_*(X_p)$ the *pushforward* of the vector X_p at p. This notion does not extend in general to vector fields, since if X is a vector field on N and $z = F(p) = F(q)$ for two distinct points $p, q \in N$, then X_p and X_q are both pushed forward to tangent vectors at $z \in M$, but there is no reason why $F_*(X_p)$ and $F_*(X_q)$ should be equal (see Figure 14.2).

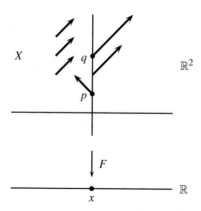

Fig. 14.2. The vector field X cannot be pushed forward under the first projection $F : \mathbb{R}^2 \to \mathbb{R}$.

In one important special case, the pushforward F_*X of any vector field X on N always makes sense, namely, when $F : N \to M$ is a diffeomorphism. In this case, since F is injective, there is no ambiguity about the meaning of $(F_*X)_{F(p)} = F_{*,p}(X_p)$, and since F is surjective, F_*X is defined everywhere on M.

14.6 Related Vector Fields

Under a C^∞ map $F : N \to M$, although in general a vector field on N cannot be pushed forward to a vector field on M, there is nonetheless a useful notion of *related vector fields*, which we now define.

Definition 14.14. Let $F : N \to M$ be a smooth map of manifolds. A vector field X on N is *F-related* to a vector field $\bar{X}$ on M if for all $p \in N$,

$$F_{*,p}(X_p) = \bar{X}_{F(p)}. \tag{14.11}$$

Example 14.15 (*Pushforward by a diffeomorphism*). If $F : N \to M$ is a diffeomorphism and X is a vector field on N, then the pushforward $F_* X$ is defined. By definition, the vector field X on N is F-related to the vector field $F_* X$ on M. In Subsection 16.5, we will see examples of vector fields related by a map F that is not a diffeomorphism.

We may reformulate condition (14.11) for F-relatedness as follows.

Proposition 14.16. *Let $F : N \to M$ be a smooth map of manifolds. A vector field X on N and a vector field $\bar{X}$ on M are F-related if and only if for all $g \in C^\infty(M)$,*

$$X(g \circ F) = (\bar{X}g) \circ F.$$

Proof.
($\Rightarrow$) Suppose X on N and $\bar{X}$ on M are F-related. By (14.11), for any $g \in C^\infty(M)$ and $p \in N$,

$$
\begin{aligned}
F_{*,p}(X_p)g &= \bar{X}_{F(p)}g && \text{(definition of F-relatedness)}, \\
X_p(g \circ F) &= (\bar{X}g)(F(p)) && \text{(definitions of F_* and $\bar{X}g$)}, \\
(X(g \circ F))(p) &= (\bar{X}g)(F(p)).
\end{aligned}
$$

Since this is true for all $p \in N$,

$$X(g \circ F) = (\bar{X}g) \circ F.$$

($\Leftarrow$) Reversing the set of equations above proves the converse.   □

Proposition 14.17. *Let $F : N \to M$ be a smooth map of manifolds. If the C^∞ vector fields X and Y on N are F-related to the C^∞ vector fields $\bar{X}$ and $\bar{Y}$, respectively, on M, then the Lie bracket $[X,Y]$ on N is F-related to the Lie bracket $[\bar{X},\bar{Y}]$ on M.*

Proof. For any $g \in C^\infty(M)$,

$$
\begin{aligned}
[X,Y](g \circ F) &= XY(g \circ F) - YX(g \circ F) && \text{(definition of $[X,Y]$)} \\
&= X((\bar{Y}g) \circ F) - Y((\bar{X}g) \circ F) && \text{(Proposition 14.16)} \\
&= (\bar{X}\bar{Y}g) \circ F - (\bar{Y}\bar{X}g) \circ F && \text{(Proposition 14.16)} \\
&= ((\bar{X}\bar{Y} - \bar{Y}\bar{X})g) \circ F \\
&= ([\bar{X},\bar{Y}]g) \circ F.
\end{aligned}
$$

By Proposition 14.16 again, this proves that $[X,Y]$ on N and $[\bar{X},\bar{Y}]$ on M are F-related.   □

Problems

14.1.* Equality of vector fields
Show that two C^∞ vector fields X and Y on a manifold M are equal if and only if for every C^∞ function f on M, we have $Xf = Yf$.

14.2. Vector field on an odd sphere
Let $x^1, y^1, \dots, x^n, y^n$ be the standard coordinates on $\mathbb{R}^{2n}$. The unit sphere S^{2n-1} in $\mathbb{R}^{2n}$ is defined by the equation $\sum_{i=1}^{n}(x^i)^2 + (y^i)^2 = 1$. Show that

$$X = \sum_{i=1}^{n} -y^i \frac{\partial}{\partial x^i} + x^i \frac{\partial}{\partial y^i}$$

is a nowhere-vanishing smooth vector field on S^{2n-1}. Since all spheres of the same dimension are diffeomorphic, this proves that on every odd-dimensional sphere there is a nowhere-vanishing smooth vector field. It is a classical theorem of differential and algebraic topology that on an even-dimensional sphere every continuous vector field must vanish somewhere (see [28, Section 5, p. 31] or [16, Theorem 16.5, p. 70]). (*Hint:* Use Problem 11.1 to show that X is tangent to S^{2n-1}.)

14.3. Maximal integral curve on a punctured line
Let M be $\mathbb{R} - \{0\}$ and let X be the vector field d/dx on M (Figure 14.3). Find the maximal integral curve of X starting at $x = 1$.

Fig. 14.3. The vector field d/dx on $\mathbb{R} - \{0\}$.

14.4. Integral curves in the plane
Find the integral curves of the vector field

$$X_{(x,y)} = x\frac{\partial}{\partial x} - y\frac{\partial}{\partial y} = \begin{bmatrix} x \\ -y \end{bmatrix} \quad \text{on } \mathbb{R}^2.$$

14.5. Maximal integral curve in the plane
Find the maximal integral curve $c(t)$ starting at the point $(a,b) \in \mathbb{R}^2$ of the vector field $X_{(x,y)} = \partial/\partial x + x\partial/\partial y$ on $\mathbb{R}^2$.

14.6. Integral curve starting at a zero of a vector field

(a)* Suppose the smooth vector field X on a manifold M vanishes at a point $p \in M$. Show that the integral curve of X with initial point p is the constant curve $c(t) \equiv p$.

(b) Show that if X is the zero vector field on a manifold M, and $c_t(p)$ is the maximal integral curve of X starting at p, then the one-parameter group of diffeomorphisms $c \colon \mathbb{R} \to \text{Diff}(M)$ is the constant map $c(t) \equiv \mathbb{1}_M$.

14.7. Maximal integral curve

Let X be the vector field $x\,d/dx$ on $\mathbb{R}$. For each p in $\mathbb{R}$, find the maximal integral curve $c(t)$ of X starting at p.

14.8. Maximal integral curve

Let X be the vector field $x^2\,d/dx$ on the real line $\mathbb{R}$. For each $p > 0$ in $\mathbb{R}$, find the maximal integral curve of X with initial point p.

14.9. Reparametrization of an integral curve

Suppose $c\colon\,]a,b[\, \to M$ is an integral curve of the smooth vector field X on M. Show that for any real number s, the map

$$c_s\colon\,]a+s,b+s[\, \to M, \quad c_s(t) = c(t-s),$$

is also an integral curve of X.

14.10. Lie bracket of vector fields

If f and g are C^∞ functions and X and Y are C^∞ vector fields on a manifold M, show that

$$[fX,gY] = fg[X,Y] + f(Xg)Y - g(Yf)X.$$

14.11. Lie bracket of vector fields on $\mathbb{R}^2$

Compute the Lie bracket

$$\left[-y\frac{\partial}{\partial x} + x\frac{\partial}{\partial y}, \frac{\partial}{\partial x} \right]$$

on $\mathbb{R}^2$.

14.12. Lie bracket in local coordinates

Consider two C^∞ vector fields X, Y on $\mathbb{R}^n$:

$$X = \sum a^i \frac{\partial}{\partial x^i}, \qquad Y = \sum b^j \frac{\partial}{\partial x^j},$$

where a^i, b^j are C^∞ functions on $\mathbb{R}^n$. Since $[X,Y]$ is also a C^∞ vector field on $\mathbb{R}^n$,

$$[X,Y] = \sum c^k \frac{\partial}{\partial x^k}$$

for some C^∞ functions c^k. Find the formula for c^k in terms of a^i and b^j.

14.13. Vector field under a diffeomorphism

Let $F\colon N \to M$ be a C^∞ diffeomorphism of manifolds. Prove that if g is a C^∞ function and X a C^∞ vector field on N, then

$$F_*(gX) = (g \circ F^{-1})F_*X.$$

14.14. Lie bracket under a diffeomorphism

Let $F\colon N \to M$ be a C^∞ diffeomorphism of manifolds. Prove that if X and Y are C^∞ vector fields on N, then

$$F_*[X,Y] = [F_*X, F_*Y].$$

Chapter 4

Lie Groups and Lie Algebras

A Lie group is a manifold that is also a group such that the group operations are smooth. Classical groups such as the general and special linear groups over $\mathbb{R}$ and over $\mathbb{C}$, orthogonal groups, unitary groups, and symplectic groups are all Lie groups.

A Lie group is a homogeneous space in the sense that left translation by a group element g is a diffeomorphism of the group onto itself that maps the identity element to g. Therefore, locally the group looks the same around any point. To study the local structure of a Lie group, it is enough to examine a neighborhood of the identity element. It is not surprising that the tangent space at the identity of a Lie group should play a key role.

The tangent space at the identity of a Lie group G turns out to have a canonical bracket operation $[\ ,\ ]$ that makes it into a Lie algebra. The tangent space T_eG with the bracket is called the *Lie algebra* of the Lie group G. The Lie algebra of a Lie group encodes within it much information about the group.

In a series of papers in the decade from 1874 to 1884, the Norwegian mathematician Sophus Lie initiated the study of Lie groups and Lie algebras. At first his work gained little notice, possibly because at the time he wrote mostly in Norwegian. In 1886, Lie became a professor in Leipzig, Germany, and his theory began to attract attention, especially after the publication of the three-volume treatise *Theorie der Transformationsgruppen* that he wrote in collaboration with his assistant Friedrich Engel.

Lie's original motivation was to study the group of transformations of a space as a continuous analogue of the group of permutations of a finite set. Indeed, a diffeomorphism of a manifold M can be viewed as a permutation of the points of M. The interplay of group theory, topology, and linear algebra makes the theory of Lie groups and Lie algebras

Sophus Lie

(1842–1899)

L. W. Tu, *An Introduction to Manifolds*, Universitext, DOI 10.1007/978-1-4419-7400-6_4,
© Springer Science+Business Media, LLC 2011

a particularly rich and vibrant branch of mathematics. In this chapter we can but scratch the surface of this vast creation. For us, Lie groups serve mainly as an important class of manifolds, and Lie algebras as examples of tangent spaces.

§15 Lie Groups

We begin with several examples of matrix groups, subgroups of the general linear group over a field. The goal is to exhibit a variety of methods for showing that a group is a Lie group and for computing the dimension of a Lie group. These examples become templates for investigating other matrix groups. A powerful tool, which we state but do not prove, is the closed subgroup theorem. According to this theorem, an abstract subgroup that is a closed subset of a Lie group is itself a Lie group. In many instances, the closed subgroup theorem is the easiest way to prove that a group is a Lie group.

The matrix exponential gives rise to curves in a matrix group with a given initial vector. It is useful in computing the differential of a map on a matrix group. As an example, we compute the differential of the determinant map on the general linear group over $\mathbb{R}$.

15.1 Examples of Lie Groups

We recall here the definition of a Lie group, which first appeared in Subsection 6.5.

Definition 15.1. A *Lie group* is a C^∞ manifold G that is also a group such that the two group operations, multiplication

$$\mu : G \times G \to G, \quad \mu(a,b) = ab,$$

and inverse

$$\iota : G \to G, \quad \iota(a) = a^{-1},$$

are C^∞.

For $a \in G$, denote by $\ell_a : G \to G$, $\ell_a(x) = \mu(a,x) = ax$, the operation of *left multiplication* by a, and by $r_a : G \to G$, $r_a(x) = xa$, the operation of *right multiplication* by a. We also call left and right multiplications *left* and *right translations*.

Exercise 15.2 (Left multiplication).* For an element a in a Lie group G, prove that the left multiplication $\ell_a : G \to G$ is a diffeomorphism.

Definition 15.3. A map $F : H \to G$ between two Lie groups H and G is a *Lie group homomorphism* if it is a C^∞ map and a group homomorphism.

The group homomorphism condition means that for all $h, x \in H$,

$$F(hx) = F(h)F(x). \tag{15.1}$$

This may be rewritten in functional notation as

$$F \circ \ell_h = \ell_{F(h)} \circ F \quad \text{for all } h \in H. \tag{15.2}$$

Let e_H and e_G be the identity elements of H and G, respectively. Taking h and x in (15.1) to be the identity e_H, it follows that $F(e_H) = e_G$. So a group homomorphism always maps the identity to the identity.

NOTATION. We use capital letters to denote matrices, but generally lowercase letters to denote their entries. Thus, the (i,j)-entry of the matrix AB is $(AB)_{ij} = \sum_k a_{ik} b_{kj}$.

Example 15.4 (*General linear group*). In Example 6.21, we showed that the general linear group

$$GL(n, \mathbb{R}) = \{A \in \mathbb{R}^{n \times n} \mid \det A \neq 0\}$$

is a Lie group.

Example 15.5 (*Special linear group*). The special linear group $SL(n, \mathbb{R})$ is the subgroup of $GL(n, \mathbb{R})$ consisting of matrices of determinant 1. By Example 9.13, $SL(n, \mathbb{R})$ is a regular submanifold of dimension $n^2 - 1$ of $GL(n, \mathbb{R})$. By Example 11.16, the multiplication map

$$\bar{\mu} : SL(n, \mathbb{R}) \times SL(n, \mathbb{R}) \to SL(n, \mathbb{R})$$

is C^∞.

To see that the inverse map

$$\bar{\iota} : SL(n, \mathbb{R}) \to SL(n, \mathbb{R})$$

is C^∞, let $i : SL(n, \mathbb{R}) \to GL(n, \mathbb{R})$ be the inclusion map and $\iota : GL(n, \mathbb{R}) \to GL(n, \mathbb{R})$ the inverse map of $GL(n, \mathbb{R})$. As the composite of two C^∞ maps,

$$\iota \circ i : SL(n, \mathbb{R}) \xrightarrow{i} GL(n, \mathbb{R}) \xrightarrow{\iota} GL(n, \mathbb{R})$$

is a C^∞ map. Since its image is contained in the regular submanifold $SL(n, \mathbb{R})$, the induced map $\bar{\iota} : SL(n, \mathbb{R}) \to SL(n, \mathbb{R})$ is C^∞ by Theorem 11.15. Thus, $SL(n, \mathbb{R})$ is a Lie group.

An entirely analogous argument proves that the complex special linear group $SL(n, \mathbb{C})$ is also a Lie group.

Example 15.6 (*Orthogonal group*). Recall that the orthogonal group $O(n)$ is the subgroup of $GL(n, \mathbb{R})$ consisting of all matrices A satisfying $A^T A = I$. Thus, $O(n)$ is the inverse image of I under the map $f(A) = A^T A$.

In Example 11.3 we showed that $f : GL(n, \mathbb{R}) \to GL(n, \mathbb{R})$ has constant rank. By the constant-rank level set theorem, $O(n)$ is a regular submanifold of $GL(n, \mathbb{R})$.

One drawback of this approach is that it does not tell us what the rank of f is, and so the dimension of $O(n)$ remains unknown.

In this example we will apply the regular level set theorem to prove that $O(n)$ is a regular submanifold of $GL(n,\mathbb{R})$. This will at the same time determine the dimension of $O(n)$. To accomplish this, we must first redefine the target space of f. Since $A^T A$ is a symmetric matrix, the image of f lies in S_n, the vector space of all $n \times n$ real symmetric matrices. The space S_n is a proper subspace of $\mathbb{R}^{n \times n}$ as soon as $n \geq 2$.

Exercise 15.7 (Space of symmetric matrices).* Show that the vector space S_n of $n \times n$ real symmetric matrices has dimension $(n^2 + n)/2$.

Consider the map $f \colon GL(n,\mathbb{R}) \to S_n$, $f(A) = A^T A$. The tangent space of S_n at any point is canonically isomorphic to S_n itself, because S_n is a vector space. Thus, the image of the differential

$$f_{*,A} \colon T_A(GL(n,\mathbb{R})) \to T_{f(A)}(S_n) \simeq S_n$$

lies in S_n. While it is true that f also maps $GL(n,\mathbb{R})$ to $GL(n,\mathbb{R})$ or $\mathbb{R}^{n \times n}$, if we had taken $GL(n,\mathbb{R})$ or $\mathbb{R}^{n \times n}$ as the target space of f, the differential $f_{*,A}$ would never be surjective for any $A \in GL(n,\mathbb{R})$ when $n \geq 2$, since $f_{*,A}$ factors through the proper subspace S_n of $\mathbb{R}^{n \times n}$. This illustrates a general principle: for the differential $f_{*,A}$ to be surjective, the target space of f should be as small as possible.

To show that the differential of

$$f \colon GL(n,\mathbb{R}) \to S_n, \quad f(A) = A^T A,$$

is surjective, we compute explicitly the differential $f_{*,A}$. Since $GL(n,\mathbb{R})$ is an open subset of $\mathbb{R}^{n \times n}$, its tangent space at any $A \in GL(n,\mathbb{R})$ is

$$T_A(GL(n,\mathbb{R})) = T_A(\mathbb{R}^{n \times n}) = \mathbb{R}^{n \times n}.$$

For any matrix $X \in \mathbb{R}^{n \times n}$, there is a curve $c(t)$ in $GL(n,\mathbb{R})$ with $c(0) = A$ and $c'(0) = X$ (Proposition 8.16). By Proposition 8.18,

$$
\begin{aligned}
f_{*,A}(X) &= \frac{d}{dt} f(c(t)) \Big|_{t=0} \\
&= \frac{d}{dt} c(t)^T c(t) \Big|_{t=0} \\
&= (c'(t)^T c(t) + c(t)^T c'(t))|_{t=0} \quad \text{(by Problem 15.2)} \\
&= X^T A + A^T X.
\end{aligned}
$$

The surjectivity of $f_{*,A}$ becomes the following question: if $A \in O(n)$ and B is any symmetric matrix in S_n, does there exist an $n \times n$ matrix X such that

$$X^T A + A^T X = B?$$

Note that since $(X^T A)^T = A^T X$, it is enough to solve

$$A^T X = \frac{1}{2} B, \tag{15.3}$$

for then

$$X^T A + A^T X = \frac{1}{2} B^T + \frac{1}{2} B = B.$$

Equation (15.3) clearly has a solution: $X = \frac{1}{2}(A^T)^{-1} B$. So $f_{*,A} \colon T_A \mathrm{GL}(n, \mathbb{R}) \to S_n$ is surjective for all $A \in \mathrm{O}(n)$, and $\mathrm{O}(n)$ is a regular level set of f. By the regular level set theorem, $\mathrm{O}(n)$ is a regular submanifold of $\mathrm{GL}(n, \mathbb{R})$ of dimension

$$\dim \mathrm{O}(n) = n^2 - \dim S_n = n^2 - \frac{n^2 + n}{2} = \frac{n^2 - n}{2}. \tag{15.4}$$

15.2 Lie Subgroups

Definition 15.8. A *Lie subgroup* of a Lie group G is (i) an abstract subgroup H that is (ii) an *immersed* submanifold via the inclusion map such that (iii) the group operations on H are C^∞.

An "abstract subgroup" simply means a subgroup in the algebraic sense, in contrast to a "Lie subgroup." The group operations on the subgroup H are the restrictions of the multiplication map μ and the inverse map ι from G to H. For an explanation of why a Lie subgroup is defined to be an immersed submanifold instead of a regular submanifold, see Remark 16.15. Because a Lie subgroup is an immersed submanifold, it need not have the relative topology. However, being an immersion, the inclusion map $i \colon H \hookrightarrow G$ of a Lie subgroup H is of course C^∞. It follows that the composite

$$\mu \circ (i \times i) \colon H \times H \to G \times G \to G$$

is C^∞. If H were defined to be a regular submanifold of G, then by Theorem 11.15, the multiplication map $H \times H \to H$ and similarly the inverse map $H \to H$ would automatically be C^∞, and condition (iii) in the definition of a Lie subgroup would be redundant. Since a Lie subgroup is defined to be an immersed submanifold, it is necessary to impose condition (iii) on the group operations on H.

Example 15.9 (*Lines with irrational slope in a torus*).   Let G be the torus $\mathbb{R}^2/\mathbb{Z}^2$ and L a line through the origin in $\mathbb{R}^2$. The torus can also be represented by the unit square with the opposite edges identified. The image H of L under the projection $\pi \colon \mathbb{R}^2 \to \mathbb{R}^2/\mathbb{Z}^2$ is a closed curve if and only if the line L goes through another lattice point, say $(m, n) \in \mathbb{Z}^2$. This is the case if and only if the slope of L is n/m, a rational number or ∞; then H is the image of finitely many line segments on the unit square. It is a closed curve diffeomorphic to a circle and is a regular submanifold of $\mathbb{R}^2/\mathbb{Z}^2$ (Figure 15.1).

If the slope of L is irrational, then its image H on the torus will never close up. In this case the restriction to L of the projection map, $f = \pi|_L \colon L \to \mathbb{R}^2/\mathbb{Z}^2$, is a one-to-one immersion. We give H the topology and manifold structure induced from f. It

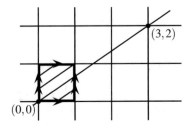

Fig. 15.1. An embedded Lie subgroup of the torus.

can be shown that H is a dense subset of the torus [3, Example III.6.15, p. 86]. Thus, H is an immersed submanifold but not a regular submanifold of the torus $\mathbb{R}^2/\mathbb{Z}^2$.

Whatever the slope of L, its image H in $\mathbb{R}^2/\mathbb{Z}^2$ is an abstract subgroup of the torus, an immersed submanifold, and a Lie group. Therefore, H is a Lie subgroup of the torus.

Exercise 15.10 (Induced topology versus subspace topology).* Suppose $H \subset \mathbb{R}^2/\mathbb{Z}^2$ is the image of a line L with irrational slope in $\mathbb{R}^2$. We call the topology on H induced from the bijection $f\colon L \xrightarrow{\sim} H$ the *induced topology* and the topology on H as a subset of $\mathbb{R}^2/\mathbb{Z}^2$ the *subspace topology*. Compare these two topologies: is one a subset of the other?

Proposition 15.11. *If H is an abstract subgroup and a regular submanifold of a Lie group G, then it is a Lie subgroup of G.*

Proof. Since a regular submanifold is the image of an embedding (Theorem 11.14), it is also an immersed submanifold.

Let $\mu\colon G \times G \to G$ be the multiplication map on G. Since H is an immersed submanifold of G, the inclusion map $i\colon H \hookrightarrow G$ is C^∞. Hence, the inclusion map $i \times i\colon H \times H \hookrightarrow G \times G$ is C^∞, and the composition $\mu \circ (i \times i)\colon H \times H \to G$ is C^∞. By Theorem 11.15, because H is a regular submanifold of G, the induced map $\bar{\mu}\colon H \times H \to H$ is C^∞.

The smoothness of the inverse map $\bar{\iota}\colon H \to H$ can be deduced from the smoothness of $\iota\colon G \to G$ just as in Example 15.5. $\square$

A subgroup H as in Proposition 15.11 is called an *embedded Lie subgroup*, because the inclusion map $i\colon H \to G$ of a regular submanifold is an embedding (Theorem 11.14).

Example. We showed in Examples 15.5 and 15.6 that the subgroups $\mathrm{SL}(n,\mathbb{R})$ and $\mathrm{O}(n)$ of $\mathrm{GL}(n,\mathbb{R})$ are both regular submanifolds. By Proposition 15.11 they are embedded Lie subgroups.

We state without proof an important theorem about Lie subgroups. If G is a Lie group, then an abstract subgroup that is a closed subset in the topology of G is called a *closed subgroup*.

Theorem 15.12 (Closed subgroup theorem). *A closed subgroup of a Lie group is an embedded Lie subgroup.*

For a proof of the closed subgroup theorem, see [38, Theorem 3.42, p. 110].

Examples.

(i) A line with irrational slope in the torus $\mathbb{R}^2/\mathbb{Z}^2$ is not a closed subgroup, since it is not the whole torus, but being dense, its closure is.
(ii) The special linear group $SL(n,\mathbb{R})$ and the orthogonal group $O(n)$ are the zero sets of polynomial equations on $GL(n,\mathbb{R})$. As such, they are closed subsets of $GL(n,\mathbb{R})$. By the closed subgroup theorem, $SL(n,\mathbb{R})$ and $O(n)$ are embedded Lie subgroups of $GL(n,\mathbb{R})$.

15.3 The Matrix Exponential

To compute the differential of a map on a subgroup of $GL(n,\mathbb{R})$, we need a curve of nonsingular matrices. Because the matrix exponential is always nonsingular, it is uniquely suited for this purpose.

A *norm* on a vector space V is a real-valued function $\|\cdot\| : V \to \mathbb{R}$ satisfying the following three properties: for all $r \in \mathbb{R}$ and $v, w \in V$,

(i) (positive-definiteness) $\|v\| \geq 0$ with equality if and only if $v = 0$,
(ii) (positive homogeneity) $\|rv\| = |r|\,\|v\|$,
(iii) (subadditivity) $\|v + w\| \leq \|v\| + \|w\|$.

A vector space V together with a norm $\|\cdot\|$ is called a *normed vector space*. The vector space $\mathbb{R}^{n \times n} \simeq \mathbb{R}^{n^2}$ of all $n \times n$ real matrices can be given the Euclidean norm: for $X = [x_{ij}] \in \mathbb{R}^{n \times n}$,

$$\|X\| = \left(\sum x_{ij}^2 \right)^{1/2}.$$

The *matrix exponential* e^X of a matrix $X \in \mathbb{R}^{n \times n}$ is defined by the same formula as the exponential of a real number:

$$e^X = I + X + \frac{1}{2!}X^2 + \frac{1}{3!}X^3 + \cdots, \tag{15.5}$$

where I is the $n \times n$ identity matrix. For this formula to make sense, we need to show that the series on the right converges in the normed vector space $\mathbb{R}^{n \times n} \simeq \mathbb{R}^{n^2}$.

A *normed algebra* V is a normed vector space that is also an algebra over $\mathbb{R}$ satisfying the submultiplicative property: for all $v, w \in V$, $\|vw\| \leq \|v\|\,\|w\|$. Matrix multiplication makes the normed vector space $\mathbb{R}^{n \times n}$ into a normed algebra.

Proposition 15.13. *For $X, Y \in \mathbb{R}^{n \times n}$, $\|XY\| \leq \|X\|\,\|Y\|$.*

Proof. Write $X = [x_{ij}]$ and $Y = [y_{ij}]$ and fix a pair of subscripts (i, j). By the Cauchy–Schwarz inequality,

$$(XY)_{ij}^2 = \left(\sum_k x_{ik} y_{kj} \right)^2 \leq \left(\sum_k x_{ik}^2 \right) \left(\sum_k y_{kj}^2 \right) = a_i b_j,$$

where we set $a_i = \sum_k x_{ik}^2$ and $b_j = \sum_k y_{kj}^2$. Then

$$\|XY\|^2 = \sum_{i,j} (XY)_{ij}^2 \leq \sum_{i,j} a_i b_j = \left(\sum_i a_i \right) \left(\sum_j b_j \right)$$
$$= \left(\sum_{i,k} x_{ik}^2 \right) \left(\sum_{j,k} y_{kj}^2 \right) = \|X\|^2 \|Y\|^2. \qquad \square$$

In a normed algebra, multiplication distributes over a finite sum. When the sum is infinite as in a convergent series, the distributivity of multiplication over the sum requires a proof.

Proposition 15.14. *Let V be a normed algebra.*

(i) *If $a \in V$ and s_m is a sequence in V that converges to s, then as_m converges to as.*
(ii) *If $a \in V$ and $\sum_{k=0}^{\infty} b_k$ is a convergent series in V, then $a \sum_k b_k = \sum_k a b_k$.*

Exercise 15.15 (Distributivity over a convergent series).* Prove Proposition 15.14.

In a normed vector space V a series $\sum a_k$ is said to *converge absolutely* if the series $\sum \|a_k\|$ of norms converges in $\mathbb{R}$. The normed vector space V is said to be *complete* if every Cauchy sequence in V converges to a point in V. For example, $\mathbb{R}^{n \times n}$ is a complete normed vector space.[1] It is easy to show that in a complete normed vector space, absolute convergence implies convergence [26, Theorem 2.9.3, p. 126]. Thus, to show that a series $\sum Y_k$ of matrices converges, it is enough to show that the series $\sum \|Y_k\|$ of real numbers converges.

For any $X \in \mathbb{R}^{n \times n}$ and $k > 0$, repeated applications of Proposition 15.13 give $\|X^k\| \leq \|X\|^k$. So the series $\sum_{k=0}^{\infty} \|X^k/k!\|$ is bounded term by term in absolute value by the convergent series

$$\sqrt{n} + \|X\| + \frac{1}{2!} \|X\|^2 + \frac{1}{3!} \|X\|^3 + \cdots = (\sqrt{n} - 1) + e^{\|X\|}.$$

By the comparison test for series of real numbers, the series $\sum_{k=0}^{\infty} \|X^k/k!\|$ converges. Therefore, the series (15.5) converges absolutely for any $n \times n$ matrix X.

NOTATION. Following standard convention we use the letter e both for the exponential map and for the identity element of a general Lie group. The context should prevent any confusion. We sometimes write $\exp(X)$ for e^X.

---

[1]A complete normed vector space is also called a *Banach space*, named after the Polish mathematician Stefan Banach, who introduced the concept in 1920–1922. Correspondingly, a complete normed algebra is called a *Banach algebra*.

Unlike the exponential of real numbers, when A and B are $n \times n$ matrices with $n > 1$, it is not necessarily true that

$$e^{A+B} = e^A e^B.$$

Exercise 15.16 (Exponentials of commuting matrices). Prove that if A and B are commuting $n \times n$ matrices, then

$$e^A e^B = e^{A+B}.$$

Proposition 15.17. *For $X \in \mathbb{R}^{n \times n}$,*

$$\frac{d}{dt} e^{tX} = X e^{tX} = e^{tX} X.$$

Proof. Because each (i,j)-entry of the series for the exponential function e^{tX} is a power series in t, it is possible to differentiate term by term [35, Theorem 8.1, p. 173]. Hence,

$$\frac{d}{dt} e^{tX} = \frac{d}{dt} \left(I + tX + \frac{1}{2!} t^2 X^2 + \frac{1}{3!} t^3 X^3 + \cdots \right)$$

$$= X + tX^2 + \frac{1}{2!} t^2 X^3 + \cdots$$

$$= X \left(I + tX + \frac{1}{2!} t^2 X^2 + \cdots \right) = X e^{tX} \qquad \text{(Proposition 15.14(ii))}.$$

In the second equality above, one could have factored out X as the second factor:

$$\frac{d}{dt} e^{tX} = X + tX^2 + \frac{1}{2!} t^2 X^3 + \cdots$$

$$= \left(I + tX + \frac{1}{2!} t^2 X^2 + \cdots \right) X = e^{tX} X. \qquad \square$$

The definition of the matrix exponential e^X makes sense even if X is a complex matrix. All the arguments so far carry over word for word; one merely has to replace the Euclidean norm $\|X\|^2 = \sum x_{ij}^2$ by the Hermitian norm $\|X\|^2 = \sum |x_{ij}|^2$, where $|x_{ij}|$ is the modulus of a complex number x_{ij}.

15.4 The Trace of a Matrix

Define the *trace* of an $n \times n$ matrix X to be the sum of its diagonal entries:

$$\text{tr}(X) = \sum_{i=1}^{n} x_{ii}.$$

Lemma 15.18.
(i) *For any two matrices $X, Y \in \mathbb{R}^{n \times n}$, $\text{tr}(XY) = \text{tr}(YX)$.*
(ii) *For $X \in \mathbb{R}^{n \times n}$ and $A \in \text{GL}(n, \mathbb{R})$, $\text{tr}(AXA^{-1}) = \text{tr}(X)$.*

Proof.
(i)

$$\text{tr}(XY) = \sum_i (XY)_{ii} = \sum_i \sum_k x_{ik} y_{ki},$$

$$\text{tr}(YX) = \sum_k (YX)_{kk} = \sum_k \sum_i y_{ki} x_{ik}.$$

(ii) Set $B = XA^{-1}$ in (i).   □

The eigenvalues of an $n \times n$ matrix X are the roots of the polynomial equation $\det(\lambda I - X) = 0$. Over the field of complex numbers, which is algebraically closed, such an equation necessarily has n roots, counted with multiplicity. Thus, the advantage of allowing complex numbers is that every $n \times n$ matrix, real or complex, has n complex eigenvalues, counted with multiplicity, whereas a real matrix need not have any real eigenvalue.

Example. The real matrix

$$\begin{bmatrix} 0 & -1 \\ 1 & 0 \end{bmatrix}$$

has no real eigenvalues. It has two complex eigenvalues, $\pm i$.

The following two facts about eigenvalues are immediate from the definitions:

(i) Two similar matrices X and AXA^{-1} have the same eigenvalues, because

$$\det(\lambda I - AXA^{-1}) = \det\left(A(\lambda I - X)A^{-1}\right) = \det(\lambda I - X).$$

(ii) The eigenvalues of a triangular matrix are its diagonal entries, because

$$\det\left(\lambda I - \begin{bmatrix} \lambda_1 & & * \\ & \ddots & \\ 0 & & \lambda_n \end{bmatrix}\right) = \prod_{i=1}^n (\lambda - \lambda_i).$$

By a theorem from algebra [19, Th. 6.4.1, p. 286], any complex square matrix X can be triangularized; more precisely, there exists a nonsingular complex square matrix A such that AXA^{-1} is upper triangular. Since the eigenvalues $\lambda_1, \ldots, \lambda_n$ of X are the same as the eigenvalues of AXA^{-1}, the triangular matrix AXA^{-1} must have the eigenvalues of X along its diagonal:

$$\begin{bmatrix} \lambda_1 & & * \\ & \ddots & \\ 0 & & \lambda_n \end{bmatrix}.$$

A real matrix X, viewed as a complex matrix, can also be triangularized, but of course the triangularizing matrix A and the triangular matrix AXA^{-1} are in general complex.

Proposition 15.19. *The trace of a matrix, real or complex, is equal to the sum of its complex eigenvalues.*

Proof. Suppose X has complex eigenvalues $\lambda_1, \ldots, \lambda_n$. Then there exists a nonsingular matrix $A \in \mathrm{GL}(n, \mathbb{C})$ such that

$$
AXA^{-1} = \begin{bmatrix} \lambda_1 & & * \\ & \ddots & \\ 0 & & \lambda_n \end{bmatrix}.
$$

By Lemma 15.18,

$$
\mathrm{tr}(X) = \mathrm{tr}(AXA^{-1}) = \sum \lambda_i. \qquad \square
$$

Proposition 15.20. *For any $X \in \mathbb{R}^{n \times n}$, $\det(e^X) = e^{\mathrm{tr}X}$.*

Proof.
Case 1. Assume that X is upper triangular:

$$
X = \begin{bmatrix} \lambda_1 & & * \\ & \ddots & \\ 0 & & \lambda_n \end{bmatrix}.
$$

Then

$$
e^X = \sum \frac{1}{k!} X^k = \sum \frac{1}{k!} \begin{bmatrix} \lambda_1^k & & * \\ & \ddots & \\ 0 & & \lambda_n^k \end{bmatrix} = \begin{bmatrix} e^{\lambda_1} & & * \\ & \ddots & \\ 0 & & e^{\lambda_n} \end{bmatrix}.
$$

Hence, $\det e^X = \prod e^{\lambda_i} = e^{\sum \lambda_i} = e^{\mathrm{tr}X}$.

Case 2. Given a general matrix X, with eigenvalues $\lambda_1, \ldots, \lambda_n$, we can find a nonsingular complex matrix A such that

$$
AXA^{-1} = \begin{bmatrix} \lambda_1 & & * \\ & \ddots & \\ 0 & & \lambda_n \end{bmatrix},
$$

an upper triangular matrix. Then

$$
e^{AXA^{-1}} = I + AXA^{-1} + \frac{1}{2!}(AXA^{-1})^2 + \frac{1}{3!}(AXA^{-1})^3 + \cdots
$$

$$
= I + AXA^{-1} + A\left(\frac{1}{2!}X^2\right)A^{-1} + A\left(\frac{1}{3!}X^3\right)A^{-1} + \cdots
$$

$$
= Ae^X A^{-1} \quad \text{(by Proposition 15.14(ii))}.
$$

Hence,

$$\det e^X = \det(Ae^X A^{-1}) = \det(e^{AXA^{-1}})$$
$$= e^{\operatorname{tr}(AXA^{-1})} \quad \text{(by Case 1, since } AXA^{-1} \text{ is upper triangular)}$$
$$= e^{\operatorname{tr}X} \qquad \text{(by Lemma 15.18).} \qquad\qquad \square$$

It follows from this proposition that the matrix exponential e^X is always non-singular, because $\det(e^X) = e^{\operatorname{tr}X}$ is never 0. This is one reason why the matrix exponential is so useful, for it allows us to write down explicitly a curve in $\operatorname{GL}(n, \mathbb{R})$ with a given initial point and a given initial velocity. For example, $c(t) = e^{tX} : \mathbb{R} \to \operatorname{GL}(n, \mathbb{R})$ is a curve in $\operatorname{GL}(n, \mathbb{R})$ with initial point I and initial velocity X, since

$$c(0) = e^{0X} = e^0 = I \quad \text{and} \quad c'(0) = \frac{d}{dt} e^{tX}\Big|_{t=0} = Xe^{tX}\Big|_{t=0} = X. \qquad (15.6)$$

Similarly, $c(t) = Ae^{tX} : \mathbb{R} \to \operatorname{GL}(n, \mathbb{R})$ is a curve in $\operatorname{GL}(n, \mathbb{R})$ with initial point A and initial velocity AX.

15.5 The Differential of det at the Identity

Let $\det : \operatorname{GL}(n, \mathbb{R}) \to \mathbb{R}$ be the determinant map. The tangent space $T_I \operatorname{GL}(n, \mathbb{R})$ to $\operatorname{GL}(n, \mathbb{R})$ at the identity matrix I is the vector space $\mathbb{R}^{n \times n}$ and the tangent space $T_1 \mathbb{R}$ to $\mathbb{R}$ at 1 is $\mathbb{R}$. So

$$\det_{*,I} : \mathbb{R}^{n \times n} \to \mathbb{R}.$$

Proposition 15.21. *For any $X \in \mathbb{R}^{n \times n}$, $\det_{*,I}(X) = \operatorname{tr}X$.*

Proof. We use a curve at I to compute the differential (Proposition 8.18). As a curve $c(t)$ with $c(0) = I$ and $c'(0) = X$, choose the matrix exponential $c(t) = e^{tX}$. Then

$$\det_{*,I}(X) = \frac{d}{dt} \det(e^{tX})\Big|_{t=0} = \frac{d}{dt} e^{t \operatorname{tr}X}\Big|_{t=0}$$
$$= (\operatorname{tr}X)e^{t \operatorname{tr}X}\Big|_{t=0} = \operatorname{tr}X. \qquad\qquad \square$$

Problems

15.1. Matrix exponential
For $X \in \mathbb{R}^{n \times n}$, define the partial sum $s_m = \sum_{k=0}^{m} X^k / k!$.

(a) Show that for $\ell \geq m$,

$$\|s_\ell - s_m\| \leq \sum_{k=m+1}^{\ell} \|X\|^k / k!.$$

(b) Conclude that s_m is a Cauchy sequence in $\mathbb{R}^{n \times n}$ and therefore converges to a matrix, which we denote by e^X. This gives another way of showing that $\sum_{k=0}^{\infty} X^k / k!$ is convergent, without using the comparison test or the theorem that absolute convergence implies convergence in a complete normed vector space.

15.2. Product rule for matrix-valued functions

Let $]a,b[$ be an open interval in $\mathbb{R}$. Suppose $A\colon]a,b[\to \mathbb{R}^{m\times n}$ and $B\colon]a,b[\to \mathbb{R}^{n\times p}$ are $m\times n$ and $n\times p$ matrices respectively whose entries are differentiable functions of $t \in]a,b[$. Prove that for $t \in]a,b[$,

$$\frac{d}{dt}A(t)B(t) = A'(t)B(t) + A(t)B'(t),$$

where $A'(t) = (dA/dt)(t)$ and $B'(t) = (dB/dt)(t)$.

15.3. Identity component of a Lie group

The *identity component* G_0 of a Lie group G is the connected component of the identity element e in G. Let μ and ι be the multiplication map and the inverse map of G.

(a) For any $x \in G_0$, show that $\mu(\{x\} \times G_0) \subset G_0$. (*Hint*: Apply Proposition A.43.)
(b) Show that $\iota(G_0) \subset G_0$.
(c) Show that G_0 is an open subset of G. (*Hint*: Apply Problem A.16.)
(d) Prove that G_0 is itself a Lie group.

15.4.* Open subgroup of a connected Lie group

Prove that an open subgroup H of a connected Lie group G is equal to G.

15.5. Differential of the multiplication map

Let G be a Lie group with multiplication map $\mu\colon G \times G \to G$, and let $\ell_a\colon G \to G$ and $r_b\colon G \to G$ be left and right multiplication by a and $b \in G$, respectively. Show that the differential of μ at $(a,b) \in G \times G$ is

$$\mu_{*,(a,b)}(X_a, Y_b) = (r_b)_*(X_a) + (\ell_a)_*(Y_b) \quad \text{for } X_a \in T_a(G), \quad Y_b \in T_b(G).$$

15.6. Differential of the inverse map

Let G be a Lie group with multiplication map $\mu\colon G \times G \to G$, inverse map $\iota\colon G \to G$, and identity element e. Show that the differential of the inverse map at $a \in G$,

$$\iota_{*,a}\colon T_a G \to T_{a^{-1}} G,$$

is given by

$$\iota_{*,a}(Y_a) = -(r_{a^{-1}})_*(\ell_{a^{-1}})_* Y_a,$$

where $(r_{a^{-1}})_* = (r_{a^{-1}})_{*,e}$ and $(\ell_{a^{-1}})_* = (\ell_{a^{-1}})_{*,a}$. (The differential of the inverse at the identity was calculated in Problem 8.8(b).)

15.7.* Differential of the determinant map at A

Show that the differential of the determinant: $\det\colon GL(n,\mathbb{R}) \to \mathbb{R}$ at $A \in GL(n,\mathbb{R})$ is given by

$$\det{}_{*,A}(AX) = (\det A)\operatorname{tr} X \quad \text{for } X \in \mathbb{R}^{n\times n}. \tag{15.7}$$

15.8.* Special linear group

Use Problem 15.7 to show that 1 is a regular value of the determinant map. This gives a quick proof that the special linear group $SL(n,\mathbb{R})$ is a regular submanifold of $GL(n,\mathbb{R})$.

15.9. Structure of a general linear group

(a) For $r \in \mathbb{R}^\times := \mathbb{R} - \{0\}$, let M_r be the $n \times n$ matrix

$$M_r = \begin{bmatrix} r & & & \\ & 1 & & \\ & & \ddots & \\ & & & 1 \end{bmatrix} = [re_1 \; e_2 \; \cdots \; e_n],$$

where $e_1, \ldots, e_n$ is the standard basis for $\mathbb{R}^n$. Prove that the map

$$f: \mathrm{GL}(n, \mathbb{R}) \to \mathrm{SL}(n, \mathbb{R}) \times \mathbb{R}^\times,$$

$$A \mapsto \left(AM_{1/\det A}, \det A \right),$$

is a diffeomorphism.
(b) The *center* $Z(G)$ of a group G is the subgroup of elements $g \in G$ that commute with all elements of G:

$$Z(G) := \{g \in G \mid gx = xg \text{ for all } x \in G\}.$$

Show that the center of $\mathrm{GL}(2, \mathbb{R})$ is isomorphic to $\mathbb{R}^\times$, corresponding to the subgroup of scalar matrices, and that the center of $\mathrm{SL}(2, \mathbb{R}) \times \mathbb{R}^\times$ is isomorphic to $\{\pm 1\} \times \mathbb{R}^\times$. The group $\mathbb{R}^\times$ has two elements of order 2, while the group $\{\pm 1\} \times \mathbb{R}^\times$ has four elements of order 2. Since their centers are not isomorphic, $\mathrm{GL}(2, \mathbb{R})$ and $\mathrm{SL}(2, \mathbb{R}) \times \mathbb{R}^\times$ are not isomorphic as groups.
(c) Show that

$$h: \mathrm{GL}(3, \mathbb{R}) \to \mathrm{SL}(3, \mathbb{R}) \times \mathbb{R}^\times,$$

$$A \mapsto \left((\det A)^{1/3} A, \det A \right),$$

is a Lie group isomorphism.

The same arguments as in (b) and (c) prove that for n even, the two Lie groups $\mathrm{GL}(n, \mathbb{R})$ and $\mathrm{SL}(n, \mathbb{R}) \times \mathbb{R}^\times$ are not isomorphic as groups, while for n odd, they are isomorphic as Lie groups.

15.10. Orthogonal group
Show that the orthogonal group $\mathrm{O}(n)$ is compact by proving the following two statements.

(a) $\mathrm{O}(n)$ is a closed subset of $\mathbb{R}^{n \times n}$.
(b) $\mathrm{O}(n)$ is a bounded subset of $\mathbb{R}^{n \times n}$.

15.11. Special orthogonal group SO(2)
The *special orthogonal group* $\mathrm{SO}(n)$ is defined to be the subgroup of $\mathrm{O}(n)$ consisting of matrices of determinant 1. Show that every matrix $A \in \mathrm{SO}(2)$ can be written in the form

$$A = \begin{bmatrix} a & c \\ b & d \end{bmatrix} = \begin{bmatrix} \cos\theta & -\sin\theta \\ \sin\theta & \cos\theta \end{bmatrix}$$

for some real number θ. Then prove that $\mathrm{SO}(2)$ is diffeomorphic to the circle S^1.

15.12. Unitary group
The *unitary group* $\mathrm{U}(n)$ is defined to be

$$\mathrm{U}(n) = \{A \in \mathrm{GL}(n, \mathbb{C}) \mid \bar{A}^T A = I\},$$

where $\bar{A}$ denotes the complex conjugate of A, the matrix obtained from A by conjugating every entry of A: $(\bar{A})_{ij} = \overline{a_{ij}}$. Show that $U(n)$ is a regular submanifold of $GL(n, \mathbb{C})$ and that $\dim U(n) = n^2$.

15.13. Special unitary group SU(2)
The *special unitary group* $SU(n)$ is defined to be the subgroup of $U(n)$ consisting of matrices of determinant 1.

(a) Show that $SU(2)$ can also be described as the set

$$SU(2) = \left\{ \begin{bmatrix} a & -\bar{b} \\ b & \bar{a} \end{bmatrix} \in \mathbb{C}^{2 \times 2} \ \Big| \ a\bar{a} + b\bar{b} = 1 \right\}.$$

(*Hint*: Write out the condition $A^{-1} = \bar{A}^T$ in terms of the entries of A.)
(b) Show that $SU(2)$ is diffeomorphic to the three-dimensional sphere

$$S^3 = \left\{ (x_1, x_2, x_3, x_4) \in \mathbb{R}^4 \ | \ x_1^2 + x_2^2 + x_3^2 + x_4^2 = 1 \right\}.$$

15.14. A matrix exponential
Compute $\exp \begin{bmatrix} 0 & 1 \\ 1 & 0 \end{bmatrix}$.

15.15. Symplectic group
This problem requires a knowledge of quaternions as in Appendix E. Let $\mathbb{H}$ be the skew field of quaternions. The *symplectic group* $Sp(n)$ is defined to be

$$Sp(n) = \{ A \in GL(n, \mathbb{H}) \ | \ \bar{A}^T A = I \},$$

where $\bar{A}$ denotes the quaternionic conjugate of A. Show that $Sp(n)$ is a regular submanifold of $GL(n, \mathbb{H})$ and compute its dimension.

15.16. Complex symplectic group
Let J be the $2n \times 2n$ matrix

$$J = \begin{bmatrix} 0 & I_n \\ -I_n & 0 \end{bmatrix},$$

where I_n denotes the $n \times n$ identity matrix. The *complex symplectic group* $Sp(2n, \mathbb{C})$ is defined to be

$$Sp(2n, \mathbb{C}) = \{ A \in GL(2n, \mathbb{C}) \ | \ A^T JA = J \}.$$

Show that $Sp(2n, \mathbb{C})$ is a regular submanifold of $GL(2n, \mathbb{C})$ and compute its dimension. (*Hint*: Mimic Example 15.6. It is crucial to choose the correct target space for the map $f(A) = A^T JA$.)

§16 Lie Algebras

In a Lie group G, because left translation by an element $g \in G$ is a diffeomorphism that maps a neighborhood of the identity to a neighborhood of g, all the local information about the group is concentrated in a neighborhood of the identity, and the tangent space at the identity assumes a special importance.

Moreover, one can give the tangent space T_eG a Lie bracket $[\ ,\ ]$, so that in addition to being a vector space, it becomes a Lie algebra, called the *Lie algebra* of the Lie group. This Lie algebra encodes in it much information about the Lie group. The goal of this section is to define the Lie algebra structure on T_eG and to identity the Lie algebras of a few classical groups.

The Lie bracket on the tangent space T_eG is defined using a canonical isomorphism between the tangent space at the identity and the vector space of left-invariant vector fields on G. With respect to this Lie bracket, the differential of a Lie group homomorphism becomes a Lie algebra homomorphism. We thus obtain a functor from the category of Lie groups and Lie group homomorphisms to the category of Lie algebras and Lie algebra homomorphisms. This is the beginning of a rewarding program, to understand the structure and representations of Lie groups through a study of their Lie algebras.

16.1 Tangent Space at the Identity of a Lie Group

Because of the existence of a multiplication, a Lie group is a very special kind of manifold. In Exercise 15.2, we learned that for any $g \in G$, left translation $\ell_g : G \to G$ by g is a diffeomorphism with inverse $\ell_{g^{-1}}$. The diffeomorphism ℓ_g takes the identity element e to the element g and induces an isomorphism of tangent spaces

$$\ell_{g*} = (\ell_g)_{*,e} : T_e(G) \to T_g(G).$$

Thus, if we can describe the tangent space $T_e(G)$ at the identity, then $\ell_{g*}T_e(G)$ will give a description of the tangent space $T_g(G)$ at any point $g \in G$.

Example 16.1 (*The tangent space to* $\mathrm{GL}(n,\mathbb{R})$ *at* I). In Example 8.19, we identified the tangent space $\mathrm{GL}(n,\mathbb{R})$ at any point $g \in \mathrm{GL}(n,\mathbb{R})$ as $\mathbb{R}^{n \times n}$, the vector space of all $n \times n$ real matrices. We also identified the isomorphism $\ell_{g*} : T_I(\mathrm{GL}(n,\mathbb{R})) \to T_g(\mathrm{GL}(n,\mathbb{R}))$ as left multiplication by $g : X \mapsto gX$.

Example 16.2 (*The tangent space to* $\mathrm{SL}(n,\mathbb{R})$ *at* I). We begin by finding a condition that a tangent vector X in $T_I(\mathrm{SL}(n,\mathbb{R}))$ must satisfy. By Proposition 8.16 there is a curve $c : \ ]-\varepsilon, \varepsilon[\ \to \mathrm{SL}(n,\mathbb{R})$ with $c(0) = I$ and $c'(0) = X$. Being in $\mathrm{SL}(n,\mathbb{R})$, this curve satisfies

$$\det c(t) = 1$$

for all t in the domain $]-\varepsilon, \varepsilon[$. We now differentiate both sides with respect to t and evaluate at $t = 0$. On the left-hand side, we have

$$\frac{d}{dt}\det(c(t))\Big|_{t=0} = (\det \circ c)_* \left(\frac{d}{dt}\Big|_0\right)$$

$$= \det_{*,I}\left(c_* \frac{d}{dt}\Big|_0\right) \quad \text{(by the chain rule)}$$

$$= \det_{*,I}(c'(0))$$

$$= \det_{*,I}(X)$$

$$= \text{tr}(X) \quad \text{(by Proposition 15.21)}.$$

Thus,

$$\text{tr}(X) = \frac{d}{dt}1\Big|_{t=0} = 0.$$

So the tangent space $T_I(\text{SL}(n,\mathbb{R}))$ is contained in the subspace V of $\mathbb{R}^{n\times n}$ defined by

$$V = \{X \in \mathbb{R}^{n\times n} \mid \text{tr}\,X = 0\}.$$

Since $\dim V = n^2 - 1 = \dim T_I(\text{SL}(n,\mathbb{R}))$, the two spaces must be equal.

Proposition 16.3. *The tangent space* $T_I(\text{SL}(n,\mathbb{R}))$ *at the identity of the special linear group* $\text{SL}(n,\mathbb{R})$ *is the subspace of* $\mathbb{R}^{n\times n}$ *consisting of all* $n \times n$ *matrices of trace* 0.

Example 16.4 (*The tangent space to* O(n) *at I*).   Let X be a tangent vector to the orthogonal group O(n) at the identity I. Choose a curve $c(t)$ in O(n) defined on a small interval containing 0 such that $c(0) = I$ and $c'(0) = X$. Since $c(t)$ is in O(n),

$$c(t)^T c(t) = I.$$

Differentiating both sides with respect to t using the matrix product rule (Problem 15.2) gives

$$c'(t)^T c(t) + c(t)^T c'(t) = 0.$$

Evaluating at $t = 0$ gives

$$X^T + X = 0.$$

Thus, X is a skew-symmetric matrix.

Let K_n be the space of all $n \times n$ real skew-symmetric matrices. For example, for $n = 3$, these are matrices of the form

$$\begin{bmatrix} 0 & a & b \\ -a & 0 & c \\ -b & -c & 0 \end{bmatrix}, \quad \text{where } a,b,c, \in \mathbb{R}.$$

The diagonal entries of such a matrix are all 0 and the entries below the diagonal are determined by those above the diagonal. So

$$\dim K_n = \frac{n^2 - \#\,\text{diagonal entries}}{2} = \frac{1}{2}(n^2 - n).$$

We have shown that

$$T_I(O(n)) \subset K_n. \tag{16.1}$$

By an earlier computation (see (15.4)),

$$\dim T_I(O(n)) = \dim O(n) = \frac{n^2 - n}{2}.$$

Since the two vector spaces in (16.1) have the same dimension, equality holds.

Proposition 16.5. *The tangent space $T_I(O(n))$ of the orthogonal group $O(n)$ at the identity is the subspace of $\mathbb{R}^{n \times n}$ consisting of all $n \times n$ skew-symmetric matrices.*

16.2 Left-Invariant Vector Fields on a Lie Group

Let X be a vector field on a Lie group G. We do not assume X to be C^∞. For any $g \in G$, because left multiplication $\ell_g \colon G \to G$ is a diffeomorphism, the pushforward $\ell_{g*}X$ is a well-defined vector field on G. We say that the vector field X is *left-invariant* if

$$\ell_{g*}X = X$$

for every $g \in G$; this means for any $h \in G$,

$$\ell_{g*}(X_h) = X_{gh}.$$

In other words, a vector field X is left-invariant if and only if it is ℓ_g-related to itself for all $g \in G$.

Clearly, a left-invariant vector field X is completely determined by its value X_e at the identity, since

$$X_g = \ell_{g*}(X_e). \tag{16.2}$$

Conversely, given a tangent vector $A \in T_e(G)$ we can define a vector field $\tilde{A}$ on G by (16.2): $(\tilde{A})_g = \ell_{g*}A$. So defined, the vector field $\tilde{A}$ is left-invariant, since

$$\begin{aligned}
\ell_{g*}(\tilde{A}_h) &= \ell_{g*}\ell_{h*}A \\
&= (\ell_g \circ \ell_h)_*A \quad \text{(by the chain rule)} \\
&= (\ell_{gh})_*(A) \\
&= \tilde{A}_{gh}.
\end{aligned}$$

We call $\tilde{A}$ the *left-invariant vector field on G generated by $A \in T_e G$.* Let $L(G)$ be the vector space of all left-invariant vector fields on G. Then there is a one-to-one correspondence

$$\begin{aligned}
T_e(G) &\leftrightarrow L(G), \tag{16.3} \\
X_e &\leftrightarrow X, \\
A &\mapsto \tilde{A}.
\end{aligned}$$

It is easy to show that this correspondence is in fact a vector space isomorphism.

Example 16.6 (*Left-invariant vector fields on* $\mathbb{R}$). On the Lie group $\mathbb{R}$, the group operation is addition and the identity element is 0. So "left multiplication" ℓ_g is actually addition:

$$\ell_g(x) = g + x.$$

Let us compute $\ell_{g*}(d/dx|_0)$. Since $\ell_{g*}(d/dx|_0)$ is a tangent vector at g, it is a scalar multiple of $d/dx|_g$:

$$\ell_{g*}\left(\left.\frac{d}{dx}\right|_0\right) = a\left.\frac{d}{dx}\right|_g. \tag{16.4}$$

To evaluate a, apply both sides of (16.4) to the function $f(x) = x$:

$$a = a\left.\frac{d}{dx}\right|_g f = \ell_{g*}\left(\left.\frac{d}{dx}\right|_0\right)f = \left.\frac{d}{dx}\right|_0 f\circ\ell_g = \left.\frac{d}{dx}\right|_0 (g + x) = 1.$$

Thus,

$$\ell_{g*}\left(\left.\frac{d}{dx}\right|_0\right) = \left.\frac{d}{dx}\right|_g.$$

This shows that d/dx is a left-invariant vector field on $\mathbb{R}$. Therefore, the left-invariant vector fields on $\mathbb{R}$ are constant multiples of d/dx.

Example 16.7 (*Left-invariant vector fields on* $\mathrm{GL}(n,\mathbb{R})$). Since $\mathrm{GL}(n,\mathbb{R})$ is an open subset of $\mathbb{R}^{n\times n}$, at any $g \in \mathrm{GL}(n,\mathbb{R})$ there is a canonical identification of the tangent space $T_g(\mathrm{GL}(n,\mathbb{R}))$ with $\mathbb{R}^{n\times n}$, under which a tangent vector corresponds to an $n \times n$ matrix:

$$\sum a_{ij}\left.\frac{\partial}{\partial x_{ij}}\right|_g \longleftrightarrow [a_{ij}]. \tag{16.5}$$

We use the same letter B to denote alternately a tangent vector $B = \sum b_{ij}\partial/\partial x_{ij}|_I \in T_I(G(n,\mathbb{R}))$ at the identity and a matrix $B = [b_{ij}]$. Let $B = \sum b_{ij}\partial/\partial x_{ij}|_I \in T_I(\mathrm{GL}(n,\mathbb{R}))$ and let $\tilde{B}$ be the left-invariant vector field on $\mathrm{GL}(n,\mathbb{R})$ generated by B. By Example 8.19,

$$\tilde{B}_g = (\ell_g)_* B \longleftrightarrow gB$$

under the identification (16.5). In terms of the standard basis $\partial/\partial x_{ij}|_g$,

$$\tilde{B}_g = \sum_{i,j}(gB)_{ij}\left.\frac{\partial}{\partial x_{ij}}\right|_g = \sum_{i,j}\left(\sum_k g_{ik}b_{kj}\right)\left.\frac{\partial}{\partial x_{ij}}\right|_g.$$

Proposition 16.8. *Any left-invariant vector field X on a Lie group G is C^∞.*

Proof. By Proposition 14.3 it suffices to show that for any C^∞ function f on G, the function Xf is also C^∞. Choose a C^∞ curve $c\colon I \to G$ defined on some interval I containing 0 such that $c(0) = e$ and $c'(0) = X_e$. If $g \in G$, then $gc(t)$ is a curve starting at g with initial vector X_g, since $gc(0) = ge = g$ and

$$(gc)'(0) = \ell_{g*}c'(0) = \ell_{g*}X_e = X_g.$$

By Proposition 8.17,

$$(Xf)(g) = X_g f = \left.\frac{d}{dt}\right|_{t=0} f(gc(t)).$$

Now the function $f(gc(t))$ is a composition of C^∞ functions

$$G \times I \xrightarrow{\mathbb{1} \times c} G \times G \xrightarrow{\mu} G \xrightarrow{f} \mathbb{R},$$
$$(g,t) \longmapsto (g,c(t)) \mapsto gc(t) \mapsto f(gc(t));$$

as such, it is C^∞. Its derivative with respect to t,

$$F(g,t) := \frac{d}{dt} f(gc(t)),$$

is therefore also C^∞. Since $(Xf)(g)$ is a composition of C^∞ functions,

$$G \to G \times I \xrightarrow{F} \mathbb{R},$$
$$g \mapsto (g,0) \mapsto F(g,0) = \left.\frac{d}{dt}\right|_{t=0} f(gc(t)),$$

it is a C^∞ function on G. This proves that X is a C^∞ vector field on G.   □

It follows from this proposition that the vector space $L(G)$ of left-invariant vector fields on G is a subspace of the vector space $\mathfrak{X}(G)$ of all C^∞ vector fields on G.

Proposition 16.9. *If X and Y are left-invariant vector fields on G, then so is $[X,Y]$.*

Proof. For any g in G, X is ℓ_g-related to itself, and Y is ℓ_g-related to itself. By Proposition 14.17, $[X,Y]$ is ℓ_g-related to itself.   □

16.3 The Lie Algebra of a Lie Group

Recall that a *Lie algebra* is a vector space $\mathfrak{g}$ together with a *bracket*, i.e., an anticommutative bilinear map $[\ ,\ ]: \mathfrak{g} \times \mathfrak{g} \to \mathfrak{g}$ that satisfies the Jacobi identity (Definition 14.12). A *Lie subalgebra* of a Lie algebra $\mathfrak{g}$ is a vector subspace $\mathfrak{h} \subset \mathfrak{g}$ that is closed under the bracket $[\ ,\ ]$. By Proposition 16.9, the space $L(G)$ of left-invariant vector fields on a Lie group G is closed under the Lie bracket $[\ ,\ ]$ and is therefore a Lie subalgebra of the Lie algebra $\mathfrak{X}(G)$ of all C^∞ vector fields on G.

As we will see in the next few subsections, the linear isomorphism $\varphi: T_eG \simeq L(G)$ in (16.3) is mutually beneficial to the two vector spaces, for each space has something that the other one lacks. The vector space $L(G)$ has a natural Lie algebra structure given by the Lie bracket of vector fields, while the tangent space at the identity has a natural notion of pushforward, given by the differential of a Lie group homomorphism. The linear isomorphism $\varphi: T_eG \simeq L(G)$ allows us to define a Lie bracket on T_eG and to push forward left-invariant vector fields under a Lie group homomorphism.

We begin with the Lie bracket on T_eG. Given $A, B \in T_eG$, we first map them via φ to the left-invariant vector fields $\tilde{A}, \tilde{B}$, take the Lie bracket $[\tilde{A}, \tilde{B}] = \tilde{A}\tilde{B} - \tilde{B}\tilde{A}$, and then map it back to T_eG via φ^{-1}. Thus, the definition of the Lie bracket $[A, B] \in T_eG$ should be

$$[A, B] = [\tilde{A}, \tilde{B}]_e. \tag{16.6}$$

Proposition 16.10. *If $A, B \in T_eG$ and $\tilde{A}, \tilde{B}$ are the left-invariant vector fields they generate, then*

$$[\tilde{A}, \tilde{B}] = [A, B]\tilde{}.$$

Proof. Applying $(\ )\tilde{}$ to both sides of (16.6) gives

$$[A, B]\tilde{} = ([\tilde{A}, \tilde{B}]_e)\tilde{} = [\tilde{A}, \tilde{B}],$$

since $(\ )\tilde{}$ and $(\ )_e$ are inverse to each other. $\qquad\square$

With the Lie bracket $[\ ,\ ]$, the tangent space $T_e(G)$ becomes a Lie algebra, called the *Lie algebra* of the Lie group G. As a Lie algebra, $T_e(G)$ is usually denoted by $\mathfrak{g}$.

16.4 The Lie Bracket on $\mathfrak{gl}(n,\mathbb{R})$

For the general linear group $\mathrm{GL}(n,\mathbb{R})$, the tangent space at the identity I can be identified with the vector space $\mathbb{R}^{n\times n}$ of all $n \times n$ real matrices. We identified a tangent vector in $T_I(\mathrm{GL}(n,\mathbb{R}))$ with a matrix $A \in \mathbb{R}^{n\times n}$ via

$$\sum a_{ij} \frac{\partial}{\partial x_{ij}}\bigg|_I \longleftrightarrow [a_{ij}]. \tag{16.7}$$

The tangent space $T_I\,\mathrm{GL}(n,\mathbb{R})$ with its Lie algebra structure is denoted by $\mathfrak{gl}(n,\mathbb{R})$. Let $\tilde{A}$ be the left-invariant vector field on $\mathrm{GL}(n,\mathbb{R})$ generated by A. Then on the Lie algebra $\mathfrak{gl}(n,\mathbb{R})$ we have the Lie bracket $[A, B] = [\tilde{A}, \tilde{B}]_I$ coming from the Lie bracket of left-invariant vector fields. In the next proposition, we identify the Lie bracket in terms of matrices.

Proposition 16.11. *Let*

$$A = \sum a_{ij} \frac{\partial}{\partial x_{ij}}\bigg|_I, \qquad B = \sum b_{ij} \frac{\partial}{\partial x_{ij}}\bigg|_I \in T_I(\mathrm{GL}(n,\mathbb{R})).$$

If

$$[A, B] = [\tilde{A}, \tilde{B}]_I = \sum c_{ij} \frac{\partial}{\partial x_{ij}}\bigg|_I, \tag{16.8}$$

then

$$c_{ij} = \sum_k a_{ik}b_{kj} - b_{ik}a_{kj}.$$

Thus, if derivations are identified with matrices via (16.7), then

$$[A, B] = AB - BA.$$

Proof. Applying both sides of (16.8) to x_{ij}, we get

$$c_{ij} = [\tilde{A}, \tilde{B}]_I x_{ij} = \tilde{A}_I \tilde{B} x_{ij} - \tilde{B}_I \tilde{A} x_{ij}$$
$$= A\tilde{B} x_{ij} - B\tilde{A} x_{ij} \quad \text{(because } \tilde{A}_I = A, \ \tilde{B}_I = B),$$

so it is necessary to find a formula for the function $\tilde{B} x_{ij}$.

In Example 16.7 we found that the left-invariant vector field $\tilde{B}$ on $GL(n, \mathbb{R})$ is given by

$$\tilde{B}_g = \sum_{i,j} (gB)_{ij} \left. \frac{\partial}{\partial x_{ij}} \right|_g \quad \text{at } g \in GL(n, \mathbb{R}).$$

Hence,

$$\tilde{B}_g x_{ij} = (gB)_{ij} = \sum_k g_{ik} b_{kj} = \sum_k b_{kj} x_{ik}(g).$$

Since this formula holds for all $g \in GL(n, \mathbb{R})$, the function $\tilde{B} x_{ij}$ is

$$\tilde{B} x_{ij} = \sum_k b_{kj} x_{ik}.$$

It follows that

$$A\tilde{B} x_{ij} = \sum_{p,q} a_{pq} \left. \frac{\partial}{\partial x_{pq}} \right|_I \left(\sum_k b_{kj} x_{ik} \right) = \sum_{p,q,k} a_{pq} b_{kj} \delta_{ip} \delta_{kq}$$
$$= \sum_k a_{ik} b_{kj} = (AB)_{ij}.$$

Interchanging A and B gives

$$B\tilde{A} x_{ij} = \sum_k b_{ik} a_{kj} = (BA)_{ij}.$$

Therefore,

$$c_{ij} = \sum_k a_{ik} b_{kj} - b_{ik} a_{kj} = (AB - BA)_{ij}. \qquad \square$$

16.5 The Pushforward of Left-Invariant Vector Fields

As we noted in Subsection 14.5, if $F: N \to M$ is a C^∞ map of manifolds and X is a C^∞ vector field on N, the pushforward $F_* X$ is in general not defined except when F is a diffeomorphism. In the case of Lie groups, however, because of the correspondence between left-invariant vector fields and tangent vectors at the identity, it is possible to push forward left-invariant vector fields under a Lie group homomorphism.

Let $F: H \to G$ be a Lie group homomorphism. A left-invariant vector field X on H is generated by its value $A = X_e \in T_e H$ at the identity, so that $X = \tilde{A}$. Since a Lie group homomorphism $F: H \to G$ maps the identity of H to the identity of G, its differential $F_{*,e}$ at the identity is a linear map from $T_e H$ to $T_e G$. The diagrams

$$T_eH \xrightarrow{F_{*,e}} T_eG \qquad A \longmapsto F_{*,e}A$$

$$\downarrow{\simeq} \qquad \downarrow{\simeq} \qquad \downarrow \qquad \downarrow$$

$$L(H) \dashrightarrow L(G), \qquad \tilde{A} \dashrightarrow (F_{*,e}A)^{\sim}$$

show clearly the existence of an induced linear map $F_*: L(H) \to L(G)$ on left-invariant vector fields as well as a way to define it.

Definition 16.12.   Let $F: H \to G$ be a Lie group homomorphism.   Define $F_*: L(H) \to L(G)$ by

$$F_*(\tilde{A}) = (F_{*,e}A)^{\sim}$$

for all $A \in T_eH$.

Proposition 16.13. *If* $F: H \to G$ *is a Lie group homomorphism and* X *is a left-invariant vector field on* H, *then the left-invariant vector field* F_*X *on* G *is* F-*related to the left-invariant vector field* X.

Proof. For each $h \in H$, we need to verify that

$$F_{*,h}(X_h) = (F_*X)_{F(h)}. \tag{16.9}$$

The left-hand side of (16.9) is

$$F_{*,h}(X_h) = F_{*,h}(\ell_{h*,e}X_e) = (F \circ \ell_h)_{*,e}(X_e),$$

while the right-hand side of (16.9) is

$$
\begin{aligned}
(F_*X)_{F(h)} &= (F_{*,e}X_e)^{\sim}_{F(h)} & \text{(definition of } F_*X) \\
&= \ell_{F(h)*}F_{*,e}(X_e) & \text{(definition of left invariance)} \\
&= (\ell_{F(h)} \circ F)_{*,e}(X_e) & \text{(chain rule).}
\end{aligned}
$$

Since F is a Lie group homomorphism, we have $F \circ \ell_h = \ell_{F(h)} \circ F$, so the two sides of (16.9) are equal. $\qquad\square$

If $F: H \to G$ is a Lie group homomorphism and X is a left-invariant vector field on H, we will call F_*X the *pushforward of X under F*.

16.6 The Differential as a Lie Algebra Homomorphism

Proposition 16.14. *If* $F: H \to G$ *is a Lie group homomorphism, then its differential at the identity,*

$$F_* = F_{*,e}: T_eH \to T_eG,$$

is a Lie algebra homomorphism, i.e., a linear map such that for all $A, B \in T_eH$,

$$F_*[A,B] = [F_*A, F_*B].$$

Proof. By Proposition 16.13, the vector field $F_*\tilde{A}$ on G is F-related to the vector field $\tilde{A}$ on H, and the vector field $F_*\tilde{B}$ is F-related to $\tilde{B}$ on H. Hence, the bracket $[F_*\tilde{A}, F_*\tilde{B}]$ on G is F-related to the bracket $[\tilde{A}, \tilde{B}]$ on H (Proposition 14.17). This means that

$$F_* \left([\tilde{A}, \tilde{B}]_e \right) = [F_*\tilde{A}, F_*\tilde{B}]_{F(e)} = [F_*\tilde{A}, F_*\tilde{B}]_e.$$

The left-hand side of this equality is $F_*[A, B]$, while the right-hand side is

$$\begin{aligned}
[F_*\tilde{A}, F_*\tilde{B}]_e &= [(F_*A)\tilde{\ }, (F_*B)\tilde{\ }]_e \quad \text{(definition of } F_*\tilde{A}) \\
&= [F_*A, F_*B] \quad \text{(definition of } [\,,\,] \text{ on } T_eG).
\end{aligned}$$

Equating the two sides gives

$$F_*[A, B] = [F_*A, F_*B]. \qquad \square$$

Suppose H is a Lie subgroup of a Lie group G, with inclusion map $i: H \to G$. Since i is an immersion, its differential

$$i_*: T_eH \to T_eG$$

is injective. To distinguish the Lie bracket on T_eH from the Lie bracket on T_eG, we temporarily attach subscripts T_eH and T_eG to the two Lie brackets respectively. By Proposition 16.14, for $X, Y \in T_eH$,

$$i_* \left([X, Y]_{T_eH} \right) = [i_*X, i_*Y]_{T_eG}. \tag{16.10}$$

This shows that if T_eH is identified with a subspace of T_eG via i_*, then the bracket on T_eH is the restriction of the bracket on T_eG to T_eH. Thus, the Lie algebra of a Lie subgroup H may be identified with a Lie subalgebra of the Lie algebra of G.

In general, the Lie algebras of the classical groups are denoted by gothic letters. For example, the Lie algebras of $GL(n, \mathbb{R})$, $SL(n, \mathbb{R})$, $O(n)$, and $U(n)$ are denoted by $\mathfrak{gl}(n, \mathbb{R})$, $\mathfrak{sl}(n, \mathbb{R})$, $\mathfrak{o}(n)$, and $\mathfrak{u}(n)$, respectively. By (16.10) and Proposition 16.11, the Lie algebra structures on $\mathfrak{sl}(n, \mathbb{R})$, $\mathfrak{o}(n)$, and $\mathfrak{u}(n)$ are given by

$$[A, B] = AB - BA,$$

as on $\mathfrak{gl}(n, \mathbb{R})$.

Remark 16.15. A fundamental theorem in Lie group theory asserts the existence of a one-to-one correspondence between the connected Lie subgroups of a Lie group G and the Lie subalgebras of its Lie algebra $\mathfrak{g}$ [38, Theorem 3.19, Corollary (a), p. 95]. For the torus $\mathbb{R}^2/\mathbb{Z}^2$, the Lie algebra $\mathfrak{g}$ has $\mathbb{R}^2$ as the underlying vector space and the one-dimensional Lie subalgebras are all the lines through the origin. Each line through the origin in $\mathbb{R}^2$ is a subgroup of $\mathbb{R}^2$ under addition. Its image under the quotient map $\mathbb{R}^2 \to \mathbb{R}^2/\mathbb{Z}^2$ is a subgroup of the torus $\mathbb{R}^2/\mathbb{Z}^2$. If a line has rational slope, then its image is a regular submanifold of the torus. If a line has irrational slope, then its image is only an immersed submanifold of the torus. According to the correspondence theorem just quoted, the one-dimensional connected Lie subgroups of the torus are the images of all the lines through the origin. Note that if a Lie subgroup had been defined as a subgroup that is also a *regular* submanifold, then one would have to exclude all the lines with irrational slopes as Lie subgroups of the torus, and it would not be possible to have a one-to-one correspondence between the connected subgroups of a Lie group and the Lie subalgebras of its Lie algebra. It is because of our desire for such a correspondence that a Lie subgroup of a Lie group is defined to be a subgroup that is also an *immersed* submanifold.

Problems

In the following problems the word "dimension" refers to the dimension as a real vector space or as a manifold.

16.1. Skew-Hermitian matrices
A complex matrix $X \in \mathbb{C}^{n \times n}$ is said to be *skew-Hermitian* if its conjugate transpose $\bar{X}^T$ is equal to $-X$. Let V be the vector space of $n \times n$ skew-Hermitian matrices. Show that $\dim V = n^2$.

16.2. Lie algebra of a unitary group
Show that the tangent space at the identity I of the unitary group $U(n)$ is the vector space of $n \times n$ skew-Hermitian matrices.

16.3. Lie algebra of a symplectic group
Refer to Problem 15.15 for the definition and notation concerning the symplectic group $Sp(n)$. Show that the tangent space at the identity I of the symplectic group $Sp(n) \subset GL(n, \mathbb{H})$ is the vector space of all $n \times n$ quaternionic matrices X such that $\bar{X}^T = -X$.

16.4. Lie algebra of a complex symplectic group
(a) Show that the tangent space at the identity I of $Sp(2n, \mathbb{C}) \subset GL(2n, \mathbb{C})$ is the vector space of all $2n \times 2n$ complex matrices X such that JX is symmetric.
(b) Calculate the dimension of $Sp(2n, \mathbb{C})$.

16.5. Left-invariant vector fields on $\mathbb{R}^n$
Find the left-invariant vector fields on $\mathbb{R}^n$.

16.6. Left-invariant vector fields on a circle
Find the left-invariant vector fields on S^1.

16.7. Integral curves of a left-invariant vector field
Let $A \in \mathfrak{gl}(n, \mathbb{R})$ and let $\tilde{A}$ be the left-invariant vector field on $GL(n, \mathbb{R})$ generated by A. Show that $c(t) = e^{tA}$ is the integral curve of $\tilde{A}$ starting at the identity matrix I. Find the integral curve of $\tilde{A}$ starting at $g \in GL(n, \mathbb{R})$.

16.8. Parallelizable manifolds
A manifold whose tangent bundle is trivial is said to be *parallelizable*. If M is a manifold of dimension n, show that parallelizability is equivalent to the existence of a smooth frame $X_1, \dots, X_n$ on M.

16.9. Parallelizability of a Lie group
Show that every Lie group is parallelizable.

16.10.* The pushforward of left-invariant vector fields
Let $F\colon H \to G$ be a Lie group homomorphism and let X and Y be left-invariant vector fields on H. Prove that $F_*[X,Y] = [F_*X, F_*Y]$.

16.11. The adjoint representation
Let G be a Lie group of dimension n with Lie algebra $\mathfrak{g}$.

(a) For each $a \in G$, the differential at the identity of the conjugation map $c_a := \ell_a \circ r_{a^{-1}}\colon$ $G \to G$ is a linear isomorphism $c_{a*}\colon \mathfrak{g} \to \mathfrak{g}$. Hence, $c_{a*} \in \mathrm{GL}(\mathfrak{g})$. Show that the map $\mathrm{Ad}\colon G \to \mathrm{GL}(\mathfrak{g})$ defined by $\mathrm{Ad}(a) = c_{a*}$ is a group homomorphism. It is called the *adjoint representation* of the Lie group G.
(b) Show that $\mathrm{Ad}\colon G \to \mathrm{GL}(\mathfrak{g})$ is C^∞.

16.12. A Lie algebra structure on $\mathbb{R}^3$
The Lie algebra $\mathfrak{o}(n)$ of the orthogonal group $\mathrm{O}(n)$ is the Lie algebra of $n \times n$ skew-symmetric real matrices, with Lie bracket $[A,B] = AB - BA$. When $n = 3$, there is a vector space isomorphism $\varphi\colon \mathfrak{o}(3) \to \mathbb{R}^3$,

$$\varphi(A) = \varphi \left(\begin{bmatrix} 0 & a_1 & a_2 \\ -a_1 & 0 & a_3 \\ -a_2 & -a_3 & 0 \end{bmatrix} \right) = \begin{bmatrix} a_1 \\ -a_2 \\ a_3 \end{bmatrix} = a.$$

Prove that $\varphi([A,B]) = \varphi(A) \times \varphi(B)$. Thus, $\mathbb{R}^3$ with the cross product is a Lie algebra.

Chapter 5

Differential Forms

Differential forms are generalizations of real-valued functions on a manifold. Instead of assigning to each point of the manifold a number, a differential k-form assigns to each point a k-covector on its tangent space. For $k = 0$ and 1, differential k-forms are functions and covector fields respectively.

Élie Cartan

(1869–1951)

Differential forms play a crucial role in manifold theory. First and foremost, they are intrinsic objects associated to any manifold, and so can be used to construct diffeomorphism invariants of a manifold. In contrast to vector fields, which are also intrinsic to a manifold, differential forms have a far richer algebraic structure. Due to the existence of the wedge product, a grading, and the exterior derivative, the set of smooth forms on a manifold is both a graded algebra and a differential complex. Such an algebraic structure is called a *differential graded algebra*. Moreover, the differential complex of smooth forms on a manifold can be pulled back under a smooth map, making the complex into a contravariant functor called the *de Rham complex* of the manifold. We will eventually construct the de Rham cohomology of a manifold from the de Rham complex.

Because integration of functions on a Euclidean space depends on a choice of coordinates and is not invariant under a change of coordinates, it is not possible to integrate functions on a manifold. The highest possible degree of a differential form is the dimension of the manifold. Among differential forms, those of top degree turn out to transform correctly under a change of coordinates and are precisely the objects that can be integrated. The theory of integration on a manifold would not be possible without differential forms.

Very loosely speaking, differential forms are whatever appears under an integral sign. In this sense, differential forms are as old as calculus, and many theorems in

L. W. Tu, *An Introduction to Manifolds,* Universitext, DOI 10.1007/978-1-4419-7400-6_5,

calculus such as Cauchy's integral theorem or Green's theorem can be interpreted as statements about differential forms. Although it is difficult to say who first gave differential forms an independent meaning, Henri Poincaré [32] and Élie Cartan [5] are generally both regarded as pioneers in this regard. In the paper [5] published in 1899, Cartan defined formally the algebra of differential forms on $\mathbb{R}^n$ as the anticommutative graded algebra over C^∞ functions generated by $dx^1, \ldots, dx^n$ in degree 1. In the same paper one finds for the first time the exterior derivative on differential forms. The modern definition of a differential form as a section of an exterior power of the cotangent bundle appeared in the late forties [6], after the theory of fiber bundles came into being.

In this chapter we give an introduction to differential forms from the vector bundle point of view. For simplicity we start with 1-forms, which already have many of the properties of k-forms. We give various characterizations of smooth forms, and show how to multiply, differentiate, and pull back these forms. In addition to the exterior derivative, we also introduce the Lie derivative and interior multiplication, two other intrinsic operations on a manifold.

§17 Differential 1-Forms

Let M be a smooth manifold and p a point in M. The *cotangent space* of M at p, denoted by $T_p^*(M)$ or T_p^*M, is defined to be the dual space of the tangent space T_pM:

$$T_p^*M = (T_pM)^\vee = \operatorname{Hom}(T_pM, \mathbb{R}).$$

An element of the cotangent space T_p^*M is called a *covector* at p. Thus, a covector ω_p at p is a linear function

$$\omega_p \colon T_pM \to \mathbb{R}.$$

A *covector field*, a *differential 1-form*, or more simply a 1-*form* on M, is a function ω that assigns to each point p in M a covector ω_p at p. In this sense it is dual to a vector field on M, which assigns to each point in M a tangent vector at p. There are many reasons for the great utility of differential forms in manifold theory, among which is the fact that they can be pulled back under a map. This is in contrast to vector fields, which in general cannot be pushed forward under a map.

Covector fields arise naturally even when one is interested only in vector fields. For example, if X is a C^∞ vector field on $\mathbb{R}^n$, then at each point $p \in \mathbb{R}^n$, $X_p = \sum a^i \partial/\partial x^i|_p$. The coefficient a^i depends on the vector X_p. It is in fact a linear function: $T_p\mathbb{R}^n \to \mathbb{R}$, i.e., a covector at p. As p varies over $\mathbb{R}^n$, a^i becomes a covector field on $\mathbb{R}^n$. Indeed, it is none other than the 1-form dx^i that picks out the ith coefficient of a vector field relative to the standard frame $\partial/\partial x^1, \ldots, \partial/\partial x^n$.

17.1 The Differential of a Function

Definition 17.1. If f is a C^∞ real-valued function on a manifold M, its *differential* is defined to be the 1-form df on M such that for any $p \in M$ and $X_p \in T_pM$,

$$(df)_p(X_p) = X_p f.$$

Instead of $(df)_p$, we also write $df|_p$ for the value of the 1-form df at p. This is parallel to the two notations for a tangent vector: $(d/dt)_p = d/dt|_p$.

In Subsection 8.2 we encountered another notion of the differential, denoted by f_*, for a map f between manifolds. Let us compare the two notions of the differential.

Proposition 17.2. *If* $f \colon M \to \mathbb{R}$ *is a* C^∞ *function, then for* $p \in M$ *and* $X_p \in T_pM$,

$$f_*(X_p) = (df)_p(X_p) \left. \frac{d}{dt} \right|_{f(p)}.$$

Proof. Since $f_*(X_p) \in T_{f(p)}\mathbb{R}$, there is a real number a such that

$$f_*(X_p) = a \left. \frac{d}{dt} \right|_{f(p)}. \tag{17.1}$$

To evaluate a, apply both sides of (17.1) to x:

$$a = f_*(X_p)(t) = X_p(t \circ f) = X_p f = (df)_p(X_p). \qquad \square$$

This proposition shows that under the canonical identification of the tangent space $T_{f(p)}\mathbb{R}$ with $\mathbb{R}$ via

$$a \left. \frac{d}{dt} \right|_{f(p)} \longleftrightarrow a,$$

f_* is the same as df. For this reason, we are justified in calling both of them the *differential* of f. In terms of the differential df, a C^∞ function $f \colon M \to \mathbb{R}$ has a critical point at $p \in M$ if and only if $(df)_p = 0$.

17.2 Local Expression for a Differential 1-Form

Let $(U, \phi) = (U, x^1, \dots, x^n)$ be a coordinate chart on a manifold M. Then the differentials $dx^1, \dots, dx^n$ are 1-forms on U.

Proposition 17.3. *At each point* $p \in U$, *the covectors* $(dx^1)_p, \dots, (dx^n)_p$ *form a basis for the cotangent space* T_p^*M *dual to the basis* $\partial/\partial x^1|_p, \dots, \partial/\partial x^n|_p$ *for the tangent space* T_pM.

Proof. The proof is just like that in the Euclidean case (Proposition 4.1):

$$(dx^i)_p \left(\frac{\partial}{\partial x^j}\bigg|_p \right) = \frac{\partial}{\partial x^j}\bigg|_p x^i = \delta^i_j. \qquad \Box$$

Thus, every 1-form ω on U can be written as a linear combination

$$\omega = \sum a_i \, dx^i,$$

where the coefficients a_i are functions on U. In particular, if f is a C^∞ function on M, then the restriction of the 1-form df to U must be a linear combination

$$df = \sum a_i \, dx^i.$$

To find a_j, we apply the usual trick of evaluating both sides on $\partial/\partial x^j$:

$$(df) \left(\frac{\partial}{\partial x^j} \right) = \sum_i a_i \, dx^i \left(\frac{\partial}{\partial x^j} \right) \quad \Longrightarrow \quad \frac{\partial f}{\partial x^j} = \sum_i a_i \delta^i_j = a_j.$$

This gives a local expression for df:

$$df = \sum \frac{\partial f}{\partial x^i} \, dx^i. \tag{17.2}$$

17.3 The Cotangent Bundle

The underlying set of the *cotangent bundle* T^*M of a manifold M is the union of the cotangent spaces at all the points of M:

$$T^*M := \bigcup_{p \in M} T^*_p M. \tag{17.3}$$

Just as in the case of the tangent bundle, the union (17.3) is a disjoint union and there is a natural map $\pi \colon T^*M \to M$ given by $\pi(\alpha) = p$ if $\alpha \in T^*_p M$. Mimicking the construction of the tangent bundle, we give T^*M a topology as follows. If $(U, \phi) = (U, x^1, \dots, x^n)$ is a chart on M and $p \in U$, then each $\alpha \in T^*_p M$ can be written uniquely as a linear combination

$$\alpha = \sum c_i(\alpha) \, dx^i|_p.$$

This gives rise to a bijection

$$\tilde{\phi} \colon T^*U \to \phi(U) \times \mathbb{R}^n, \tag{17.4}$$
$$\alpha \mapsto (\phi(p), c_1(\alpha), \dots, c_n(\alpha)) = (\phi \circ \pi, c_1, \dots, c_n)(\alpha).$$

Using this bijection, we can transfer the topology of $\phi(U) \times \mathbb{R}^n$ to T^*U.

Now for each domain U of a chart in the maximal atlas of M, let $\mathcal{B}_U$ be the collection of all open subsets of T^*U, and let $\mathcal{B}$ be the union of the $\mathcal{B}_U$. As in Subsection 12.1, $\mathcal{B}$ satisfies the conditions for a collection of subsets of T^*M to be a basis. We give T^*M the topology generated by the basis $\mathcal{B}$. As for the tangent bundle, with the maps $\tilde{\phi} = (x^1 \circ \pi, \ldots, x^n \circ \pi, c_1, \ldots, c_n)$ of (17.4) as coordinate maps, T^*M becomes a C^∞ manifold, and the projection map $\pi \colon T^*M \to M$ becomes a vector bundle of rank n over M, justifying the "bundle" in the name "cotangent bundle." If $x^1, \ldots, x^n$ are coordinates on $U \subset M$, then $\pi^* x^1, \ldots, \pi^* x^n, c_1, \ldots, c_n$ are coordinates on $\pi^{-1} U \subset T^*M$. Properly speaking, the *cotangent bundle* of a manifold M is the triple (T^*M, M, π), while T^*M and M are the *total space* and the *base space* of the cotangent bundle respectively, but by abuse of language, it is customary to call T^*M the cotangent bundle of M.

In terms of the cotangent bundle, a 1-form on M is simply a section of the cotangent bundle T^*M; i.e., it is a map $\omega \colon M \to T^*M$ such that $\pi \circ \omega = \mathbb{1}_M$, the identity map on M. We say that a 1-form ω is C^∞ if it is C^∞ as a map $M \to T^*M$.

Example 17.4 (*Liouville form on the cotangent bundle*). If a manifold M has dimension n, then the total space T^*M of its cotangent bundle $\pi \colon T^*M \to M$ is a manifold of dimension $2n$. Remarkably, on T^*M there is a 1-form λ, called the *Liouville form* (or the *Poincaré form* in some books), defined independently of charts as follows. A point in T^*M is a covector $\omega_p \in T_p^*M$ at some point $p \in M$. If X_{ω_p} is a tangent vector to T^*M at ω_p, then the pushforward $\pi_*\left(X_{\omega_p}\right)$ is a tangent vector to M at p. Therefore, one can pair up ω_p and $\pi_*\left(X_{\omega_p}\right)$ to obtain a real number $\omega_p\left(\pi_*\left(X_{\omega_p}\right)\right)$. Define

$$\lambda_{\omega_p}\left(X_{\omega_p}\right) = \omega_p\left(\pi_*\left(X_{\omega_p}\right)\right).$$

The cotangent bundle and the Liouville form on it play an important role in the mathematical theory of classical mechanics [1, p. 202].

17.4 Characterization of C^∞ 1-Forms

We define a 1-form ω on a manifold M to be *smooth* if $\omega \colon M \to T^*M$ is smooth as a section of the cotangent bundle $\pi \colon T^*M \to M$. The set of all smooth 1-forms on M has the structure of a vector space, denoted by $\Omega^1(M)$. In a coordinate chart $(U, \phi) = (U, x^1, \ldots, x^n)$ on M, the value of the 1-form ω at $p \in U$ is a linear combination

$$\omega_p = \sum a_i(p) dx^i |_p.$$

As p varies in U, the coefficients a_i become functions on U. We will now derive smoothness criteria for a 1-form in terms of the coefficient functions a_i. The development is parallel to that of smoothness criteria for a vector field in Subsection 14.1.

By Subsection 17.3, the chart (U, ϕ) on M induces a chart

$$(T^*U, \tilde{\phi}) = (T^*U, \bar{x}^1, \ldots, \bar{x}^n, c_1, \ldots, c_n)$$

on T^*M, where $\bar{x}^i = \pi^* x^i = x^i \circ \pi$ and the c_i are defined by

$$\alpha = \sum c_i(\alpha) dx^i |_p, \quad \alpha \in T_p^*M.$$

Comparing the coefficients in

$$\omega_p = \sum a_i(p)\, dx^i|_p = \sum c_i(\omega_p)\, dx^i|_p,$$

we get $a_i = c_i \circ \omega$, where ω is viewed as a map from U to T^*U. Being coordinate functions, the c_i are smooth on T^*U. Thus, if ω is smooth, then the coefficients a_i of $\omega = \sum a_i dx^i$ relative to the frame dx^i are smooth on U. The converse is also true, as indicated in the following lemma.

Lemma 17.5. *Let $(U, \phi) = (U, x^1, \dots, x^n)$ be a chart on a manifold M. A 1-form $\omega = \sum a_i dx^i$ on U is smooth if and only if the coefficient functions a_i are all smooth.*

Proof. This lemma is a special case of Proposition 12.12, with E the cotangent bundle T^*M and s_j the coordinate 1-forms dx^j. However, a direct proof is also possible (cf. Lemma 14.1).

Since $\tilde{\phi} : T^*U \to U \times \mathbb{R}^n$ is a diffeomorphism, $\omega : U \to T^*M$ is smooth if and only if $\tilde{\phi} \circ \omega : U \to U \times \mathbb{R}^n$ is smooth. For $p \in U$,

$$(\tilde{\phi} \circ \omega)(p) = \tilde{\phi}(\omega_p) = \left(x^1(p), \dots, x^n(p), c_1(\omega_p), \dots, c_n(\omega_p) \right)$$
$$= \left(x^1(p), \dots, x^n(p), a_1(p), \dots, a_n(p) \right).$$

As coordinate functions, $x^1, \dots, x^n$ are smooth on U. Therefore, by Proposition 6.13, $\tilde{\phi} \circ \omega$ is smooth on U if and only if all a_i are smooth on U. ☐

Proposition 17.6 (Smoothness of a 1-form in terms of coefficients). *Let ω be a 1-form on a manifold M. The following are equivalent:*

(i) *The 1-form ω is smooth on M.*
(ii) *The manifold M has an atlas such that on any chart $(U, x^1, \dots, x^n)$ of the atlas, the coefficients a_i of $\omega = \sum a_i dx^i$ relative to the frame dx^i are all smooth.*
(iii) *On any chart $(U, x^1, \dots, x^n)$ on the manifold, the coefficients a_i of $\omega = \sum a_i dx^i$ relative to the frame dx^i are all smooth.*

Proof. The proof is omitted, since it is virtually identical to that of Proposition 14.2. ☐

Corollary 17.7. *If f is a C^∞ function on a manifold M, then its differential df is a C^∞ 1-form on M.*

Proof. On any chart $(U, x^1, \dots, x^n)$ on M, the equality $df = \sum (\partial f / \partial x^i)\, dx^i$ holds. Since the coefficients $\partial f / \partial x^i$ are all C^∞, by Proposition 17.6(iii), the 1-form df is C^∞. ☐

If ω is a 1-form and X is a vector field on a manifold M, we define a function $\omega(X)$ on M by the formula

$$\omega(X)_p = \omega_p(X_p) \in \mathbb{R}, \quad p \in M.$$

Proposition 17.8 (Linearity of a 1-form over functions). *Let ω be a 1-form on a manifold M. If f is a function and X is a vector field on M, then $\omega(fX) = f\omega(X)$.*

Proof. At each point $p \in M$,

$$\omega(fX)_p = \omega_p(f(p)X_p) = f(p)\omega_p(X_p) = (f\,\omega(X))_p,$$

because $\omega(X)$ is defined pointwise, and at each point, ω_p is $\mathbb{R}$-linear in its argument.

$\square$

Proposition 17.9 (Smoothness of a 1-form in terms of vector fields). *A 1-form ω on a manifold M is C^∞ if and only if for every C^∞ vector field X on M, the function $\omega(X)$ is C^∞ on M.*

Proof.
($\Rightarrow$) Suppose ω is a C^∞ 1-form and X is a C^∞ vector field on M. On any chart $(U, x^1, \ldots, x^n)$ on M, by Propositions 14.2 and 17.6, $\omega = \sum a_i\, dx^i$ and $X = \sum b^j \partial/\partial x^j$ for C^∞ functions a_i, b^j. By the linearity of 1-forms over functions (Proposition 17.8),

$$\omega(X) = \left(\sum a_i\, dx^i\right)\left(\sum b^j \frac{\partial}{\partial x^j}\right) = \sum_{i,j} a_i b^j \delta_j^i = \sum a_i b^i,$$

a C^∞ function on U. Since U is an arbitrary chart on M, the function $\omega(X)$ is C^∞ on M.

($\Leftarrow$) Suppose ω is a 1-form on M such that the function $\omega(X)$ is C^∞ for every C^∞ vector field X on M. Given $p \in M$, choose a coordinate neighborhood $(U, x^1, \ldots, x^n)$ about p. Then $\omega = \sum a_i\, dx^i$ on U for some functions a_i.

Fix an integer j, $1 \le j \le n$. By Proposition 14.4, we can extend the C^∞ vector field $X = \partial/\partial x^j$ on U to a C^∞ vector field $\bar{X}$ on M that agrees with $\partial/\partial x^j$ in a neighborhood V_p^j of p in U. Restricted to the open set V_p^j,

$$\omega(\bar{X}) = \left(\sum a_i\, dx^i\right)\left(\frac{\partial}{\partial x^j}\right) = a_j.$$

This proves that a_j is C^∞ on the coordinate chart $(V_p^j, x^1, \ldots, x^n)$. On the intersection $V_p := \bigcap_j V_p^j$, all a_j are C^∞. By Lemma 17.5, the 1-form ω is C^∞ on V_p. So for each $p \in M$, we have found a coordinate neighborhood V_p on which ω is C^∞. It follows that ω is a C^∞ map from M to T^*M.

$\square$

Let $\mathcal{F} = C^\infty(M)$ be the ring of all C^∞ functions on M. By Proposition 17.9, a 1-form ω on M defines a map $\mathfrak{X}(M) \to \mathcal{F}, X \mapsto \omega(X)$. According to Proposition 17.8, this map is both $\mathbb{R}$-linear and $\mathcal{F}$-linear.

17.5 Pullback of 1-Forms

If $F: N \to M$ is a C^∞ map of manifolds, then at each point $p \in N$ the differential

$$F_{*,p}: T_pN \to T_{F(p)}M$$

is a linear map that pushes forward vectors at p from N to M. The *codifferential*, i.e., the dual of the differential,

$$(F_{*,p})^\vee : T^*_{F(p)}M \to T^*_p N,$$

reverses the arrow and pulls back a covector at $F(p)$ from M to N. Another notation for the codifferential is $F^* = (F_{*,p})^\vee$. By the definition of the dual, if $\omega_{F(p)} \in T^*_{F(p)}M$ is a covector at $F(p)$ and $X_p \in T_p N$ is a tangent vector at p, then

$$F^*\left(\omega_{F(p)}\right)(X_p) = \left((F_{*,p})^\vee \omega_{F(p)}\right)(X_p) = \omega_{F(p)}(F_{*,p}X_p).$$

We call $F^*\left(\omega_{F(p)}\right)$ the *pullback* of the covector $\omega_{F(p)}$ by F. Thus, the pullback of covectors is simply the codifferential.

Unlike vector fields, which in general cannot be pushed forward under a C^∞ map, every covector field can be pulled back by a C^∞ map. If ω is a 1-form on M, its *pullback* $F^*\omega$ is the 1-form on N defined pointwise by

$$(F^*\omega)_p = F^*\left(\omega_{F(p)}\right), \quad p \in N.$$

This means that

$$(F^*\omega)_p(X_p) = \omega_{F(p)}(F_*(X_p))$$

for all $X_p \in T_p N$. Recall that functions can also be pulled back: if F is a C^∞ map from N to M and $g \in C^\infty(M)$, then $F^*g = g \circ F \in C^\infty(N)$.

This difference in the behavior of vector fields and forms under a map can be traced to a basic asymmetry in the concept of a function—every point in the domain maps to only one image point in the range, but a point in the range can have several preimage points in the domain.

Now that we have defined the pullback of a 1-form under a map, a question naturally suggests itself. Is the pullback of a C^∞ 1-form under a C^∞ map C^∞? To answer this question, we first need to establish three commutation properties of the pullback: its commutation with the differential, sum, and product.

Proposition 17.10 (Commutation of the pullback with the differential). *Let $F : N \to M$ be a C^∞ map of manifolds. For any $h \in C^\infty(M)$, $F^*(dh) = d(F^*h)$.*

Proof. It suffices to check that for any point $p \in N$ and any tangent vector $X_p \in T_p N$,

$$(F^*dh)_p(X_p) = (dF^*h)_p(X_p). \tag{17.5}$$

The left-hand side of (17.5) is

$$\begin{aligned}
(F^*dh)_p(X_p) &= (dh)_{F(p)}(F_*(X_p)) && \text{(definition of the pullback of a 1-form)} \\
&= (F_*(X_p))h && \text{(definition of the differential } dh) \\
&= X_p(h \circ F) && \text{(definition of } F_*).
\end{aligned}$$

The right-hand side of (17.5) is

$$\begin{aligned}
(dF^*h)_p(X_p) &= X_p(F^*h) && \text{(definition of } d \text{ of a function)} \\
&= X_p(h \circ F) && \text{(definition of } F^* \text{ of a function).} \quad \square
\end{aligned}$$

Pullback of functions and 1-forms respects addition and scalar multiplication.

Proposition 17.11 (Pullback of a sum and a product). *Let $F: N \to M$ be a C^∞ map of manifolds. Suppose $\omega,\ \tau \in \Omega^1(M)$ and $g \in C^\infty(M)$. Then*

(i) $F^*(\omega + \tau) = F^*\omega + F^*\tau$,
(ii) $F^*(g\omega) = (F^*g)(F^*\omega)$.

Proof. Problem 17.5.

Proposition 17.12 (Pullback of a C^∞ 1-form). *The pullback $F^*\omega$ of a C^∞ 1-form ω on M under a C^∞ map $F: N \to M$ is C^∞ 1-form on N.*

Proof. Given $p \in N$, choose a chart $(V, \psi) = (V, y^1, \ldots, y^n)$ in M about $F(p)$. By the continuity of F, there is a chart $(U, \phi) = (U, x^1, \ldots, x^n)$ about p in N such that $F(U) \subset V$. On V, $\omega = \sum a_i \, dy^i$ for some $a_i \in C^\infty(V)$. On U,

$$
\begin{aligned}
F^*\omega &= \sum (F^*a_i) F^*(dy^i) &&\text{(Proposition 17.11)}\\
&= \sum (F^*a_i)\, dF^*y^i &&\text{(Proposition 17.10)}\\
&= \sum (a_i \circ F)\, d(y^i \circ F) &&\text{(definition of } F^* \text{ of a function)}\\
&= \sum_{i,j} (a_i \circ F) \frac{\partial F^i}{\partial x^j}\, dx^j &&\text{(equation (17.2)).}
\end{aligned}
$$

Since the coefficients $(a_i \circ F)\, \partial F^i / \partial x^j$ are all C^∞, by Proposition 17.5 the 1-form $F^*\omega$ is C^∞ on U and therefore at p. Since p was an arbitrary point in N, the pullback $F^*\omega$ is C^∞ on N. $\qquad\square$

Example 17.13 (*Liouville form on the cotangent bundle*). Let M be a manifold. In terms of the pullback, the Liouville form λ on the cotangent bundle T^*M introduced in Example 17.4 can be expressed as $\lambda_{\omega_p} = \pi^*(\omega_p)$ at any $\omega_p \in T^*M$.

17.6 Restriction of 1-Forms to an Immersed Submanifold

Let $S \subset M$ be an immersed submanifold and $i: S \to M$ the inclusion map. At any $p \in S$, since the differential $i_*: T_pS \to T_pM$ is injective, one may view the tangent space T_pS as a subspace of T_pM. If ω is a 1-form on M, then the *restriction* of ω to S is the 1-form $\omega|_S$ defined by

$$
(\omega|_S)_p (v) = \omega_p(v) \quad \text{for all } p \in S \text{ and } v \in T_pS.
$$

Thus, the restriction $\omega|_S$ is the same as ω except that its domain has been restricted from M to S and for each $p \in S$, the domain of $(\omega|_S)_p$ has been restricted from T_pM to T_pS. The following proposition shows that the restriction of 1-forms is simply the pullback of the inclusion i.

Proposition 17.14. *If $i: S \hookrightarrow M$ is the inclusion map of an immersed submanifold S and ω is a 1-form on M, then $i^*\omega = \omega|_S$.*

Proof. For $p \in S$ and $v \in T_pS$,

$$(i^*\omega)_p(v) = \omega_{i(p)}(i_*v) \quad \text{(definition of pullback)}$$
$$= \omega_p(v) \quad \text{(both } i \text{ and } i_* \text{ are inclusions)}$$
$$= (\omega|_S)_p(v) \quad \text{(definition of } \omega|_S\text{).} \qquad \square$$

To avoid too cumbersome a notation, we sometimes write ω to mean $\omega|_S$, relying on the context to make clear that it is the restriction of ω to S.

Example 17.15 (*A* 1-*form on the circle*). The velocity vector field of the unit circle $c(t) = (x, y) = (\cos t, \sin t)$ in $\mathbb{R}^2$ is

$$c'(t) = (-\sin t, \cos t) = (-y, x).$$

Thus,

$$X = -y\frac{\partial}{\partial x} + x\frac{\partial}{\partial y}$$

is a C^∞ vector field on the unit circle S^1. What this notation means is that if x, y are the standard coordinates on $\mathbb{R}^2$ and $i: S^1 \hookrightarrow \mathbb{R}^2$ is the inclusion map, then at a point $p = (x, y) \in S^1$, one has $i_*X_p = -y\partial/\partial x|_p + x\partial/\partial y|_p$, where $\partial/\partial x|_p$ and $\partial/\partial y|_p$ are tangent vectors at p in $\mathbb{R}^2$. Find a 1-form $\omega = a\,dx + b\,dy$ on S^1 such that $\omega(X) \equiv 1$.

Solution. Here ω is viewed as the restriction to S^1 of the 1-form $a\,dx + b\,dy$ on $\mathbb{R}^2$. We calculate in $\mathbb{R}^2$, where dx, dy are dual to $\partial/\partial x, \partial/\partial y$:

$$\omega(X) = (a\,dx + b\,dy)\left(-y\frac{\partial}{\partial x} + x\frac{\partial}{\partial y}\right) = -ay + bx = 1. \qquad (17.6)$$

Since $x^2 + y^2 = 1$ on S^1, $a = -y$ and $b = x$ is a solution to (17.6). So $\omega = -y\,dx + x\,dy$ is one such 1-form. Since $\omega(X) \equiv 1$, the form ω is nowhere vanishing on the circle.

Remark. In the notation of Problem 11.2, ω should be written $-\bar{y}\,d\bar{x} + \bar{x}\,d\bar{y}$, since x, y are functions on $\mathbb{R}^2$ and $\bar{x}, \bar{y}$ are their restrictions to S^1. However, one generally uses the same notation for a form on $\mathbb{R}^n$ and for its restriction to a submanifold. Since $i^*x = \bar{x}$ and $i^*dx = d\bar{x}$, there is little possibility of confusion in omitting the bar while dealing with the restriction of forms on $\mathbb{R}^n$. This is in contrast to the situation for vector fields, where $i_*(\partial/\partial\bar{x}|_p) \neq \partial/\partial x|_p$.

Example 17.16 (*Pullback of a* 1-*form*). Let $h: \mathbb{R} \to S^1 \subset \mathbb{R}^2$ be given by $h(t) = (x, y) = (\cos t, \sin t)$. If ω is the 1-form $-y\,dx + x\,dy$ on S^1, compute the pullback $h^*\omega$.

Solution.

$$h^*(-y\,dx + x\,dy) = -(h^*y)\,d(h^*x) + (h^*x)\,d(h^*y) \quad \text{(by Proposition 17.11)}$$
$$= -(\sin t)\,d(\cos t) + (\cos t)\,d(\sin t)$$
$$= \sin^2 t\,dt + \cos^2 t\,dt = dt.$$

Problems

17.1. A 1-form on $\mathbb{R}^2 - \{(0,0)\}$
Denote the standard coordinates on $\mathbb{R}^2$ by x, y, and let

$$X = -y\frac{\partial}{\partial x} + x\frac{\partial}{\partial y} \quad \text{and} \quad Y = x\frac{\partial}{\partial x} + y\frac{\partial}{\partial y}$$

be vector fields on $\mathbb{R}^2$. Find a 1-form ω on $\mathbb{R}^2 - \{(0,0)\}$ such that $\omega(X) = 1$ and $\omega(Y) = 0$.

17.2. Transition formula for 1-forms
Suppose $(U, x^1, \ldots, x^n)$ and $(V, y^1, \ldots, y^n)$ are two charts on M with nonempty overlap $U \cap V$. Then a C^∞ 1-form ω on $U \cap V$ has two different local expressions:

$$\omega = \sum a_j \, dx^j = \sum b_i \, dy^i.$$

Find a formula for a_j in terms of b_i.

17.3. Pullback of a 1-form on S^1
Multiplication in the unit circle S^1, viewed as a subset of the complex plane, is given by

$$e^{it} \cdot e^{iu} = e^{i(t+u)}, \quad t, u \in \mathbb{R}.$$

In terms of real and imaginary parts,

$$(\cos t + i\sin t)(x + iy) = ((\cos t)x - (\sin t)y) + i((\sin t)x + (\cos t)y).$$

Hence, if $g = (\cos t, \sin t) \in S^1 \subset \mathbb{R}^2$, then the left multiplication $\ell_g : S^1 \to S^1$ is given by

$$\ell_g(x, y) = ((\cos t)x - (\sin t)y, (\sin t)x + (\cos t)y).$$

Let $\omega = -y \, dx + x \, dy$ be the 1-form found in Example 17.15. Prove that $\ell_g^* \omega = \omega$ for all $g \in S^1$.

17.4. Liouville form on the cotangent bundle
(a) Let $(U, \phi) = (U, x^1, \ldots, x^n)$ be a chart on a manifold M, and let

$$(\pi^{-1}U, \tilde{\phi}) = (\pi^{-1}U, \bar{x}^1, \ldots, \bar{x}^n, c_1, \ldots, c_n)$$

be the induced chart on the cotangent bundle T^*M. Find a formula for the Liouville form λ on $\pi^{-1}U$ in terms of the coordinates $\bar{x}^1, \ldots, \bar{x}^n, c_1, \ldots, c_n$.
(b) Prove that the Liouville form λ on T^*M is C^∞. (*Hint:* Use (a) and Proposition 17.6.)

17.5. Pullback of a sum and a product
Prove Proposition 17.11 by verifying both sides of each equality on a tangent vector X_p at a point p.

17.6. Construction of the cotangent bundle
Let M be a manifold of dimension n. Mimicking the construction of the tangent bundle in Section 12, write out a detailed proof that $\pi : T^*M \to M$ is a C^∞ vector bundle of rank n.

§18 Differential k-Forms

We now generalize the construction of 1-forms on a manifold to k-forms. After defining k-forms on a manifold, we show that locally they look no different from k-forms on $\mathbb{R}^n$. In parallel to the construction of the tangent and cotangent bundles on a manifold, we construct the kth exterior power $\bigwedge^k(T^*M)$ of the cotangent bundle. A differential k-form is seen to be a section of the bundle $\bigwedge^k(T^*M)$. This gives a natural notion of smoothness of differential forms: a differential k-form is smooth if and only if it is smooth as a section of the vector bundle $\bigwedge^k(T^*M)$. The pullback and the wedge product of differential forms are defined pointwise. As examples of differential forms, we consider left-invariant forms on a Lie group.

18.1 Differential Forms

Recall that a k-*tensor* on a vector space V is a k-linear function

$$f: V \times \cdots \times V \to \mathbb{R}.$$

The k-tensor f is *alternating* if for any permutation $\sigma \in S_k$,

$$f(v_{\sigma(1)}, \ldots, v_{\sigma(k)}) = (\operatorname{sgn}\sigma) f(v_1, \ldots, v_k). \tag{18.1}$$

When $k = 1$, the only element of the permutation group S_1 is the identity permutation. So for 1-tensors the condition (18.1) is vacuous and all 1-tensors are alternating (and symmetric too). An alternating k-tensor on V is also called a k-*covector* on V.

For any vector space V, denote by $A_k(V)$ the vector space of alternating k-tensors on V. Another common notation for the space $A_k(V)$ is $\bigwedge^k(V^\vee)$. Thus,

$$\bigwedge^0(V^\vee) = A_0(V) = \mathbb{R},$$
$$\bigwedge^1(V^\vee) = A_1(V) = V^\vee,$$
$$\bigwedge^2(V^\vee) = A_2(V), \quad \text{and so on.}$$

In fact, there is a purely algebraic construction $\bigwedge^k(V)$, called the kth *exterior power* of the vector space V, with the property that $\bigwedge^k(V^\vee)$ is isomorphic to $A_k(V)$. To delve into this construction would lead us too far afield, so in this book $\bigwedge^k(V^\vee)$ will simply be an alternative notation for $A_k(V)$.

We apply the functor $A_k(\ )$ to the tangent space T_pM of a manifold M at a point p. The vector space $A_k(T_pM)$, usually denoted by $\bigwedge^k(T_p^*M)$, is the space of all alternating k-tensors on the tangent space T_pM. A k-*covector field* on M is a function ω that assigns to each point $p \in M$ a k-covector $\omega_p \in \bigwedge^k(T_p^*M)$. A k-covector field is also called a *differential k-form*, a *differential form of degree k*, or simply a *k-form*. A *top form* on a manifold is a differential form whose degree is the dimension of the manifold.

If ω is a k-form on a manifold M and $X_1,\ldots,X_k$ are vector fields on M, then $\omega(X_1,\ldots,X_k)$ is the function on M defined by

$$(\omega(X_1,\ldots,X_k))(p) = \omega_p((X_1)_p,\ldots,(X_k)_p).$$

Proposition 18.1 (Multilinearity of a form over functions). *Let ω be a k-form on a manifold M. For any vector fields $X_1,\ldots,X_k$ and any function h on M,*

$$\omega(X_1,\ldots,hX_i,\ldots,X_k) = h\omega(X_1,\ldots,X_i,\ldots,X_k).$$

Proof. The proof is essentially the same as that of Proposition 17.8. $\square$

Example 18.2. Let $(U,x^1,\ldots,x^n)$ be a coordinate chart on a manifold. At each point $p \in U$, a basis for the tangent space T_pU is

$$\left.\frac{\partial}{\partial x^1}\right|_p, \ldots, \left.\frac{\partial}{\partial x^n}\right|_p.$$

As we saw in Proposition 17.3, the dual basis for the cotangent space T_p^*U is

$$(dx^1)_p,\ldots,(dx^n)_p.$$

As p varies over points in U, we get differential 1-forms $dx^1,\ldots,dx^n$ on U.

By Proposition 3.29 a basis for the alternating k-tensors in $\bigwedge^k(T_p^*U)$ is

$$(dx^{i_1})_p \wedge \cdots \wedge (dx^{i_k})_p, \quad 1 \le i_1 < \cdots < i_k \le n.$$

If ω is a k-form on $\mathbb{R}^n$, then at each point $p \in \mathbb{R}^n$, ω_p is a linear combination

$$\omega_p = \sum a_{i_1\cdots i_k}(p)\,(dx^{i_1})_p \wedge \cdots \wedge (dx^{i_k})_p.$$

Omitting the point p, we write

$$\omega = \sum a_{i_1\cdots i_k}\,dx^{i_1} \wedge \cdots \wedge dx^{i_k}.$$

In this expression the coefficients $a_{i_1\cdots i_k}$ are functions on U because they vary with the point p. To simplify the notation, we let

$$\mathcal{I}_{k,n} = \{I = (i_1,\ldots,i_k) \mid 1 \le i_1 < \cdots < i_k \le n\}$$

be the set of all strictly ascending multi-indices between 1 and n of length k, and write

$$\omega = \sum_{I \in \mathcal{I}_{k,n}} a_I\,dx^I,$$

where dx^I stands for $dx^{i_1} \wedge \cdots \wedge dx^{i_k}$.

18.2 Local Expression for a k-Form

By Example 18.2, on a coordinate chart $(U,x^1,\ldots,x^n)$ of a manifold M, a k-form on U is a linear combination $\omega = \sum a_I dx^I$, where $I \in \mathcal{I}_{k,n}$ and the a_I are functions on U. As a shorthand, we write $\partial_i = \partial/\partial x^i$ for the ith coordinate vector field. Evaluating pointwise as in Lemma 3.28, we obtain the following equality on U for $I,J \in \mathcal{I}_{k,n}$:

$$dx^I(\partial_{j_1},\ldots,\partial_{j_k}) = \delta_J^I = \begin{cases} 1 & \text{for } I = J, \\ 0 & \text{for } I \neq J. \end{cases} \tag{18.2}$$

Proposition 18.3 (A wedge of differentials in local coordinates). *Let* $(U,x^1,\ldots,x^n)$ *be a chart on a manifold and* $f^1,\ldots,f^k$ *smooth functions on* U. *Then*

$$df^1 \wedge \cdots \wedge df^k = \sum_{I \in \mathcal{I}_{k,n}} \frac{\partial(f^1,\ldots,f^k)}{\partial(x^{i_1},\ldots,x^{i_k})} dx^{i_1} \wedge \cdots \wedge dx^{i_k}.$$

Proof. On U,

$$df^1 \wedge \cdots \wedge df^k = \sum_{J \in \mathcal{I}_{k,n}} c_J\, dx^{j_1} \wedge \cdots \wedge dx^{j_k} \tag{18.3}$$

for some functions c_J. By the definition of the differential, $df^i(\partial/\partial x^j) = \partial f^i/\partial x^j$. Applying both sides of (18.3) to the list of coordinate vectors $\partial_{i_1},\ldots,\partial_{i_k}$, we get

$$\text{LHS} = (df^1 \wedge \cdots \wedge df^k)(\partial_{i_1},\ldots,\partial_{i_k}) = \det\left[\frac{\partial f^i}{\partial x^{i_j}}\right] \quad \text{by Proposition 3.27}$$

$$= \frac{\partial(f^1,\ldots,f^k)}{\partial(x^{i_1},\ldots,x^{i_k})},$$

$$\text{RHS} = \sum_J c_J\, dx^J(\partial_{i_1},\ldots,\partial_{i_k}) = \sum_J c_J \delta_I^J = c_I \quad \text{by Lemma 18.2.}$$

Hence, $c_I = \partial(f^1,\ldots,f^k)/\partial(x^{i_1},\ldots,x^{i_k})$. $\qquad\square$

If $(U,x^1,\ldots,x^n)$ and $(V,y^1,\ldots,y^n)$ are two overlapping charts on a manifold, then on the intersection $U \cap V$, Proposition 18.3 becomes the transition formula for k-forms:

$$dy^J = \sum_I \frac{\partial(y^{j_1},\ldots,y^{j_k})}{\partial(x^{i_1},\ldots,x^{i_k})} dx^I.$$

Two cases of Proposition 18.3 are of special interest:

Corollary 18.4. *Let* $(U,x^1,\ldots,x^n)$ *be a chart on a manifold, and let* $f, f^1,\ldots,f^n$ *be* C^∞ *functions on* U. *Then*

(i) *(1-forms)* $df = \sum(\partial f/\partial x^i)\, dx^i$,
(ii) *(top forms)* $df^1 \wedge \cdots \wedge df^n = \det[\partial f^j/\partial x^i]\, dx^1 \wedge \cdots \wedge dx^n$.

Case (i) of the corollary agrees with the formula we derived in (17.2).

Exercise 18.5 (Transition formula for a 2-form).* If $(U, x^1, \ldots, x^n)$ and $(V, y^1, \ldots, y^n)$ are two overlapping coordinate charts on M, then a C^∞ 2-form ω on $U \cap V$ has two local expressions:

$$\omega = \sum_{i<j} a_{ij}\, dx^i \wedge dx^j = \sum_{k<\ell} b_{k\ell}\, dy^k \wedge dy^\ell.$$

Find a formula for a_{ij} in terms of $b_{k\ell}$ and the coordinate functions $x^1, \ldots, x^n, y^1, \ldots, y^n$.

18.3 The Bundle Point of View

Let M be a manifold of dimension n. To better understand differential forms, we mimic the construction of the tangent and cotangent bundles and form the set

$$\textstyle\bigwedge^k(T^*M) := \bigcup_{p \in M} \bigwedge^k(T_p^*M) = \bigcup_{p \in M} A_k(T_p M)$$

of all alternating k-tensors at all points of the manifold M. This set is called the kth *exterior power* of the cotangent bundle. There is a projection map $\pi \colon \bigwedge^k(T^*M) \to M$ given by $\pi(\alpha) = p$ if $\alpha \in \bigwedge^k(T_p^*M)$.

If (U, ϕ) is a coordinate chart on M, then there is a bijection

$$\textstyle\bigwedge^k(T^*U) = \bigcup_{p \in U} \bigwedge^k(T_p^*U) \simeq \phi(U) \times \mathbb{R}^{\binom{n}{k}},$$

$$\textstyle\alpha \in \bigwedge^k(T_p^*U) \mapsto (\phi(p), \{c_I(\alpha)\}_I),$$

where $\alpha = \sum c_I(\alpha)\, dx^I|_p \in \bigwedge^k(T_p^*U)$ and $I = (1 \le i_1 < \cdots < i_k \le n)$. In this way we can give $\bigwedge^k(T^*U)$ and hence $\bigwedge^k(T^*M)$ a topology and even a differentiable structure. The details are just like those for the construction of the tangent bundle, so we omit them. The upshot is that the projection map $\pi \colon \bigwedge^k(T^*M) \to M$ is a C^∞ vector bundle of rank $\binom{n}{k}$ and that a differential k-form is simply a section of this bundle. As one might expect, we define a k-form to be C^∞ if it is C^∞ as a section of the bundle $\pi \colon \bigwedge^k(T^*M) \to M$.

NOTATION. If $E \to M$ is a C^∞ vector bundle, then the vector space of C^∞ sections of E is denoted by $\Gamma(E)$ or $\Gamma(M, E)$. The vector space of all C^∞ k-forms on M is usually denoted by $\Omega^k(M)$. Thus,

$$\textstyle\Omega^k(M) = \Gamma\left(\bigwedge^k(T^*M)\right) = \Gamma\left(M, \bigwedge^k(T^*M)\right).$$

18.4 Smooth k-Forms

There are several equivalent characterizations of a smooth k-form. Since the proofs are similar to those for 1-forms (Lemma 17.5 and Propositions 17.6 and 17.9), we omit them.

Lemma 18.6 (Smoothness of a k-form on a chart). *Let $(U, x^1, \ldots, x^n)$ be a chart on a manifold M. A k-form $\omega = \sum a_I\, dx^I$ on U is smooth if and only if the coefficient functions a_I are all smooth on U.*

Proposition 18.7 (Characterization of a smooth k-form). *Let ω be a k-form on a manifold M. The following are equivalent:*

(i) *The k-form ω is C^∞ on M.*
(ii) *The manifold M has an atlas such that on every chart $(U,\phi) = (U,x^1,\ldots,x^n)$ in the atlas, the coefficients a_I of $\omega = \sum a_I\, dx^I$ relative to the coordinate frame $\{dx^I\}_{I\in\mathcal{J}_{k,n}}$ are all C^∞.*
(iii) *On every chart $(U,\phi) = (U,x^1,\ldots,x^n)$ on M, the coefficients a_I of $\omega = \sum a_I\, dx^I$ relative to the coordinate frame $\{dx^I\}_{I\in\mathcal{J}_{k,n}}$ are all C^∞.*
(iv) *For any k smooth vector fields $X_1,\ldots,X_k$ on M, the function $\omega(X_1,\ldots,X_k)$ is C^∞ on M.*

We defined the 0-tensors and the 0-covectors to be the constants, that is, $L_0(V) = A_0(V) = \mathbb{R}$. Therefore, the bundle $\bigwedge^0(T^*M)$ is simply $M \times \mathbb{R}$ and a 0-form on M is a function on M. A C^∞ 0-form on M is thus the same as a C^∞ function on M. In our new notation,

$$\Omega^0(M) = \Gamma\big(\textstyle\bigwedge^0(T^*M)\big) = \Gamma(M \times \mathbb{R}) = C^\infty(M).$$

Proposition 13.2 on C^∞ extensions of functions has a generalization to differential forms.

Proposition 18.8 (C^∞ extension of a form). *Suppose τ is a C^∞ differential form defined on a neighborhood U of a point p in a manifold M. Then there is a C^∞ form $\tilde{\tau}$ on M that agrees with τ on a possibly smaller neighborhood of p.*

The proof is identical to that of Proposition 13.2. We leave it as an exercise. Of course, the extension $\tilde{\tau}$ is not unique. In the proof it depends on p and on the choice of a bump function at p.

18.5 Pullback of k-Forms

We have defined the pullback of 0-forms and 1-forms under a C^∞ map $F\colon N \to M$. For a C^∞ 0-form on M, i.e., a C^∞ function on M, the pullback F^*f is simply the composition

$$N \xrightarrow{F} M \xrightarrow{f} \mathbb{R}, \quad F^*(f) = f \circ F \in \Omega^0(N).$$

To generalize the pullback to k-forms for all $k \geq 1$, we first recall the pullback of k-covectors from Subsection 10.3. A linear map $L\colon V \to W$ of vector spaces induces a pullback map $L^*\colon A_k(W) \to A_k(V)$ by

$$(L^*\alpha)(v_1,\ldots,v_k) = \alpha(L(v_1),\ldots,L(v_k))$$

for $\alpha \in A_k(W)$ and $v_1,\ldots,v_k \in V$.

Now suppose $F\colon N \to M$ is a C^∞ map of manifolds. At each point $p \in N$, the differential

$$F_{*,p}\colon T_pN \to T_{F(p)}M$$

is a linear map of tangent spaces, and so by the preceding paragraph there is a pull-back map

$$(F_{*,p})^* : A_k(T_{F(p)}M) \to A_k(T_pN).$$

This ugly notation is usually simplified to F^*. Thus, if $\omega_{F(p)}$ is a k-covector at $F(p)$ in M, then its *pullback* $F^* \left(\omega_{F(p)} \right)$ is the k-covector at p in N given by

$$F^* \left(\omega_{F(p)} \right)(v_1, \ldots, v_k) = \omega_{F(p)}(F_{*,p}v_1, \ldots, F_{*,p}v_k), \quad v_i \in T_pN.$$

Finally, if ω is a k-form on M, then its *pullback* $F^*\omega$ is the k-form on N defined pointwise by $(F^*\omega)_p = F^* \left(\omega_{F(p)} \right)$ for all $p \in N$. Equivalently,

$$(F^*\omega)_p(v_1, \ldots, v_k) = \omega_{F(p)}(F_{*,p}v_1, \ldots, F_{*,p}v_k), \quad v_i \in T_pN. \tag{18.4}$$

When $k = 1$, this formula specializes to the definition of the pullback of a 1-form in Subsection 17.5. The pullback of a k-form (18.4) can be viewed as a composition

$$T_pN \times \cdots \times T_pN \xrightarrow{F_* \times \cdots \times F_*} T_{F(p)}M \times \cdots \times T_{F(p)}M \xrightarrow{\omega_{F(p)}} \mathbb{R}.$$

Proposition 18.9 (Linearity of the pullback). *Let $F : N \to M$ be a C^∞ map. If ω, τ are k-forms on M and a is a real number, then*

(i) $F^*(\omega + \tau) = F^*\omega + F^*\tau$;
(ii) $F^*(a\omega) = aF^*\omega$.

Proof. Problem 18.2. $\qquad\qquad\qquad\qquad\qquad\qquad\qquad\qquad\qquad\qquad\qquad\qquad\square$

At this point, we still do not know, other than for $k = 0, 1$, whether the pullback of a C^∞ k-form under a C^∞ map remains C^∞. This very basic question will be answered in Subsection 19.5.

18.6 The Wedge Product

We learned in Section 3 that if α and β are alternating tensors of degree k and ℓ respectively on a vector space V, then their wedge product $\alpha \wedge \beta$ is the alternating $(k+\ell)$-tensor on V defined by

$$(\alpha \wedge \beta)(v_1, \ldots, v_{k+\ell}) = \sum (\text{sgn } \sigma)\alpha(v_{\sigma(1)}, \ldots, v_{\sigma(k)})\beta(v_{\sigma(k+1)}, \ldots, v_{\sigma(k+\ell)}),$$

where $v_i \in V$ and σ runs over all (k, ℓ)-shuffles of $1, \ldots, k+\ell$. For example, if α and β are 1-covectors, then

$$(\alpha \wedge \beta)(v_1, v_2) = \alpha(v_1)\beta(v_2) - \alpha(v_2)\beta(v_1).$$

The wedge product extends pointwise to differential forms on a manifold: for a k-form ω and an ℓ-form τ on M, define their *wedge product* $\omega \wedge \tau$ to be the $(k+\ell)$-form on M such that

$$(\omega \wedge \tau)_p = \omega_p \wedge \tau_p$$

at all $p \in M$.

Proposition 18.10. *If ω and τ are C^∞ forms on M, then $\omega \wedge \tau$ is also C^∞.*

Proof. Let $(U, x^1, \ldots, x^n)$ be a chart on M. On U,

$$\omega = \sum a_I \, dx^I, \quad \tau = \sum b_J \, dx^J$$

for C^∞ function a_I, b_J on U. Their wedge product on U is

$$\omega \wedge \tau = \left(\sum a_I \, dx^I \right) \wedge \left(\sum b_J \, dx^J \right) = \sum a_I b_J \, dx^I \wedge dx^J.$$

In this sum, $dx^I \wedge dx^J = 0$ if I and J have an index in common. If I and J are disjoint, then $dx^I \wedge dx^J = \pm dx^K$, where $K = I \cup J$ but reordered as an increasing sequence. Thus,

$$\omega \wedge \tau = \sum_K \left(\sum_{\substack{I \cup J = K \\ I, J \text{ disjoint}}} \pm a_I b_J \right) dx^K.$$

Since the coefficients of dx^K are C^∞ on U, by Proposition 18.7, $\omega \wedge \tau$ is C^∞.   □

Proposition 18.11 (Pullback of a wedge product). *If $F \colon N \to M$ is a C^∞ map of manifolds and ω and τ are differential forms on M, then*

$$F^*(\omega \wedge \tau) = F^* \omega \wedge F^* \tau.$$

Proof. Problem 18.3.   □

Define the vector space $\Omega^*(M)$ of C^∞ differential forms on a manifold M of dimension n to be the direct sum

$$\Omega^*(M) = \bigoplus_{k=0}^{n} \Omega^k(M).$$

What this means is that each element of $\Omega^*(M)$ is uniquely a sum $\sum_{k=0}^{n} \omega_k$, where $\omega_k \in \Omega^k(M)$. With the wedge product, the vector space $\Omega^*(M)$ becomes a graded algebra, the grading being the degree of differential forms.

18.7 Differential Forms on a Circle

Consider the map

$$h \colon \mathbb{R} \to S^1, \quad h(t) = (\cos t, \sin t).$$

Since the derivative $\dot{h}(t) = (-\sin t, \cos t)$ is nonzero for all t, the map $h \colon \mathbb{R} \to S^1$ is a submersion. By Problem 18.8, the pullback map $h^* \colon \Omega^*(S^1) \to \Omega^*(\mathbb{R})$ on smooth differential forms is injective. This will allow us to identify the differential forms on S^1 with a subspace of differential forms on $\mathbb{R}$.

Let $\omega = -y \, dx + x \, dy$ be the nowhere-vanishing form on S^1 from Example 17.15. In Example 17.16, we showed that $h^* \omega = dt$. Since ω is nowhere vanishing, it is a frame for the cotangent bundle $T^* S^1$ over S^1, and every C^∞ 1-form α on S^1 can be written as $\alpha = f\omega$ for some function f on S^1. By Proposition 12.12, the function f is C^∞. Its pullback $\bar{f} := h^* f$ is a C^∞ function on $\mathbb{R}$. Since pulling back preserves multiplication (Proposition 18.11),

$$h^* \alpha = (h^* f)(h^* \omega) = \bar{f} \, dt. \tag{18.5}$$

We say that a function g or a 1-form $g \, dt$ on $\mathbb{R}$ is *periodic* of *period a* if $g(t+a) = g(t)$ for all $t \in \mathbb{R}$.

Proposition 18.12. *For $k = 0, 1$, under the pullback map $h^*: \Omega^*(S^1) \to \Omega^*(\mathbb{R})$, smooth k-forms on S^1 are identified with smooth periodic k-forms of period 2π on $\mathbb{R}$.*

Proof. If $f \in \Omega^0(S^1)$, then since $h: \mathbb{R} \to S^1$ is periodic of period 2π, the pullback $h^* f = f \circ h \in \Omega^0(\mathbb{R})$ is periodic of period 2π.

Conversely, suppose $\bar{f} \in \Omega^0(\mathbb{R})$ is periodic of period 2π. For $p \in S^1$, let s be the C^∞ inverse in a neighborhood U of p of the local diffeomorphism h and define $f = \bar{f} \circ s$ on U. To show that f is well defined, let s_1 and s_2 be two inverses of h over U. By the periodic properties of sine and cosine, $s_1 = s_2 + 2\pi n$ for some $n \in \mathbb{Z}$. Because $\bar{f}$ is periodic of period 2π, we have $\bar{f} \circ s_1 = \bar{f} \circ s_2$. This proves that f is well defined on U. Moreover,

$$\bar{f} = f \circ s^{-1} = f \circ h = h^* f \quad \text{on } h^{-1}(U).$$

As p varies over S^1, we obtain a well-defined C^∞ function f on S^1 such that $\bar{f} = h^* f$. Thus, the image of $h^*: \Omega^0(S^1) \to \Omega^0(\mathbb{R})$ consists precisely of the C^∞ periodic functions of period 2π on $\mathbb{R}$.

As for 1-forms, note that $\Omega^1(S^1) = \Omega^0(S^1)\omega$ and $\Omega^1(\mathbb{R}) = \Omega^0(\mathbb{R})dt$. The pullback $h^*: \Omega^1(S^1) \to \Omega^1(\mathbb{R})$ is given by $h^*(f\omega) = (h^* f)dt$, so the image of $h^*: \Omega^1(S^1) \to \Omega^1(\mathbb{R})$ consists of C^∞ periodic 1-forms of period 2π. $\qquad \square$

18.8 Invariant Forms on a Lie Group

Just as there are left-invariant vector fields on a Lie group G, so also are there left-invariant differential forms. For $g \in G$, let $\ell_g: G \to G$ be left multiplication by g. A k-form ω on G is said to be *left-invariant* if $\ell_g^* \omega = \omega$ for all $g \in G$. This means that for all $g, x \in G$,

$$\ell_g^*(\omega_{gx}) = \omega_x.$$

Thus, a left-invariant k-form is uniquely determined by its value at the identity, since for any $g \in G$,

$$\omega_g = \ell_{g^{-1}}^*(\omega_e). \tag{18.6}$$

Example 18.13 (*A left-invariant 1-form on S^1*). By Problem 17.3, $\omega = -y \, dx + x \, dy$ is a left-invariant 1-form on S^1.

We have the following analogue of Proposition 16.8.

Proposition 18.14. *Every left-invariant k-form ω on a Lie group G is C^∞.*

Proof. By Proposition 18.7(iii), it suffices to prove that for any k smooth vector fields $X_1, \ldots, X_k$ on G, the function $\omega(X_1, \ldots, X_k)$ is C^∞ on G. Let $(Y_1)_e, \ldots, (Y_n)_e$ be a basis for the tangent space $T_e G$ and $Y_1, \ldots, Y_n$ the left-invariant vector fields they generate. Then $Y_1, \ldots, Y_n$ is a C^∞ frame on G (Proposition 16.8). Each X_j can be written as a linear combination $X_j = \sum a^i_j Y_i$. By Proposition 12.12, the functions a^i_j are C^∞. Hence, to prove that ω is C^∞, it suffices to show that $\omega(Y_{i_1}, \ldots, Y_{i_k})$ is C^∞ for the *left-invariant* vector fields $Y_{i_1}, \ldots, Y_{i_k}$. But

$$
\begin{aligned}
(\omega(Y_{i_1}, \ldots, Y_{i_k}))(g) &= \omega_g((Y_{i_1})_g, \ldots, (Y_{i_k})_g) \\
&= (\ell^*_{g^{-1}}(\omega_e)) \left(\ell_{g*}(Y_{i_1})_e, \ldots, \ell_{g*}(Y_{i_k})_e \right) \\
&= \omega_e((Y_{i_1})_e, \ldots, (Y_{i_k})_e),
\end{aligned}
$$

which is a constant, independent of g. Being a constant function, $\omega(Y_{i_1}, \ldots, Y_{i_k})$ is C^∞ on G. $\qquad\square$

Similarly, a k-form ω on G is said to be *right-invariant* if $r^*_g \omega = \omega$ for all $g \in G$. The analogue of Proposition 18.14, that every right-invariant form on a Lie group is C^∞, is proven in the same way.

Let $\Omega^k(G)^G$ denote the vector space of left-invariant k-forms on G. The linear map

$$
\Omega^k(G)^G \to \bigwedge^k(\mathfrak{g}^\vee), \quad \omega \mapsto \omega_e,
$$

has an inverse defined by (18.6) and is therefore an isomorphism. It follows that $\dim \Omega^k(G)^G = \binom{n}{k}$.

Problems

18.1. Characterization of a smooth k-form
Write out a proof of Proposition 18.7(i)⇔(iv).

18.2. Linearity of the pullback
Prove Proposition 18.9.

18.3. Pullback of a wedge product
Prove Proposition 18.11.

18.4.* Support of a sum or product
Generalizing the support of a function, we define the *support of a k-form* $\omega \in \Omega^k(M)$ to be

$$
\operatorname{supp} \omega = \text{closure of } \{p \in M \mid \omega_p \neq 0\} = \overline{Z(\omega)^c},
$$

where $Z(\omega)^c$ is the complement of the zero set $Z(\omega)$ of ω in M. Let ω and τ be differential forms on a manifold M. Prove that

(a) $\operatorname{supp}(\omega + \tau) \subset \operatorname{supp} \omega \cup \operatorname{supp} \tau$,
(b) $\operatorname{supp}(\omega \wedge \tau) \subset \operatorname{supp} \omega \cap \operatorname{supp} \tau$.

18.5. Support of a linear combination

Prove that if the k-forms $\omega^1,\ldots,\omega^r \in \Omega^k(M)$ are linearly independent at every point of a manifold M and $a_1,\ldots,a_r$ are C^∞ functions on M, then

$$\operatorname{supp} \sum_{i=1}^r a_i \omega^i = \bigcup_{i=1}^r \operatorname{supp} a_i.$$

18.6.* Locally finite collection of supports

Let $\{\rho_\alpha\}_{\alpha \in A}$ be a collection of functions on M and ω a C^∞ k-form with compact support on M. If the collection $\{\operatorname{supp} \rho_\alpha\}_{\alpha \in A}$ of supports is locally finite, prove that $\rho_\alpha \omega \equiv 0$ for all but finitely many α.

18.7. Locally finite sums

We say that a sum $\sum \omega_\alpha$ of differential k-forms on a manifold M is *locally finite* if the collection $\{\operatorname{supp} \omega_\alpha\}$ of supports is locally finite. Suppose $\sum \omega_\alpha$ and $\sum \tau_\alpha$ are locally finite sums and f is a C^∞ function on M.

(a) Show that every point $p \in M$ has a neighborhood U on which $\sum \omega_\alpha$ is a finite sum.
(b) Show that $\sum \omega_\alpha + \tau_\alpha$ is a locally finite sum and

$$\sum \omega_\alpha + \tau_\alpha = \sum \omega_\alpha + \sum \tau_\alpha.$$

(c) Show that $\sum f \omega_\alpha$ is a locally finite sum and

$$\sum f \cdot \omega_\alpha = f \cdot \left(\sum \omega_\alpha \right).$$

18.8.* Pullback by a surjective submersion

In Subsection 19.5, we will show that the pullback of a C^∞ form is C^∞. Assuming this fact for now, prove that if $\pi \colon \tilde{M} \to M$ is a surjective submersion, then the pullback map $\pi^* \colon \Omega^*(M) \to \Omega^*(\tilde{M})$ is an injective algebra homomorphism.

18.9. Bi-invariant top forms on a compact, connected Lie group

Suppose G is a compact, connected Lie group of dimension n with Lie algebra $\mathfrak{g}$. This exercise proves that every left-invariant n-form on G is right-invariant.

(a) Let ω be a left-invariant n-form on G. For any $a \in G$, show that $r_a^* \omega$ is also left-invariant, where $r_a \colon G \to G$ is right multiplication by a.
(b) Since $\dim \Omega^n(G)^G = \dim \bigwedge^n(\mathfrak{g}^\vee) = 1$, $r_a^* \omega = f(a) \omega$ for some nonzero real number $f(a)$ depending on $a \in G$. Show that $f \colon G \to \mathbb{R}^\times$ is a group homomorphism.
(c) Show that $f \colon G \to \mathbb{R}^\times$ is C^∞. (*Hint*: Note that $f(a) \omega_e = (r_a^* \omega)_e = r_a^*(\omega_a) = r_a^* \ell_{a^{-1}}^*(\omega_e)$. Thus, $f(a)$ is the pullback of the map $\operatorname{Ad}(a^{-1}) \colon \mathfrak{g} \to \mathfrak{g}$. See Problem 16.11.)
(d) As the continuous image of a compact connected set G, the set $f(G) \subset \mathbb{R}^\times$ is compact and connected. Prove that $f(G) = 1$. Hence, $r_a^* \omega = \omega$ for all $a \in G$.

§19 The Exterior Derivative

In contrast to undergraduate calculus, where the basic objects of study are functions, the basic objects in calculus on manifolds are differential forms. Our program now is to learn how to integrate and differentiate differential forms.

Recall that an *antiderivation* on a graded algebra $A = \bigoplus_{k=0}^{\infty} A^k$ is an $\mathbb{R}$-linear map $D\colon A \to A$ such that

$$D(\omega \cdot \tau) = (D\omega) \cdot \tau + (-1)^k \omega \cdot D\tau$$

for $\omega \in A^k$ and $\tau \in A^\ell$. In the graded algebra A, an element of A^k is called a *homogeneous element* of degree k. The antiderivation is *of degree m* if

$$\deg D\omega = \deg \omega + m$$

for all homogeneous elements $\omega \in A$.

Let M be a manifold and $\Omega^*(M)$ the graded algebra of C^∞ differential forms on M. On the graded algebra $\Omega^*(M)$ there is a uniquely and intrinsically defined antiderivation called the *exterior derivative*. The process of applying the exterior derivative is called *exterior differentiation*.

Definition 19.1. An *exterior derivative* on a manifold M is an $\mathbb{R}$-linear map

$$D\colon \Omega^*(M) \to \Omega^*(M)$$

such that

  (i) D is an antiderivation of degree 1,
 (ii) $D \circ D = 0$,
(iii) if f is a C^∞ function and X a C^∞ vector field on M, then $(Df)(X) = Xf$.

Condition (iii) says that on 0-forms an exterior derivative agrees with the differential df of a function f. Hence, by (17.2), on a coordinate chart $(U, x^1, \ldots, x^n)$,

$$Df = df = \sum \frac{\partial f}{\partial x^i}\, dx^i.$$

In this section we prove the existence and uniqueness of an exterior derivative on a manifold. Using its three defining properties, we then show that the exterior derivative commutes with the pullback. This will finally allow us to prove that the pullback of a C^∞ form by a C^∞ map is C^∞.

19.1 Exterior Derivative on a Coordinate Chart

We showed in Subsection 4.4 the existence and uniqueness of an exterior derivative on an open subset of $\mathbb{R}^n$. The same proof carries over to any coordinate chart on a manifold.

More precisely, suppose $(U, x^1, \ldots, x^n)$ is a coordinate chart on a manifold M. Then any k-form ω on U is uniquely a linear combination

$$\omega = \sum a_I \, dx^I, \quad a_I \in C^\infty(U).$$

If D is an exterior derivative on U, then

$$
\begin{aligned}
D\omega &= \sum (Da_I) \wedge dx^I + \sum a_I D\, dx^I && \text{(by (i))} \\
&= \sum (Da_I) \wedge dx^I && \text{(by (iii) and (ii), } Dd = D^2 = 0) \\
&= \sum_I \sum_j \frac{\partial a_I}{\partial x^j} dx^j \wedge dx^I && \text{(by (iii)).} && (19.1)
\end{aligned}
$$

Hence, if an exterior derivative D exists on U, then it is uniquely defined by (19.1).

To show existence, we define D by the formula (19.1). The proof that D satisfies (i), (ii), and (iii) is the same as in the case of $\mathbb{R}^n$ in Proposition 4.7. We will denote the unique exterior derivative on a chart (U, ϕ) by d_U.

Like the derivative of a function on $\mathbb{R}^n$, an antiderivation D on $\Omega^*(M)$ has the property that for a k-form ω, the value of $D\omega$ at a point p depends only on the values of ω in a neighborhood of p. To explain this, we make a digression on local operators.

19.2 Local Operators

An endomorphism of a vector space W is often called an *operator* on W. For example, if $W = C^\infty(\mathbb{R})$ is the vector space of C^∞ functions on $\mathbb{R}$, then the derivative d/dx is an operator on W:

$$\frac{d}{dx} f(x) = f'(x).$$

The derivative has the property that the value of $f'(x)$ at a point p depends only on the values of f in a small neighborhood of p. More precisely, if $f = g$ on an open set U in $\mathbb{R}$, then $f' = g'$ on U. We say that the derivative is a *local operator on* $C^\infty(\mathbb{R})$.

Definition 19.2. An operator $D: \Omega^*(M) \to \Omega^*(M)$ is said to be *local* if for all $k \geq 0$, whenever a k-form $\omega \in \Omega^k(M)$ restricts to 0 on an open set U in M, then $D\omega \equiv 0$ on U.

Here by restricting to 0 on U, we mean that $\omega_p = 0$ at every point p in U, and the symbol "$\equiv 0$" means "is identically zero": $(D\omega)_p = 0$ at every point p in U. An equivalent criterion for an operator D to be local is that for all $k \geq 0$, whenever two k-forms $\omega, \tau \in \Omega^k(M)$ agree on an open set U, then $D\omega \equiv D\tau$ on U.

Example. Define the integral operator

$$I: C^\infty([a,b]) \to C^\infty([a,b])$$

by

$$I(f) = \int_a^b f(t)dt.$$

Here $I(f)$ is a number, which we view as a constant function on $[a,b]$. The integral is not a local operator, since the value of $I(f)$ at any point p depends on the values of f over the entire interval $[a,b]$.

Proposition 19.3. *Any antiderivation D on $\Omega^*(M)$ is a local operator.*

Proof. Suppose $\omega \in \Omega^k(M)$ and $\omega \equiv 0$ on an open subset U. Let p be an arbitrary point in U. It suffices to prove that $(D\omega)_p = 0$.

Choose a C^∞ bump function f at p supported in U. In particular, $f \equiv 1$ in a neighborhood of p in U. Then $f\omega \equiv 0$ on M, since if a point q is in U, then $\omega_q = 0$, and if q is not in U, then $f(q) = 0$. Applying the antiderivation property of D to $f\omega$, we get

$$0 = D(0) = D(f\omega) = (Df) \wedge \omega + (-1)^0 f \wedge (D\omega).$$

Evaluating the right-hand side at p, noting that $\omega_p = 0$ and $f(p) = 1$, gives $0 = (D\omega)_p$. $\qquad\square$

Remark. The same proof shows that a derivation on $\Omega^*(M)$ is also a local operator.

19.3 Existence of an Exterior Derivative on a Manifold

To define an exterior derivative on a manifold M, let ω be a k-form on M and $p \in M$. Choose a chart $(U,x^1,\ldots,x^n)$ about p. Suppose $\omega = \sum a_I\, dx^I$ on U. In Subsection 19.1 we showed the existence of an exterior derivative d_U on U with the property

$$d_U\omega = \sum da_I \wedge dx^I \quad \text{on } U. \tag{19.2}$$

Define $(d\omega)_p = (d_U\omega)_p$. We now show that $(d_U\omega)_p$ is independent of the chart U containing p. If $(V,y^1,\ldots,y^n)$ is another chart about p and $\omega = \sum b_J dy^J$ on V, then on $U \cap V$,

$$\sum a_I\, dx^I = \sum b_J\, dy^J.$$

On $U \cap V$ there is a unique exterior derivative

$$d_{U\cap V}: \Omega^*(U \cap V) \to \Omega^*(U \cap V).$$

By the properties of the exterior derivative, on $U \cap V$

$$d_{U\cap V}\left(\sum a_I\, dx^I\right) = d_{U\cap V}\left(\sum b_J\, dy^J\right),$$

$$\text{or} \quad \sum da_I \wedge dx^I = \sum db_J \wedge dy^J.$$

In particular,

$$\left(\sum da_I \wedge dx^I\right)_p = \left(\sum db_J \wedge dy^J\right)_p.$$

Thus, $(d\omega)_p = (d_U\omega)_p$ is well defined, independently of the chart $(U, x^1, \ldots, x^n)$. As p varies over all points of M, this defines an operator

$$d: \Omega^*(M) \to \Omega^*(M).$$

To check properties (i), (ii), and (iii), it suffices to check them at each point $p \in M$. As in Subsection 19.1, the verification reduces to the same calculation as for the exterior derivative on $\mathbb{R}^n$ in Proposition 4.7.

19.4 Uniqueness of the Exterior Derivative

Suppose $D: \Omega^*(M) \to \Omega^*(M)$ is an exterior derivative. We will show that D coincides with the exterior derivative d defined in Subsection 19.3.

If f is a C^∞ function and X a C^∞ vector field on M, then by condition (iii) of Definition 19.1,

$$(Df)(X) = Xf = (df)(X).$$

Therefore, $Df = df$ on functions $f \in \Omega^0(M)$.

Next consider a wedge product of exact 1-forms $df^1 \wedge \cdots \wedge df^k$:

$$
\begin{aligned}
D(df^1 \wedge \cdots \wedge df^k) & \\
= D(Df^1 \wedge \cdots \wedge Df^k) & \qquad \text{(because } Df^i = df^i) \\
= \sum_{i=1}^{k} (-1)^{i-1} Df^1 \wedge \cdots \wedge DDf^i \wedge \cdots \wedge Df^k & \quad \text{(D is an antiderivation)} \\
= 0 & \qquad (D^2 = 0).
\end{aligned}
$$

Finally, we show that D agrees with d on any k-form $\omega \in \Omega^k(M)$. Fix $p \in M$. Choose a chart $(U, x^1, \ldots, x^n)$ about p and suppose $\omega = \sum a_I \, dx^I$ on U. Extend the functions $a_I, x^1, \ldots, x^n$ on U to C^∞ functions $\tilde{a}_I, \tilde{x}^1, \ldots, \tilde{x}^n$ on M that agree with $a_I, x^1, \ldots, x^n$ on a neighborhood of V of p (by Proposition 18.8). Define

$$\tilde{\omega} = \sum \tilde{a}_I \, d\tilde{x}^I \in \Omega^k(M).$$

Then

$$\omega \equiv \tilde{\omega} \quad \text{on } V.$$

Since D is a local operator,

$$D\omega = D\tilde{\omega} \quad \text{on } V.$$

Thus,

$$(D\omega)_p = (D\tilde{\omega})_p = \left(D\sum \tilde{a}_I \, d\tilde{x}^I\right)_p$$
$$= \left(\sum D\tilde{a}_I \wedge d\tilde{x}^I + \sum \tilde{a}_I \wedge D \, d\tilde{x}^I\right)_p$$
$$= \left(\sum d\tilde{a}_I \wedge d\tilde{x}^I\right)_p \qquad (\text{because } D \, d\tilde{x}^I = DD\tilde{x} = 0)$$
$$= \left(\sum da_I \wedge dx^I\right)_p \qquad (\text{since } D \text{ is a local operator})$$
$$= (d\omega)_p.$$

We have proven the following theorem.

Theorem 19.4. *On any manifold M there exists an exterior derivative $d \colon \Omega^*(M) \to \Omega^*(M)$ characterized uniquely by the three properties of Definition 19.1.*

19.5 Exterior Differentiation Under a Pullback

The pullback of differential forms commutes with the exterior derivative. This fact, together with Proposition 18.11 that the pullback preserves the wedge product, is a cornerstone of calculations involving the pullback. Using these two properties, we will finally be in a position to prove that the pullback of a C^∞ form under a C^∞ map is C^∞.

Proposition 19.5 (Commutation of the pullback with d). *Let $F \colon N \to M$ be a smooth map of manifolds. If $\omega \in \Omega^k(M)$, then $dF^*\omega = F^*d\omega$.*

Proof. The case $k = 0$, when ω is a C^∞ function on M, is Proposition 17.10. Next consider the case $k \geq 1$. It suffices to verify $dF^*\omega = F^*d\omega$ at an arbitrary point $p \in N$. This reduces the proof to a local computation, i.e., computation in a coordinate chart. If $(V, y^1, \dots, y^m)$ is a chart on M about $F(p)$, then on V,

$$\omega = \sum a_I \, dy^{i_1} \wedge \cdots \wedge dy^{i_k}, \quad I = (i_1 < \cdots < i_k),$$

for some C^∞ functions a_I on V and

$$F^*\omega = \sum (F^*a_I) F^* dy^{i_1} \wedge \cdots \wedge F^* dy^{i_k} \quad (\text{Proposition 18.11})$$
$$= \sum (a_I \circ F) \, dF^{i_1} \wedge \cdots \wedge dF^{i_k} \qquad (F^* dy^i = dF^* y^i = d(y^i \circ F) = dF^i).$$

So

$$dF^*\omega = \sum d(a_I \circ F) \wedge dF^{i_1} \wedge \cdots \wedge dF^{i_k}.$$

On the other hand,

$$F^*d\omega = F^*\left(\sum da_I \wedge dy^{i_1} \wedge \cdots \wedge dy^{i_k}\right)$$
$$= \sum F^* da_I \wedge F^* dy^{i_1} \wedge \cdots \wedge F^* dy^{i_k}$$
$$= \sum d(F^*a_I) \wedge dF^{i_1} \wedge \cdots \wedge dF^{i_k} \quad (\text{by the case } k = 0)$$
$$= \sum d(a_I \circ F) \wedge dF^{i_1} \wedge \cdots \wedge dF^{i_k}.$$

Therefore,

$$dF^*\omega = F^*d\omega. \qquad \square$$

Corollary 19.6. *If U is an open subset of a manifold M and $\omega \in \Omega^k(M)$, then $(d\omega)|_U = d(\omega|_U)$.*

Proof. Let $i: U \hookrightarrow M$ be the inclusion map. Then $\omega|_U = i^*\omega$, so the corollary is simply a restatement of the commutativity of d with i^*. $\qquad\square$

Example. Let U be the open set $]0, \infty[\times]0, 2\pi[$ in the (r, θ)-plane $\mathbb{R}^2$. Define $F: U \subset \mathbb{R}^2 \to \mathbb{R}^2$ by

$$F(r, \theta) = (r\cos\theta, r\sin\theta).$$

If x, y are the standard coordinates on the target $\mathbb{R}^2$, compute the pullback $F^*(dx \wedge dy)$.

Solution. We first compute F^*dx:

$$
\begin{aligned}
F^*dx &= dF^*x && \text{(Proposition 19.5)}\\
&= d(x \circ F) && \text{(definition of the pullback of a function)}\\
&= d(r\cos\theta)\\
&= (\cos\theta)\,dr - r\sin\theta\,d\theta.
\end{aligned}
$$

Similarly,

$$F^*dy = dF^*y = d(r\sin\theta) = (\sin\theta)\,dr + r\cos\theta\,d\theta.$$

Since the pullback commutes with the wedge product (Proposition 18.11),

$$
\begin{aligned}
F^*(dx \wedge dy) &= (F^*dx) \wedge (F^*dy)\\
&= ((\cos\theta)\,dr - r\sin\theta\,d\theta) \wedge ((\sin\theta)\,dr + r\cos\theta\,d\theta)\\
&= (r\cos^2\theta + r\sin^2\theta)\,dr \wedge d\theta \quad \text{(because } d\theta \wedge dr = -dr \wedge d\theta)\\
&= r\,dr \wedge d\theta. \qquad\qquad\qquad\qquad\qquad\qquad\qquad\qquad\qquad\quad\square
\end{aligned}
$$

Proposition 19.7. *If $F: N \to M$ is a C^∞ map of manifolds and ω is a C^∞ k-form on M, then $F^*\omega$ is a C^∞ k-form on N.*

Proof. It is enough to show that every point in N has a neighborhood on which $F^*\omega$ is C^∞. Fix $p \in N$ and choose a chart $(V, y^1, \dots, y^m)$ on M about $F(p)$. Let $F^i = y^i \circ F$ be the ith coordinate of the map F in this chart. By the continuity of F, there is a chart $(U, x^1, \dots, x^n)$ on N about p such that $F(U) \subset V$. Because ω is C^∞, on V,

$$\omega = \sum_I a_I\,dy^{i_1} \wedge \cdots \wedge dy^{i_k}$$

for some C^∞ functions $a_I \in C^\infty(V)$ (Proposition 18.7(i)$\Rightarrow$(ii)). By properties of the pullback,

$$
\begin{aligned}
F^*\omega &= \sum (F^*a_I) F^*(dy^{i_1}) \wedge \cdots F^*(dy^{i_k}) && \text{(Propositions 18.9 and 18.11)}\\
&= \sum (F^*a_I)\,dF^*y^{i_1} \wedge \cdots \wedge dF^*y^{i_k} && \text{(Proposition 19.5)}\\
&= \sum (a_I \circ F)\,dF^{i_1} \wedge \cdots \wedge dF^{i_k} && (F^*y^i = y^i \circ F = F^i)\\
&= \sum_{I,J} (a_I \circ F)\,\frac{\partial(F^{i_1}, \dots, F^{i_k})}{\partial(x^{j_1}, \dots, x^{j_k})}\,dx^J && \text{(Proposition 18.3)}.
\end{aligned}
$$

Since the $a_I \circ F$ and $\partial(F^{i_1}, \ldots, F^{i_k})/\partial(x^{j_1}, \ldots, x^{j_k})$ are all C^∞, $F^*\omega$ is C^∞ by Proposition 18.7(iii)$\Rightarrow$(i). $\square$

In summary, if $F \colon N \to M$ is a C^∞ map of manifolds, then the pullback map $F^* \colon \Omega^*(M) \to \Omega^*(N)$ is a morphism of differential graded algebras, i.e., a degree-preserving algebra homomorphism that commutes with the differential.

19.6 Restriction of k-Forms to a Submanifold

The restriction of a k-form to an immersed submanifold is just like the restriction of a 1-form, but with k arguments. Let S be a regular submanifold of a manifold M. If ω is a k-form on M, then the *restriction* of ω to S is the k-form $\omega|_S$ on S defined by

$$(\omega|_S)_p(v_1, \ldots, v_k) = \omega_p(v_1, \ldots, v_k)$$

for $v_1, \ldots, v_k \in T_pS \subset T_pM$. Thus, $(\omega|_S)_p$ is obtained from ω_p by restricting the domain of ω_p to $T_pS \times \cdots \times T_pS$ (k times). As in Proposition 17.14, the restriction of k-forms is the same as the pullback under the inclusion map $i \colon S \hookrightarrow M$.

A nonzero form on M may restrict to the zero form on a submanifold S. For example, if S is a smooth curve in $\mathbb{R}^2$ defined by the nonconstant function $f(x, y)$, then $df = (\partial f/\partial x)\,dx + (\partial f/\partial y)\,dy$ is a nonzero 1-form on $\mathbb{R}^2$, but since f is identically zero on S, the differential df is also identically zero on S. Thus, $(df)|_S \equiv 0$. Another example is Problem 19.9.

One should distinguish between a *nonzero* form and a *nowhere-zero* or *nowhere-vanishing* form. For example, $x\,dy$ is a nonzero form on $\mathbb{R}^2$, meaning that it is not identically zero. However, it is not nowhere-zero, because it vanishes on the y-axis. On the other hand, dx and dy are nowhere-zero 1-forms on $\mathbb{R}^2$.

NOTATION. Since pullback and exterior differentiation commute, $(df)|_S = d(f|_S)$, so one may write $df|_S$ to mean either expression.

19.7 A Nowhere-Vanishing 1-Form on the Circle

In Example 17.15 we found a nowhere-vanishing 1-form $-y\,dx + x\,dy$ on the unit circle. As an application of the exterior derivative, we will construct in a different way a nowhere-vanishing 1-form on the circle. One advantage of the new method is that it generalizes to the construction of a nowhere-vanishing top form on a *smooth hypersurface* in $\mathbb{R}^{n+1}$, a regular level set of a smooth function $f \colon \mathbb{R}^{n+1} \to \mathbb{R}$. As we will see in Section 21, the existence of a nowhere-vanishing top form is intimately related to orientations on a manifold.

Example 19.8. Let S^1 be the unit circle defined by $x^2 + y^2 = 1$ in $\mathbb{R}^2$. The 1-form dx restricts from $\mathbb{R}^2$ to a 1-form on S^1. At each point $p \in S^1$, the domain of $(dx|_{S^1})_p$ is $T_p(S^1)$ instead of $T_p(\mathbb{R}^2)$:

$$(dx|_{S^1})_p \colon T_p(S^1) \to \mathbb{R}.$$

At $p = (1,0)$, a basis for the tangent space $T_p(S^1)$ is $\partial/\partial y$ (Figure 19.1). Since

$$(dx)_p \left(\frac{\partial}{\partial y} \right) = 0,$$

we see that although dx is a nowhere-vanishing 1-form on $\mathbb{R}^2$, it vanishes at $(1,0)$ when restricted to S^1.

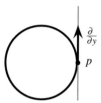

Fig. 19.1. The tangent space to S^1 at $p = (1,0)$.

To find a nowhere-vanishing 1-form on S^1, we take the exterior derivative of both sides of the equation

$$x^2 + y^2 = 1.$$

Using the antiderivation property of d, we get

$$2x\,dx + 2y\,dy = 0. \tag{19.3}$$

Of course, this equation is valid only at a point $(x,y) \in S^1$. Let

$$U_x = \{(x,y) \in S^1 \mid x \neq 0\} \quad \text{and} \quad U_y = \{(x,y) \in S^1 \mid y \neq 0\}.$$

By (19.3), on $U_x \cap U_y$,

$$\frac{dy}{x} = -\frac{dx}{y}.$$

Define a 1-form ω on S^1 by

$$\omega = \begin{cases} \dfrac{dy}{x} & \text{on } U_x, \\[2mm] -\dfrac{dx}{y} & \text{on } U_y. \end{cases} \tag{19.4}$$

Since these two 1-forms agree on $U_x \cap U_y$, ω is a well-defined 1-form on $S^1 = U_x \cup U_y$. To show that ω is C^∞ and nowhere-vanishing, we need charts. Let

$$U_x^+ = \{(x,y) \in S^1 \mid x > 0\}.$$

We define similarly U_x^-, U_y^+, U_y^- (Figure 19.2). On U_x^+, y is a local coordinate, and so dy is a basis for the cotangent space $T_p^*(S^1)$ at each point $p \in U_x^+$. Since $\omega = dy/x$ on U_x^+, ω is C^∞ and nowhere zero on U_x^+. A similar argument applies to dy/x on U_x^- and $-dx/y$ on U_y^+ and U_y^-. Hence, ω is C^∞ and nowhere vanishing on S^1.

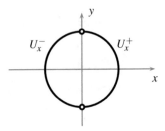

Fig. 19.2. Two charts on the unit circle.

Problems

19.1. Pullback of a differential form

Let U be the open set $]0,\infty[\times]0,\pi[\times]0,2\pi[$ in the (ρ,ϕ,θ)-space $\mathbb{R}^3$. Define $F: U \to \mathbb{R}^3$ by

$$F(\rho,\phi,\theta) = (\rho\sin\phi\cos\theta, \rho\sin\phi\sin\theta, \rho\cos\phi).$$

If x,y,z are the standard coordinates on the target $\mathbb{R}^3$, show that

$$F^*(dx \wedge dy \wedge dz) = \rho^2\sin\phi\, d\rho \wedge d\phi \wedge d\theta.$$

19.2. Pullback of a differential form

Let $F: \mathbb{R}^2 \to \mathbb{R}^2$ be given by

$$F(x,y) = (x^2+y^2, xy).$$

If u,v are the standard coordinates on the target $\mathbb{R}^2$, compute $F^*(u\,du + v\,dv)$.

19.3. Pullback of a differential form by a curve

Let τ be the 1-form $\tau = (-y\,dx + x\,dy)/(x^2+y^2)$ on $\mathbb{R}^2 - \{\mathbf{0}\}$. Define $\gamma: \mathbb{R} \to \mathbb{R}^2 - \{\mathbf{0}\}$ by $\gamma(t) = (\cos t, \sin t)$. Compute $\gamma^*\tau$. (This problem is related to Example 17.16 in that if $i: S^1 \hookrightarrow \mathbb{R}^2 - \{\mathbf{0}\}$ is the inclusion, then $\gamma = i \circ c$ and $\omega = i^*\tau$.)

19.4. Pullback of a restriction

Let $F: N \to M$ be a C^∞ map of manifolds, U an open subset of M, and $F|_{F^{-1}(U)}: F^{-1}(U) \to U$ the restriction of F to $F^{-1}(U)$. Prove that if $\omega \in \Omega^k(M)$, then

$$\left(F|_{F^{-1}(U)}\right)^* (\omega|_U) = (F^*\omega)|_{F^{-1}(U)}.$$

19.5. Coordinate functions and differential forms

Let $f^1, \ldots, f^n$ be C^∞ functions on a neighborhood U of a point p in a manifold of dimension n. Show that there is a neighborhood W of p on which $f^1, \ldots, f^n$ form a coordinate system if and only if $(df^1 \wedge \cdots \wedge df^n)_p \neq 0$.

19.6. Local operators

An operator $L: \Omega^*(M) \to \Omega^*(M)$ is *support-decreasing* if $\operatorname{supp} L(\omega) \subset \operatorname{supp}\omega$ for every k-form $\omega \in \Omega^*(M)$ for all $k \geq 0$. Show that an operator on $\Omega^*(M)$ is local if and only if it is support-decreasing.

19.7. Derivations of C^∞ functions are local operators

Let M be a smooth manifold. The definition of a *local operator* D on $C^\infty(M)$ is similar to that of a local operator on $\Omega^*(M)$: D is *local* if whenever a function $f \in C^\infty(M)$ vanishes identically on an open subset U, then $Df \equiv 0$ on U. Prove that a derivation of $C^\infty(M)$ is a local operator on $C^\infty(M)$.

19.8. Nondegenerate 2-forms

A 2-covector α on a $2n$-dimensional vector space V is said to be *nondegenerate* if $\alpha^n :=$ $\alpha \wedge \cdots \wedge \alpha$ (n times) is not the zero $2n$-covector. A 2-form ω on a $2n$-dimensional manifold M is said to be *nondegenerate* if at every point $p \in M$, the 2-covector ω_p is nondegenerate on the tangent space T_pM.

(a) Prove that on $\mathbb{C}^n$ with real coordinates $x^1, y^1, \ldots, x^n, y^n$, the 2-form

$$\omega = \sum_{j=1}^{n} dx^j \wedge dy^j$$

is nondegenerate.
(b) Prove that if λ is the Liouville form on the total space T^*M of the cotangent bundle of an n-dimensional manifold M, then $d\lambda$ is a nondegenerate 2-form on T^*M.

19.9.* Vertical planes

Let x, y, z be the standard coordinates on $\mathbb{R}^3$. A plane in $\mathbb{R}^3$ is *vertical* if it is defined by $ax + by = 0$ for some $(a, b) \neq (0, 0) \in \mathbb{R}^2$. Prove that restricted to a vertical plane, $dx \wedge dy = 0$.

19.10. Nowhere-vanishing form on S^1

Prove that the nowhere-vanishing form ω on S^1 constructed in Example 19.8 is the form $-y\,dx + x\,dy$ of Example 17.15. (*Hint*: Consider U_x and U_y separately. On U_x, substitute $dx = -(y/x)\,dy$ into $-y\,dx + x\,dy$.)

19.11. A C^∞ nowhere-vanishing form on a smooth hypersurface

(a) Let $f(x, y)$ be a C^∞ function on $\mathbb{R}^2$ and assume that 0 is a regular value of f. By the regular level set theorem, the zero set M of $f(x, y)$ is a one-dimensional submanifold of $\mathbb{R}^2$. Construct a C^∞ nowhere-vanishing 1-form on M.
(b) Let $f(x, y, z)$ be a C^∞ function on $\mathbb{R}^3$ and assume that 0 is a regular value of f. By the regular level set theorem, the zero set M of $f(x, y, z)$ is a two-dimensional submanifold of $\mathbb{R}^3$. Let f_x, f_y, f_z be the partial derivatives of f with respect to x, y, z, respectively. Show that the equalities

$$\frac{dx \wedge dy}{f_z} = \frac{dy \wedge dz}{f_x} = \frac{dz \wedge dx}{f_y}$$

hold on M whenever they make sense, and therefore the three 2-forms piece together to give a C^∞ nowhere-vanishing 2-form on M.
(c) Generalize this problem to a regular level set of $f(x^1, \ldots, x^{n+1})$ in $\mathbb{R}^{n+1}$.

19.12. Vector fields as derivations of C^∞ functions

In Subsection 14.1 we showed that a C^∞ vector field X on a manifold M gives rise to a derivation of $C^\infty(M)$. We will now show that every derivation of $C^\infty(M)$ arises from one and only one vector field, as promised earlier. To distinguish the vector field from the derivation, we will temporarily denote the derivation arising from X by $\varphi(X)$. Thus, for any $f \in C^\infty(M)$,

$$(\varphi(X)f)(p) = X_p f \quad \text{for all } p \in M.$$

(a) Let $\mathcal{F} = C^\infty(M)$. Prove that $\varphi\colon \mathfrak{X}(M) \to \mathrm{Der}(C^\infty(M))$ is an $\mathcal{F}$-linear map.

(b) Show that φ is injective.

(c) If D is a derivation of $C^\infty(M)$ and $p \in M$, define $D_p\colon C_p^\infty(M) \to C_p^\infty(M)$ by

$$D_p[f] = [D\tilde{f}] \in C_p^\infty(M),$$

where $[f]$ is the germ of f at p and $\tilde{f}$ is a global extension of f, such as those given by Proposition 18.8. Show that $D_p[f]$ is well defined. (*Hint:* Apply Problem 19.7.)

(d) Show that D_p is a derivation of $C_p^\infty(M)$.

(e) Prove that $\varphi\colon \mathfrak{X}(M) \to \mathrm{Der}(C^\infty(M))$ is an isomorphism of $\mathcal{F}$-modules.

19.13. Twentieth-century formulation of Maxwell's equations

In Maxwell's theory of electricity and magnetism, developed in the late nineteenth century, the electric field $\mathbf{E} = \langle E_1, E_2, E_3 \rangle$ and the magnetic field $\mathbf{B} = \langle B_1, B_2, B_3 \rangle$ in a vacuum $\mathbb{R}^3$ with no charge or current satisfy the following equations:

$$\nabla \times \mathbf{E} = -\frac{\partial \mathbf{B}}{\partial t}, \qquad \nabla \times \mathbf{B} = \frac{\partial \mathbf{E}}{\partial t},$$
$$\mathrm{div}\,\mathbf{E} = 0, \qquad \mathrm{div}\,\mathbf{B} = 0.$$

By the correspondence in Subsection 4.6, the 1-form E on $\mathbb{R}^3$ corresponding to the vector field $\mathbf{E}$ is

$$E = E_1\,dx + E_2\,dy + E_3\,dz$$

and the 2-form B on $\mathbb{R}^3$ corresponding to the vector field $\mathbf{B}$ is

$$B = B_1\,dy \wedge dz + B_2\,dz \wedge dx + B_3\,dx \wedge dy.$$

Let $\mathbb{R}^4$ be space-time with coordinates (x, y, z, t). Then both E and B can be viewed as differential forms on $\mathbb{R}^4$. Define F to be the 2-form

$$F = E \wedge dt + B$$

on space-time. Decide which two of Maxwell's equations are equivalent to the equation

$$dF = 0.$$

Prove your answer. (The other two are equivalent to $d * F = 0$ for a star-operator $*$ defined in differential geometry. See [2, Section 19.1, p. 689].)

§20 The Lie Derivative and Interior Multiplication

The only portion of this section necessary for the remainder of the book is Subsection 20.4 on interior multiplication. The rest may be omitted on first reading.

The construction of exterior differentiation in Section 19 is local and depends on a choice of coordinates: if $\omega = \sum a_I \, dx^I$, then

$$d\omega = \sum \frac{\partial a_I}{\partial x^j} \, dx^j \wedge dx^I.$$

It turns out, however, that this d is in fact global and intrinsic to the manifold, i.e., independent of the choice of local coordinates. Indeed, for a C^∞ 1-form ω and C^∞ vector fields X, Y on a manifold M, one has the formula

$$(d\omega)(X,Y) = X\omega(Y) - Y\omega(X) - \omega([X,Y]).$$

In this section we will derive a global intrinsic formula like this for the exterior derivative of a k-form.

The proof uses the Lie derivative and interior multiplication, two other intrinsic operations on a manifold. The Lie derivative is a way of differentiating a vector field or a differential form on a manifold along another vector field. For any vector field X on a manifold, the interior multiplication ι_X is an antiderivation of degree -1 on differential forms. Being intrinsic operators on a manifold, both the Lie derivative and interior multiplication are important in their own right in differential topology and geometry.

20.1 Families of Vector Fields and Differential Forms

A collection $\{X_t\}$ or $\{\omega_t\}$ of vector fields or differential forms on a manifold is said to be a 1-*parameter family* if the parameter t runs over some subset of the real line. Let I be an open interval in $\mathbb{R}$ and let M be a manifold. Suppose $\{X_t\}$ is a 1-parameter family of vector fields on M defined for all $t \in I$ except at $t_0 \in I$. We say that the *limit* $\lim_{t \to t_0} X_t$ exists if every point $p \in M$ has a coordinate neighborhood $(U, x^1, \ldots, x^n)$ on which $X_t|_p = \sum a^i(t,p)\partial/\partial x^i|_p$ and $\lim_{t \to t_0} a^i(t,p)$ exists for all i. In this case, we set

$$\lim_{t \to t_0} X_t|_p = \sum_{i=1}^n \lim_{t \to t_0} a^i(t,p) \left. \frac{\partial}{\partial x^i} \right|_p. \tag{20.1}$$

In Problem 20.1 we ask the reader to show that this definition of the limit of X_t as $t \to t_0$ is independent of the choice of the coordinate neighborhood $(U, x^1, \ldots, x^n)$.

A 1-parameter family $\{X_t\}_{t \in I}$ of smooth vector fields on M is said to *depend smoothly* on t if every point in M has a coordinate neighborhood $(U, x^1, \ldots, x^n)$ on which

$$(X_t)_p = \sum a^i(t,p) \left. \frac{\partial}{\partial x^i} \right|_p, \qquad (t,p) \in I \times U, \tag{20.2}$$

for some C^∞ functions a^i on $I \times U$. In this case we also say that $\{X_t\}_{t \in I}$ is a *smooth family of vector fields* on M.

For a smooth family of vector fields on M, one can define its derivative with respect to t at $t = t_0$ by

$$\left(\frac{d}{dt}\bigg|_{t=t_0} X_t \right)_p = \sum \frac{\partial a^i}{\partial t}(t_0, p) \frac{\partial}{\partial x^i}\bigg|_p \tag{20.3}$$

for $(t_0, p) \in I \times U$. It is easy to check that this definition is independent of the chart $(U, x^1, \ldots, x^n)$ containing p (Problem 20.3). Clearly, the derivative $d/dt|_{t=t_0} X_t$ is a smooth vector field on M.

Similarly, a 1-parameter family $\{\omega_t\}_{t \in I}$ of smooth k-forms on M is said to *depend smoothly* on t if every point of M has a coordinate neighborhood $(U, x^1, \ldots, x^n)$ on which

$$(\omega_t)_p = \sum b_J(t, p) \, dx^J|_p, \qquad (t, p) \in I \times U,$$

for some C^∞ functions b_J on $I \times U$. We also call such a family $\{\omega_t\}_{t \in I}$ a *smooth family of k-forms* on M and define its derivative with respect to t to be

$$\left(\frac{d}{dt}\bigg|_{t=t_0} \omega_t \right)_p = \sum \frac{\partial b_J}{\partial t}(t_0, p) \, dx^J|_p.$$

As for vector fields, this definition is independent of the chart and defines a C^∞ k-form $d/dt|_{t=t_0} \omega_t$ on M.

NOTATION. We write d/dt for the derivative of a smooth family of vector fields or differential forms, but $\partial/\partial t$ for the partial derivative of a function of several variables.

Proposition 20.1 (Product rule for d/dt). *If $\{\omega_t\}$ and $\{\tau_t\}$ are smooth families of k-forms and ℓ-forms respectively on a manifold M, then*

$$\frac{d}{dt}(\omega_t \wedge \tau_t) = \left(\frac{d}{dt}\omega_t \right) \wedge \tau_t + \omega_t \wedge \frac{d}{dt}\tau_t.$$

Proof. Written out in local coordinates, this reduces to the usual product rule in calculus. We leave the details as an exercise (Problem 20.4). $\qquad \square$

Proposition 20.2 (Commutation of $d/dt|_{t=t_0}$ with d). *If $\{\omega_t\}_{t \in I}$ is a smooth family of differential forms on a manifold M, then*

$$\frac{d}{dt}\bigg|_{t=t_0} d\omega_t = d\left(\frac{d}{dt}\bigg|_{t=t_0} \omega_t \right).$$

Proof. In this proposition, there are three operations—exterior differentiation, differentiation with respect to t, and evaluation at $t = t_0$. We will first show that d and d/dt commute:

$$\frac{d}{dt}(d\omega_t) = d\left(\frac{d}{dt}\omega_t\right). \tag{20.4}$$

It is enough to check the equality at an arbitrary point $p \in M$. Let $(U, x^1, \ldots, x^n)$ be a neighborhood of p such that $\omega = \sum_J b_J \, dx^J$ for some C^∞ functions b_J on $I \times U$. On U,

$$\frac{d}{dt}(d\omega_t) = \frac{d}{dt}\sum_{J,i}\frac{\partial b_J}{\partial x^i}\,dx^i \wedge dx^J \quad \text{(note that there is no } dt \text{ term)}$$

$$= \sum_{i,J}\frac{\partial}{\partial x^i}\left(\frac{\partial b_J}{\partial t}\right)dx^i \wedge dx^J \quad \text{(since } b_J \text{ is } C^\infty)$$

$$= d\left(\sum_J\frac{\partial b_J}{\partial t}\,dx^J\right) = d\left(\frac{d}{dt}\omega_t\right).$$

Evaluation at $t = t_0$ commutes with d, because d involves only partial derivatives with respect to the x^i variables. Explicitly,

$$\left(d\left(\frac{d}{dt}\omega_t\right)\right)\Big|_{t=t_0} = \left(\sum_{i,J}\frac{\partial}{\partial x^i}\frac{\partial}{\partial t}b_J\,dx^i \wedge dx^J\right)\Big|_{t=t_0}$$

$$= \sum_{i,J}\frac{\partial}{\partial x^i}\left(\frac{\partial}{\partial t}\Big|_{t=t_0}b_J\right)dx^i \wedge dx^J = d\left(\frac{\partial}{\partial t}\Big|_{t_0}\omega_t\right).$$

Evaluating both sides of (20.4) at $t = t_0$ completes the proof of the proposition. $\square$

20.2 The Lie Derivative of a Vector Field

In a first course on calculus, one defines the derivative of a real-valued function f on $\mathbb{R}$ at a point $p \in \mathbb{R}$ as

$$f'(p) = \lim_{t \to 0}\frac{f(p+t) - f(p)}{t}.$$

The problem in generalizing this definition to the derivative of a vector field Y on a manifold M is that at two nearby points p and q in M, the tangent vectors Y_p and Y_q are in different vector spaces T_pM and T_qM and so it is not possible to compare them by subtracting one from the other. One way to get around this difficulty is to use the local flow of another vector field X to transport Y_q to the tangent space T_pM at p. This leads to the definition of the Lie derivative of a vector field.

Recall from Subsection 14.3 that for any smooth vector field X on M and point p in M, there is a neighborhood U of p on which the vector field X has a *local flow*; this means that there exist a real number $\varepsilon > 0$ and a map

$$\varphi : \,]-\varepsilon, \varepsilon[\,\times U \to M$$

such that if we set $\varphi_t(q) = \varphi(t, q)$, then

$$\frac{\partial}{\partial t}\varphi_t(q) = X_{\varphi_t(q)}, \quad \varphi_0(q) = q \quad \text{for } q \in U. \tag{20.5}$$

In other words, for each q in U, the curve $\varphi_t(q)$ is an integral curve of X with initial point q. By definition, $\varphi_0 \colon U \to U$ is the identity map. The local flow satisfies the property

$$\varphi_s \circ \varphi_t = \varphi_{s+t}$$

whenever both sides are defined (see (14.10)). Consequently, for each t the map $\varphi_t \colon U \to \varphi_t(U)$ is a diffeomorphism onto its image, with a C^∞ inverse φ_{-t}:

$$\varphi_{-t} \circ \varphi_t = \varphi_0 = \mathbb{1}, \quad \varphi_t \circ \varphi_{-t} = \varphi_0 = \mathbb{1}.$$

Let Y be a C^∞ vector field on M. To compare the values of Y at $\varphi_t(p)$ and at p, we use the diffeomorphism $\varphi_{-t} \colon \varphi_t(U) \to U$ to push $Y_{\varphi_t(p)}$ into T_pM (Figure 20.1).

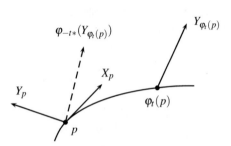

Fig. 20.1. Comparing the values of Y at nearby points.

Definition 20.3. For $X, Y \in \mathfrak{X}(M)$ and $p \in M$, let $\varphi \colon]-\varepsilon, \varepsilon[\times U \to M$ be a local flow of X on a neighborhood U of p and define the *Lie derivative* $\mathcal{L}_X Y$ of Y with respect to X at p to be the vector

$$(\mathcal{L}_X Y)_p = \lim_{t \to 0} \frac{\varphi_{-t*}\left(Y_{\varphi_t(p)}\right) - Y_p}{t} = \lim_{t \to 0} \frac{(\varphi_{-t*}Y)_p - Y_p}{t} = \frac{d}{dt}\Big|_{t=0} (\varphi_{-t*}Y)_p.$$

In this definition the limit is taken in the finite-dimensional vector space T_pM. For the derivative to exist, it suffices that $\{\varphi_{-t*}Y\}$ be a smooth family of vector fields on M. To show the smoothness of the family $\{\varphi_{-t*}Y\}$, we write out $\varphi_{-t*}Y$ in local coordinates $x^1, \ldots, x^n$ in a chart. Let φ_t^i and φ^i be the ith components of φ_t and φ respectively. Then

$$(\varphi_t)^i(p) = \varphi^i(t, p) = (x^i \circ \varphi)(t, p).$$

By Proposition 8.11, relative to the frame $\{\partial/\partial x^j\}$, the differential φ_{t*} at p is represented by the Jacobian matrix $[\partial(\varphi_t)^i/\partial x^j(p)] = [\partial\varphi^i/\partial x^j(t, p)]$. This means that

$$\varphi_{t*}\left(\left.\frac{\partial}{\partial x^j}\right|_p\right) = \sum_i \frac{\partial \varphi^i}{\partial x^j}(t,p)\left.\frac{\partial}{\partial x^i}\right|_{\varphi_t(p)}.$$

Thus, if $Y = \sum b^j \partial/\partial x^j$, then

$$\varphi_{-t*}\left(Y_{\varphi_t(p)}\right) = \sum_j b^j(\varphi(t,p))\varphi_{-t*}\left(\left.\frac{\partial}{\partial x^j}\right|_{\varphi_t(p)}\right)$$

$$= \sum_{i,j} b^j(\varphi(t,p))\frac{\partial \varphi^i}{\partial x^j}(-t,p)\left.\frac{\partial}{\partial x^i}\right|_p. \tag{20.6}$$

When X and Y are C^∞ vector fields on M, both φ^i and b^j are C^∞ functions. The formula (20.6) then shows that $\{\varphi_{-t*}Y\}$ is a smooth family of vector fields on M. It follows that the Lie derivative $\mathcal{L}_X Y$ exists and is given in local coordinates by

$$(\mathcal{L}_X Y)_p = \left.\frac{d}{dt}\right|_{t=0}\varphi_{-t*}\left(Y_{\varphi_t(p)}\right)$$

$$= \sum_{i,j}\left.\frac{\partial}{\partial t}\right|_{t=0}\left(b^j(\varphi(t,p))\frac{\partial \varphi^i}{\partial x^j}(-t,p)\right)\left.\frac{\partial}{\partial x^i}\right|_p. \tag{20.7}$$

It turns out that the Lie derivative of a vector field gives nothing new.

Theorem 20.4. *If X and Y are C^∞ vector fields on a manifold M, then the Lie derivative $\mathcal{L}_X Y$ coincides with the Lie bracket $[X,Y]$.*

Proof. It suffices to check the equality $\mathcal{L}_X Y = [X,Y]$ at every point. To do this, we expand both sides in local coordinates. Suppose a local flow for X is φ: $]-\varepsilon,\varepsilon[\times U \to M$, where U is a coordinate chart with coordinates $x^1,\ldots,x^n$. Let $X = \sum a^i \partial/\partial x^i$ and $Y = \sum b^j \partial/\partial x^j$ on U. The condition (20.5) that $\varphi_t(p)$ be an integral curve of X translates into the equations

$$\frac{\partial \varphi^i}{\partial t}(t,p) = a^i(\varphi(t,p)), \qquad i=1,\ldots,n, \quad (t,p)\in]-\varepsilon,\varepsilon[\times U.$$

At $t=0$, $\partial \varphi^i/\partial t(0,p) = a^i(\varphi(0,p)) = a^i(p)$.

By Problem 14.12, the Lie bracket in local coordinates is

$$[X,Y] = \sum_{i,k}\left(a^k\frac{\partial b^i}{\partial x^k} - b^k\frac{\partial a^i}{\partial x^k}\right)\frac{\partial}{\partial x^i}.$$

Expanding (20.7) by the product rule and the chain rule, we get

$$(\mathcal{L}_X Y)_p = \left[\sum_{i,j,k}\left(\frac{\partial b^j}{\partial x^k}(\varphi(t,p))\frac{\partial \varphi^k}{\partial t}(t,p)\frac{\partial \varphi^i}{\partial x^j}(-t,p)\right)\frac{\partial}{\partial x^i}\right.$$

$$\left.-\sum_{i,j}\left(b^j(\varphi(t,p))\frac{\partial}{\partial x^j}\frac{\partial \varphi^i}{\partial t}(-t,p)\right)\frac{\partial}{\partial x^i}\right]_{t=0}$$

$$= \sum_{i,j,k}\left(\frac{\partial b^j}{\partial x^k}(p)a^k(p)\frac{\partial \varphi^i}{\partial x^j}(0,p)\right)\frac{\partial}{\partial x^i} - \sum_{i,j}\left(b^j(p)\frac{\partial a^i}{\partial x^j}(p)\right)\frac{\partial}{\partial x^i}. \tag{20.8}$$

Since $\varphi(0,p) = p$, φ_0 is the identity map and hence its Jacobian matrix is the identity matrix. Thus,

$$\frac{\partial \varphi^i}{\partial x^j}(0,p) = \delta^i_j, \quad \text{the Kronecker delta.}$$

So (20.8) simplifies to

$$\mathcal{L}_X Y = \sum_{i,k} \left(a^k \frac{\partial b^i}{\partial x^k} - b^k \frac{\partial a^i}{\partial x^k} \right) \frac{\partial}{\partial x^i} = [X,Y]. \qquad \square$$

Although the Lie derivative of a vector field gives us nothing new, in conjunction with the Lie derivative of differential forms it turns out to be a tool of great utility, for example, in the proof of the global formula for the exterior derivative in Theorem 20.14.

20.3 The Lie Derivative of a Differential Form

Let X be a smooth vector field and ω a smooth k-form on a manifold M. Fix a point $p \in M$ and let $\varphi_t : U \to M$ be a flow of X in a neighborhood U of p. The definition of the Lie derivative of a differential form is similar to that of the Lie derivative of a vector field. However, instead of pushing a vector at $\varphi_t(p)$ to p via $(\varphi_{-t})_*$, we now pull the k-covector $\omega_{\varphi_t(p)}$ back to p via φ_t^*.

Definition 20.5. For X a smooth vector field and ω a smooth k-form on a manifold M, the *Lie derivative* $\mathcal{L}_X \omega$ at $p \in M$ is

$$(\mathcal{L}_X \omega)_p = \lim_{t \to 0} \frac{\varphi_t^* \left(\omega_{\varphi_t(p)} \right) - \omega_p}{t} = \lim_{t \to 0} \frac{(\varphi_t^* \omega)_p - \omega_p}{t} = \frac{d}{dt}\Big|_{t=0} (\varphi_t^* \omega)_p.$$

By an argument similar to that for the existence of the Lie derivative $\mathcal{L}_X Y$ in Section 20.2, one shows that $\{\varphi_t^* \omega\}$ is a smooth family of k-forms on M by writing it out in local coordinates. The existence of $(\mathcal{L}_X \omega)_p$ follows.

Proposition 20.6. *If f is a C^∞ function and X a C^∞ vector field on M, then $\mathcal{L}_X f = Xf$.*

Proof. Fix a point p in M and let $\varphi_t : U \to M$ be a local flow of X as above. Then

$$\begin{aligned}
(\mathcal{L}_X f)_p &= \frac{d}{dt}\Big|_{t=0} (\varphi_t^* f)_p && \text{(definition of } \mathcal{L}_X f) \\
&= \frac{d}{dt}\Big|_{t=0} (f \circ \varphi_t)(p) && \text{(definition of } \varphi_t^* f) \\
&= X_p f && \text{(Proposition 8.17),}
\end{aligned}$$

since $\varphi_t(p)$ is a curve through p with initial vector X_p. $\qquad \square$

20.4 Interior Multiplication

We first define interior multiplication on a vector space. If β is a k-covector on a vector space V and $v \in V$, for $k \geq 2$ the *interior multiplication* or *contraction* of β with v is the $(k-1)$-covector $\iota_v \beta$ defined by

$$(\iota_v \beta)(v_2, \ldots, v_k) = \beta(v, v_2, \ldots, v_k), \quad v_2, \ldots, v_k \in V.$$

We define $\iota_v \beta = \beta(v) \in \mathbb{R}$ for a 1-covector β on V and $\iota_v \beta = 0$ for a 0-covector β (a constant) on V.

Proposition 20.7. *For* 1-*covectors* $\alpha^1, \ldots, \alpha^k$ *on a vector space* V *and* $v \in V$,

$$\iota_v(\alpha^1 \wedge \cdots \wedge \alpha^k) = \sum_{i=1}^{k} (-1)^{i-1} \alpha^i(v) \alpha^1 \wedge \cdots \wedge \widehat{\alpha^i} \wedge \cdots \wedge \alpha^k,$$

where the caret $\widehat{}$ *over* α^i *means that* α^i *is omitted from the wedge product.*

Proof.

$$\left(\iota_v\left(\alpha^1 \wedge \cdots \wedge \alpha^k\right)\right)(v_2, \ldots, v_k)$$

$$= \left(\alpha^1 \wedge \cdots \wedge \alpha^k\right)(v, v_2, \ldots, v_k)$$

$$= \det \begin{bmatrix} \alpha^1(v) & \alpha^1(v_2) & \cdots & \alpha^1(v_k) \\ \alpha^2(v) & \alpha^2(v_2) & \cdots & \alpha^2(v_k) \\ \vdots & \vdots & \ddots & \vdots \\ \alpha^k(v) & \alpha^k(v_2) & \cdots & \alpha^k(v_k) \end{bmatrix} \qquad \text{(Proposition 3.27)}$$

$$= \sum_{i=1}^{k} (-1)^{i+1} \alpha^i(v) \det[\alpha^\ell(v_j)]_{\substack{1 \leq \ell \leq k, \ell \neq i \\ 2 \leq j \leq k}} \quad \text{(expansion along first column)}$$

$$= \sum_{i=1}^{k} (-1)^{i+1} \alpha^i(v) \left(\alpha^1 \wedge \cdots \wedge \widehat{\alpha^i} \wedge \cdots \wedge \alpha^k\right)(v_2, \ldots, v_k) \quad \text{(Proposition 3.27)}.$$

$\square$

Proposition 20.8. *For* v *in a vector space* V, *let* $\iota_v \colon \bigwedge^*(V^\vee) \to \bigwedge^{*-1}(V^\vee)$ *be interior multiplication by* v. *Then*

(i) $\iota_v \circ \iota_v = 0$,
(ii) *for* $\beta \in \bigwedge^k(V^\vee)$ *and* $\gamma \in \bigwedge^\ell(V^\vee)$,

$$\iota_v(\beta \wedge \gamma) = (\iota_v \beta) \wedge \gamma + (-1)^k \beta \wedge \iota_v \gamma.$$

In other words, ι_v *is an antiderivation of degree* -1 *whose square is zero.*

Proof. (i) Let $\beta \in \bigwedge^k(V^\vee)$. By the definition of interior multiplication,

$$(\iota_v(\iota_v \beta))(v_3, \ldots, v_k) = (\iota_v \beta)(v, v_3, \ldots, v_k) = \beta(v, v, v_3, \ldots, v_k) = 0,$$

because β is alternating and there is a repeated variable v among its arguments.

(ii) Since both sides of the equation are linear in β and in γ, we may assume that

$$\beta = \alpha^1 \wedge \cdots \wedge \alpha^k, \quad \gamma = \alpha^{k+1} \wedge \cdots \wedge \alpha^{k+\ell},$$

where the α^i are all 1-covectors. Then

$$
\begin{aligned}
\iota_v(\beta \wedge \gamma) \\
&= \iota_v(\alpha^1 \wedge \cdots \wedge \alpha^{k+\ell}) \\
&= \left(\sum_{i=1}^{k} (-1)^{i-1} \alpha^i(v) \alpha^1 \wedge \cdots \wedge \widehat{\alpha^i} \wedge \cdots \wedge \alpha^k \right) \wedge \alpha^{k+1} \wedge \cdots \wedge \alpha^{k+\ell} \\
&\quad + (-1)^k \alpha^1 \wedge \cdots \wedge \alpha^k \wedge \sum_{i=1}^{k} (-1)^{i+1} \alpha^{k+i}(v) \alpha^{k+1} \wedge \cdots \wedge \widehat{\alpha^{k+i}} \wedge \cdots \wedge \alpha^{k+\ell}
\end{aligned}
$$

(by Proposition 20.7)

$$= (\iota_v \beta) \wedge \gamma + (-1)^k \beta \wedge \iota_v \gamma. \qquad \square$$

Interior multiplication on a manifold is defined pointwise. If X is a smooth vector field on M and $\omega \in \Omega^k(M)$, then $\iota_X \omega$ is the $(k-1)$-form defined by $(\iota_X \omega)_p = \iota_{X_p} \omega_p$ for all $p \in M$. The form $\iota_X \omega$ on M is smooth because for any smooth vector fields $X_2, \ldots, X_k$ on M,

$$(\iota_X \omega)(X_2, \ldots, X_k) = \omega(X, X_2, \ldots, X_k)$$

is a smooth function on M (Proposition 18.7(iii)$\Rightarrow$(i)). Of course, $\iota_X \omega = \omega(X)$ for a 1-form ω and $\iota_X f = 0$ for a function f on M. By the properties of interior multiplication at each point $p \in M$ (Proposition 20.8), the map $\iota_X : \Omega^*(M) \to \Omega^*(M)$ is an antiderivation of degree -1 such that $\iota_X \circ \iota_X = 0$.

Let $\mathcal{F}$ be the ring $C^\infty(M)$ of C^∞ functions on the manifold M. Because $\iota_X \omega$ is a point operator—that is, its value at p depends only on X_p and ω_p—it is $\mathcal{F}$-linear in either argument. This means that $\iota_X \omega$ is additive in each argument and moreover, for any $f \in \mathcal{F}$,

(i) $\iota_{fX} \omega = f \iota_X \omega$;
(ii) $\iota_X(f\omega) = f \iota_X \omega$.

Explicitly, the proof of (i) goes as follows. For any $p \in M$,

$$(\iota_{fX} \omega)_p = \iota_{f(p)X_p} \omega_p = f(p) \iota_{X_p} \omega_p = (f \iota_X \omega)_p.$$

Hence, $\iota_{fX} \omega = f \iota_X \omega$. The proof of (ii) is similar. Additivity is more or less obvious.

Example 20.9 (*Interior multiplication on* $\mathbb{R}^2$). Let $X = x\partial/\partial x + y\partial/\partial y$ be the radial vector field and $\alpha = dx \wedge dy$ the area 2-form on the plane $\mathbb{R}^2$. Compute the contraction $\iota_X \alpha$.

Solution. We first compute $\iota_X \, dx$ and $\iota_X \, dy$:

$$\iota_X dx = dx(X) = dx\left(x\frac{\partial}{\partial x}+y\frac{\partial}{\partial y}\right) = x,$$

$$\iota_X dy = dy(X) = dy\left(x\frac{\partial}{\partial x}+y\frac{\partial}{\partial y}\right) = y.$$

By the antiderivation property of ι_X,

$$\iota_X \alpha = \iota_X(dx \wedge dy) = (\iota_X dx)dy - dx(\iota_X dy) = x\,dy - y\,dx,$$

which restricts to the nowhere-vanishing 1-form ω on the circle S^1 in Example 17.15.

20.5 Properties of the Lie Derivative

In this section we state and prove several basic properties of the Lie derivative. We also relate the Lie derivative to two other intrinsic operators on differential forms on a manifold: the exterior derivative and interior multiplication. The interplay of these three operators results in some surprising formulas.

Theorem 20.10. *Assume X to be a C^∞ vector field on a manifold M.*

(i) *The Lie derivative $\mathcal{L}_X : \Omega^*(M) \to \Omega^*(M)$ is a derivation: it is an $\mathbb{R}$-linear map and if $\omega \in \Omega^k(M)$ and $\tau \in \Omega^\ell(M)$, then*

$$\mathcal{L}_X(\omega \wedge \tau) = (\mathcal{L}_X \omega) \wedge \tau + \omega \wedge (\mathcal{L}_X \tau).$$

(ii) *The Lie derivative $\mathcal{L}_X$ commutes with the exterior derivative d.*
(iii) *(Cartan homotopy formula) $\mathcal{L}_X = d\iota_X + \iota_X d$.*
(iv) *("Product" formula) For $\omega \in \Omega^k(M)$ and $Y_1,\ldots,Y_k \in \mathfrak{X}(M)$,*

$$\mathcal{L}_X(\omega(Y_1,\ldots,Y_k)) = (\mathcal{L}_X\omega)(Y_1,\ldots,Y_k) + \sum_{i=1}^{k} \omega(Y_1,\ldots,\mathcal{L}_X Y_i,\ldots,Y_k).$$

Proof. In the proof let $p \in M$ and let $\varphi_t : U \to M$ be a local flow of the vector field X in a neighborhood U of p.
(i) Since the Lie derivative $\mathcal{L}_X$ is d/dt of a vector-valued function of t, the derivation property of $\mathcal{L}_X$ is really just the product rule for d/dt (Proposition 20.1). More precisely,

$$
\begin{aligned}
(\mathcal{L}_X(\omega \wedge \tau))_p &= \frac{d}{dt}\bigg|_{t=0} (\varphi_t^*(\omega \wedge \tau))_p \\
&= \frac{d}{dt}\bigg|_{t=0} (\varphi_t^* \omega)_p \wedge (\varphi_t^* \tau)_p \\
&= \left(\frac{d}{dt}\bigg|_{t=0} (\varphi_t^* \omega)_p\right) \wedge \tau_p + \omega_p \wedge \frac{d}{dt}\bigg|_{t=0} (\varphi_t^* \tau)_p \\
&\qquad \text{(product rule for } d/dt) \\
&= (\mathcal{L}_X \omega)_p \wedge \tau_p + \omega_p \wedge (\mathcal{L}_X \tau)_p.
\end{aligned}
$$

(ii)

$$\mathcal{L}_X d\omega = \frac{d}{dt}\bigg|_{t=0} \varphi_t^* d\omega \qquad \text{(definition of } \mathcal{L}_X)$$

$$= \frac{d}{dt}\bigg|_{t=0} d\varphi_t^* \omega \qquad \text{(}d \text{ commutes with pullback)}$$

$$= d\left(\frac{d}{dt}\bigg|_{t=0} \varphi_t^* \omega\right) \quad \text{(by Proposition 20.2)}$$

$$= d\mathcal{L}_X \omega.$$

(iii) We make two observations that reduce the problem to a simple case. First, for any $\omega \in \Omega^k(M)$, to prove the equality $\mathcal{L}_X \omega = (d\iota_X + \iota_X d)\omega$ it suffices to check it at any point p, which is a local problem. In a coordinate neighborhood $(U, x^1, \ldots, x^n)$ about p, we may assume by linearity that ω is a wedge product $\omega = f\, dx^{i_1} \wedge \cdots \wedge dx^{i_k}$.

Second, on the left-hand side of the Cartan homotopy formula, by (i) and (ii), $\mathcal{L}_X$ is a derivation that commutes with d. On the right-hand side, since d and ι_X are antiderivations, $d\iota_X + \iota_X d$ is a derivation by Problem 4.7. It clearly commutes with d. Thus, both sides of the Cartan homotopy formula are derivations that commute with d. Consequently, if the formula holds for two differential forms ω and τ, then it holds for the wedge product $\omega \wedge \tau$ as well as for $d\omega$. These observations reduce the verification of (iii) to checking

$$\mathcal{L}_X f = (d\iota_X + \iota_X d)f \quad \text{for } f \in C^\infty(U).$$

This is quite easy:

$$(d\iota_X + \iota_X d)f = \iota_X df \qquad \text{(because } \iota_X f = 0)$$

$$= (df)(X) \qquad \text{(definition of } \iota_X)$$

$$= Xf = \mathcal{L}_X f \quad \text{(Proposition 20.6)}.$$

(iv) We call this the "product" formula, even though there is no product in $\omega(Y_1, \ldots, Y_k)$, because this formula can be best remembered as though the juxtaposition of symbols were a product. In fact, even its proof resembles that of the product formula in calculus. To illustrate this, consider the case $k = 2$. Let $\omega \in \Omega^2(M)$ and $X, Y, Z \in \mathfrak{X}(M)$. The proof looks forbidding, but the idea is quite simple. To compare the values of $\omega(Y, Z)$ at the two points $\varphi_t(p)$ and p, we subtract the value at p from the value at $\varphi_t(p)$. The trick is to add and subtract terms so that each time only one of the three variables ω, Y, and Z moves from one point to the other. By the definitions of the Lie derivative and the pullback of a function,

$$(\mathcal{L}_X(\omega(Y, Z)))_p = \lim_{t \to 0} \frac{(\varphi_t^*(\omega(Y, Z)))_p - (\omega(Y, Z))_p}{t}$$

$$= \lim_{t \to 0} \frac{\omega_{\varphi_t(p)}\left(Y_{\varphi_t(p)}, Z_{\varphi_t(p)}\right) - \omega_p(Y_p, Z_p)}{t}$$

$$= \lim_{t \to 0} \frac{\omega_{\varphi_t(p)}\left(Y_{\varphi_t(p)}, Z_{\varphi_t(p)}\right) - \omega_p\left(\varphi_{-t*}\left(Y_{\varphi_t(p)}\right), \varphi_{-t*}\left(Z_{\varphi_t(p)}\right)\right)}{t}$$

$$(20.9)$$

$$+ \lim_{t \to 0} \frac{\omega_p\left(\varphi_{-t*}\left(Y_{\varphi_t(p)}\right), \varphi_{-t*}\left(Z_{\varphi_t(p)}\right)\right) - \omega_p\left(Y_p, \varphi_{-t*}\left(Z_{\varphi_t(p)}\right)\right)}{t}$$

$$(20.10)$$

$$+ \lim_{t \to 0} \frac{\omega_p\left(Y_p, \varphi_{-t*}\left(Z_{\varphi_t(p)}\right)\right) - \omega_p(Y_p, Z_p)}{t}.$$

$$(20.11)$$

In this sum the quotient in the first limit (20.9) is

$$\frac{\left(\varphi_t^* \omega_{\varphi_t(p)}\right)\left(\varphi_{-t*}\left(Y_{\varphi_t(p)}\right), \varphi_{-t*}\left(Z_{\varphi_t(p)}\right)\right) - \omega_p\left(\varphi_{-t*}\left(Y_{\varphi_t(p)}\right), \varphi_{-t*}\left(Z_{\varphi_t(p)}\right)\right)}{t}$$

$$= \frac{\varphi_t^*\left(\omega_{\varphi_t(p)}\right) - \omega_p}{t} \left(\varphi_{-t*}\left(Y_{\varphi_t(p)}\right), \varphi_{-t*}\left(Z_{\varphi_t(p)}\right)\right).$$

On the right-hand side of this equality, the difference quotient has a limit at $t = 0$, namely the Lie derivative $(\mathcal{L}_X \omega)_p$, and by (20.6) the two arguments of the difference quotient are C^∞ functions of t. Therefore, the right-hand side is a continuous function of t and its limit as t goes to 0 is $(\mathcal{L}_X \omega)_p(Y_p, Z_p)$ (by Problem 20.2).

By the bilinearity of ω_p, the second term (20.10) is

$$\lim_{t \to 0} \omega_p \left(\frac{\varphi_{-t*}\left(Y_{\varphi_t(p)}\right) - Y_p}{t}, \varphi_{-t*}\left(Z_{\varphi_t(p)}\right) \right) = \omega_p((\mathcal{L}_X Y)_p, Z_p).$$

Similarly, the third term (20.11) is $\omega_p(Y_p, (\mathcal{L}_X Z)_p)$.

Thus

$$\mathcal{L}_X(\omega(Y, Z)) = (\mathcal{L}_X \omega)(Y, Z) + \omega(\mathcal{L}_X Y, Z) + \omega(Y, \mathcal{L}_X Z).$$

The general case is similar. □

Remark. Unlike interior multiplication, the Lie derivative $\mathcal{L}_X \omega$ is not $\mathcal{F}$-linear in either argument. By the derivation property of the Lie derivative (Theorem 20.10(i)),

$$\mathcal{L}_X(f\omega) = (\mathcal{L}_X f)\omega + f\mathcal{L}_X \omega = (Xf)\omega + f\mathcal{L}_X \omega.$$

We leave the problem of expanding $\mathcal{L}_{fX}\omega$ as an exercise (Problem 20.7).

Theorem 20.10 can be used to calculate the Lie derivative of a differential form.

Example 20.11 (*The Lie derivative on a circle*). Let ω be the 1-form $-y\,dx + x\,dy$ and let X be the tangent vector field $-y\partial/\partial x + x\partial/\partial y$ on the unit circle S^1 from Example 17.15. Compute the Lie derivative $\mathcal{L}_X \omega$.

Solution. By Proposition 20.6,

$$\mathcal{L}_X(x) = Xx = \left(-y\frac{\partial}{\partial x} + x\frac{\partial}{\partial y}\right)x = -y,$$

$$\mathcal{L}_X(y) = Xy = \left(-y\frac{\partial}{\partial x} + x\frac{\partial}{\partial y}\right)y = x.$$

Next we compute $\mathcal{L}_X(-y\,dx)$:

$$
\begin{aligned}
\mathcal{L}_X(-y\,dx) &= -(\mathcal{L}_X y)\,dx - y\mathcal{L}_X\,dx && (\mathcal{L}_X \text{ is a derivation})\\
&= -(\mathcal{L}_X y)\,dx - y\,d\mathcal{L}_X x && (\mathcal{L}_X \text{ commutes with } d)\\
&= -x\,dx + y\,dy.
\end{aligned}
$$

Similarly, $\mathcal{L}_X(x\,dy) = -y\,dy + x\,dx$. Hence, $\mathcal{L}_X \omega = \mathcal{L}_X(-y\,dx + x\,dy) = 0$.

20.6 Global Formulas for the Lie and Exterior Derivatives

The definition of the Lie derivative $\mathcal{L}_X \omega$ is local, since it makes sense only in a neighborhood of a point. The product formula in Theorem 20.10(iv), however, gives a global formula for the Lie derivative.

Theorem 20.12 (Global formula for the Lie derivative). *For a smooth k-form ω and smooth vector fields $X, Y_1, \ldots, Y_k$ on a manifold M,*

$$(\mathcal{L}_X \omega)(Y_1, \ldots, Y_k) = X(\omega(Y_1, \ldots, Y_k)) - \sum_{i=1}^{k} \omega(Y_1, \ldots, [X, Y_i], \ldots, Y_k).$$

Proof. In Theorem 20.10(iv), $\mathcal{L}_X(\omega(Y_1, \ldots, Y_k)) = X(\omega(Y_1, \ldots, Y_k))$ by Proposition 20.6 and $\mathcal{L}_X Y_i = [X, Y_i]$ by Theorem 20.4. $\qquad\square$

The definition of the exterior derivative d is also local. Using the Lie derivative, we obtain a very useful global formula for the exterior derivative. We first derive the formula for the exterior derivative of a 1-form, the case most useful in differential geometry.

Proposition 20.13. *If ω is a C^∞ 1-form and X and Y are C^∞ vector fields on a manifold M, then*

$$d\omega(X, Y) = X\omega(Y) - Y\omega(X) - \omega([X, Y]).$$

Proof. It is enough to check the formula in a chart $(U, x^1, \ldots, x^n)$, so we may assume $\omega = \sum a_i \, dx^i$. Since both sides of the equation are $\mathbb{R}$-linear in ω, we may further assume that $\omega = f \, dg$, where $f, g \in C^\infty(U)$.

In this case, $d\omega = d(f\,dg) = df \wedge dg$ and

$$
\begin{aligned}
d\omega(X, Y) &= df(X)\,dg(Y) - df(Y)\,dg(X) = (Xf)Yg - (Yf)Xg,\\
X\omega(Y) &= X(f\,dg(Y)) = X(fYg) = (Xf)Yg + fXYg,\\
Y\omega(X) &= Y(f\,dg(X)) = Y(fXg) = (Yf)Xg + fYXg,\\
\omega([X, Y]) &= f\,dg([X, Y]) = f(XY - YX)g.
\end{aligned}
$$

It follows that

$$X\omega(Y) - Y\omega(X) - \omega([X,Y]) = (Xf)Yg - (Yf)Xg = d\omega(X,Y). \qquad \square$$

Theorem 20.14 (Global formula for the exterior derivative). *Assume* $k \geq 1$. *For a smooth k-form* ω *and smooth vector fields* $Y_0, Y_1, \ldots, Y_k$ *on a manifold* M,

$$(d\omega)(Y_0, \ldots, Y_k) = \sum_{i=0}^{k} (-1)^i Y_i \omega(Y_0, \ldots, \widehat{Y_i}, \ldots, Y_k)$$

$$+ \sum_{0 \leq i < j \leq k} (-1)^{i+j} \omega([Y_i, Y_j], Y_0, \ldots, \widehat{Y_i}, \ldots, \widehat{Y_j}, \ldots, Y_k).$$

Proof. When $k = 1$, the formula is proven in Proposition 20.13.

Assuming the formula for forms of degree $k - 1$, we can prove it by induction for a form ω of degree k. By the definition of ι_{Y_0} and Cartan's homotopy formula (Theorem 20.10(iii)),

$$(d\omega)(Y_0, Y_1, \ldots, Y_k) = (\iota_{Y_0} d\omega)(Y_1, \ldots, Y_k)$$

$$= (\mathcal{L}_{Y_0} \omega)(Y_1, \ldots, Y_k) - (d\iota_{Y_0} \omega)(Y_1, \ldots, Y_k).$$

The first term of this expression can be computed using the global formula for the Lie derivative $\mathcal{L}_{Y_0}\omega$, while the second term can be computed using the global formula for d of a form of degree $k - 1$. This kind of verification is best done by readers on their own. We leave it as an exercise (Problem 20.6). $\qquad \square$

Problems

20.1. The limit of a family of vector fields
Let I be an open interval, M a manifold, and $\{X_t\}$ a 1-parameter family of vector fields on M defined for all $t \neq t_0 \in I$. Show that the definition of $\lim_{t \to t_0} X_t$ in (20.1), if the limit exists, is independent of coordinate charts.

20.2. Limits of families of vector fields and differential forms
Let I be an open interval containing 0. Suppose $\{\omega_t\}_{t \in I}$ and $\{Y_t\}_{t \in I}$ are 1-parameter families of 1-forms and vector fields respectively on a manifold M. Prove that if $\lim_{t \to 0} \omega_t = \omega_0$ and $\lim_{t \to 0} Y_t = Y_0$, then $\lim_{t \to 0} \omega_t(Y_t) = \omega_0(Y_0)$. (*Hint:* Expand in local coordinates.) By the same kind of argument, one can show that there is a similar formula for a family $\{\omega_t\}$ of 2-forms: $\lim_{t \to 0} \omega_t(Y_t, Z_t) = \omega_0(Y_0, Z_0)$.

20.3.* Derivative of a smooth family of vector fields
Show that the definition (20.3) of the derivative of a smooth family of vector fields on M is independent of the chart $(U, x^1, \ldots, x^n)$ containing p.

20.4. Product rule for d/dt
Prove that if $\{\omega_t\}$ and $\{\tau_t\}$ are smooth families of k-forms and ℓ-forms respectively on a manifold M, then

$$\frac{d}{dt}(\omega_t \wedge \tau_t) = \left(\frac{d}{dt}\omega_t\right) \wedge \tau_t + \omega_t \wedge \frac{d}{dt}\tau_t.$$

20.5. Smooth families of forms and vector fields

If $\{\omega_t\}_{t\in I}$ is a smooth family of 2-forms and $\{Y_t\}_{t\in I}$ and $\{Z_t\}_{t\in I}$ are smooth families of vector fields on a manifold M, prove that $\omega_t(X_t, Y_t)$ is a C^∞ function on $I \times M$.

20.6.* Global formula for the exterior derivative

Complete the proof of Theorem 20.14.

20.7. $\mathcal{F}$-Linearity and the Lie Derivative

Let ω be a differential form, X a vector field, and f a smooth function on a manifold. The Lie derivative $\mathcal{L}_X\omega$ is not $\mathcal{F}$-linear in either variable, but prove that it satisfies the following identity:

$$\mathcal{L}_{fX}\omega = f\mathcal{L}_X\omega + df \wedge \iota_X\omega.$$

(*Hint*: Start with Cartan's homotopy formula $\mathcal{L}_X = d\iota_X + \iota_X d$.)

20.8. Bracket of the Lie Derivative and Interior Multiplication

If X and Y are smooth vector fields on a manifold M, prove that on differential forms on M

$$\mathcal{L}_X\iota_Y - \iota_Y\mathcal{L}_X = \iota_{[X,Y]}.$$

(*Hint*: Let $\omega \in \Omega^k(M)$ and $Y, Y_1, \ldots, Y_{k-1} \in \mathfrak{X}(M)$. Apply the global formula for $\mathcal{L}_X$ to

$$(\iota_Y\mathcal{L}_X\,\omega)(Y_1, \ldots, Y_{k-1}) = (\mathcal{L}_X\,\omega)(Y, Y_1, \ldots, Y_{k-1}).)$$

20.9. Interior multiplication on $\mathbb{R}^n$

Let $\omega = dx^1 \wedge \cdots \wedge dx^n$ be the volume form and $X = \sum x^i \partial/\partial x^i$ the radial vector field on $\mathbb{R}^n$. Compute the contraction $\iota_X\omega$.

20.10. The Lie derivative on the 2-sphere

Let $\omega = x\,dy \wedge dz - y\,dx \wedge dy + z\,dx \wedge dy$ and $X = -y\partial/\partial x + x\partial/\partial y$ on the unit 2-sphere S^2 in $\mathbb{R}^3$. Compute the Lie derivative $\mathcal{L}_X\omega$.

Chapter 6

Integration

On a manifold one integrates not functions as in calculus on $\mathbb{R}^n$ but differential forms. There are actually two theories of integration on manifolds, one in which the integration is over a submanifold and the other in which the integration is over what is called a *singular chain*. Singular chains allow one to integrate over an object such as a closed rectangle in $\mathbb{R}^2$:

$$[a,b] \times [c,d] := \{(x,y) \in \mathbb{R}^2 \mid a \leq x \leq b, \, c \leq y \leq d\},$$

which is not a submanifold of $\mathbb{R}^2$ because of its corners.

For simplicity we will discuss only integration of smooth forms over a submanifold. For integration of noncontinuous forms over more general sets, the reader may consult the many excellent references in the bibliography, for example [3, Section VI.2], [7, Section 8.2], or [25, Chapter 14].

For integration over a manifold to be well defined, the manifold needs to be oriented. We begin the chapter with a discussion of orientations on a manifold. We then enlarge the category of manifolds to include manifolds with boundary. Our treatment of integration culminates in Stokes's theorem for an n-dimensional manifold. Stokes's theorem for a surface with boundary in $\mathbb{R}^3$ was first published as a question in the Smith's Prize Exam that Stokes set at the University of Cambridge in 1854. It is not known whether any student solved the problem. According to [21, p. 150], the same theorem had appeared four years earlier in a letter of Lord Kelvin to Stokes, which only goes to confirm that the attribution of credit in mathematics is fraught with pitfalls. Stokes's theorem for a general manifold resulted from the work of many mathematicians, including Vito Volterra (1889), Henri Poincaré (1899), Edouard Goursat (1917), and Élie Cartan (1899 and 1922). First there were many special cases, then a general statement in terms of coordinates, and finally a general statement in terms of differential forms. Cartan was the master of differential forms par excellence, and it was in his work that the differential form version of Stokes's theorem found its clearest expression.

L. W. Tu, *An Introduction to Manifolds,* Universitext, DOI 10.1007/978-1-4419-7400-6_6,

§21 Orientations

It is a familiar fact from vector calculus that line and surface integrals depend on the orientation of the curve or surface over which the integration takes place: reversing the orientation changes the sign of the integral. The goal of this section is to define orientation for n-dimensional manifolds and to investigate various equivalent characterizations of orientation.

We assume all vector spaces in this section to be finite-dimensional and real. An orientation of a finite-dimensional real vector space is simply an equivalence class of ordered bases, two ordered bases being equivalent if and only if their transition matrix has positive determinant. By its alternating nature, a multicovector of top degree turns out to represent perfectly an orientation of a vector space.

An orientation on a manifold is a choice of an orientation for each tangent space satisfying a continuity condition. Globalizing n-covectors over a manifold, we obtain differential n-forms. An orientation on an n-manifold can also be given by an equivalence class of C^∞ nowhere-vanishing n-forms, two such forms being equivalent if and only if one is a multiple of the other by a positive function. Finally, a third way to represent an orientation on a manifold is through an *oriented atlas*, an atlas in which any two overlapping charts are related by a transition function with everywhere positive Jacobian determinant.

21.1 Orientations of a Vector Space

On $\mathbb{R}^1$ an orientation is one of two directions (Figure 21.1).

Fig. 21.1. Orientations of a line.

On $\mathbb{R}^2$ an orientation is either counterclockwise or clockwise (Figure 21.2).

Fig. 21.2. Orientations of a plane.

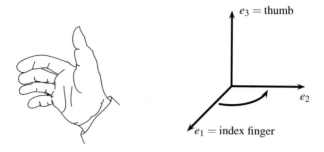

Fig. 21.3. Right-handed orientation (e_1, e_2, e_3) of $\mathbb{R}^3$.

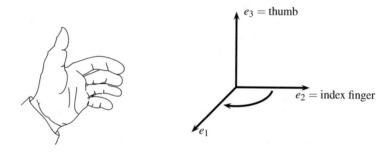

Fig. 21.4. Left-handed orientation (e_2, e_1, e_3) of $\mathbb{R}^3$.

On $\mathbb{R}^3$ an orientation is either right-handed (Figure 21.3) or left-handed (Figure 21.4). The right-handed orientation of $\mathbb{R}^3$ is the choice of a Cartesian coordinate system such that if you hold out your right hand with the index finger curling from the vector e_1 in the x-axis to the vector e_2 in the y-axis, then your thumb points in the direction of of the vector e_3 in the z-axis.

How should one define an orientation for $\mathbb{R}^4$, $\mathbb{R}^5$, and beyond? If we analyze the three examples above, we see that an orientation can be specified by an ordered basis for $\mathbb{R}^n$. Let $e_1, \ldots, e_n$ be the standard basis for $\mathbb{R}^n$. For $\mathbb{R}^1$ an orientation could be given by either e_1 or $-e_1$. For $\mathbb{R}^2$ the counterclockwise orientation is (e_1, e_2), while the clockwise orientation is (e_2, e_1). For $\mathbb{R}^3$ the right-handed orientation is (e_1, e_2, e_3), and the left-handed orientation is (e_2, e_1, e_3).

For any two ordered bases (u_1, u_2) and (v_1, v_2) for $\mathbb{R}^2$, there is a unique nonsingular 2×2 matrix $A = [a_j^i]$ such that

$$u_j = \sum_{i=1}^{2} v_i a_j^i, \quad j = 1, 2,$$

called the *change-of-basis matrix* from (v_1, v_2) to (u_1, u_2). In matrix notation, if we write ordered bases as row vectors, for example, $[u_1 \, u_2]$ for the basis (u_1, u_2), then

$$[u_1 \, u_2] = [v_1 \, v_2]A.$$

We say that two ordered bases are *equivalent* if the change-of-basis matrix A has positive determinant. It is easy to check that this is indeed an equivalence relation on the set of all ordered bases for $\mathbb{R}^2$. It therefore partitions ordered bases into two equivalence classes. Each equivalence class is called an *orientation* of $\mathbb{R}^2$. The equivalence class containing the ordered basis (e_1, e_2) is the counterclockwise orientation and the equivalence class of (e_2, e_1) is the clockwise orientation.

The general case is similar. We assume all vector spaces in this section to be finite-dimensional. Two ordered bases $u = [u_1 \, \cdots \, u_n]$ and $v = [v_1 \, \cdots \, v_n]$ of a vector space V are said to be *equivalent*, written $u \sim v$, if $u = vA$ for an $n \times n$ matrix A with positive determinant. An *orientation* of V is an equivalence class of ordered bases. Any finite-dimensional vector space has two orientations. If μ is an orientation of a finite-dimensional vector space V, we denote the other orientation by $-\mu$ and call it the *opposite* of the orientation μ.

The zero-dimensional vector space $\{0\}$ is a special case because it does not have a basis. We define an orientation on $\{0\}$ to be one of the two signs $+$ and $-$.

NOTATION. A basis for a vector space is normally written $v_1, \ldots, v_n$, without parentheses, brackets, or braces. If it is an *ordered* basis, then we enclose it in parentheses: $(v_1, \ldots, v_n)$. In matrix notation, we also write an ordered basis as a row vector $[v_1 \, \cdots \, v_n]$. An orientation is an equivalence class of ordered bases, so the notation is $[(v_1, \ldots, v_n)]$, where the brackets now stand for equivalence class.

21.2 Orientations and *n*-Covectors

Instead of using an ordered basis, we can also use an *n*-covector to specify an orientation of an *n*-dimensional vector space V. This approach to orientations is based on the fact that the space $\bigwedge^n(V^\vee)$ of *n*-covectors on V is one-dimensional.

Lemma 21.1. *Let $u_1, \ldots, u_n$ and $v_1, \ldots, v_n$ be vectors in a vector space V. Suppose*

$$u_j = \sum_{i=1}^n v_i a^i_j, \quad j = 1, \ldots, n,$$

for a matrix $A = [a^i_j]$ of real numbers. If β is an n-covector on V, then

$$\beta(u_1, \ldots, u_n) = (\det A)\, \beta(v_1, \ldots, v_n).$$

Proof. By hypothesis,

$$u_j = \sum_i v_i a^i_j.$$

Since β is *n*-linear,

$$\beta(u_1,\ldots,u_n) = \beta\left(\sum v_{i_1}a_1^{i_1},\ldots,\sum v_{i_n}a_n^{i_n}\right) = \sum a_1^{i_1}\cdots a_n^{i_n}\beta(v_{i_1},\ldots,v_{i_n}).$$

For $\beta(v_{i_1},\ldots,v_{i_n})$ to be nonzero, the subscripts $i_1,\ldots,i_n$ must be all distinct. An ordered n-tuple $I = (i_1,\ldots,i_n)$ with distinct components corresponds to a permutation σ_I of $1,\ldots,n$ with $\sigma_I(j) = i_j$ for $j = 1,\ldots,n$. Since β is an alternating n-tensor,

$$\beta(v_{i_1},\ldots,v_{i_n}) = (\operatorname{sgn}\sigma_I)\beta(v_1,\ldots,v_n).$$

Thus,

$$\beta(u_1,\ldots,u_n) = \sum_{\sigma_I \in S_n} (\operatorname{sgn}\sigma_I)a_1^{i_1}\cdots a_n^{i_n}\beta(v_1,\ldots,v_n) = (\det A)\beta(v_1,\ldots,v_n). \quad \square$$

As a corollary, if $u_1,\ldots,u_n$ and $v_1,\ldots,v_n$ are ordered bases of a vector space V, then

$$\beta(u_1,\ldots,u_n) \text{ and } \beta(v_1,\ldots,v_n) \text{ have the same sign}$$
$$\Longleftrightarrow \quad \det A > 0$$
$$\Longleftrightarrow \quad u_1,\ldots,u_n \text{ and } v_1,\ldots,v_n \text{ are equivalent ordered bases.}$$

We say that the n-covector β *determines* or *specifies* the orientation $(v_1,\ldots,v_n)$ if $\beta(v_1,\ldots,v_n) > 0$. By the preceding corollary, this is a well-defined notion, independent of the choice of ordered basis for the orientation. Moreover, two n-covectors β and β' on V determine the same orientation if and only if $\beta = a\beta'$ for some positive real number a. We define an equivalence relation on the nonzero n-covectors on the n-dimensional vector space V by setting

$$\beta \sim \beta' \quad \Longleftrightarrow \quad \beta = a\beta' \text{ for some } a > 0.$$

Thus, in addition to an equivalence class of ordered bases, an orientation of V is also given by an equivalence class of nonzero n-covectors on V.

A linear isomorphism $\bigwedge^n(V^\vee) \simeq \mathbb{R}$ identifies the set of nonzero n-covectors on V with $\mathbb{R} - \{0\}$, which has two connected components. Two nonzero n-covectors β and β' on V are in the same component if and only if $\beta = a\beta'$ for some real number $a > 0$. Thus, each connected component of $\bigwedge^n(V^\vee) - \{0\}$ determines an orientation of V.

Example. Let e_1,e_2 be the standard basis for $\mathbb{R}^2$ and α^1,α^2 its dual basis. Then the 2-covector $\alpha^1 \wedge \alpha^2$ determines the counterclockwise orientation of $\mathbb{R}^2$, since

$$(\alpha^1 \wedge \alpha^2)(e_1,e_2) = 1 > 0.$$

Example. Let $\partial/\partial x|_p, \partial/\partial y|_p$ be the standard basis for the tangent space $T_p(\mathbb{R}^2)$, and $(dx)_p,(dy)_p$ its dual basis. Then $(dx \wedge dy)_p$ determines the counterclockwise orientation of $T_p(\mathbb{R}^2)$.

21.3 Orientations on a Manifold

Recall that every vector space of dimension n has two orientations, corresponding to the two equivalence classes of ordered bases or the two equivalence classes of nonzero n-covectors. To orient a manifold M, we orient the tangent space at each point in M, but of course this has to be done in a "coherent" way so that the orientation does not change abruptly anywhere.

As we learned in Subsection 12.5, a *frame* on an open set $U \subset M$ is an n-tuple $(X_1, \ldots, X_n)$ of possibly discontinuous vector fields on U such that at every point $p \in U$, the n-tuple $(X_{1,p}, \ldots, X_{n,p})$ of vectors is an ordered basis for the tangent space T_pM. A *global frame* is a frame defined on the entire manifold M, while a *local frame* about $p \in M$ is a frame defined on some neighborhood of p. We introduce an equivalence relation on frames on U:

$$(X_1, \ldots, X_n) \sim (Y_1, \ldots, Y_n) \iff (X_{1,p}, \ldots, X_{n,p}) \sim (Y_{1,p}, \ldots, Y_{n,p}) \text{ for all } p \in U.$$

In other words, if $Y_j = \sum_i a_j^i X_i$, then two frames $(X_1, \ldots, X_n)$ and $(Y_1, \ldots, Y_n)$ are equivalent if and only if the change-of-basis matrix $A = [a_j^i]$ has positive determinant at every point in U.

A *pointwise orientation* on a manifold M assigns to each $p \in M$ an orientation μ_p of the tangent space T_pM. In terms of frames, a pointwise orientation on M is simply an equivalence class of possibly discontinuous frames on M. A pointwise orientation μ on M is said to be *continuous at* $p \in M$ if p has a neighborhood U on which μ is represented by a *continuous frame*; i.e., there exist continuous vector fields $Y_1, \ldots, Y_n$ on U such that $\mu_q = \left[(Y_{1,q}, \ldots, Y_{n,q}) \right]$ for all $q \in U$. The pointwise orientation μ is *continuous on* M if it is continuous at every point $p \in M$. Note that a continuous pointwise orientation need not be represented by a continuous global frame; it suffices that it be locally representable by a continuous local frame. A continuous pointwise orientation on M is called an *orientation* on M. A manifold is said to be *orientable* if it has an orientation. A manifold together with an orientation is said to be *oriented*.

Example. The Euclidean space $\mathbb{R}^n$ is orientable with orientation given by the continuous global frame $(\partial/\partial r^1, \ldots, \partial/\partial r^n)$.

Example 21.2 (*The open Möbius band*). Let R be the rectangle

$$R = \{(x, y) \in \mathbb{R}^2 \mid 0 \le x \le 1, \ -1 < y < 1\}.$$

The open Möbius band M (Figures 21.5 and 21.6) is the quotient of the rectangle R by the equivalence relation generated by

$$(0, y) \sim (1, -y). \tag{21.1}$$

The interior of R is the open rectangle

$$U = \{(x, y) \in \mathbb{R}^2 \mid 0 < x < 1, \ -1 < y < 1\}.$$

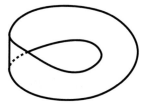

Fig. 21.5. Möbius band.

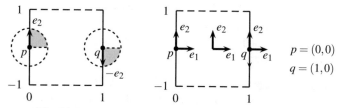

$$p = (0,0)$$
$$q = (1,0)$$

Fig. 21.6. Nonorientability of the Möbius band.

Suppose the Möbius band M is orientable. An orientation on M restricts to an orientation on U. To avoid confusion with an ordered pair of numbers, in this example we write an ordered basis without the parentheses. For the sake of definiteness, we first assume the orientation on U to be given by e_1, e_2. By continuity the orientations at the points $(0,0)$ and $(1,0)$ are also given by e_1, e_2. But under the identification (21.1), the ordered basis e_1, e_2 at $(1,0)$ maps to $e_1, -e_2$ at $(0,0)$. Thus, at $(0,0)$ the orientation has to be given by both e_1, e_2 and $e_1, -e_2$, a contradiction. Assuming the orientation on U to be given by e_2, e_1 also leads to a contradiction. This proves that the Möbius band is not orientable.

Proposition 21.3. *A connected orientable manifold M has exactly two orientations.*

Proof. Let μ and v be two orientations on M. At any point $p \in M$, μ_p and v_p are orientations of T_pM. They either are the same or are opposite orientations. Define a function $f: M \to \{\pm 1\}$ by

$$f(p) = \begin{cases} 1 & \text{if } \mu_p = v_p, \\ -1 & \text{if } \mu_p = -v_p. \end{cases}$$

Fix a point $p \in M$. By continuity, there exists a connected neighborhood U of p on which $\mu = [(X_1, \ldots, X_n)]$ and $v = [(Y_1, \ldots, Y_n)]$ for some continuous vector fields X_i and Y_j on U. Then there is a matrix-valued function $A = [a^i_j]: U \to \mathrm{GL}(n, \mathbb{R})$ such that $Y_j = \sum_i a^i_j X_i$. By Proposition 12.12 and Remark 12.13, the entries a^i_j are continuous, so that the determinant $\det A: U \to \mathbb{R}^\times$ is continuous also. By the intermediate value theorem, the continuous nowhere-vanishing function $\det A$ on the connected set U is everywhere positive or everywhere negative. Hence, $\mu = v$ or $\mu = -v$ on

U. This proves that the function $f: M \to \{\pm 1\}$ is locally constant. Since a locally constant function on a connected set is constant (Problem 21.1), $\mu = \nu$ or $\mu = -\nu$ on M.   □

21.4 Orientations and Differential Forms

While the definition of an orientation on a manifold as a continuous pointwise orientation is geometrically intuitive, in practice it is easier to manipulate the nowhere-vanishing top forms that specify a pointwise orientation. In this section we show that the continuity condition on pointwise orientations translates to a C^∞ condition on nowhere-vanishing top forms.

If f is a real-valued function on a set M, we use the notation $f > 0$ to mean that f is everywhere positive on M.

Lemma 21.4. *A pointwise orientation $[(X_1, \ldots, X_n)]$ on a manifold M is continuous if and only if each point $p \in M$ has a coordinate neighborhood $(U, x^1, \ldots, x^n)$ on which the function $(dx^1 \wedge \cdots \wedge dx^n)(X_1, \ldots, X_n)$ is everywhere positive.*

Proof.
($\Rightarrow$) Assume that the pointwise orientation $\mu = [(X_1, \ldots, X_n)]$ on M is continuous. This does not mean that the global frame $(X_1, \ldots, X_n)$ is continuous. What it means is that every point $p \in M$ has a neighborhood W on which μ is represented by a continuous frame $(Y_1, \ldots, Y_n)$. Choose a connected coordinate neighborhood $(U, x^1, \ldots, x^n)$ of p contained in W and let $\partial_i = \partial/\partial x^i$. Then $Y_j = \sum_i b^i_j \partial_i$ for a continuous matrix function $[b^i_j]: U \to \mathrm{GL}(n, \mathbb{R})$, the change-of-basis matrix at each point. By Lemma 21.1,

$$\left(dx^1 \wedge \cdots \wedge dx^n\right)(Y_1, \ldots, Y_n) = \left(\det[b^i_j]\right)\left(dx^1 \wedge \cdots \wedge dx^n\right)(\partial_1, \ldots, \partial_n) = \det[b^i_j],$$

which is never zero, because $[b^i_j]$ is nonsingular. As a continuous nowhere-vanishing real-valued function on a connected set, $(dx^1 \wedge \cdots \wedge dx^n)(Y_1, \ldots, Y_n)$ is everywhere positive or everywhere negative on U. If it is negative, then by setting $\tilde{x}^1 = -x^1$, we have on the chart $(U, \tilde{x}^1, x^2, \ldots, x^n)$ that

$$\left(d\tilde{x}^1 \wedge dx^2 \wedge \cdots \wedge dx^n\right)(Y_1, \ldots, Y_n) > 0.$$

Renaming $\tilde{x}^1$ as x^1, we may assume that on the coordinate neighborhood $(U, x^1, \ldots, x^n)$ of p, the function $(dx^1 \wedge \cdots \wedge dx^n)(Y_1, \ldots, Y_n)$ is always positive.

Since $\mu = [(X_1, \ldots, X_n)] = [(Y_1, \ldots, Y_n)]$ on U, the change-of-basis matrix $C = [c^i_j]$ such that $X_j = \sum_i c^i_j Y_i$ has positive determinant. By Lemma 21.1 again, on U,

$$\left(dx^1 \wedge \cdots \wedge dx^n\right)(X_1, \ldots, X_n) = (\det C)\left(dx^1 \wedge \cdots \wedge dx^n\right)(Y_1, \ldots, Y_n) > 0.$$

($\Leftarrow$) On the chart $(U, x^1, \ldots, x^n)$, suppose $X_j = \sum a^i_j \partial_i$. As before,

$$\left(dx^1 \wedge \cdots \wedge dx^n\right)(X_1, \ldots, X_n) = \left(\det[a^i_j]\right)\left(dx^1 \wedge \cdots \wedge dx^n\right)(\partial_1, \ldots, \partial_n) = \det[a^i_j].$$

By hypothesis, the left-hand side of the equalities above is positive. Therefore, on U, $\det[a^i_j] > 0$ and $[(X_1,\dots,X_n)] = [(\partial_1,\dots,\partial_n)]$, which proves that the pointwise orientation μ is continuous at p. Since p was arbitrary, μ is continuous on M. $\square$

Theorem 21.5. *A manifold M of dimension n is orientable if and only if there exists a C^∞ nowhere-vanishing n-form on M.*

Proof.
($\Rightarrow$) Suppose $[(X_1,\dots,X_n)]$ is an orientation on M. By Lemma 21.4, each point p has a coordinate neighborhood $(U,x^1,\dots,x^n)$ on which

$$\left(dx^1 \wedge \cdots \wedge dx^n\right)(X_1,\dots,X_n) > 0. \tag{21.2}$$

Let $\{(U_\alpha, x^1_\alpha, \dots, x^n_\alpha)\}$ be a collection of these charts that covers M, and let $\{\rho_\alpha\}$ be a C^∞ partition of unity subordinate to the open cover $\{U_\alpha\}$. Being a locally finite sum, the n-form $\omega = \sum_\alpha \rho_\alpha dx^1_\alpha \wedge \cdots \wedge dx^n_\alpha$ is well defined and C^∞ on M. Fix $p \in M$. Since $\rho_\alpha(p) \geq 0$ for all α and $\rho_\alpha(p) > 0$ for at least one α, by (21.2),

$$\omega_p(X_{1,p},\dots,X_{n,p}) = \sum_\alpha \rho_\alpha(p)\left(dx^1_\alpha \wedge \cdots \wedge dx^n_\alpha\right)_p (X_{1,p},\dots,X_{n,p}) > 0.$$

Therefore, ω is a C^∞ nowhere-vanishing n-form on M.
($\Leftarrow$) Suppose ω is a C^∞ nowhere-vanishing n-form on M. At each point $p \in M$, choose an ordered basis $(X_{1,p},\dots,X_{n,p})$ for T_pM such that $\omega_p(X_{1,p},\dots,X_{n,p}) > 0$. Fix $p \in M$ and let $(U,x^1,\dots,x^n)$ be a connected coordinate neighborhood of p. On U, $\omega = f\, dx^1 \wedge \cdots \wedge dx^n$ for a C^∞ nowhere-vanishing function f. Being continuous and nowhere vanishing on a connected set, f is everywhere positive or everywhere negative on U. If $f > 0$, then on the chart $(U,x^1,\dots,x^n)$,

$$\left(dx^1 \wedge \cdots \wedge dx^n\right)(X_1,\dots,X_n) > 0.$$

If $f < 0$, then on the chart $(U,-x^1,x^2,\dots,x^n)$,

$$\left(d(-x^1) \wedge dx^2 \wedge \cdots \wedge dx^n\right)(X_1,\dots,X_n) > 0.$$

In either case, by Lemma 21.4, $\mu = [(X_1,\dots,X_n)]$ is a continuous pointwise orientation on M. $\square$

Example 21.6 (*Orientability of a regular zero set*). By the regular level set theorem, if 0 is a regular value of a C^∞ function $f(x,y,z)$ on $\mathbb{R}^3$, then the zero set $f^{-1}(0)$ is a C^∞ manifold. In Problem 19.11 we constructed a nowhere-vanishing 2-form on the regular zero set of a C^∞ function. It then follows from Theorem 21.5 that the regular zero set of a C^∞ function on $\mathbb{R}^3$ is orientable.

As an example, the unit sphere S^2 in $\mathbb{R}^3$ is orientable. As another example, since an open Möbius band is not orientable (Example 21.2), it cannot be realized as the regular zero set of a C^∞ function on $\mathbb{R}^3$.

According to a classical theorem from algebraic topology, a continuous vector field on an even-dimensional sphere must vanish somewhere [18, Theorem 2.28, p. 135]. Thus, although the sphere S^2 has a continuous pointwise orientation, any global frame (X_1,X_2) that represents the orientation is necessarily discontinuous.

If ω and ω' are two nowhere-vanishing C^∞ n-forms on a manifold M of dimension n, then $\omega = f\omega'$ for some nowhere-vanishing function f on M. Locally, on a chart $(U, x^1, \ldots, x^n)$, $\omega = h\,dx^1 \wedge \cdots \wedge dx^n$ and $\omega' = g\,dx^1 \wedge \cdots \wedge dx^n$, where h and g are C^∞ nowhere-vanishing functions on U. Therefore, $f = h/g$ is also a C^∞ nowhere-vanishing function on U. Since U is an arbitrary chart, f is C^∞ and nowhere vanishing on M. On a *connected* manifold M, such a function f is either everywhere positive or everywhere negative. In this way the nowhere-vanishing C^∞ n-forms on a connected orientable manifold M are partitioned into two equivalence classes by the equivalence relation

$$\omega \sim \omega' \iff \omega = f\omega' \text{ with } f > 0.$$

To each orientation $\mu = [(X_1, \ldots, X_n)]$ on a connected orientable manifold M, we associate the equivalence class of a C^∞ nowhere-vanishing n-form ω on M such that $\omega(X_1, \ldots, X_n) > 0$. (Such an ω exists by the proof of Theorem 21.5.) If $\mu \mapsto [\omega]$, then $-\mu \mapsto [-\omega]$. On a connected orientable manifold, this sets up a one-to-one correspondence

$$\{\text{orientations on } M\} \quad \longleftrightarrow \quad \left\{ \begin{array}{l} \text{equivalence classes of} \\ C^\infty \text{ nowhere-vanishing} \\ n\text{-forms on } M \end{array} \right\}, \qquad (21.3)$$

each side being a set of two elements. By considering one connected component at a time, we see that the bijection (21.3) still holds for an arbitrary orientable manifold, each connected component having two possible orientations and two equivalence classes of C^∞ nowhere-vanishing n-forms. If ω is a C^∞ nowhere-vanishing n-form such that $\omega(X_1, \ldots, X_n) > 0$, we say that ω *determines* or *specifies* the orientation $[(X_1, \ldots, X_n)]$ and we call ω an *orientation form* on M. An oriented manifold can be described by a pair $(M, [\omega])$, where $[\omega]$ is the equivalence class of an orientation form on M. We sometimes write M, instead of $(M, [\omega])$, for an oriented manifold if it is clear from the context what the orientation is. For example, unless otherwise specified, $\mathbb{R}^n$ is oriented by $dx^1 \wedge \cdots \wedge dx^n$.

Remark 21.7 (*Orientations on zero-dimensional manifolds*). A connected manifold of dimension 0 is a point. The equivalence class of a nowhere-vanishing 0-form on a point is either $[-1]$ or $[1]$. Hence, a connected zero-dimensional manifold is always orientable. Its two orientations are specified by the two numbers ± 1. A general zero-dimensional manifold M is a countable discrete set of points (Example 5.13), and an orientation on M is given by a function that assigns to each point of M either 1 or -1.

A diffeomorphism $F : (N, [\omega_N]) \to (M, [\omega_M])$ of oriented manifolds is said to be *orientation-preserving* if $[F^*\omega_M] = [\omega_N]$; it is *orientation-reversing* if $[F^*\omega_M] = [-\omega_N]$.

Proposition 21.8. *Let U and V be open subsets of $\mathbb{R}^n$, both with the standard orientation inherited from $\mathbb{R}^n$. A diffeomorphism $F : U \to V$ is orientation-preserving if and only if the Jacobian determinant $\det[\partial F^i / \partial x^j]$ is everywhere positive on U.*

Proof. Let $x^1,\ldots,x^n$ and $y^1,\ldots,y^n$ be the standard coordinates on $U \subset \mathbb{R}^n$ and $V \subset \mathbb{R}^n$. Then

$$
\begin{aligned}
F^*(dy^1 \wedge \cdots \wedge dy^n) &= d(F^* y^1) \wedge \cdots \wedge d(F^* y^n) \quad \text{(Propositions 18.11 and 19.5)} \\
&= d(y^1 \circ F) \wedge \cdots \wedge d(y^n \circ F) \quad \text{(definition of pullback)} \\
&= dF^1 \wedge \cdots \wedge dF^n \\
&= \det\left[\frac{\partial F^i}{\partial x^j}\right] dx^1 \wedge \cdots \wedge dx^n \quad \text{(by Corollary 18.4(ii))}.
\end{aligned}
$$

Thus, F is orientation-preserving if and only if $\det[\partial F^i / \partial x^j]$ is everywhere positive on U. $\qquad\square$

21.5 Orientations and Atlases

Using the characterization of an orientation-preserving diffeomorphism by the sign of its Jacobian determinant, we can describe orientability of manifolds in terms of atlases.

Definition 21.9. An atlas on M is said to be *oriented* if for any two overlapping charts $(U,x^1,\ldots,x^n)$ and $(V,y^1,\ldots,y^n)$ of the atlas, the Jacobian determinant $\det[\partial y^i / \partial x^j]$ is everywhere positive on $U \cap V$.

Theorem 21.10. *A manifold M is orientable if and only if it has an oriented atlas.*

Proof.
($\Rightarrow$) Let $\mu = [(X_1,\ldots,X_n)]$ be an orientation on the manifold M. By Lemma 21.4, each point $p \in M$ has a coordinate neighborhood $(U,x^1,\ldots,x^n)$ on which

$$
\left(dx^1 \wedge \cdots \wedge dx^n\right)(X_1,\ldots,X_n) > 0.
$$

We claim that the collection $\mathfrak{U} = \{(U,x^1,\ldots,x^n)\}$ of these charts is an oriented atlas.

If $(U,x^1,\ldots,x^n)$ and $(V,y^1,\ldots,y^n)$ are two overlapping charts from $\mathfrak{U}$, then on $U \cap V$,

$$
\left(dx^1 \wedge \cdots \wedge dx^n\right)(X_1,\ldots,X_n) > 0 \quad \text{and} \quad \left(dy^1 \wedge \cdots \wedge dy^n\right)(X_1,\ldots,X_n) > 0. \quad (21.4)
$$

Since $dy^1 \wedge \cdots \wedge dy^n = \left(\det[\partial y^i / \partial x^j]\right) dx^1 \wedge \cdots \wedge dx^n$, it follows from (21.4) that $\det[\partial y^i / \partial x^j] > 0$ on $U \cap V$. Therefore, $\mathfrak{U}$ is an oriented atlas.

($\Rightarrow$) Suppose $\{(U,x^1,\ldots,x^n)\}$ is an oriented atlas. For each $p \in (U,x^1,\ldots,x^n)$, define μ_p to be the equivalence class of the ordered basis $(\partial/\partial x^1|_p,\ldots,\partial/\partial x^n|_p)$ for T_pM. If two charts $(U,x^1,\ldots,x^n)$ and $(V,y^1,\ldots,y^n)$ in the oriented atlas contain p, then by the orientability of the atlas, $\det[\partial y^i / \partial x^j] > 0$, so that $(\partial/\partial x^1|_p,\ldots,\partial/\partial x^n|_p)$ is equivalent to $(\partial/\partial y^1|_p,\ldots,\partial/\partial y^n|_p)$. This proves that μ is a well-defined pointwise orientation on M. It is continuous because every point p has a coordinate neighborhood $(U,x^1,\ldots,x^n)$ on which $\mu = [(\partial/\partial x^1,\ldots,\partial/\partial x^n)]$ is represented by a continuous frame. $\qquad\square$

Definition 21.11. Two oriented atlases $\{(U_\alpha, \phi_\alpha)\}$ and $\{(V_\beta, \psi_\beta)\}$ on a manifold M are said to be *equivalent* if the transition functions

$$\phi_\alpha \circ \psi_\beta^{-1} : \psi_\beta(U_\alpha \cap V_\beta) \to \phi_\alpha(U_\alpha \cap V_\beta)$$

have positive Jacobian determinant for all α, β.

It is not difficult to show that this is an equivalence relation on the set of oriented atlases on a manifold M (Problem 21.3).

In the proof of Theorem 21.10, an oriented atlas $\{(U, x^1, \ldots, x^n)\}$ on a manifold M determines an orientation $U \ni p \mapsto [(\partial/\partial x^1|_p, \ldots, \partial/\partial x^n|_p)]$ on M, and conversely, an orientation $[(X_1, \ldots, X_n)]$ on M gives rise to an oriented atlas $\{(U, x^1, \ldots, x^n)\}$ on M such that $(dx^1 \wedge \cdots \wedge dx^n)(X_1, \ldots, X_n) > 0$ on U. We leave it as an exercise to show that for an orientable manifold M, the two induced maps

$$\left\{ \begin{array}{l} \text{equivalence classes of} \\ \text{oriented atlases on } M \end{array} \right\} \quad \underset{\longleftarrow}{\longrightarrow} \quad \{\text{orientations on } M\}$$

are well defined and inverse to each other. Therefore, one can also specify an orientation on an orientable manifold by an equivalence class of oriented atlases.

For an oriented manifold M, we denote by $-M$ the same manifold but with the opposite orientation. If $\{(U, \phi)\} = \{(U, x^1, x^2, \ldots, x^n)\}$ is an oriented atlas specifying the orientation of M, then an oriented atlas specifying the orientation of $-M$ is $\{(U, \tilde{\phi})\} = \{(U, -x^1, x^2, \ldots, x^n)\}$.

Problems

21.1.* Locally constant map on a connected space
A map $f : S \to Y$ between two topological spaces is *locally constant* if for every $p \in S$ there is a neighborhood U of p such that f is constant on U. Show that a locally constant map $f : S \to Y$ on a nonempty connected space S is constant. (*Hint:* Show that for every $y \in Y$, the inverse image $f^{-1}(y)$ is open. Then $S = \bigcup_{y \in Y} f^{-1}(y)$ exhibits S as a disjoint union of open subsets.)

21.2. Continuity of pointwise orientations
Prove that a pointwise orientation $[(X_1, \ldots, X_n)]$ on a manifold M is continuous if and only if every point $p \in M$ has a coordinate neighborhood $(U, \phi) = (U, x^1, \ldots, x^n)$ such that for all $q \in U$, the differential $\phi_{*,q} : T_q M \to T_{f(q)} \mathbb{R}^n \simeq \mathbb{R}^n$ carries the orientation of $T_q M$ to the standard orientation of $\mathbb{R}^n$ in the following sense: $(\phi_* X_{1,q}, \ldots, \phi_* X_{n,q}) \sim (\partial/\partial r^1, \ldots, \partial/\partial r^n)$.

21.3. Equivalence of oriented atlases
Show that the relation in Definition 21.11 is an equivalence relation.

21.4. Orientation-preserving diffeomorphisms
Let $F : (N, [\omega_N]) \to (M, [\omega_M])$ be an orientation-preserving diffeomorphism. If $\{(V_\alpha, \psi_\alpha)\} = \{(V_\alpha, y_\alpha^1, \ldots, y_\alpha^n)\}$ is an oriented atlas on M that specifies the orientation of M, show that $\{(F^{-1}V_\alpha, F^*\psi_\alpha)\} = \{(F^{-1}V_\alpha, F_\alpha^1, \ldots, F_\alpha^n)\}$ is an oriented atlas on N that specifies the orientation of N, where $F_\alpha^i = y_\alpha^i \circ F$.

21.5. Orientation-preserving or orientation-reversing diffeomorphisms

Let U be the open set $(0,\infty) \times (0,2\pi)$ in the (r,θ)-plane $\mathbb{R}^2$. We define $F: U \subset \mathbb{R}^2 \to \mathbb{R}^2$ by $F(r,\theta) = (r\cos\theta, r\sin\theta)$. Decide whether F is orientation-preserving or orientation-reversing as a diffeomorphism onto its image.

21.6. Orientability of a regular level set in $\mathbb{R}^{n+1}$

Suppose $f(x^1,\ldots,x^{n+1})$ is a C^∞ function on $\mathbb{R}^{n+1}$ with 0 as a regular value. Show that the zero set of f is an orientable submanifold of $\mathbb{R}^{n+1}$. In particular, the unit n-sphere S^n in $\mathbb{R}^{n+1}$ is orientable.

21.7. Orientability of a Lie group

Show that every Lie group G is orientable by constructing a nowhere-vanishing top form on G.

21.8. Orientability of a parallelizable manifold

Show that a parallelizable manifold is orientable. (In particular, this shows again that every Lie group is orientable.)

21.9. Orientability of the total space of the tangent bundle

Let M be a smooth manifold and $\pi: TM \to M$ its tangent bundle. Show that if $\{(U,\phi)\}$ is any atlas on M, then the atlas $\{(TU,\tilde\phi)\}$ on TM, with $\tilde\phi$ defined in equation (12.1), is oriented. This proves that the total space TM of the tangent bundle is always orientable, regardless of whether M is orientable.

21.10. Oriented atlas on a circle

In Example 5.16 we found an atlas $\mathfrak{U} = \{(U_i,\phi_i)\}_{i=1}^4$ on the unit circle S^1. Is $\mathfrak{U}$ an oriented atlas? If not, alter the coordinate functions ϕ_i to make $\mathfrak{U}$ into an oriented atlas.

§22 Manifolds with Boundary

The prototype of a manifold with boundary is the *closed upper half-space*

$$\mathcal{H}^n = \{(x^1,\ldots,x^n) \in \mathbb{R}^n \mid x^n \geq 0\},$$

with the subspace topology inherited from $\mathbb{R}^n$. The points $(x^1,\ldots,x^n)$ in $\mathcal{H}^n$ with $x^n > 0$ are called the *interior points* of $\mathcal{H}^n$, and the points with $x^n = 0$ are called the *boundary points* of $\mathcal{H}^n$. These two sets are denoted by $(\mathcal{H}^n)^\circ$ and $\partial(\mathcal{H}^n)$, respectively (Figure 22.1).

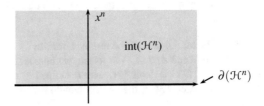

Fig. 22.1. Upper half-space.

In the literature the upper half-space often means the open set

$$\{(x^1,\ldots,x^n) \in \mathbb{R}^n \mid x^n > 0\}.$$

We require that $\mathcal{H}^n$ include the boundary in order for it to serve as a model for manifolds with boundary.

If M is a manifold with boundary, then its boundary ∂M turns out to be a manifold of dimension one less without boundary. Moreover, an orientation on M induces an orientation on ∂M. The choice of the induced orientation on the boundary is a matter of convention, guided by the desire to make Stokes's theorem sign-free. Of the various ways to describe the boundary orientation, two stand out for their simplicity: (1) contraction of an orientation form on M with an outward-pointing vector field on ∂M and (2) "outward vector first."

22.1 Smooth Invariance of Domain in $\mathbb{R}^n$

To discuss C^∞ functions on a manifold with boundary, we need to extend the definition of a C^∞ function to allow nonopen domains.

Definition 22.1. Let $S \subset \mathbb{R}^n$ be an arbitrary subset. A function $f \colon S \to \mathbb{R}^m$ is *smooth at a point p in S* if there exist a neighborhood U of p in $\mathbb{R}^n$ and a C^∞ function $\tilde{f} \colon U \to \mathbb{R}^m$ such that $\tilde{f} = f$ on $U \cap S$. The function is *smooth on S* if it is smooth at each point of S.

With this definition it now makes sense to speak of an arbitrary subset $S \subset \mathbb{R}^n$ being diffeomorphic to an arbitrary subset $T \subset \mathbb{R}^m$; this will be the case if and only if there are smooth maps $f \colon S \to T \subset \mathbb{R}^m$ and $g \colon T \to S \subset \mathbb{R}^n$ that are inverse to each other.

Exercise 22.2 (Smooth functions on a nonopen set).* Using a partition of unity, show that a function $f \colon S \to \mathbb{R}^m$ is C^∞ on $S \subset \mathbb{R}^n$ if and only if there exist an open set U in $\mathbb{R}^n$ containing S and a C^∞ function $\tilde{f} \colon U \to \mathbb{R}^m$ such that $f = \tilde{f}|_S$.

The following theorem is the C^∞ analogue of a classical theorem in the continuous category. We will use it to show that interior points and boundary points are invariant under diffeomorphisms of open subsets of $\mathcal{H}^n$.

Theorem 22.3 (Smooth invariance of domain). *Let $U \subset \mathbb{R}^n$ be an open subset, $S \subset \mathbb{R}^n$ an arbitrary subset, and $f \colon U \to S$ a diffeomorphism. Then S is open in $\mathbb{R}^n$.*

More succinctly, a diffeomorphism between an open subset U of $\mathbb{R}^n$ and an arbitrary subset S of $\mathbb{R}^n$ forces S to be open in $\mathbb{R}^n$. The theorem is not automatic. A diffeomorphism $f \colon \mathbb{R}^n \supset U \to S \subset \mathbb{R}^n$ takes an open subset of U to an open subset of S. Thus, a priori we know only that $f(U)$ is open in S, not that $f(U)$, which is S, is open in $\mathbb{R}^n$. It is crucial that the two Euclidean spaces be of the same dimension. For example, there are a diffeomorphism between the open interval $]0,1[$ in $\mathbb{R}^1$ and the open segment $S =]0,1[\times \{0\}$ in $\mathbb{R}^2$, but S is not open in $\mathbb{R}^2$.

Proof. Let $f(p)$ be an arbitrary point in S, with $p \in U$. Since $f \colon U \to S$ is a diffeomorphism, there are an open set $V \subset \mathbb{R}^n$ containing S and a C^∞ map $g \colon V \to \mathbb{R}^n$ such that $g|_S = f^{-1}$. Thus,

$$U \xrightarrow{f} V \xrightarrow{g} \mathbb{R}^n$$

satisfies

$$g \circ f = \mathbb{1}_U \colon U \to U \subset \mathbb{R}^n,$$

the identity map on U. By the chain rule,

$$g_{*,f(p)} \circ f_{*,p} = \mathbb{1}_{T_pU} \colon T_pU \to T_pU \simeq T_p(\mathbb{R}^n),$$

the identity map on the tangent space T_pU. Hence, $f_{*,p}$ is injective. Since U and V have the same dimension, $f_{*,p} \colon T_pU \to T_{f(p)}V$ is invertible. By the inverse function theorem, f is locally invertible at p. This means that there are open neighborhoods U_p of p in U and $V_{f(p)}$ of $f(p)$ in V such that $f \colon U_p \to V_{f(p)}$ is a diffeomorphism. It follows that

$$f(p) \in V_{f(p)} = f(U_p) \subset f(U) = S.$$

Since V is open in $\mathbb{R}^n$ and $V_{f(p)}$ is open in V, the set $V_{f(p)}$ is open in $\mathbb{R}^n$. By the local criterion for openness (Lemma A.2), S is open in $\mathbb{R}^n$. $\quad\square$

Proposition 22.4. *Let U and V be open subsets of the upper half-space $\mathcal{H}^n$ and $f \colon U \to V$ a diffeomorphism. Then f maps interior points to interior points and boundary points to boundary points.*

Proof. Let $p \in U$ be an interior point. Then p is contained in an open ball B, which is actually open in $\mathbb{R}^n$ (not just in $\mathcal{H}^n$). By smooth invariance of domain, $f(B)$ is open in $\mathbb{R}^n$ (again not just in $\mathcal{H}^n$). Therefore, $f(B) \subset (\mathcal{H}^n)^\circ$. Since $f(p) \in f(B)$, $f(p)$ is an interior point of $\mathcal{H}^n$.

If p is a boundary point in $U \cap \partial\mathcal{H}^n$, then $f^{-1}(f(p)) = p$ is a boundary point. Since $f^{-1}: V \to U$ is a diffeomorphism, by what has just been proven, $f(p)$ cannot be an interior point. Thus, $f(p)$ is a boundary point. □

Remark 22.5. Replacing Euclidean spaces by manifolds throughout this subsection, one can prove in exactly the same way smooth invariance of domain for manifolds: if there is a diffeomorphism between an open subset U of an n-dimensional manifold N and an arbitrary subset S of another n-dimensional manifold M, then S is open in M.

22.2 Manifolds with Boundary

In the upper half-space $\mathcal{H}^n$ one may distinguish two kinds of open subsets, depending on whether the set is disjoint from the boundary or intersects the boundary (Figure 22.2). Charts on a manifold are homeomorphic to only the first kind of open sets.

Fig. 22.2. Two types of open subsets of $\mathcal{H}^n$.

A manifold with boundary generalizes the definition of a manifold by allowing both kinds of open sets. We say that a topological space M is *locally* $\mathcal{H}^n$ if every point $p \in M$ has a neighborhood U homeomorphic to an open subset of $\mathcal{H}^n$.

Definition 22.6.   A *topological n-manifold with boundary* is a second countable, Hausdorff topological space that is locally $\mathcal{H}^n$.

Let M be a topological n-manifold with boundary. For $n \geq 2$, a *chart* on M is defined to be a pair (U, ϕ) consisting of an open set U in M and a homeomorphism

$$\phi: U \to \phi(U) \subset \mathcal{H}^n$$

of U with an open subset $\phi(U)$ of $\mathcal{H}^n$. As Example 22.9 (p. 254) will show, a slight modification is necessary when $n = 1$: we need to allow two local models, the *right half-line* $\mathcal{H}^1$ and the *left half-line*

$$\mathcal{L}^1 := \{x \in \mathbb{R} \mid x \leq 0\}.$$

A chart (U, ϕ) in dimension 1 consists of an open set U in M and a homeomorphism ϕ of U with an open subset of $\mathcal{H}^1$ or $\mathcal{L}^1$. With this convention, if $(U, x^1, x^2, \ldots, x^n)$ is a chart of an n-dimensional manifold with boundary, then so is $(U, -x^1, x^2, \ldots, x^n)$ for any $n \geq 1$. A manifold with boundary has dimension at least 1, since a manifold of dimension 0, being a discrete set of points, necessarily has empty boundary.

A collection $\{(U, \phi)\}$ of charts is a C^∞ *atlas* if for any two charts (U, ϕ) and (V, ψ), the transition map

$$\psi \circ \phi^{-1} : \phi(U \cap V) \to \psi(U \cap V) \subset \mathcal{H}^n$$

is a diffeomorphism. A C^∞ *manifold with boundary* is a topological manifold with boundary together with a maximal C^∞ atlas.

A point p of M is called an *interior point* if in some chart (U, ϕ), the point $\phi(p)$ is an interior point of $\mathcal{H}^n$. Similarly, p is a *boundary point* of M if $\phi(p)$ is a boundary point of $\mathcal{H}^n$. These concepts are well defined, independent of the charts, because if (V, ψ) is another chart, then the diffeomorphism $\psi \circ \phi^{-1}$ maps $\phi(p)$ to $\psi(p)$, and so by Proposition 22.4, $\phi(p)$ and $\psi(p)$ are either both interior points or both boundary points (Figure 22.3). The set of boundary points of M is denoted by ∂M.

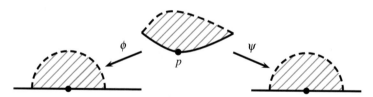

Fig. 22.3. Boundary charts.

Most of the concepts introduced for a manifold extend word for word to a manifold with boundary, the only difference being that now a chart can be either of two types and the local model is $\mathcal{H}^n$ (or $\mathcal{L}^1$). For example, a function $f : M \to \mathbb{R}$ is C^∞ at a boundary point $p \in \partial M$ if there is a chart (U, ϕ) about p such that $f \circ \phi^{-1}$ is C^∞ at $\phi(p) \in \mathcal{H}^n$. This in turn means that $f \circ \phi^{-1}$ has a C^∞ extension to a neighborhood of $\phi(p)$ in $\mathbb{R}^n$.

In point-set topology there are other notions of interior and boundary, defined for a subset A of a topological space S. A point $p \in S$ is said to be an *interior point* of A if there exists an open subset U of S such that

$$p \in U \subset S.$$

The point $p \in S$ is an *exterior point* of A if there exists an open subset U of S such that

$$p \in U \subset S - A.$$

Finally, $p \in S$ is a *boundary point* of A if every neighborhood of p contains both a point in A and a point not in A. We denote by $\text{int}(A)$, $\text{ext}(A)$, and $\text{bd}(A)$ the sets of interior, exterior, and boundary points respectively of A in S. Clearly, the topological space S is the disjoint union

$$S = \text{int}(A) \ \amalg \ \text{ext}(A) \ \amalg \ \text{bd}(A).$$

In case the subset $A \subset S$ is a manifold with boundary, we call $\text{int}(A)$ the *topological interior* and $\text{bd}(A)$ the *topological boundary* of A, to distinguish them from the *manifold interior* A° and the *manifold boundary* ∂A. Note that the topological interior and the topological boundary of a set depend on an ambient space, while the manifold interior and the manifold boundary are intrinsic.

Example 22.7 (*Topological boundary versus manifold boundary*). Let A be the open unit disk in $\mathbb{R}^2$:

$$A = \{x \in \mathbb{R}^2 \mid \|x\| < 1\}.$$

Then its topological boundary $\text{bd}(A)$ in $\mathbb{R}^2$ is the unit circle, while its manifold boundary ∂A is the empty set (Figure 22.4).

If B is the closed unit disk in $\mathbb{R}^2$, then its topological boundary $\text{bd}(B)$ and its manifold boundary ∂B coincide; both are the unit circle.

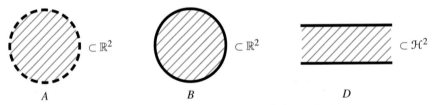

Fig. 22.4. Interiors and boundaries.

Example 22.8 (*Topological interior versus manifold interior*). Let S be the upper half-plane $\mathcal{H}^2$ and let D be the subset (Figure 22.4)

$$D = \{(x,y) \in \mathcal{H}^2 \mid y \le 1\}.$$

The topological interior of D is the set

$$\text{int}(D) = \{(x,y) \in \mathcal{H}^2 \mid 0 \le y < 1\},$$

containing the x-axis, while the manifold interior of D is the set

$$D^\circ = \{(x,y) \in \mathcal{H}^2 \mid 0 < y < 1\},$$

not containing the x-axis.

To indicate the dependence of the topological interior of a set A on its ambient space S, we might denote it by $\text{int}_S(A)$ instead of $\text{int}(A)$. Then in this example, the topological interior $\text{int}_{\mathcal{H}^2}(D)$ of D in $\mathcal{H}^2$ is as above, but the topological interior $\text{int}_{\mathbb{R}^2}(D)$ of D in $\mathbb{R}^2$ coincides with D°.

22.3 The Boundary of a Manifold with Boundary

Let M be a manifold of dimension n with boundary ∂M. If (U, ϕ) is a chart on M, we denote by $\phi' = \phi|_{U \cap \partial M}$ the restriction of the coordinate map ϕ to the boundary. Since ϕ maps boundary points to boundary points,

$$\phi' : U \cap \partial M \to \partial \mathcal{H}^n = \mathbb{R}^{n-1}.$$

Moreover, if (U, ϕ) and (V, ψ) are two charts on M, then

$$\psi' \circ (\phi')^{-1} : \phi'(U \cap V \cap \partial M) \to \psi'(U \cap V \cap \partial M)$$

is C^∞. Thus, an atlas $\{(U_\alpha, \phi_\alpha)\}$ for M induces an atlas $\{(U_\alpha \cap \partial M, \phi_\alpha|_{U_\alpha \cap \partial M})\}$ for ∂M, making ∂M into a manifold of dimension $n - 1$ without boundary.

22.4 Tangent Vectors, Differential Forms, and Orientations

Let M be a manifold with boundary and let $p \in \partial M$. As in Subsection 2.2, two C^∞ functions $f : U \to \mathbb{R}$ and $g : V \to \mathbb{R}$ defined on neighborhoods U and V of p in M are said to be *equivalent* if they agree on some neighborhood W of p contained in $U \cap V$. A *germ* of C^∞ functions at p is an equivalence class of such functions. With the usual addition, multiplication, and scalar multiplication of germs, the set $C_p^\infty(M)$ of germs of C^∞ functions at p is an $\mathbb{R}$-algebra. The *tangent space* $T_p M$ at p is then defined to be the vector space of all point-derivations on the algebra $C_p^\infty(M)$.

For example, for p in the boundary of the upper half-plane $\mathcal{H}^2$, $\partial/\partial x|_p$ and $\partial/\partial y|_p$ are both derivations on $C_p^\infty(\mathcal{H}^2)$. The tangent space $T_p(\mathcal{H}^2)$ is represented by a 2-dimensional vector space with the origin at p. Since $\partial/\partial y|_p$ is a tangent vector to $\mathcal{H}^2$ at p, its negative $-\partial/\partial y|_p$ is also a tangent vector at p (Figure 22.5), although there is no curve through p in $\mathcal{H}^2$ with initial velocity $-\partial/\partial y|_p$.

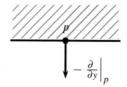

Fig. 22.5. A tangent vector at the boundary.

The *cotangent space* $T_p^* M$ is defined to be the dual of the tangent space:

$$T_p^* M = \operatorname{Hom}(T_p M, \mathbb{R}).$$

Differential k-forms on M are defined as before, as sections of the vector bundle $\bigwedge^k(T^* M)$. A differential k-form is C^∞ if it is C^∞ as a section of the vector bundle

$\bigwedge^k(T^*M)$. For example, $dx \wedge dy$ is a C^∞ 2-form on $\mathcal{H}^2$. An *orientation* on an *n*-manifold M with boundary is again a continuous pointwise orientation on M.

The discussion in Section 21 on orientations goes through word for word for manifolds with boundary. Thus, the orientability of a manifold with boundary is equivalent to the existence of a C^∞ nowhere-vanishing top form and to the existence of an oriented atlas. At one point in the proof of Lemma 21.4, it was necessary to replace the chart $(U, x^1, x^2, \ldots, x^n)$ by $(U, -x^1, x^2, \ldots, x^n)$. This would not have been possible for $n = 1$ if we had not allowed the left half-line $\mathcal{L}^1$ as a local model in the definition of a chart on a 1-dimensional manifold with boundary.

Example 22.9. The closed interval $[0,1]$ is a C^∞ manifold with boundary. It has an atlas with two charts (U_1, ϕ_1) and (U_2, ϕ_2), where $U_1 = [0,1[$, $\phi_1(x) = x$, and $U_2 =]0,1]$, $\phi_2(x) = 1 - x$. With d/dx as a continuous pointwise orientation, $[0,1]$ is an oriented manifold with boundary. However, $\{(U_1, \phi_1), (U_2, \phi_2)\}$ is not an oriented atlas, because the Jacobian determinant of the transition function $(\phi_2 \circ \phi_1^{-1})(x) = 1 - x$ is negative. If we change the sign of ϕ_2, then $\{(U_1, \phi_1), (U_2, -\phi_2)\}$ is an oriented atlas. Note that $-\phi_2(x) = x - 1$ maps $]0,1]$ into the left half-line $\mathcal{L}^1 \subset \mathbb{R}$. If we had allowed only $\mathcal{H}^1$ as a local model for a 1-dimensional manifold with boundary, the closed interval $[0,1]$ would not have an oriented atlas.

22.5 Outward-Pointing Vector Fields

Let M be a manifold with boundary and $p \in \partial M$. We say that a tangent vector $X_p \in T_p(M)$ is *inward-pointing* if $X_p \notin T_p(\partial M)$ and there are a positive real number ε and a curve $c \colon [0, \varepsilon[\to M$ such that $c(0) = p$, $c(]0, \varepsilon[) \subset M^\circ$, and $c'(0) = X_p$. A vector $X_p \in T_p(M)$ is *outward-pointing* if $-X_p$ is inward-pointing. For example, on the upper half-plane $\mathcal{H}^2$, the vector $\partial/\partial y|_p$ is inward-pointing and the vector $-\partial/\partial y|_p$ is outward-pointing at a point p on the x-axis.

A vector field *along* ∂M is a function X that assigns to each point p in ∂M a vector X_p in the tangent space $T_p M$ (as opposed to $T_p(\partial M)$). In a coordinate neighborhood $(U, x^1, \ldots, x^n)$ of p in M, such a vector field X can be written as a linear combination

$$X_q = \sum_i a^i(q) \left. \frac{\partial}{\partial x^i} \right|_q, \quad q \in \partial M.$$

The vector field X along ∂M is said to be *smooth at* $p \in M$ if there exists a coordinate neighborhood of p for which the functions a^i on ∂M are C^∞ at p; it is said to be *smooth* if it is smooth at every point p. In terms of local coordinates, a vector X_p is outward-pointing if and only if $a^n(p) < 0$ (see Figure 22.5 and Problem 22.3).

Proposition 22.10. *On a manifold M with boundary ∂M, there is a smooth outward-pointing vector field along ∂M.*

Proof. Cover ∂M with coordinate open sets $(U_\alpha, x_\alpha^1, \ldots, x_\alpha^n)$ in M. On each U_α the vector field $X_\alpha = -\partial/\partial x_\alpha^n$ along $U_\alpha \cap \partial M$ is smooth and outward-pointing. Choose a partition of unity $\{\rho_\alpha\}_{\alpha \in A}$ on ∂M subordinate to the open cover $\{U_\alpha \cap \partial M\}_{\alpha \in A}$. Then one can check that $X := \sum \rho_\alpha X_\alpha$ is a smooth outward-pointing vector field along ∂M (Problem 22.4). $\square$

22.6 Boundary Orientation

In this section we show that the boundary of an orientable manifold M with boundary is an orientable manifold (without boundary, by Subsection 22.3). We will designate one of the orientations on the boundary as the boundary orientation. It is easily described in terms of an orientation form or of a pointwise orientation on ∂M.

Proposition 22.11. *Let M be an oriented n-manifold with boundary. If ω is an orientation form on M and X is a smooth outward-pointing vector field on ∂M, then $\iota_X \omega$ is a smooth nowhere-vanishing $(n-1)$-form on ∂M. Hence, ∂M is orientable.*

Proof. Since ω and X are both smooth on ∂M, so is the contraction $\iota_X \omega$ (Subsection 20.4). We will now prove by contradiction that $\iota_X \omega$ is nowhere-vanishing on ∂M. Suppose $\iota_X \omega$ vanishes at some $p \in \partial M$. This means that $(\iota_X \omega)_p (v_1, \ldots, v_{n-1}) = 0$ for all $v_1, \ldots, v_{n-1} \in T_p(\partial M)$. Let $e_1, \ldots, e_{n-1}$ be a basis for $T_p(\partial M)$. Then $X_p, e_1, \ldots, e_{n-1}$ is a basis for $T_p M$, and

$$\omega_p (X_p, e_1, \ldots, e_{n-1}) = (\iota_X \omega)_p (e_1, \ldots, e_{n-1}) = 0.$$

By Problem 3.9, $\omega_p \equiv 0$ on $T_p M$, a contradiction. Therefore, $\iota_X \omega$ is nowhere vanishing on ∂M. By Theorem 21.5, ∂M is orientable. $\qquad\square$

In the notation of the preceding proposition, we define the *boundary orientation* on ∂M to be the orientation with orientation form $\iota_X \omega$. For the boundary orientation to be well defined, we need to check that it is independent of the choice of the orientation form ω and of the outward-pointing vector field X. The verification is not difficult (see Problem 22.5).

Proposition 22.12. *Suppose M is an oriented n-manifold with boundary. Let p be a point of the boundary ∂M and let X_p be an outward-pointing vector in $T_p M$. An ordered basis $(v_1, \ldots, v_{n-1})$ for $T_p(\partial M)$ represents the boundary orientation at p if and only if the ordered basis $(X_p, v_1, \ldots, v_{n-1})$ for $T_p M$ represents the orientation on M at p.*

To make this rule easier to remember, we summarize it under the rubric "outward vector first."

Proof. For p in ∂M, let $(v_1, \ldots, v_{n-1})$ be an ordered basis for the tangent space $T_p(\partial M)$. Then

$(v_1, \ldots, v_{n-1})$ represents the boundary orientation on ∂M at p

$\iff \quad (\iota_{X_p} \omega_p)(v_1, \ldots, v_{n-1}) > 0$

$\iff \quad \omega_p(X_p, v_1, \ldots, v_{n-1}) > 0$

$\iff \quad (X_p, v_1, \ldots, v_{n-1})$ represents the orientation on M at p. $\qquad\square$

Example 22.13 (*The boundary orientation on* $\partial\mathcal{H}^n$). An orientation form for the standard orientation on the upper half-space $\mathcal{H}^n$ is $\omega = dx^1 \wedge \cdots \wedge dx^n$. A smooth outward-pointing vector field on $\partial\mathcal{H}^n$ is $-\partial/\partial x^n$. By definition, an orientation form for the boundary orientation on $\partial\mathcal{H}^n$ is given by the contraction

$$\iota_{-\partial/\partial x^n}(\omega) = -\iota_{\partial/\partial x^n}(dx^1 \wedge \cdots \wedge dx^{n-1} \wedge dx^n)$$
$$= -(-1)^{n-1} dx^1 \wedge \cdots \wedge dx^{n-1} \wedge \iota_{\partial/\partial x^n}(dx^n)$$
$$= (-1)^n dx^1 \wedge \cdots \wedge dx^{n-1}.$$

Thus, the boundary orientation on $\partial\mathcal{H}^1 = \{0\}$ is given by -1, the boundary orientation on $\partial\mathcal{H}^2$, given by dx^1, is the usual orientation on the real line $\mathbb{R}$ (Figure 22.6(a)), and the boundary orientation on $\partial\mathcal{H}^3$, given by $-dx^1 \wedge dx^2$, is the clockwise orientation in the (x_1, x_2)-plane $\mathbb{R}^2$ (Figure 22.6(b)).

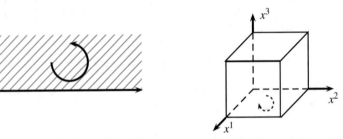

(a) Boundary orientation on $\partial\mathcal{H}^2 = \mathbb{R}$.   (b) Boundary orientation on $\partial\mathcal{H}^3 = \mathbb{R}^2$.

Fig. 22.6. Boundary orientations.

Example. The closed interval $[a,b]$ in the real line with coordinate x has a standard orientation given by the vector field d/dx, with orientation form dx. At the right endpoint b, an outward vector is d/dx. Hence, the boundary orientation at b is given by $\iota_{d/dx}(dx) = +1$. Similarly, the boundary orientation at the left endpoint a is given by $\iota_{-d/dx}(dx) = -1$.

Example. Suppose $c\colon [a,b] \to M$ is a C^∞ immersion whose image is a 1-dimensional manifold C with boundary. An orientation on $[a,b]$ induces an orientation on C via the differential $c_{*,p}\colon T_p([a,b]) \to T_pC$ at each point $p \in [a,b]$. In a situation like this, we give C the orientation induced from the standard orientation on $[a,b]$. The boundary orientation on the boundary of C is given by $+1$ at the endpoint $c(b)$ and -1 at the initial point $c(a)$.

Problems

22.1. Topological boundary versus manifold boundary
Let M be the subset $[0,1[\cup \{2\}$ of the real line. Find its topological boundary $\mathrm{bd}(M)$ and its manifold boundary ∂M.

22.2. Topological boundary of an intersection

Let A and B be two subsets of a topological space S. Prove that

$$\operatorname{bd}(A \cap B) \subset \operatorname{bd}(A) \cup \operatorname{bd}(B).$$

22.3.* Inward-pointing vectors at the boundary

Let M be a manifold with boundary and let $p \in \partial M$. Show that $X_p \in T_p M$ is inward-pointing if and only if in any coordinate chart $(U, x^1, \ldots, x^n)$ centered at p, the coefficient of $(\partial/\partial x^n)_p$ in X_p is positive.

22.4.* Smooth outward-pointing vector field along the boundary

Show that the vector field $X = \sum \rho_\alpha X_\alpha$ defined in the proof of Proposition 22.10 is a smooth outward-pointing vector field along ∂M.

22.5. Boundary orientation

Let M be an oriented manifold with boundary, ω an orientation form for M, and X a C^∞ outward-pointing vector field along ∂M.

(a) If τ is another orientation form on M, then $\tau = f\omega$ for a C^∞ everywhere-positive function f on M. Show that $\iota_X \tau = f \iota_X \omega$ and therefore, $\iota_X \tau \sim \iota_X \omega$ on ∂M. (Here "$\sim$" is the equivalence relation defined in Subsection 21.4.)

(b) Prove that if Y is another C^∞ outward-pointing vector field along ∂M, then $\iota_X \omega \sim \iota_Y \omega$ on ∂M.

22.6.* Induced atlas on the boundary

Assume $n \geq 2$ and let (U, ϕ) and (V, ψ) be two charts in an oriented atlas of an orientable n-manifold M with boundary. Prove that if $U \cap V \cap \partial M \neq \varnothing$, then the restriction of the transition function $\psi \circ \phi^{-1}$ to the boundary $B := \phi(U \cap V) \cap \partial \mathcal{H}^n$,

$$(\psi \circ \phi^{-1})|_B : \phi(U \cap V) \cap \partial \mathcal{H}^n \to \psi(U \cap V) \cap \partial \mathcal{H}^n,$$

has positive Jacobian determinant. (*Hint:* Let $\phi = (x^1, \ldots, x^n)$ and $\psi = (y^1, \ldots, y^n)$. Show that the Jacobian matrix of $\psi \circ \phi^{-1}$ in local coordinates is block triangular with $J(\psi \circ \phi^{-1})|_B$ and $\partial y^n/\partial x^n$ as the diagonal blocks, and that $\partial y^n/\partial x^n > 0$.)

Thus, if $\{(U_\alpha, \phi_\alpha)\}$ is an oriented atlas for a manifold M with boundary, then the induced atlas $\{(U_\alpha \cap \partial M, \phi_\alpha|_{U_\alpha \cap \partial M})\}$ for ∂M is oriented.

22.7.* Boundary orientation of the left half-space

Let M be the left half-space

$$\{(x^1, \ldots, x^n) \in \mathbb{R}^n \mid x^1 \leq 0\},$$

with orientation form $dx^1 \wedge \cdots \wedge dx^n$. Show that an orientation form for the boundary orientation on $\partial M = \{(0, x^2, \ldots, x^n) \in \mathbb{R}^n\}$ is $dx^2 \wedge \cdots \wedge dx^n$.

Unlike the upper half-space $\mathcal{H}^n$, whose boundary orientation takes on a sign (Example 22.13), this exercise shows that the boundary orientation for the left half-space has no sign. For this reason some authors use the left half-space as the model of a manifold with boundary, e.g., [7].

22.8. Boundary orientation on a cylinder

Let M be the cylinder $S^1 \times [0, 1]$ with the counterclockwise orientation when viewed from the exterior (Figure 22.7(a)). Describe the boundary orientation on $C_0 = S^1 \times \{0\}$ and $C_1 = S^1 \times \{1\}$.

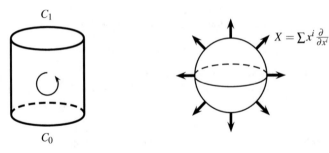

$$X = \sum x^i \frac{\partial}{\partial x^i}$$

(a) Oriented cylinder.   (b) Radial vector field on a sphere.

Fig. 22.7. Boundary orientations.

22.9. Boundary orientation on a sphere

Orient the unit sphere S^n in $\mathbb{R}^{n+1}$ as the boundary of the closed unit ball. Show that an orientation form on S^n is

$$\omega = \sum_{i=1}^{n+1} (-1)^{i-1} x^i \, dx^1 \wedge \cdots \wedge \widehat{dx^i} \wedge \cdots \wedge dx^{n+1},$$

where the caret $\widehat{}$ over dx^i indicates that dx^i is to be omitted. (*Hint:* An outward-pointing vector field on S^n is the radial vector field $X = \sum x^i \, \partial/\partial x^i$ as in Figure 22.7(b).)

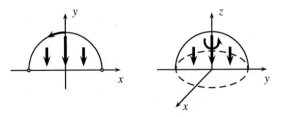

Fig. 22.8. Projection of the upper hemisphere to a disk.

22.10. Orientation on the upper hemisphere of a sphere

Orient the unit sphere S^n in $\mathbb{R}^{n+1}$ as the boundary of the closed unit ball. Let U be the upper hemisphere

$$U = \{x \in S^n \mid x^{n+1} > 0\}.$$

It is a coordinate chart on the sphere with coordinates $x^1, \ldots, x^n$.

(a) Find an orientation form on U in terms of $dx^1, \ldots, dx^n$.

(b) Show that the projection map $\pi \colon U \to \mathbb{R}^n$,

$$\pi(x^1, \ldots, x^n, x^{n+1}) = (x^1, \ldots, x^n),$$

is orientation-preserving if and only if n is even (Figure 22.8).

22.11. Antipodal map on a sphere and the orientability of $\mathbb{R}P^n$

(a) The antipodal map $a\colon S^n \to S^n$ on the n-sphere is defined by

$$a(x^1, \ldots, x^{n+1}) = (-x^1, \ldots, -x^{n+1}).$$

Show that the antipodal map is orientation-preserving if and only if n is odd.

(b) Use part (a) and Problem 21.6 to prove that an odd-dimensional real projective space $\mathbb{R}P^n$ is orientable.

§23 Integration on Manifolds

In this chapter we first recall Riemann integration for a function over a closed rectangle in Euclidean space. By Lebesgue's theorem, this theory can be extended to integrals over domains of integration, bounded subsets of $\mathbb{R}^n$ whose boundary has measure zero.

The integral of an *n*-form with compact support in an open set of $\mathbb{R}^n$ is defined to be the Riemann integral of the corresponding function. Using a partition of unity, we define the integral of an *n*-form with compact support on a manifold by writing the form as a sum of forms each with compact support in a coordinate chart. We then prove the general Stokes theorem for an oriented manifold and show how it generalizes the fundamental theorem for line integrals as well as Green's theorem from calculus.

23.1 The Riemann Integral of a Function on $\mathbb{R}^n$

We assume that the reader is familiar with the theory of Riemann integration in $\mathbb{R}^n$, as in [26] or [35]. What follows is a brief synopsis of the Riemann integral of a bounded function over a bounded set in $\mathbb{R}^n$.

A *closed rectangle* in $\mathbb{R}^n$ is a Cartesian product $R = [a^1, b^1] \times \cdots \times [a^n, b^n]$ of closed intervals in $\mathbb{R}$, where $a^i, b^i \in \mathbb{R}$. Let $f \colon R \to \mathbb{R}$ be a bounded function defined on a closed rectangle R. The *volume* vol(R) of the closed rectangle R is defined to be

$$\text{vol}(R) := \prod_{i=1}^{n}(b_i - a_i). \tag{23.1}$$

A *partition* of the closed interval $[a, b]$ is a set of real numbers $\{p_0, \ldots, p_n\}$ such that

$$a = p_0 < p_1 < \cdots < p_n = b.$$

A *partition* of the rectangle R is a collection $P = \{P_1, \ldots, P_n\}$, where each P_i is a partition of $[a^i, b^i]$. The partition P divides the rectangle R into closed subrectangles, which we denote by R_j (Figure 23.1).

We define the *lower sum* and the *upper sum* of f with respect to the partition P to be

$$L(f, P) := \sum (\inf_{R_j} f)\, \text{vol}(R_j), \quad U(f, P) := \sum (\sup_{R_j} f)\, \text{vol}(R_j),$$

where each sum runs over all subrectangles of the partition P. For any partition P, clearly $L(f, P) \leq U(f, P)$. In fact, more is true: for any two partitions P and P' of the rectangle R,

$$L(f, P) \leq U(f, P'),$$

which we show next.

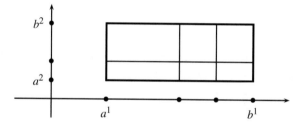

Fig. 23.1. A partition of a closed rectangle.

A partition $P' = \{P'_1,\ldots,P'_n\}$ is a *refinement* of the partition $P = \{P_1,\ldots,P_n\}$ if $P_i \subset P'_i$ for all $i = 1,\ldots,n$. If P' is a refinement of P, then each subrectangle R_j of P is subdivided into subrectangles R'_{jk} of P', and it is easily seen that

$$L(f,P) \leq L(f,P'),\tag{23.2}$$

because if $R'_{jk} \subset R_j$, then $\inf_{R_j} f \leq \inf_{R'_{jk}} f$. Similarly, if P' is a refinement of P, then

$$U(f,P') \leq U(f,P).\tag{23.3}$$

Any two partitions P and P' of the rectangle R have a common refinement $Q = \{Q_1,\ldots,Q_n\}$ with $Q_i = P_i \cup P'_i$. By (23.2) and (23.3),

$$L(f,P) \leq L(f,Q) \leq U(f,Q) \leq U(f,P').$$

It follows that the supremum of the lower sum $L(f,P)$ over all partitions P of R is less than or equal to the infimum of the upper sum $U(f,P)$ over all partitions P of R. We define these two numbers to be the *lower integral* $\underline{\int}_R f$ and the *upper integral* $\overline{\int}_R f$, respectively:

$$\underline{\int}_R f := \sup_P L(f,P),\qquad \overline{\int}_R f := \inf_P L(f,P).$$

Definition 23.1. Let R be a closed rectangle in $\mathbb{R}^n$. A bounded function $f\colon R \to \mathbb{R}$ is said to be *Riemann integrable* if $\underline{\int}_R f = \overline{\int}_R f$; in this case, the Riemann integral of f is this common value, denoted by $\int_R f(x)\,dx^1 \cdots dx^n$, where $x^1,\ldots,x^n$ are the standard coordinates on $\mathbb{R}^n$.

Remark. When we speak of a rectangle $[a^1,b^1] \times \cdots \times [a^n,b^n]$ in $\mathbb{R}^n$, we have already tacitly chosen n coordinates axes, with coordinates $x^1,\ldots,x^n$. Thus, the definition of a Riemann integral depends on the coordinates $x^1,\ldots,x^n$.

If $f\colon A \subset \mathbb{R}^n \to \mathbb{R}$, then the *extension of f by zero* is the function $\tilde{f}\colon \mathbb{R}^n \to \mathbb{R}$ such that

$$\tilde{f}(x) = \begin{cases} f(x) & \text{for } x \in A, \\ 0 & \text{for } x \notin A. \end{cases}$$

Now suppose $f \colon A \to \mathbb{R}$ is a bounded function on a bounded set A in $\mathbb{R}^n$. Enclose A in a closed rectangle R and define the Riemann integral of f over A to be

$$\int_A f(x)\, dx^1 \cdots dx^n = \int_R \tilde{f}(x)\, dx^1 \cdots dx^n$$

if the right-hand side exists. In this way we can deal with the integral of a bounded function whose domain is an arbitrary bounded set in $\mathbb{R}^n$.

The *volume* vol(A) of a subset $A \subset \mathbb{R}^n$ is defined to be the integral $\int_A 1\, dx^1 \cdots dx^n$ if the integral exists. This concept generalizes the volume of a closed rectangle defined in (23.1).

23.2 Integrability Conditions

In this section we describe some conditions under which a function defined on an open subset of $\mathbb{R}^n$ is Riemann integrable.

Definition 23.2. A set $A \subset \mathbb{R}^n$ is said to have *measure zero* if for every $\varepsilon > 0$, there is a countable cover $\{R_i\}_{i=1}^{\infty}$ of A by closed rectangles R_i such that $\sum_{i=1}^{\infty} \text{vol}(R_i) < \varepsilon$.

The most useful integrability criterion is the following theorem of Lebesgue [26, Theorem 8.3.1, p. 455].

Theorem 23.3 (Lebesgue's theorem). *A bounded function $f \colon A \to \mathbb{R}$ on a bounded subset $A \subset \mathbb{R}^n$ is Riemann integrable if and only if the set* Disc$(\tilde{f})$ *of discontinuities of the extended function $\tilde{f}$ has measure zero.*

Proposition 23.4. *If a continuous function $f \colon U \to \mathbb{R}$ defined on an open subset U of $\mathbb{R}^n$ has compact support, then f is Riemann integrable on U.*

Proof. Being continuous on a compact set, the function f is bounded. Being compact, the set supp f is closed and bounded in $\mathbb{R}^n$. We claim that the extension $\tilde{f}$ is continuous.

Since $\tilde{f}$ agrees with f on U, the extended function $\tilde{f}$ is continuous on U. It remains to show that $\tilde{f}$ is continuous on the complement of U in $\mathbb{R}^n$ as well. If $p \notin U$, then $p \notin$ supp f. Since supp f is a closed subset of $\mathbb{R}^n$, there is an open ball B containing p and disjoint from supp f. On this open ball, $\tilde{f} \equiv 0$, which implies that $\tilde{f}$ is continuous at $p \notin U$. Thus, $\tilde{f}$ is continuous on $\mathbb{R}^n$. By Lebesgue's theorem, f is Riemann integrable on U. $\square$

Example 23.5. The continuous function $f \colon\,]-1, 1[\, \to \mathbb{R}$, $f(x) = \tan(\pi x/2)$, is defined on an open subset of finite length in $\mathbb{R}$, but is not bounded (Figure 23.2). The support of f is the open interval $]-1, 1[$, which is not compact. Thus, the function f does not satisfy the hypotheses of either Lebesgue's theorem or Proposition 23.4. Note that it is not Riemann integrable.

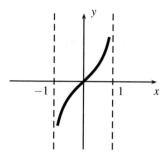

Fig. 23.2. The function $f(x) = \tan(\pi x/2)$ on $]-1,1[$.

Remark. The support of a real-valued function is the closure *in its domain* of the subset where the function is not zero. In Example 23.5, the support of f is the open interval $]-1,1[$, not the closed interval $[-1,1]$, because the domain of f is $]-1,1[$, not $\mathbb{R}$.

Definition 23.6. A subset $A \subset \mathbb{R}^n$ is called a *domain of integration* if it is bounded and its topological boundary $\mathrm{bd}(A)$ is a set of measure zero.

Familiar plane figures such as triangles, rectangles, and circular disks are all domains of integration in $\mathbb{R}^2$.

Proposition 23.7. *Every bounded continuous function f defined on a domain of integration A in $\mathbb{R}^n$ is Riemann integrable over A.*

Proof. Let $\tilde{f}\colon \mathbb{R}^n \to \mathbb{R}$ be the extension of f by zero. Since f is continuous on A, the extension $\tilde{f}$ is necessarily continuous at all interior points of A. Clearly, $\tilde{f}$ is continuous at all exterior points of A also, because every exterior point has a neighborhood contained entirely in $\mathbb{R}^n - A$, on which $\tilde{f}$ is identically zero. Therefore, the set $\mathrm{Disc}(\tilde{f})$ of discontinuities of $\tilde{f}$ is a subset of $\mathrm{bd}(A)$, a set of measure zero. By Lebesgue's theorem, f is Riemann integrable on A. $\qquad\square$

23.3 The Integral of an *n*-Form on $\mathbb{R}^n$

Once a set of coordinates $x^1,\dots,x^n$ has been fixed on $\mathbb{R}^n$, n-forms on $\mathbb{R}^n$ can be identified with functions on $\mathbb{R}^n$, since every n-form on $\mathbb{R}^n$ can be written as $\omega = f(x)\,dx^1 \wedge \cdots \wedge dx^n$ for a unique function $f(x)$ on $\mathbb{R}^n$. In this way the theory of Riemann integration of functions on $\mathbb{R}^n$ carries over to n-forms on $\mathbb{R}^n$.

Definition 23.8. Let $\omega = f(x)\,dx^1 \wedge \cdots \wedge dx^n$ be a C^∞ n-form on an open subset $U \subset \mathbb{R}^n$, with standard coordinates $x^1,\dots,x^n$. Its *integral* over a subset $A \subset U$ is defined to be the Riemann integral of $f(x)$:

$$\int_A \omega = \int_A f(x)\,dx^1 \wedge \cdots \wedge dx^n := \int_A f(x)\,dx^1 \cdots dx^n,$$

if the Riemann integral exists.

In this definition the n-form must be written in the order $dx^1 \wedge \cdots \wedge dx^n$. To integrate, for example, $\tau = f(x) dx^2 \wedge dx^1$ over $A \subset \mathbb{R}^2$, one would write

$$\int_A \tau = \int_A -f(x) dx^1 \wedge dx^2 = -\int_A f(x) dx^1 dx^2.$$

Example. If f is a bounded continuous function defined on a domain of integration A in $\mathbb{R}^n$, the the integral $\int_A f dx^1 \wedge \cdots \wedge dx^n$ exists by Proposition 23.7.

Let us see how the integral of an n-form $\omega = f dx^1 \wedge \cdots \wedge dx^n$ on an open subset $U \subset \mathbb{R}^n$ transforms under a change of variables. A change of variables on U is given by a diffeomorphism $T : \mathbb{R}^n \supset V \to U \subset \mathbb{R}^n$. Let $x^1, \ldots, x^n$ be the standard coordinates on U and $y^1, \ldots, y^n$ the standard coordinates on V. Then $T^i := x^i \circ T = T^*(x^i)$ is the ith component of T. We will assume that U and V are connected, and write $x = (x^1, \ldots, x^n)$ and $y = (y^1, \ldots, y^n)$. Denote by $J(T)$ the Jacobian matrix $[\partial T^i / \partial y^j]$. By Corollary 18.4(ii),

$$dT^1 \wedge \cdots \wedge dT^n = \det(J(T)) dy^1 \wedge \cdots \wedge dy^n.$$

Hence,

$$\int_V T^*\omega = \int_V (T^*f) T^* dx^1 \wedge \cdots \wedge T^* dx^n \qquad \text{(Proposition 18.11)}$$

$$= \int_V (f \circ T) dT^1 \wedge \cdots \wedge dT^n \qquad \text{(because } T^*d = dT^*)$$

$$= \int_V (f \circ T) \det(J(T)) dy^1 \wedge \cdots \wedge dy^n$$

$$= \int_V (f \circ T) \det(J(T)) dy^1 \cdots dy^n. \qquad (23.4)$$

On the other hand, the change-of-variables formula from advanced calculus gives

$$\int_U \omega = \int_U f dx^1 \cdots dx^n = \int_V (f \circ T) |\det(J(T))| dy^1 \cdots dy^n, \qquad (23.5)$$

with an absolute-value sign around the Jacobian determinant. Equations (23.4) and (23.5) differ by the sign of $\det(J(T))$. Hence,

$$\int_V T^*\omega = \pm \int_U \omega, \qquad (23.6)$$

depending on whether the Jacobian determinant $\det(J(T))$ is positive or negative.

By Proposition 21.8, a diffeomorphism $T : \mathbb{R}^n \supset V \to U \subset \mathbb{R}^n$ is orientation-preserving if and only if its Jacobian determinant $\det(J(T))$ is everywhere positive on V. Equation (23.6) shows that the integral of a differential form is not invariant under all diffeomorphisms of V with U, but only under orientation-preserving diffeomorphisms.

23.4 Integral of a Differential Form over a Manifold

Integration of an n-form on $\mathbb{R}^n$ is not so different from integration of a function. Our approach to integration over a general manifold has several distinguishing features:

(i) The manifold must be oriented (in fact, $\mathbb{R}^n$ has a standard orientation).
(ii) On a manifold of dimension n, one can integrate only n-forms, not functions.
(iii) The n-forms must have compact support.

Let M be an oriented manifold of dimension n, with an oriented atlas $\{(U_\alpha, \phi_\alpha)\}$ giving the orientation of M. Denote by $\Omega_c^k(M)$ the vector space of C^∞ k-forms with compact support on M. Suppose $\{(U, \phi)\}$ is a chart in this atlas. If $\omega \in \Omega_c^n(U)$ is an n-form with compact support on U, then because $\phi : U \to \phi(U)$ is a diffeomorphism, $(\phi^{-1})^*\omega$ is an n-form with compact support on the open subset $\phi(U) \subset \mathbb{R}^n$. We define the integral of ω on U to be

$$\int_U \omega := \int_{\phi(U)} (\phi^{-1})^*\omega. \tag{23.7}$$

If (U, ψ) is another chart in the oriented atlas with the same U, then $\phi \circ \psi^{-1} : \psi(U) \to \phi(U)$ is an orientation-preserving diffeomorphism, and so

$$\int_{\phi(U)} (\phi^{-1})^*\omega = \int_{\psi(U)} (\phi \circ \psi^{-1})^*(\phi^{-1})^*\omega = \int_{\psi(U)} (\psi^{-1})^*\omega.$$

Thus, the integral $\int_U \omega$ on a chart U of the atlas is well defined, independent of the choice of coordinates on U. By the linearity of the integral on $\mathbb{R}^n$, if $\omega, \tau \in \Omega_c^n(U)$, then

$$\int_U \omega + \tau = \int_U \omega + \int_U \tau.$$

Now let $\omega \in \Omega_c^n(M)$. Choose a partition of unity $\{\rho_\alpha\}$ subordinate to the open cover $\{U_\alpha\}$. Because ω has compact support and a partition of unity has locally finite supports, all except finitely many $\rho_\alpha \omega$ are identically zero by Problem 18.6. In particular,

$$\omega = \sum_\alpha \rho_\alpha \omega$$

is a *finite* sum. Since by Problem 18.4(b),

$$\text{supp}(\rho_\alpha \omega) \subset \text{supp} \rho_\alpha \cap \text{supp} \omega,$$

$\text{supp}(\rho_\alpha \omega)$ is a closed subset of the compact set $\text{supp} \omega$. Hence, $\text{supp}(\rho_\alpha \omega)$ is compact. Since $\rho_\alpha \omega$ is an n-form with compact support in the chart U_α, its integral $\int_{U_\alpha} \rho_\alpha \omega$ is defined. Therefore, we can define the integral of ω over M to be the finite sum

$$\int_M \omega := \sum_\alpha \int_{U_\alpha} \rho_\alpha \omega. \tag{23.8}$$

For this integral to be well defined, we must show that it is independent of the choices of oriented atlas and partition of unity. Let $\{V_\beta\}$ be another oriented atlas of M specifying the orientation of M, and $\{\chi_\beta\}$ a partition of unity subordinate to $\{V_\beta\}$. Then $\{(U_\alpha \cap V_\beta, \phi_\alpha|_{U_\alpha \cap U_\beta})\}$ and $\{(U_\alpha \cap V_\beta, \psi_\beta|_{U_\alpha \cap U_\beta})\}$ are two new atlases of M specifying the orientation of M, and

$$\sum_\alpha \int_{U_\alpha} \rho_\alpha \omega = \sum_\alpha \int_{U_\alpha} \rho_\alpha \sum_\beta \chi_\beta \omega \qquad (\text{because } \sum_\beta \chi_\beta = 1)$$

$$= \sum_\alpha \sum_\beta \int_{U_\alpha} \rho_\alpha \chi_\beta \omega \qquad (\text{these are } \textit{finite} \text{ sums})$$

$$= \sum_\alpha \sum_\beta \int_{U_\alpha \cap V_\beta} \rho_\alpha \chi_\beta \omega,$$

where the last line follows from the fact that the support of $\rho_\alpha \chi_\beta$ is contained in $U_\alpha \cap V_\beta$. By symmetry, $\sum_\beta \int_{V_\beta} \chi_\beta \omega$ is equal to the same sum. Hence,

$$\sum_\alpha \int_{U_\alpha} \rho_\alpha \omega = \sum_\beta \int_{V_\beta} \chi_\beta \omega,$$

proving that the integral (23.8) is well defined.

Proposition 23.9. *Let ω be an n-form with compact support on an oriented manifold M of dimension n. If $-M$ denotes the same manifold but with the opposite orientation, then $\int_{-M} \omega = -\int_M \omega$.*

Thus, reversing the orientation of M reverses the sign of an integral over M.

Proof. By the definition of an integral ((23.7) and (23.8)), it is enough to show that for every chart $(U, \phi) = (U, x^1, \ldots, x^n)$ and differential form $\tau \in \Omega_c^n(U)$, if $(U, \bar{\phi}) = (U, -x^1, x^2, \ldots, x^n)$ is the chart with the opposite orientation, then

$$\int_{\bar{\phi}(U)} \left(\bar{\phi}^{-1}\right)^* \tau = -\int_{\phi(U)} \left(\phi^{-1}\right)^* \tau.$$

Let $r^1, \ldots, r^n$ be the standard coordinates on $\mathbb{R}^n$. Then $x^i = r^i \circ \phi$ and $r^i = x^i \circ \phi^{-1}$. With $\bar{\phi}$, the only difference is that when $i = 1$,

$$-x^1 = r^1 \circ \bar{\phi} \quad \text{and} \quad r^1 = -x^1 \circ \bar{\phi}^{-1}.$$

Suppose $\tau = f \, dx^1 \wedge \cdots \wedge dx^n$ on U. Then

$$\left(\bar{\phi}^{-1}\right)^* \tau = \left(f \circ \bar{\phi}^{-1}\right) d(x^1 \circ \bar{\phi}^{-1}) \wedge d(x^2 \circ \bar{\phi}^{-1}) \wedge \cdots \wedge d(x^n \circ \bar{\phi}^{-1})$$

$$= -\left(f \circ \bar{\phi}^{-1}\right) dr^1 \wedge dr^2 \wedge \cdots \wedge dr^n. \qquad (23.9)$$

Similarly,

$$\left(\phi^{-1}\right)^* \tau = \left(f \circ \phi^{-1}\right) dr^1 \wedge dr^2 \wedge \cdots \wedge dr^n.$$

Since $\phi \circ \bar{\phi}^{-1} : \bar{\phi}(U) \to \phi(U)$ is given by

$$(\phi \circ \bar{\phi}^{-1})(a^1, a^2, \ldots, a^n) = (-a^1, a^2, \ldots, a^n),$$

the absolute value of its Jacobian determinant is

$$|J(\phi \circ \bar{\phi}^{-1})| = |-1| = 1. \tag{23.10}$$

Therefore,

$$
\begin{aligned}
\int_{\bar{\phi}(U)} \left(\bar{\phi}^{-1}\right)^* \tau &= -\int_{\bar{\phi}(U)} \left(f \circ \bar{\phi}^{-1}\right) dr^1 \cdots dr^n \quad \text{(by (23.9))} \\
&= -\int_{\bar{\phi}(U)} \left(f \circ \phi^{-1}\right) \circ \left(\phi \circ \bar{\phi}^{-1}\right) |J(\phi \circ \bar{\phi}^{-1})| \, dr^1 \cdots dr^n \quad \text{(by (23.10))} \\
&= -\int_{\phi(U)} \left(f \circ \phi^{-1}\right) dr^1 \cdots dr^n \quad \text{(by the change-of-variables formula)} \\
&= -\int_{\phi(U)} \left(\phi^{-1}\right)^* \tau. \qquad\qquad \square
\end{aligned}
$$

The treatment of integration above can be extended almost word for word to oriented manifolds with boundary. It has the virtue of simplicity and is of great utility in proving theorems. However, it is not practical for actual computation of integrals; an n-form multiplied by a partition of unity can rarely be integrated as a closed expression. To calculate explicitly integrals over an oriented n-manifold M, it is best to consider integrals over a parametrized set.

Definition 23.10. A *parametrized set* in an oriented n-manifold M is a subset A together with a C^∞ map $F : D \to M$ from a compact domain of integration $D \subset \mathbb{R}^n$ to M such that $A = F(D)$ and F restricts to an orientation-preserving diffeomorphism from $\text{int}(D)$ to $F(\text{int}(D))$. Note that by smooth invariance of domain for manifolds (Remark 22.5), $F(\text{int}(D))$ is an open subset of M. The C^∞ map $F : D \to A$ is called a *parametrization* of A.

If A is a parametrized set in M with parametrization $F : D \to A$ and ω is a C^∞ n-form on M, not necessarily with compact support, then we define $\int_A \omega$ to be $\int_D F^* \omega$. It can be shown that the definition of $\int_A \omega$ is independent of the parametrization and that in case A is a manifold, it agrees with the earlier definition of integration over a manifold. Subdividing an oriented manifold into a union of parametrized sets can be an effective method of calculating an integral over the manifold. We will not delve into this theory of integration (see [31, Theorem 25.4, p. 213] or [25, Proposition 14.7, p. 356]), but will content ourselves with an example.

Example 23.11 (*Integral over a sphere*). In spherical coordinates, ρ is the distance $\sqrt{x^2 + y^2 + z^2}$ of the point $(x, y, z) \in \mathbb{R}^3$ to the origin, φ is the angle that the vector $\langle x, y, z \rangle$ makes with the positive z-axis, and θ is the angle that the vector $\langle x, y \rangle$ in the (x, y)-plane makes with the positive x-axis (Figure 23.3(a)). Let ω be the 2-form on the unit sphere S^2 in $\mathbb{R}^3$ given by

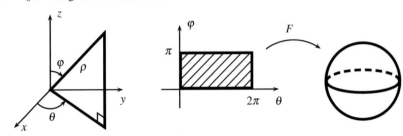

(a) Spherical coordinates in $\mathbb{R}^3$ (b) A parametrization by spherical coordinates

Fig. 23.3. The sphere as a parametrized set.

$$\omega = \begin{cases} \dfrac{dy \wedge dz}{x} & \text{for } x \neq 0, \\[2mm] \dfrac{dz \wedge dx}{y} & \text{for } y \neq 0, \\[2mm] \dfrac{dx \wedge dy}{z} & \text{for } z \neq 0. \end{cases}$$

Calculate $\displaystyle\int_{S^2} \omega$.

Up to a factor of 2, the form ω is the 2-form on S^2 from Problem 19.11(b). In Riemannian geometry, it is shown that ω is the area form of the sphere S^2 with respect to the Euclidean metric. Therefore, the integral $\displaystyle\int_{S^2} \omega$ is the surface area of the sphere.

Solution. The sphere S^2 has a parametrization by spherical coordinates (Figure 23.3(b)):

$$F(\varphi, \theta) = (\sin \varphi \cos \theta, \sin \varphi \sin \theta, \cos \varphi)$$

on $D = \{(\varphi, \theta) \in \mathbb{R}^2 \mid 0 \leq \varphi \leq \pi, \ 0 \leq \theta \leq 2\pi\}$. Since

$$F^* x = \sin \varphi \cos \theta, \quad F^* y = \sin \varphi \sin \theta, \quad \text{and} \quad F^* z = \cos \varphi,$$

we have

$$F^* dy = dF^* y = \cos \varphi \sin \theta \, d\varphi + \sin \varphi \cos \theta \, d\theta$$

and

$$F^* dz = -\sin \varphi \, d\varphi,$$

so for $x \neq 0$,

$$F^* \omega = \frac{F^* dy \wedge F^* dz}{F^* x} = \sin \varphi \, d\varphi \wedge d\theta.$$

For $y \neq 0$ and $z \neq 0$, similar calculations show that $F^*\omega$ is given by the same formula. Therefore, $F^*\omega = \sin\varphi\, d\varphi \wedge d\theta$ everywhere on D, and

$$\int_{S^2} \omega = \int_D F^*\omega = \int_0^{2\pi}\int_0^{\pi} \sin\varphi\, d\varphi\, d\theta = 2\pi\left[-\cos\varphi\right]_0^{\pi} = 4\pi. \qquad \square$$

Integration over a zero-dimensional manifold

The discussion of integration so far assumes implicitly that the manifold M has dimension $n \geq 1$. We now treat integration over a zero-dimensional manifold. A compact oriented manifold M of dimension 0 is a finite collection of points, each point oriented by $+1$ or -1. We write this as $M = \sum p_i - \sum q_j$. The integral of a 0-form $f\colon M \to \mathbb{R}$ is defined to be the sum

$$\int_M f = \sum f(p_i) - \sum f(q_j).$$

23.5 Stokes's Theorem

Let M be an oriented manifold of dimension n with boundary. We give its boundary ∂M the boundary orientation and let $i\colon \partial M \hookrightarrow M$ be the inclusion map. If ω is an $(n-1)$-form on M, it is customary to write $\int_{\partial M} \omega$ instead of $\int_{\partial M} i^*\omega$.

Theorem 23.12 (Stokes's theorem). *For any smooth $(n-1)$-form ω with compact support on the oriented n-dimensional manifold M,*

$$\int_M d\omega = \int_{\partial M} \omega.$$

Proof. Choose an atlas $\{(U_\alpha, \phi_\alpha)\}$ for M in which each U_α is diffeomorphic to either $\mathbb{R}^n$ or $\mathcal{H}^n$ via an orientation-preserving diffeomorphism. This is possible since any open disk is diffeomorphic to $\mathbb{R}^n$ and any half-disk containing its boundary diameter is diffeomorphic to $\mathcal{H}^n$ (see Problem 1.5). Let $\{\rho_\alpha\}$ be a C^∞ partition of unity subordinate to $\{U_\alpha\}$. As we showed in the preceding section, the $(n-1)$-form $\rho_\alpha\omega$ has compact support in U_α.

Suppose Stokes's theorem holds for $\mathbb{R}^n$ and for $\mathcal{H}^n$. Then it holds for all the charts U_α in our atlas, which are diffeomorphic to $\mathbb{R}^n$ or $\mathcal{H}^n$. Also, note that

$$(\partial M) \cap U_\alpha = \partial U_\alpha.$$

Therefore,

$$\int_{\partial M} \omega = \int_{\partial M} \sum_\alpha \rho_\alpha \omega \quad \left(\sum_\alpha \rho_\alpha = 1\right)$$

$$= \sum_\alpha \int_{\partial M} \rho_\alpha \omega \quad \left(\sum_\alpha \rho_\alpha \omega \text{ is a finite sum by Problem 18.6}\right)$$

$$= \sum_\alpha \int_{\partial U_\alpha} \rho_\alpha \omega \quad (\text{supp}\, \rho_\alpha \omega \subset U_\alpha)$$

$$= \sum_\alpha \int_{U_\alpha} d(\rho_\alpha \omega) \quad (\text{Stokes's theorem for } U_\alpha)$$

$$= \sum_\alpha \int_M d(\rho_\alpha \omega) \quad (\text{supp}\, d(\rho_\alpha \omega) \subset U_\alpha)$$

$$= \int_M d\left(\sum_\alpha \rho_\alpha \omega\right) \quad \left(\sum_\alpha \rho_\alpha \omega \text{ is a finite sum}\right)$$

$$= \int_M d\omega.$$

Thus, it suffices to prove Stokes's theorem for $\mathbb{R}^n$ and for $\mathcal{H}^n$. We will give a proof only for $\mathcal{H}^2$, since the general case is similar (see Problem 23.4).

Proof of Stokes's theorem for the upper half-plane $\mathcal{H}^2$. Let x,y be the coordinates on $\mathcal{H}^2$. Then the standard orientation on $\mathcal{H}^2$ is given by $dx \wedge dy$, and the boundary orientation on $\partial \mathcal{H}^2$ is given by $\iota_{-\partial/\partial y}(dx \wedge dy) = dx$.

The form ω is a linear combination

$$\omega = f(x,y)\, dx + g(x,y)\, dy \tag{23.11}$$

for C^∞ functions f, g with compact support in $\mathcal{H}^2$. Since the supports of f and g are compact, we may choose a real number $a > 0$ large enough that the supports of f and g are contained in the interior of the square $[-a, a] \times [0, a]$. We will use the notation f_x, f_y to denote the partial derivatives of f with respect to x and y, respectively. Then

$$d\omega = \left(\frac{\partial g}{\partial x} - \frac{\partial f}{\partial y}\right) dx \wedge dy = (g_x - f_y)\, dx \wedge dy,$$

and

$$\int_{\mathcal{H}^2} d\omega = \int_{\mathcal{H}^2} g_x\, dx\, dy - \int_{\mathcal{H}^2} f_y\, dx\, dy$$

$$= \int_0^\infty \int_{-\infty}^\infty g_x\, dx\, dy - \int_{-\infty}^\infty \int_0^\infty f_y\, dy\, dx$$

$$= \int_0^a \int_{-a}^a g_x\, dx\, dy - \int_{-a}^a \int_0^a f_y\, dy\, dx. \tag{23.12}$$

In this expression,

$$\int_{-a}^a g_x(x,y)\, dx = g(x,y)\big]_{x=-a}^a = 0$$

because $\text{supp}\, g$ lies in the interior of $[-a, a] \times [0, a]$. Similarly,

$$\int_0^a f_y(x,y)\,dy = f(x,y)\big]_{y=0}^a = -f(x,0)$$

because $f(x,a) = 0$. Thus, (23.12) becomes

$$\int_{\mathcal{H}^2} d\omega = \int_{-a}^a f(x,0)\,dx.$$

On the other hand, $\partial\mathcal{H}^2$ is the x-axis and $dy = 0$ on $\partial\mathcal{H}^2$. It follows from (23.11) that $\omega = f(x,0)\,dx$ when restricted to $\partial\mathcal{H}^2$ and

$$\int_{\partial\mathcal{H}^2} \omega = \int_{-a}^a f(x,0)\,dx.$$

This proves Stokes's theorem for the upper half-plane.   □

23.6 Line Integrals and Green's Theorem

We will now show how Stokes's theorem for a manifold unifies some of the theorems of vector calculus on $\mathbb{R}^2$ and $\mathbb{R}^3$. Recall the calculus notation $\mathbf{F}\cdot d\mathbf{r} = P\,dx + Q\,dy + R\,dz$ for $\mathbf{F} = \langle P,Q,R\rangle$ and $\mathbf{r} = (x,y,z)$. As in calculus, we assume in this section that functions, vector fields, and regions of integration have sufficient smoothness or regularity properties so that all the integrals are defined.

Theorem 23.13 (Fundamental theorem for line integrals). *Let C be a curve in $\mathbb{R}^3$, parametrized by $\mathbf{r}(t) = (x(t),y(t),z(t))$, $a \le t \le b$, and let $\mathbf{F}$ be a vector field on $\mathbb{R}^3$. If $\mathbf{F} = \operatorname{grad} f$ for some scalar function f, then*

$$\int_C \mathbf{F}\cdot d\mathbf{r} = f(\mathbf{r}(b)) - f(\mathbf{r}(a)).$$

Suppose in Stokes's theorem we take M to be a curve C with parametrization $\mathbf{r}(t)$, $a \le t \le b$, and ω to be the function f on C. Then

$$\int_C d\omega = \int_C df = \int_C \frac{\partial f}{\partial x}\,dx + \frac{\partial f}{\partial y}\,dy + \frac{\partial f}{\partial z}\,dz = \int_C \operatorname{grad} f \cdot d\mathbf{r}$$

and

$$\int_{\partial C} \omega = f\big]_{\mathbf{r}(a)}^{\mathbf{r}(b)} = f(\mathbf{r}(b)) - f(\mathbf{r}(a)).$$

In this case Stokes's theorem specializes to the fundamental theorem for line integrals.

Theorem 23.14 (Green's theorem). *If D is a plane region with boundary ∂D, and P and Q are C^∞ functions on D, then*

$$\int_{\partial D} P\,dx + Q\,dy = \int_D \left(\frac{\partial Q}{\partial x} - \frac{\partial P}{\partial y}\right)dA.$$

In this statement, dA is the usual calculus notation for $dx\,dy$. To obtain Green's theorem, let M be a plane region D with boundary ∂D and let ω be the 1-form $P\,dx + Q\,dy$ on D. Then

$$\int_{\partial D} \omega = \int_{\partial D} P\,dx + Q\,dy$$

and

$$\int_D d\omega = \int_D P_y\,dy \wedge dx + Q_x\,dx \wedge dy = \int_D (Q_x - P_y)\,dx \wedge dy$$
$$= \int_D (Q_x - P_y)\,dx\,dy = \int_D (Q_x - P_y)\,dA.$$

In this case Stokes's theorem is Green's theorem in the plane.

Problems

23.1. Area of an ellipse
Use the change-of-variables formula to compute the area enclosed by the ellipse

$$x^2/a^2 + y^2/b^2 = 1$$

in $\mathbb{R}^2$.

23.2. Characterization of boundedness in $\mathbb{R}^n$
Prove that a subset $A \subset \mathbb{R}^n$ is bounded if and only if its closure $\bar{A}$ in $\mathbb{R}^n$ is compact.

23.3.* Integral under a diffeomorphism
Suppose N and M are connected, oriented n-manifolds and $F: N \to M$ is a diffeomorphism. Prove that for any $\omega \in \Omega_c^k(M)$,

$$\int_N F^*\omega = \pm \int_M \omega,$$

where the sign depends on whether F is orientation-preserving or orientation-reversing.

23.4.* Stokes's theorem
Prove Stokes's theorem for $\mathbb{R}^n$ and for $\mathcal{H}^n$.

23.5. Area form on the sphere S^2
Prove that the area form ω on S^2 in Example 23.11 is equal to the orientation form

$$x\,dy \wedge dz - y\,dx \wedge dz + z\,dx \wedge dy$$

of S^2 in Problem 22.9.

Chapter 7

De Rham Theory

By the fundamental theorem for line integrals
(Theorem 23.13), if a smooth vector field $\mathbf{F}$ is the
gradient of a scalar function f, then for any two
points p and q in $\mathbb{R}^3$, the line integral $\int_C \mathbf{F} \cdot d\mathbf{r}$ over
a curve C from p to q is independent of the curve.
In this case, the line integral $\int_C \mathbf{F} \cdot d\mathbf{r}$ can be computed in terms of its values at the two endpoints as
$f(q) - f(p)$. Similarly, by the classical Stokes theorem for a surface, the surface integral of smooth a
vector field $\mathbf{F}$ over an oriented surface S with boundary C in $\mathbb{R}^3$ can be evaluated as an integral over the
curve C if $\mathbf{F}$ is the curl of another vector field. It is
thus of interest to know whether a vector field $\mathbb{R}^3$
is the gradient of a function or is the curl of another vector field. By the correspondence of Section 4.6 between vector fields and differential forms,
this translates into whether a differential form ω on
$\mathbb{R}^3$ is exact.

Henri Poincaré

(1854–1912)

Considerations such as these led Henri Poincaré to look for conditions under
which a differential form is exact on $\mathbb{R}^n$. Of course, a necessary condition is that
the form ω be closed. Poincaré proved in 1887 that for $k = 1, 2, 3$, a k-form on $\mathbb{R}^n$
is exact if and only if it is closed, a lemma that now bears his name. Vito Volterra
published in 1889 the first complete proof of the Poincaré lemma for all k.

It turns out that whether every closed form on a manifold is exact depends on
the topology of the manifold. For example, on $\mathbb{R}^2$ every closed k-form is exact for
$k > 0$, but on the punctured plane $\mathbb{R}^2 - \{(0,0)\}$ there are closed 1-forms that are not
exact. The extent to which closed forms are not exact is measured by the de Rham
cohomology, possibly the most important diffeomorphism invariant of a manifold.

L. W. Tu, *An Introduction to Manifolds*, Universitext, DOI 10.1007/978-1-4419-7400-6_7,
© Springer Science+Business Media, LLC 2011

Georges de Rham

(1903–1990)

In a series of groundbreaking papers, starting with "Analysis situs" [33] in 1895, Poincaré introduced the concept of homology and laid the foundations of modern algebraic topology. Roughly speaking, a compact submanifold with no boundary is a *cycle*, and a cycle is *homologous* to zero if it is the boundary of another manifold. The equivalence classes of cycles under the homology relation are called *homology classes*. In his doctoral thesis [8] in 1931, Georges de Rham showed that differential forms satisfy the same axioms as cycles and boundaries, in effect proving a duality between what are now called de Rham cohomology and singular homology with real coefficients. Although he did not define explicitly de Rham cohomology in this paper, it was implicit in his work. A formal definition of de Rham cohomology appeared in 1938 [9].

§24 De Rham Cohomology

In this section we define de Rham cohomology, prove some of its basic properties, and compute two elementary examples: the de Rham cohomology vector spaces of the real line and of the unit circle.

24.1 De Rham Cohomology

Suppose $\mathbf{F}(x,y) = \langle P(x,y), Q(x,y) \rangle$ is a smooth vector field representing a force on an open subset U of $\mathbb{R}^2$, and C is a parametrized curve $c(t) = (x(t), y(t))$ in U from a point p to a point q, with $a \leq t \leq b$. Then the work done by the force in moving a particle from p to q along C is given by the line integral $\int_C P\,dx + Q\,dy$.

Such a line integral is easy to compute if the vector field $\mathbf{F}$ is the gradient of a scalar function $f(x,y)$:

$$\mathbf{F} = \operatorname{grad} f = \langle f_x, f_y \rangle,$$

where $f_x = \partial f / \partial x$ and $f_y = \partial f / \partial y$. By Stokes's theorem, the line integral is simply

$$\int_C f_x\,dx + f_y\,dy = \int_C df = f(q) - f(p).$$

A necessary condition for the vector field $\mathbf{F} = \langle P, Q \rangle$ to be a gradient is that

$$P_y = f_{xy} = f_{yx} = Q_x.$$

The question is now the following: if $P_y - Q_x = 0$, is the vector field $\mathbf{F} = \langle P, Q \rangle$ on U the gradient of some scalar function $f(x,y)$ on U?

In Section 4.6 we established a one-to-one correspondence between vector fields and differential 1-forms on an open subset of $\mathbb{R}^3$. There is a similar correspondence on an open subset of any $\mathbb{R}^n$. For $\mathbb{R}^2$, it is as follows:

$$\text{vector fields} \longleftrightarrow \text{differential 1-forms,}$$
$$\mathbf{F} = \langle P, Q \rangle \longleftrightarrow \omega = P\,dx + Q\,dy,$$
$$\operatorname{grad} f = \langle f_x, f_y \rangle \longleftrightarrow df = f_x\,dx + f_y\,dy,$$
$$Q_x - P_y = 0 \longleftrightarrow d\omega = (Q_x - P_y)\,dx \wedge dy = 0.$$

In terms of differential forms the question above becomes the following: if the 1-form $\omega = P\,dx + Q\,dy$ is closed on U, is it exact? The answer to this question is sometimes yes and sometimes no, depending on the topology of U.

Just as for an open subset of $\mathbb{R}^n$, a differential form ω on a manifold M is said to be *closed* if $d\omega = 0$, and *exact* if $\omega = d\tau$ for some form τ of degree one less. Since $d^2 = 0$, every exact form is closed. In general, not every closed form is exact.

Let $Z^k(M)$ be the vector space of all closed k-forms and $B^k(M)$ the vector space of all exact k-forms on the manifold M. Because every exact form is closed, $B^k(M)$ is a subspace of $Z^k(M)$. The quotient vector space $H^k(M) := Z^k(M)/B^k(M)$ measures the extent to which closed k-forms fail to be exact, and is called the *de Rham cohomology* of M in degree k. As explained in Appendix D, the quotient vector space construction introduces an equivalence relation on $Z^k(M)$:

$$\omega' \sim \omega \quad \text{in } Z^k(M) \quad \text{iff} \quad \omega' - \omega \in B^k(M).$$

The equivalence class of a closed form ω is called its *cohomology class* and denoted by $[\omega]$. Two closed forms ω and ω' determine the same cohomology class if and only if they differ by an exact form:

$$\omega' = \omega + d\tau.$$

In this case we say that the two closed forms ω and ω' are *cohomologous*.

Proposition 24.1. *If the manifold M has r connected components, then its de Rham cohomology in degree 0 is $H^0(M) = \mathbb{R}^r$. An element of $H^0(M)$ is specified by an ordered r-tuple of real numbers, each real number representing a constant function on a connected component of M.*

Proof. Since there are no nonzero exact 0-forms,

$$H^0(M) = Z^0(M) = \{\text{closed 0-forms}\}.$$

Supposed f is a closed 0-form on M; i.e., f is a C^∞ function on M such that $df = 0$. On any chart $(U, x^1, \ldots, x^n)$,

$$df = \sum \frac{\partial f}{\partial x^i}\,dx^i.$$

Thus, $df = 0$ on U if and only if all the partial derivatives $\partial f/\partial x^i$ vanish identically on U. This in turn is equivalent to f being locally constant on U. Hence, the closed 0-forms on M are precisely the locally constant functions on M. Such a function must be constant on each connected component of M. If M has r connected components, then a locally constant function on M can be specified by an ordered set of r real numbers. Thus, $Z^0(M) = \mathbb{R}^r$. □

Proposition 24.2. *On a manifold M of dimension n, the de Rham cohomology $H^k(M)$ vanishes for $k > n$.*

Proof. At any point $p \in M$, the tangent space T_pM is a vector space of dimension n. If ω is a k-form on M, then $\omega_p \in A_k(T_pM)$, the space of alternating k-linear functions on T_pM. By Corollary 3.31, if $k > n$, then $A_k(T_pM) = 0$. Hence, for $k > n$, the only k-form on M is the zero form. □

24.2 Examples of de Rham Cohomology

Example 24.3 (*De Rham cohomology of the real line*). Since the real line $\mathbb{R}^1$ is connected, by Proposition 24.1,

$$H^0(\mathbb{R}^1) = \mathbb{R}.$$

For dimensional reasons, on $\mathbb{R}^1$ there are no nonzero 2-forms. This implies that every 1-form on $\mathbb{R}^1$ is closed. A 1-form $f(x)\,dx$ on $\mathbb{R}^1$ is exact if and only if there is a C^∞ function $g(x)$ on $\mathbb{R}^1$ such that

$$f(x)\,dx = dg = g'(x)\,dx,$$

where $g'(x)$ is the calculus derivative of g with respect to x. Such a function $g(x)$ is simply an antiderivative of $f(x)$, for example

$$g(x) = \int_0^x f(t)\,dt.$$

This proves that every 1-form on $\mathbb{R}^1$ is exact. Therefore, $H^1(\mathbb{R}^1) = 0$. In combination with Proposition 24.2, we have

$$H^k(\mathbb{R}^1) = \begin{cases} \mathbb{R} & \text{for } k = 0, \\ 0 & \text{for } k \geq 1. \end{cases}$$

Example 24.4 (*De Rham cohomology of a circle*). Let S^1 be the unit circle in the xy-plane. By Proposition 24.1, because S^1 is connected, $H^0(S^1) = \mathbb{R}$, and because S^1 is one-dimensional, $H^k(S^1) = 0$ for all $k \geq 2$. It remains to compute $H^1(S^1)$.

Recall from Subsection 18.7 the map $h \colon \mathbb{R} \to S^1$, $h(t) = (\cos t, \sin t)$. Let $i \colon [0, 2\pi] \to \mathbb{R}$ be the inclusion map. Restricting the domain of h to $[0, 2\pi]$ gives a parametrization $F := h \circ i \colon [0, 2\pi] \to S^1$ of the circle. In Examples 17.15 and 17.16, we found a nowhere-vanishing 1-form $\omega = -y\,dx + x\,dy$ on S^1 and showed that $F^*\omega = i^*h^*\omega = i^*dt = dt$. Thus,

$$\int_{S^1} \omega = \int_{F([0,2\pi])} \omega = \int_{[0,2\pi]} F^*\omega = \int_0^{2\pi} dt = 2\pi.$$

Since the circle has dimension 1, all 1-forms on S^1 are closed, so $\Omega^1(S^1) = Z^1(S^1)$. The integration of 1-forms on S^1 defines a linear map

$$\varphi: Z^1(S^1) = \Omega^1(S^1) \to \mathbb{R}, \quad \varphi(\alpha) = \int_{S^1} \alpha.$$

Because $\varphi(\omega) = 2\pi \neq 0$, the linear map $\varphi: \Omega^1(S^1) \to \mathbb{R}$ is onto.

By Stokes's theorem, the exact 1-forms on S^1 are in $\ker\varphi$. Conversely, we will show that all 1-forms in $\ker\varphi$ are exact. Suppose $\alpha = f\omega$ is a smooth 1-form on S^1 such that $\varphi(\alpha) = 0$. Let $\bar{f} = h^*f = f \circ h \in \Omega^0(\mathbb{R})$. Then $\bar{f}$ is periodic of period 2π and

$$0 = \int_{S^1} \alpha = \int_{F([0,2\pi])} \alpha = \int_{[0,2\pi]} F^*\alpha = \int_{[0,2\pi]} (i^*h^*f)(t) \cdot F^*\omega = \int_0^{2\pi} \bar{f}(t)\,dt.$$

Lemma 24.5. *Suppose $\bar{f}$ is a C^∞ periodic function of period 2π on $\mathbb{R}$ and $\int_0^{2\pi} \bar{f}(u)\,du = 0$. Then $\bar{f}\,dt = d\bar{g}$ for a C^∞ periodic function $\bar{g}$ of period 2π on $\mathbb{R}$.*

Proof. Define $\bar{g} \in \Omega^0(\mathbb{R})$ by

$$\bar{g}(t) = \int_0^t \bar{f}(u)\,du.$$

Since $\int_0^{2\pi} \bar{f}(u)\,du = 0$ and $\bar{f}$ is periodic of period 2π,

$$\bar{g}(t+2\pi) = \int_0^{2\pi} \bar{f}(u)\,du + \int_{2\pi}^{t+2\pi} \bar{f}(u)\,du$$

$$= 0 + \int_{2\pi}^{t+2\pi} \bar{f}(u)\,du = \int_0^t \bar{f}(u)\,du = \bar{g}(t).$$

Hence, $\bar{g}(t)$ is also periodic of period 2π on $\mathbb{R}$. Moreover,

$$d\bar{g} = \bar{g}'(t)\,dt = \bar{f}(t)\,dt. \qquad \square$$

Let $\bar{g}$ be the periodic function of period 2π on $\mathbb{R}$ from Lemma 24.5. By Proposition 18.12, $\bar{g} = h^*g$ for some C^∞ function g on S^1. It follows that

$$d\bar{g} = dh^*g = h^*(dg).$$

On the other hand,

$$\bar{f}(t)\,dt = (h^*f)(h^*\omega) = h^*(f\omega) = h^*\alpha.$$

Since $h^*: \Omega^1(S^1) \to \Omega^1(\mathbb{R})$ is injective, $\alpha = dg$. This proves that the kernel of φ consists of exact forms. Therefore, integration induces an isomorphism

$$H^1(S^1) = \frac{Z^1(S^1)}{B^1(S^1)} \xrightarrow{\sim} \mathbb{R}.$$

In the next section we will develop a tool, the *Mayer–Vietoris sequence*, using which the computation of the cohomology of the circle becomes more or less routine.

24.3 Diffeomorphism Invariance

For any smooth map $F: N \to M$ of manifolds, there is a *pullback map* $F^*: \Omega^*(M) \to \Omega^*(N)$ of differential forms. Moreover, the pullback F^* commutes with the exterior derivative d (Proposition 19.5).

Lemma 24.6. *The pullback map F^* sends closed forms to closed forms, and sends exact forms to exact forms.*

Proof. Suppose ω is closed. By the commutativity of F^* with d,

$$dF^*\omega = F^*d\omega = 0.$$

Hence, $F^*\omega$ is also closed.

Next suppose $\omega = d\tau$ is exact. Then

$$F^*\omega = F^*d\tau = dF^*\tau.$$

Hence, $F^*\omega$ is exact. □

It follows that F^* induces a linear map of quotient spaces, denoted by $F^\#$:

$$F^\#: \frac{Z^k(M)}{B^k(M)} \to \frac{Z^k(N)}{B^k(N)}, \quad F^\#([\omega]) = [F^*(\omega)].$$

This is a map in cohomology,

$$F^\#: H^k(M) \to H^k(N),$$

called the *pullback map in cohomology*.

Remark 24.7. The functorial properties of the pullback map F^* on differential forms easily yield the same functorial properties for the induced map in cohomology:

(i) If $\mathbb{1}_M: M \to M$ is the identity map, then $\mathbb{1}_M^\#: H^k(M) \to H^k(M)$ is also the identity map.
(ii) If $F: N \to M$ and $G: M \to P$ are smooth maps, then

$$(G \circ F)^\# = F^\# \circ G^\#.$$

It follows from (i) and (ii) that $(H^k(\ ), F^\#)$ is a contravariant functor from the category of C^∞ manifolds and C^∞ maps to the category of vector spaces and linear maps. By Proposition 10.3, if $F: N \to M$ is a diffeomorphism of manifolds, then $F^\#: H^k(M) \to H^k(N)$ is an isomorphism of vector spaces.

In fact, the usual notation for the induced map in cohomology is F^*, the same as for the pullback map on differential forms. Unless there is a possibility of confusion, henceforth we will follow this convention. It is usually clear from the context whether F^* is a map in cohomology or on forms.

24.4 The Ring Structure on de Rham Cohomology

The wedge product of differential forms on a manifold M gives the vector space $\Omega^*(M)$ of differential forms a product structure. This product structure induces a product structure in cohomology: if $[\omega] \in H^k(M)$ and $[\tau] \in H^\ell(M)$, define

$$[\omega] \wedge [\tau] = [\omega \wedge \tau] \in H^{k+\ell}(M). \tag{24.1}$$

For the product to be well defined, we need to check three things about closed forms ω and τ:

(i) The wedge product $\omega \wedge \tau$ is a closed form.
(ii) The class $[\omega \wedge \tau]$ is independent of the choice of representative for $[\tau]$. In other words, if τ is replaced by a cohomologous form $\tau' = \tau + d\sigma$, then in the equation

$$\omega \wedge \tau' = \omega \wedge \tau + \omega \wedge d\sigma,$$

we need to show that $\omega \wedge d\sigma$ is exact.
(iii) The class $[\omega \wedge \tau]$ is independent of the choice of representative for $[\omega]$.

These all follow from the antiderivation property of d. For example, in (i), since ω and τ are closed,

$$d(\omega \wedge \tau) = (d\omega) \wedge \tau + (-1)^k \omega \wedge d\tau = 0.$$

In (ii),

$$d(\omega \wedge \sigma) = (d\omega) \wedge \sigma + (-1)^k \omega \wedge d\sigma = (-1)^k \omega \wedge d\sigma \quad (\text{since } d\omega = 0),$$

which shows that $\omega \wedge d\sigma$ is exact. Item (iii) is analogous to (ii), with the roles of ω and τ reversed.

If M is a manifold of dimension n, we set

$$H^*(M) = \bigoplus_{k=0}^{n} H^k(M).$$

What this means is that an element α of $H^*(M)$ is uniquely a finite sum of cohomology classes in $H^k(M)$ for various k's:

$$\alpha = \alpha_0 + \cdots + \alpha_n, \quad \alpha_k \in H^k(M).$$

Elements of $H^*(M)$ can be added and multiplied in the same way that one would add or multiply polynomials, except here multiplication is the wedge product. It is easy to check that under addition and multiplication, $H^*(M)$ satisfies all the properties of a ring, called the *cohomology ring* of M. The ring $H^*(M)$ has a natural grading by the degree of a closed form. Recall that a ring A is *graded* if it can be written as a direct sum $A = \bigoplus_{k=0}^{\infty} A^k$ so that the ring multiplication sends $A^k \times A^\ell$ to $A^{k \times \ell}$. A graded ring $A = \bigoplus_{k=0}^{\infty} A^k$ is said to be *anticommutative* if for all $a \in A^k$ and $b \in A^\ell$,

$$a \cdot b = (-1)^{k\ell} b \cdot a.$$

In this terminology, $H^*(M)$ is an anticommutative graded ring. Since $H^*(M)$ is also a real vector space, it is in fact an anticommutative graded algebra over $\mathbb{R}$.

Suppose $F \colon N \to M$ is a C^∞ map of manifolds. Because $F^*(\omega \wedge \tau) = F^*\omega \wedge F^*\tau$ for differential forms ω and τ on M (Proposition 18.11), the linear map $F^* \colon H^*(M) \to H^*(N)$ is a ring homomorphism. By Remark 24.7, if $F \colon N \to M$ is a diffeomorphism, then the pullback $F^* \colon H^*(M) \to H^*(N)$ is a ring isomorphism.

To sum up, de Rham cohomology gives a contravariant functor from the category of C^∞ manifolds to the category of anticommutative graded rings. If M and N are diffeomorphic manifolds, then $H^*(M)$ and $H^*(N)$ are isomorphic as anticommutative graded rings. In this way the de Rham cohomology becomes a powerful diffeomorphism invariant of C^∞ manifolds.

Problems

24.1. Nowhere-vanishing 1-forms
Prove that a nowhere-vanishing 1-form on a compact manifold cannot be exact.

24.2. Cohomology in degree zero
Suppose a manifold M has infinitely many connected components. Compute its de Rham cohomology vector space $H^0(M)$ in degree 0. (*Hint*: By second countability, the number of connected components of a manifold is countable.)

§25 The Long Exact Sequence in Cohomology

A *cochain complex* $\mathcal{C}$ is a collection of vector spaces $\{C^k\}_{k\in\mathbb{Z}}$ together with a sequence of linear maps $d_k \colon C^k \to C^{k+1}$,

$$\cdots \to C^{-1} \xrightarrow{d_{-1}} C^0 \xrightarrow{d_0} C^1 \xrightarrow{d_1} C^2 \xrightarrow{d_2} \cdots,$$

such that

$$d_k \circ d_{k-1} = 0 \tag{25.1}$$

for all k. We will call the collection of linear maps $\{d_k\}$ the *differential* of the cochain complex $\mathcal{C}$.

The vector space $\Omega^*(M)$ of differential forms on a manifold M together with the exterior derivative d is a cochain complex, the *de Rham complex* of M:

$$0 \to \Omega^0(M) \xrightarrow{d} \Omega^1(M) \xrightarrow{d} \Omega^2(M) \xrightarrow{d} \cdots, \quad d \circ d = 0.$$

It turns out that many of the results on the de Rham cohomology of a manifold depend not on the topological properties of the manifold, but on the algebraic properties of the de Rham complex. To better understand de Rham cohomology, it is useful to isolate these algebraic properties. In this section we investigate the properties of a cochain complex that constitute the beginning of a subject known as *homological algebra*.

25.1 Exact Sequences

This subsection is a compendium of a few basic properties of exactness that will be used over and over again.

Definition 25.1. A sequence of homomorphisms of vector spaces

$$A \xrightarrow{f} B \xrightarrow{g} C$$

is said to be *exact at B* if $\operatorname{im} f = \ker g$. A sequence of homomorphisms

$$A^0 \xrightarrow{f_0} A^1 \xrightarrow{f_1} A^2 \xrightarrow{f_2} \cdots \xrightarrow{f_{n-1}} A^n$$

that is exact at every term except the first and the last is simply said to be an *exact sequence*. A five-term exact sequence of the form

$$0 \to A \to B \to C \to 0$$

is said to be *short exact*.

The same definition applies to homomorphisms of groups or modules, but we are mainly concerned with vector spaces.

Remark. (i) When $A = 0$, the sequence

$$0 \xrightarrow{f} B \xrightarrow{g} C$$

is exact if and only if

$$\ker g = \operatorname{im} f = 0,$$

so that g is injective.

(ii) Similarly, when $C = 0$, the sequence

$$A \xrightarrow{f} B \xrightarrow{g} 0$$

is exact if and only if

$$\operatorname{im} f = \ker g = B,$$

so that f is surjective.

The following two propositions are very useful for dealing with exact sequences.

Proposition 25.2 (A three-term exact sequence). *Suppose*

$$A \xrightarrow{f} B \xrightarrow{g} C$$

is an exact sequence. Then

(i) *the map f is surjective if and only if g is the zero map;*

(ii) *the map g is injective if and only if f is the zero map.*

Proof. Problem 25.1.   □

Proposition 25.3 (A four-term exact sequence).

(i) *The four-term sequence $0 \to A \xrightarrow{f} B \to 0$ of vector spaces is exact if and only if $f : A \to B$ is an isomorphism.*

(ii) *If*

$$A \xrightarrow{f} B \to C \to 0$$

is an exact sequence of vector spaces, then there is a linear isomorphism

$$C \simeq \operatorname{coker} f := \frac{B}{\operatorname{im} f}.$$

Proof. Problem 25.2.   □

25.2 Cohomology of Cochain Complexes

If $\mathcal{C}$ is a cochain complex, then by (25.1),

$$\operatorname{im} d_{k-1} \subset \ker d_k.$$

We can therefore form the quotient vector space

$$H^k(\mathcal{C}) := \frac{\ker d_k}{\operatorname{im} d_{k-1}},$$

which is called the *kth cohomology vector space* of the cochain complex $\mathcal{C}$. It is a measure of the extent to which the cochain complex $\mathcal{C}$ fails to be exact at C^k. Elements of the vector space C^k are called *cochains of degree k* or *k-cochains* for short. A k-cochain in $\ker d_k$ is called a *k-cocycle* and a k-cochain in $\operatorname{im} d_{k-1}$ is called a *k-coboundary*. The equivalence class $[c] \in H^k(\mathcal{C})$ of a k-cocycle $c \in \ker d_k$ is called its *cohomology class*. We denote the subspaces of k-cocycles and k-coboundaries of $\mathcal{C}$ by $Z^k(\mathcal{C})$ and $B^k(\mathcal{C})$ respectively. The letter Z for cocycles comes from *Zyklen*, the German word for cycles.

To simplify the notation we will usually omit the subscript in d_k, and write $d \circ d = 0$ instead of $d_k \circ d_{k-1} = 0$.

Example. In the de Rham complex, a cocycle is a closed form and a coboundary is an exact form.

If $\mathcal{A}$ and $\mathcal{B}$ are two cochain complexes with differentials d and d' respectively, a *cochain map* $\varphi \colon \mathcal{A} \to \mathcal{B}$ is a collection of linear maps $\varphi_k \colon A^k \to B^k$, one for each k, that commute with d and d':

$$d' \circ \varphi_k = \varphi_{k+1} \circ d.$$

In other words, the following diagram is commutative:

$$
\begin{array}{ccccccc}
\cdots \longrightarrow & A^{k-1} & \overset{d}{\longrightarrow} & A^k & \overset{d}{\longrightarrow} & A^{k+1} & \longrightarrow \cdots \\
& \Big\downarrow{\varphi_{k-1}} & & \Big\downarrow{\varphi_k} & & \Big\downarrow{\varphi_{k+1}} & \\
\cdots \longrightarrow & B^{k-1} & \underset{d'}{\longrightarrow} & B^k & \underset{d'}{\longrightarrow} & B^{k+1} & \longrightarrow \cdots .
\end{array}
$$

We will usually omit the subscript k in φ_k.

A cochain map $\varphi \colon \mathcal{A} \to \mathcal{B}$ naturally induces a linear map in cohomology

$$\varphi^* \colon H^k(\mathcal{A}) \to H^k(\mathcal{B})$$

by

$$\varphi^*[a] = [\varphi(a)]. \tag{25.2}$$

To show that this is well defined, we need to check that a cochain map takes cocycles to cocycles, and coboundaries to coboundaries:

(i) for $a \in Z^k(\mathcal{A})$, $d'(\varphi(a)) = \varphi(da) = 0$;
(ii) for $a' \in A^{k-1}$, $\varphi(da') = d'(\varphi(a'))$.

Example 25.4.
(i) For a smooth map $F: N \to M$ of manifolds, the pullback map $F^*: \Omega^*(M) \to \Omega^*(N)$ on differential forms is a cochain map, because F^* commutes with d (Proposition 19.5). By the discussion above, there is an induced map $F^*: H^*(M) \to H^*(N)$ in cohomology, as we saw once before, after Lemma 24.6.
(ii) If X is a C^∞ vector field on a manifold M, then the Lie derivative $\mathcal{L}_X: \Omega^*(M) \to \Omega^*(M)$ commutes with d (Theorem 20.10(ii)). By (25.2), $\mathcal{L}_X$ induces a linear map $\mathcal{L}_X^*: H^*(M) \to H^*(M)$ in cohomology.

25.3 The Connecting Homomorphism

A sequence of cochain complexes

$$0 \to \mathcal{A} \xrightarrow{i} \mathcal{B} \xrightarrow{j} \mathcal{C} \to 0$$

is *short exact* if i and j are cochain maps and for each k,

$$0 \to A^k \xrightarrow{i_k} B^k \xrightarrow{j_k} C^k \to 0$$

is a short exact sequence of vector spaces. Since we usually omit subscripts on cochain maps, we will write i, j instead of i_k, j_k.

Given a short exact sequence as above, we can construct a linear map $d^*: H^k(\mathcal{C}) \to H^{k+1}(\mathcal{A})$, called the *connecting homomorphism*, as follows. Consider the short exact sequences in dimensions k and $k+1$:

$$
\begin{array}{ccccccccc}
0 & \longrightarrow & A^{k+1} & \xrightarrow{i} & B^{k+1} & \xrightarrow{j} & C^{k+1} & \longrightarrow & 0 \\
 & & \big\uparrow{\scriptstyle d} & & \big\uparrow{\scriptstyle d} & & \big\uparrow{\scriptstyle d} & & \\
0 & \longrightarrow & A^k & \xrightarrow{i} & B^k & \xrightarrow{j} & C^k & \longrightarrow & 0.
\end{array}
$$

To keep the notation simple, we use the same symbol d to denote the a priori distinct differentials d_A, d_B, d_C of the three cochain complexes. Start with $[c] \in H^k(\mathcal{C})$. Since $j: B^k \to C^k$ is onto, there is an element $b \in B^k$ such that $j(b) = c$. Then $db \in B^{k+1}$ is in $\ker j$ because

$$
\begin{aligned}
jdb &= djb && \text{(by the commutativity of the diagram)} \\
&= dc = 0 && \text{(because } c \text{ is a cocycle).}
\end{aligned}
$$

By the exactness of the sequence in degree $k+1$, $\ker j = \operatorname{im} i$. This implies that $db = i(a)$ for some a in A^{k+1}. Once b is chosen, this a is unique because i is injective. The injectivity of i also implies that $da = 0$, since

$$i(da) = d(ia) = ddb = 0. \tag{25.3}$$

Therefore, a is a cocycle and defines a cohomology class $[a]$. We set

$$d^*[c] = [a] \in H^{k+1}(\mathcal{A}).$$

In defining $d^*[c]$ we made two choices: a cocycle c to represent the cohomology class $[c] \in H^k(\mathcal{C})$ and then an element $b \in B^k$ that maps to c under j. For d^* to be well defined, one must show that the cohomology class $[a] \in H^{k+1}(\mathcal{A})$ does not depend on these choices.

Exercise 25.5 (Connecting homomorphism).* Show that the connecting homomorphism

$$d^* : H^k(\mathcal{C}) \to H^{k+1}(\mathcal{A})$$

is a well-defined linear map.

The recipe for defining the connecting homomorphism d^* is best remembered as a zig-zag diagram,

$$
\begin{array}{ccc}
a & \overset{i}{\rightarrowtail} & db \\
 & & \big\uparrow{\scriptstyle d} \\
 & b & \overset{j}{\longmapsto\mkern-14mu\longrightarrow} c,
\end{array}
$$

where $a \rightarrowtail db$ means that a maps to db under an injection and $b \longmapsto\mkern-14mu\rightarrow c$ means that b maps to c under a surjection.

25.4 The Zig-Zag Lemma

The zig-zag lemma produces a long exact sequence in cohomology from a short exact sequence of cochain complexes. It is most useful when some of the terms in the long exact sequence are known to be zero, for then by exactness, the adjacent maps will be injections, surjections, or even isomorphisms. For example, if the cohomology of one of the three cochain complexes is zero, then the cohomology vector spaces of the other two cochain complexes will be isomorphic.

Theorem 25.6 (The zig-zag lemma). *A short exact sequence of cochain complexes*

$$0 \to \mathcal{A} \overset{i}{\to} \mathcal{B} \overset{j}{\to} \mathcal{C} \to 0$$

gives rise to a long exact sequence in cohomology:

$$
\begin{array}{l}
\longrightarrow H^{k+1}(\mathcal{A}) \overset{i^*}{\longrightarrow} \cdots, \\[4pt]
\overline{d^*} \\[2pt]
\longrightarrow H^k(\mathcal{A}) \overset{i^*}{\longrightarrow} H^k(\mathcal{B}) \overset{j^*}{\longrightarrow} H^k(\mathcal{C}) \longrightarrow \\[4pt]
\overline{d^*} \\[2pt]
\cdots \overset{j^*}{\longrightarrow} H^{k-1}(\mathcal{C}) \longrightarrow
\end{array}
\tag{25.4}
$$

where i^* and j^* are the maps in cohomology induced from the cochain maps i and j, and d^* is the connecting homomorphism.

To prove the theorem one needs to check exactness at $H^k(\mathcal{A})$, $H^k(\mathcal{B})$, and $H^k(\mathcal{C})$ for each k. The proof is a sequence of trivialities involving what is commonly called *diagram-chasing*. As an example, we prove exactness at $H^k(\mathcal{C})$.

Claim. $\operatorname{im} j^* \subset \ker d^*$.

Proof. Let $[b] \in H^k(\mathcal{B})$. Then

$$d^* j^*[b] = d^*[j(b)].$$

In the recipe above for d^*, we can choose the element in B^k that maps to $j(b)$ to be b. Then $db \in B^{k+1}$. Because b is a cocycle, $db = 0$. Following the zig-zag diagram

$$
\begin{array}{ccc}
0 & \xrightarrow{\ i\ } & db = 0 \\
 & & \big\uparrow{\scriptstyle d} \\
 & & b \xrightarrow{\ j\ } j(b),
\end{array}
$$

we see that since $i(0) = 0 = db$, we must have $d^*[j(b)] = [0]$. So $j^*[b] \in \ker d^*$. □

Claim. $\ker d^* \subset \operatorname{im} j^*$.

Proof. Suppose $d^*[c] = [a] = 0$, where $[c] \in H^k(\mathcal{C})$. This means that $a = da'$ for some $a' \in A^k$. The calculation of $d^*[c]$ can be represented by the zig-zag diagram

$$
\begin{array}{ccc}
a & \xrightarrow{\ i\ } & db \\
\big\uparrow{\scriptstyle d} & & \big\uparrow{\scriptstyle d} \\
a' & & b \xrightarrow{\ j\ } c,
\end{array}
$$

where b is an element in B^k with $j(b) = c$ and $i(a) = db$. Then $b - i(a')$ is a cocycle in B^k that maps to c under j:

$$d(b - i(a')) = db - di(a') = db - id(a') = db - ia = 0,$$
$$j(b - i(a')) = j(b) - ji(a') = j(b) = c.$$

Therefore,

$$j^*[b - i(a')] = [c].$$

So $[c] \in \operatorname{im} j^*$. □

These two claims together imply the exactness of (25.4) at $H^k(\mathcal{C})$. As for the exactness of the cohomology sequence (25.4) at $H^k(\mathcal{A})$ and at $H^k(\mathcal{B})$, we will leave it as an exercise (Problem 25.3).

Problems

25.1. A three-term exact sequence
Prove Proposition 25.1.

25.2. A four-term exact sequence
Prove Proposition 25.2.

25.3. Long exact cohomology sequence
Prove the exactness of the cohomology sequence (25.4) at $H^k(\mathcal{A})$ and $H^k(\mathcal{B})$.

25.4.* The snake lemma[1]
Use the zig-zag lemma to prove the following:
The snake lemma. *A commutative diagram with exact rows*

$$
\begin{array}{ccccccccc}
0 & \longrightarrow & A^1 & \longrightarrow & B^1 & \longrightarrow & C^1 & \longrightarrow & 0 \\
& & \uparrow{\scriptstyle\alpha} & & \uparrow{\scriptstyle\beta} & & \uparrow{\scriptstyle\gamma} & & \\
0 & \longrightarrow & A^0 & \longrightarrow & B^0 & \longrightarrow & C^0 & \longrightarrow & 0
\end{array}
$$

induces a long exact sequence

$$
\begin{array}{c}
 \longrightarrow \operatorname{coker}\alpha \longrightarrow \operatorname{coker}\beta \longrightarrow \operatorname{coker}\gamma \longrightarrow 0. \\[2mm]
0 \longrightarrow \ker\alpha \longrightarrow \ker\beta \longrightarrow \ker\gamma
\end{array}
$$

---

[1]The snake lemma, also called the serpent lemma, derives its name from the shape of the long exact sequence in it, usually drawn as an S. It may be the only result from homological algebra that has made its way into popular culture. In the 1980 film *It's My Turn* there is a scene in which the actress Jill Clayburgh, who plays a mathematics professor, explains the proof of the snake lemma.

§26 The Mayer–Vietoris Sequence

As the example of the cohomology of the real line $\mathbb{R}^1$ illustrates, calculating the de Rham cohomology of a manifold amounts to solving a canonically given system of differential equations on the manifold and, in case it is not solvable, to finding obstructions to its solvability. This is usually quite difficult to do directly. We introduce in this section one of the most useful tools in the calculation of de Rham cohomology, the Mayer–Vietoris sequence. Another tool, the homotopy axiom, will come in the next section.

26.1 The Mayer–Vietoris Sequence

Let $\{U,V\}$ be an open cover of a manifold M, and let $i_U : U \to M$, $i_U(p) = p$, be the inclusion map. Then the pullback

$$i_U^* : \Omega^k(M) \to \Omega^k(U)$$

is the restriction map that restricts the domain of a k-form on M to U: $i_U^* \omega = \omega|_U$. In fact, there are four inclusion maps that form a commutative diagram:

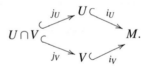

By restricting a k-form from M to U and to V, we get a homomorphism of vector spaces

$$i: \Omega^k(M) \to \Omega^k(U) \oplus \Omega^k(V),$$
$$\sigma \mapsto (i_U^* \sigma, i_V^* \sigma) = (\sigma|_U, \sigma|_V).$$

Define the map
$$j: \Omega^k(U) \oplus \Omega^k(V) \to \Omega^k(U \cap V)$$

by
$$j(\omega, \tau) = j_V^* \tau - j_U^* \omega = \tau|_{U \cap V} - \omega|_{U \cap V}. \tag{26.1}$$

If $U \cap V$ is empty, we define $\Omega^k(U \cap V) = 0$. In this case, j is simply the zero map. We call i the *restriction map* and j the *difference map*. Since the direct sum $\Omega^*(U) \oplus \Omega^*(V)$ is the de Rham complex $\Omega^*(U \amalg V)$ of the disjoint union $U \amalg V$, the exterior derivative d on $\Omega^*(U) \oplus \Omega^*(V)$ is given by $d(\omega, \tau) = (d\omega, d\tau)$.

Proposition 26.1. *Both the restriction map i and the difference map j commute with the exterior derivative d.*

Proof. This is a consequence of the commutativity of d with the pullback (Proposition 19.5). For $\sigma \in \Omega^k(M)$,

$$di\sigma = d(i_U^*\sigma, i_V^*\sigma) = (di_U^*\sigma, di_V^*\sigma) = (i_U^*d\sigma, i_V^*d\sigma) = id\sigma.$$

For $(\omega, \tau) \in \Omega^k(U) \oplus \Omega^k(V)$,

$$dj(\omega, \tau) = d(j_V^*\tau - j_U^*\omega) = j_V^*d\tau - j_U^*d\omega = jd(\omega, \tau). \qquad \square$$

Thus, i and j are cochain maps.

Proposition 26.2. *For each integer $k \geq 0$, the sequence*

$$0 \to \Omega^k(M) \xrightarrow{i} \Omega^k(U) \oplus \Omega^k(V) \xrightarrow{j} \Omega^k(U \cap V) \to 0 \qquad (26.2)$$

is exact.

Proof. Exactness at the first two terms $\Omega^k(M)$ and $\Omega^k(U) \oplus \Omega^k(V)$ is straightforward. We leave it as an exercise (Problem 26.1). We will prove exactness at $\Omega^k(U \cap V)$.

To prove the surjectivity of the difference map

$$j: \Omega^k(U) \oplus \Omega^k(V) \to \Omega^k(U \cap V),$$

it is best to consider first the case of functions on $M = \mathbb{R}^1$. Let f be a C^∞ function on $U \cap V$ as in Figure 26.1. We have to write f as the difference of a C^∞ function on V and a C^∞ function on U.

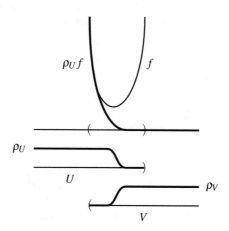

Fig. 26.1. Writing f as the difference of a C^∞ function on V and a C^∞ function on U.

Let $\{\rho_U, \rho_V\}$ be a partition of unity subordinate to the open cover $\{U, V\}$. Define $f_V: V \to \mathbb{R}$ by

$$f_V(x) = \begin{cases} \rho_U(x)f(x) & \text{for } x \in U \cap V, \\ 0 & \text{for } x \in V - (U \cap V). \end{cases}$$

Exercise 26.3 (Smooth extension of a function). Prove that f_V is a C^∞ function on V.

The function f_V is called the *extension by zero* of $\rho_U f$ from $U \cap V$ to V. Similarly, we define f_U to be the extension by zero of $\rho_V f$ from $U \cap V$ to U. Note that to "extend" the domain of f from $U \cap V$ to one of the two open sets, we multiply by the partition function of the other open set. Since

$$j(-f_U, f_V) = f_V|_{U \cap V} + f_U|_{U \cap V} = \rho_U f + \rho_V f = f \quad \text{on } U \cap V,$$

j is surjective.

For differential k-forms on a general manifold M, the formula is similar. For $\omega \in \Omega^k(U \cap V)$, define ω_U to be the extension by zero of $\rho_V \omega$ from $U \cap V$ to U, and ω_V to be the extension by zero of $\rho_U \omega$ from $U \cap V$ to V. On $U \cap V$, $(-\omega_U, \omega_V)$ restricts to $(-\rho_V \omega, \rho_U \omega)$. Hence, j maps $(-\omega_U, \omega_V) \in \Omega^k(U) \oplus \Omega^k(V)$ to

$$\rho_V \omega - (-\rho_U \omega) = \omega \in \Omega^k(U \cap V).$$

This shows that j is surjective and the sequence (26.2) is exact at $\Omega^k(U \cap V)$.   □

It follows from Proposition 26.2 that the sequence of cochain complexes

$$0 \to \Omega^*(M) \xrightarrow{i} \Omega^*(U) \oplus \Omega^*(V) \xrightarrow{j} \Omega^*(U \cap V) \to 0$$

is short exact. By the zig-zag lemma (Theorem 25.6), this short exact sequence of cochain complexes gives rise to a long exact sequence in cohomology, called the *Mayer–Vietoris sequence*:

$$
\begin{array}{l}
 \to H^{k+1}(M) \xrightarrow{i^*} \cdots . \\[2mm]
 \underset{d^*}{} \\[2mm]
 \to H^k(M) \xrightarrow{i^*} H^k(U) \oplus H^k(V) \xrightarrow{j^*} H^k(U \cap V) \qquad (26.3)\\[2mm]
 \underset{d^*}{} \\[2mm]
 \cdots \xrightarrow{j^*} H^{k-1}(U \cap V)
\end{array}
$$

In this sequence i^* and j^* are induced from i and j:

$$i^*[\sigma] = [i(\sigma)] = ([\sigma|_U], [\sigma|_V]) \in H^k(U) \oplus H^k(V),$$
$$j^*([\omega], [\tau]) = [j(\omega, \tau)] = [\tau|_{U \cap V} - \omega|_{U \cap V}] \in H^k(U \cap V).$$

By the recipe of Section 25.3, the connecting homomorphism $d^*: H^k(U \cap V) \to H^{k+1}(M)$ is obtained in three steps as in the diagrams below:

$$\Omega^{k+1}(M) \overset{i}{\rightarrowtail} \Omega^{k+1}(U) \oplus \Omega^{k+1}(V) \qquad\qquad \alpha \underset{(3)}{\overset{i}{\rightarrowtail}} (-d\zeta_U, d\zeta_V) \overset{j}{\longmapsto\!\!\!\rightarrow} 0$$

$$d \Big\uparrow \qquad\qquad\qquad\qquad\qquad d \Big\uparrow (2) \qquad d \Big\uparrow$$

$$\Omega^k(U) \oplus \Omega^k(V) \overset{j}{\longrightarrow} \Omega^k(U \cap V), \qquad (-\zeta_U, \zeta_V) \underset{(1)}{\overset{j}{\longmapsto}} \zeta.$$

(1) Starting with a closed k-form $\zeta \in \Omega^k(U \cap V)$ and using a partition of unity $\{\rho_U, \rho_V\}$ subordinate to $\{U,V\}$, one can extend $\rho_U \zeta$ by zero from $U \cap V$ to a k-form ζ_V on V and extend $\rho_V \zeta$ by zero from $U \cap V$ to a k-form ζ_U on U (see the proof of Proposition 26.2). Then

$$j(-\zeta_U, \zeta_V) = \zeta_V|_{U \cap V} + \zeta_U|_{U \cap V} = (\rho_U + \rho_V)\zeta = \zeta.$$

(2) The commutativity of the square for d and j shows that the pair $(-d\zeta_U, d\zeta_V)$ maps to 0 under j. More formally, since $jd = dj$ and since ζ is a cocycle,

$$j(-d\zeta_U, d\zeta_V) = jd(-\zeta_U, \zeta_V) = dj(-\zeta_U, \zeta_V) = d\zeta = 0.$$

It follows that the $(k+1)$-forms $-d\zeta_U$ on U and $d\zeta_V$ on V agree on $U \cap V$.

(3) Therefore, $-d\zeta_U$ on U and $d\zeta_V$ patch together to give a global $(k+1)$-form α on M. Diagram-chasing shows that α is closed (see (25.3)). By Section 25.3, $d^*[\zeta] = [\alpha] \in H^{k+1}(M)$.

Because $\Omega^k(M) = 0$ for $k \leq -1$, the Mayer–Vietoris sequence starts with

$$0 \to H^0(M) \to H^0(U) \oplus H^0(V) \to H^0(U \cap V) \to \cdots .$$

Proposition 26.4. *In the Mayer–Vietoris sequence, if U, V, and $U \cap V$ are connected and nonempty, then*

(i) *M is connected and*

$$0 \to H^0(M) \to H^0(U) \oplus H^0(V) \to H^0(U \cap V) \to 0$$

is exact;

(ii) *we may start the Mayer–Vietoris sequence with*

$$0 \to H^1(M) \overset{i^*}{\to} H^1(U) \oplus H^1(V) \overset{j^*}{\to} H^1(U \cap V) \to \cdots .$$

Proof.

(i) The connectedness of M follows from a lemma in point-set topology (Proposition A.44). It is also a consequence of the Mayer–Vietoris sequence. On a nonempty, connected open set, the de Rham cohomology in dimension 0 is simply the vector space of constant functions (Proposition 24.1). By (26.1), the map

$$j^* : H^0(U) \oplus H^0(V) \to H^0(U \cap V)$$

is given by

$$(u,v) \mapsto v - u, \quad u,v \in \mathbb{R}.$$

This map is clearly surjective. The surjectivity of j^* implies that

$$\operatorname{im} j^* = H^0(U \cap V) = \ker d^*,$$

from which we conclude that $d^* \colon H^0(U \cap V) \to H^1(M)$ is the zero map. Thus the Mayer–Vietoris sequence starts with

$$0 \to H^0(M) \xrightarrow{i^*} \mathbb{R} \oplus \mathbb{R} \xrightarrow{j^*} \mathbb{R} \xrightarrow{d^*} 0. \tag{26.4}$$

This short exact sequence shows that

$$H^0(M) \simeq \operatorname{im} i^* = \ker j^*.$$

Since

$$\ker j^* = \{(u,v) \mid v - u = 0\} = \{(u,u) \in \mathbb{R} \oplus \mathbb{R}\} \simeq \mathbb{R},$$

$H^0(M) \simeq \mathbb{R}$, which proves that M is connected.

(ii) From (i) we know that $d^* \colon H^0(U \cap V) \to H^1(M)$ is the zero map. Thus, in the Mayer–Vietoris sequence, the sequence of two maps

$$H^0(U \cap V) \xrightarrow{d^*} H^1(M) \xrightarrow{i^*} H^1(U) \oplus H^1(V)$$

may be replaced by

$$0 \to H^1(M) \xrightarrow{i^*} H^1(U) \oplus H^1(V)$$

without affecting exactness.   □

26.2 The Cohomology of the Circle

In Example 24.4 we showed that integration of 1-forms induces an isomorphism of $H^1(S^1)$ with $\mathbb{R}$. In this section we apply the Mayer–Vietoris sequence to give an alternative computation of the cohomology of the circle.

Cover the circle with two open arcs U and V as in Figure 26.2. The intersection $U \cap V$ is the disjoint union of two open arcs, which we call A and B. Since an open arc is diffeomorphic to an open interval and hence to the real line $\mathbb{R}^1$, the cohomology rings of U and V are isomorphic to that of $\mathbb{R}^1$, and the cohomology ring of $U \cap V$ to that of the disjoint union $\mathbb{R}^1 \amalg \mathbb{R}^1$. They fit into the Mayer–Vietoris sequence, which we arrange in tabular form:

		S^1	$U \amalg V$		$U \cap V$
H^2	$\to$	0	$\to$ 0	$\to$	0
H^1	$\xrightarrow{d^*}$	$H^1(S^1) \to$	0	$\to$	0
H^0	$0 \to$	$\mathbb{R}$ $\xrightarrow{i^*}$	$\mathbb{R} \oplus \mathbb{R}$	$\xrightarrow{j^*}$	$\mathbb{R} \oplus \mathbb{R}$

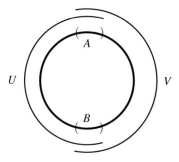

Fig. 26.2. An open cover of the circle.

From the exact sequence

$$0 \to \mathbb{R} \xrightarrow{i^*} \mathbb{R} \oplus \mathbb{R} \xrightarrow{j^*} \mathbb{R} \oplus \mathbb{R} \xrightarrow{d^*} H^1(S^1) \to 0$$

and Problem 26.2, we conclude that $\dim H^1(S^1) = 1$. Hence, the cohomology of the circle is given by

$$H^k(S^1) = \begin{cases} \mathbb{R} & \text{for } k = 0, 1, \\ 0 & \text{otherwise.} \end{cases}$$

By analyzing the maps in the Mayer–Vietoris sequence, it is possible to write down an explicit generator for $H^1(S^1)$. First, according to Proposition 24.1, an element of $H^0(U) \oplus H^0(V)$ is an ordered pair $(u, v) \in \mathbb{R} \oplus \mathbb{R}$, representing a constant function u on U and a constant function v on V. An element of $H^0(U \cap V) = H^0(A) \oplus H^0(B)$ is an ordered pair $(a, b) \in \mathbb{R} \oplus \mathbb{R}$, representing a constant function a on A and a constant function b on B. The restriction map $j_U^* : Z^0(U) \to Z^0(U \cap V)$ is the restriction of a constant function on U to the two connected components A and B of the intersection $U \cap V$:

$$j_U^*(u) = u|_{U \cap V} = (u, u) \in Z^0(A) \oplus Z^0(B).$$

Similarly,

$$j_V^*(v) = v|_{U \cap V} = (v, v) \in Z^0(A) \oplus Z^0(B).$$

By (26.1), $j : Z^0(U) \oplus Z^0(V) \to Z^0(U \cap V)$ is given by

$$j(u, v) = v|_{U \cap V} - u|_{U \cap V} = (v, v) - (u, u) = (v - u, v - u).$$

Hence, in the Mayer–Vietoris sequence, the induced map $j^* : H^0(U) \oplus H^0(V) \to H^0(U \cap V)$ is given by

$$j^*(u, v) = (v - u, v - u).$$

The image of j^* is therefore the diagonal Δ in $\mathbb{R}^2$:

$$\Delta = \{(a, a) \in \mathbb{R}^2\}.$$

Since $H^1(S^1)$ is isomorphic to $\mathbb{R}$, a generator of $H^1(S^1)$ is simply a nonzero element. Moreover, because $d^* : H^0(U \cap V) \to H^1(S^1)$ is surjective and

$$\ker d^* = \operatorname{im} j^* = \Delta,$$

such a nonzero element in $H^1(S^1)$ is the image under d^* of an element $(a,b) \in H^0(U \cap V) \simeq \mathbb{R}^2$ with $a \neq b$.

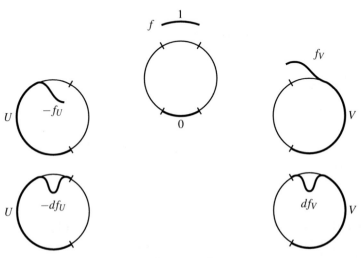

Fig. 26.3. A generator of H^1 of the circle.

So we may start with $(a,b) = (1,0) \in H^0(U \cap V)$. This corresponds to a function f with value 1 on A and 0 on B. Let $\{\rho_U, \rho_V\}$ be a partition of unity subordinate to the open cover $\{U,V\}$, and let f_U, f_V be the extensions by zero of $\rho_V f, \rho_U f$ from $U \cap V$ to U and to V, respectively. By the proof of Proposition 26.2, $j(-f_U, f_V) = f$ on $U \cap V$. From Section 25.3, $d^*(1,0)$ is represented by a 1-form on S^1 whose restriction to U is $-df_U$ and whose restriction to V is df_V. Now f_V is the function on V that is ρ_U on A and 0 on $V - A$, so df_V is a 1-form on V whose support is contained entirely in A. A similar analysis shows that $-df_U$ restricts to the same 1-form on A, because $\rho_U + \rho_V = 1$. The extension of either df_V or $-df_U$ by zero to a 1-form on S^1 represents a generator of $H^1(S^1)$. It is a bump 1-form on S^1 supported in A (Figure 26.3).

The explicit description of the map j^* gives another way to compute $H^1(S^1)$, for by the exactness of the Mayer–Vietoris sequence and the first isomorphism theorem of linear algebra, there is a sequence of vector-space isomorphisms

$$H^1(S^1) = \operatorname{im} d^* \simeq \frac{\mathbb{R} \oplus \mathbb{R}}{\ker d^*} = \frac{\mathbb{R} \oplus \mathbb{R}}{\operatorname{im} j^*} \simeq \frac{\mathbb{R} \oplus \mathbb{R}}{\operatorname{im} j^*} \simeq \frac{\mathbb{R}^2}{\mathbb{R}} \simeq \mathbb{R}.$$

26.3 The Euler Characteristic

If the cohomology vector space $H^k(M)$ of an n-manifold M is finite-dimensional for every k, we define its *Euler characteristic* to be the alternating sum

$$\chi(M) = \sum_{k=0}^{n} (-1)^k \dim H^k(M).$$

As a corollary of the Mayer–Vietoris sequence, the Euler characteristic of $U \cup V$ is computable from those of U, V, and $U \cap V$, as follows.

Exercise 26.5 (Euler characteristics in terms of an open cover). Suppose a manifold M has an open cover $\{U,V\}$ and the spaces M, U, V, and $U \cap V$ all have finite-dimensional cohomology. By applying Problem 26.2 to the Mayer–Vietoris sequence, prove that

$$\chi(M) - (\chi(U) + \chi(V)) + \chi(U \cap V) = 0.$$

Problems

26.1. Short exact Mayer–Vietoris sequence
Prove the exactness of (26.2) at $\Omega^k(M)$ and at $\Omega^k(U) \oplus \Omega^k(V)$.

26.2. Alternating sum of dimensions
Let

$$0 \to A^0 \xrightarrow{d_0} A^1 \xrightarrow{d_1} A^2 \xrightarrow{d_2} \cdots \to A^m \to 0$$

be an exact sequence of finite-dimensional vector spaces. Show that

$$\sum_{k=0}^{m} (-1)^k \dim A^k = 0.$$

(*Hint*: By the rank–nullity theorem from linear algebra,

$$\dim A^k = \dim \ker d_k + \dim \operatorname{im} d_k.$$

Take the alternating sum of these equations over k and use the fact that $\dim \ker d_k = \dim \operatorname{im} d_{k-1}$ to simplify it.)

§27 Homotopy Invariance

The homotopy axiom is a powerful tool for computing de Rham cohomology. While homotopy is normally defined in the continuous category, since we are primarily interested in smooth manifolds and smooth maps, our notion of homotopy will be *smooth homotopy*. It differs from the usual homotopy in topology only in that all the maps are assumed to be smooth. In this section we define smooth homotopy, state the homotopy axiom for de Rham cohomology, and compute a few examples. We postpone the proof of the homotopy axiom to Section 29.

27.1 Smooth Homotopy

Let M and N be manifolds. Two C^∞ maps $f, g \colon M \to N$ are *(smoothly) homotopic* if there is a C^∞ map

$$F \colon M \times \mathbb{R} \to N$$

such that

$$F(x, 0) = f(x) \quad \text{and} \quad F(x, 1) = g(x)$$

for all $x \in M$; the map F is called a *homotopy* from f to g. A homotopy F from f to g can be viewed as a smoothly varying family of maps $\{f_t \colon M \to N \mid t \in \mathbb{R}\}$, where

$$f_t(x) = F(x, t), \quad x \in M,$$

such that $f_0 = f$ and $f_1 = g$. We can think of the parameter t as time and a homotopy as an evolution through time of the map $f_0 \colon M \to N$. If f and g are homotopic, we write

$$f \sim g.$$

Since any open interval is diffeomorphic to $\mathbb{R}$ (Problem 1.3), in the definition of homotopy we could have used any open interval containing 0 and 1, instead of $\mathbb{R}$. The advantage of an open interval over the closed interval $[0, 1]$ is that an open interval is a manifold without boundary.

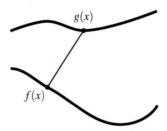

Fig. 27.1. Straight-line homotopies.

Example 27.1 (*Straight-line homotopy*). Let f and g be C^∞ maps from a manifold M to $\mathbb{R}^n$. Define $F\colon M \times \mathbb{R} \to \mathbb{R}^n$ by

$$F(x,t) = f(x) + t(g(x) - f(x)) = (1-t)f(x) + tg(x).$$

Then F is a homotopy from f to g, called the *straight-line homotopy* from f to g (Figure 27.1).

Exercise 27.2 (Homotopy). Let M and N be manifolds. Prove that homotopy is an equivalence relation on the set of all C^∞ maps from M to N.

27.2 Homotopy Type

As usual, $\mathbb{1}_M$ denotes the identity map on a manifold M.

Definition 27.3. A map $f\colon M \to N$ is a *homotopy equivalence* if it has a *homotopy inverse*, i.e., a map $g\colon N \to M$ such that $g \circ f$ is homotopic to the identity $\mathbb{1}_M$ on M and $f \circ g$ is homotopic to the identity $\mathbb{1}_N$ on N:

$$g \circ f \sim \mathbb{1}_M \quad \text{and} \quad f \circ g \sim \mathbb{1}_N.$$

In this case we say that M is *homotopy equivalent* to N, or that M and N have the same *homotopy type*.

Example. A diffeomorphism is a homotopy equivalence.

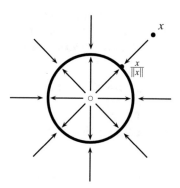

Fig. 27.2. The punctured plane retracts to the unit circle.

Example 27.4 (*Homotopy type of the punctured plane*). Let $i\colon S^1 \to \mathbb{R}^2 - \{\mathbf{0}\}$ be the inclusion map and let $r\colon \mathbb{R}^2 - \{\mathbf{0}\} \to S^1$ be the map

$$r(x) = \frac{x}{\|x\|}.$$

Then $r \circ i$ is the identity map on S^1.

We claim that

$$i \circ r \colon \mathbb{R}^2 - \{\mathbf{0}\} \to \mathbb{R}^2 - \{\mathbf{0}\}$$

is homotopic to the identity map. Note that in the definition of a smooth homotopy $F(x,t)$, the domain of t is required to be the entire real line. The straight-line homotopy

$$H(x,t) = (1-t)x + t\frac{x}{\|x\|}, \quad (x,t) \in (\mathbb{R}^2 - \{\mathbf{0}\}) \times \mathbb{R},$$

will be fine if t is restricted to the closed interval $[0,1]$. However, if t is allowed to be any real number, then $H(x,t)$ may be equal to 0. Indeed, for $t = \|x\|/(\|x\| - 1)$, $H(x,t) = 0$, and so H does not map into $\mathbb{R}^2 - \{\mathbf{0}\}$. To correct this problem, we modify the straight-line homotopy so that for all t the modified map $F(x,t)$ is always a positive multiple of x and hence never zero. Set

$$F(x,t) = (1-t)^2 x + t^2 \frac{x}{\|x\|} = \left((1-t)^2 + \frac{t^2}{\|x\|}\right)x.$$

Then

$$F(x,t) = 0 \quad \Longleftrightarrow \quad (1-t)^2 = 0 \text{ and } \frac{t^2}{\|x\|} = 0$$

$$\Longleftrightarrow \quad t = 1 = 0, \text{ a contradiction.}$$

Therefore, $F \colon (\mathbb{R}^2 - \{\mathbf{0}\}) \times \mathbb{R} \to \mathbb{R}^2 - \{\mathbf{0}\}$ provides a homotopy between the identity map on $\mathbb{R}^2 - \{\mathbf{0}\}$ and $i \circ r$ (Figure 27.2). It follows that r and i are homotopy inverse to each other, and $\mathbb{R}^2 - \{\mathbf{0}\}$ and S^1 have the same homotopy type.

Definition 27.5. A manifold is *contractible* if it has the homotopy type of a point.

In this definition, by "the homotopy type of a point" we mean the homotopy type of a set $\{p\}$ whose single element is a point. Such a set is called a *singleton set* or just a *singleton*.

Example 27.6 (*The Euclidean space $\mathbb{R}^n$ is contractible*).   Let p be a point in $\mathbb{R}^n$, $i \colon \{p\} \to \mathbb{R}^n$ the inclusion map, and $r \colon \mathbb{R}^n \to \{p\}$ the constant map. Then $r \circ i = \mathbb{1}_{\{p\}}$, the identity map on $\{p\}$. The straight-line homotopy provides a homotopy between the constant map $i \circ r \colon \mathbb{R}^n \to \mathbb{R}^n$ and the identity map on $\mathbb{R}^n$:

$$F(x,t) = (1-t)x + t r(x) = (1-t)x + tp.$$

Hence, the Euclidean space $\mathbb{R}^n$ and the set $\{p\}$ have the same homotopy type.

27.3 Deformation Retractions

Let S be a submanifold of a manifold M, with $i\colon S \to M$ the inclusion map.

Definition 27.7. A *retraction* from M to S is a map $r\colon M \to S$ that restricts to the identity map on S; in other words, $r \circ i = \mathbb{1}_S$. If there is a retraction from M to S, we say that S is a *retract* of M.

Definition 27.8. A *deformation retraction* from M to S is a map $F\colon M \times \mathbb{R} \to M$ such that for all $x \in M$,

(i) $F(x,0) = x$,
(ii) there is a retraction $r\colon M \to S$ such that $F(x,1) = r(x)$,
(iii) for all $s \in S$ and $t \in \mathbb{R}$, $F(s,t) = s$.

If there is a deformation retraction from M to S, we say that S is a *deformation retract* of M.

Setting $f_t(x) = F(x,t)$, we can think of a deformation retraction $F\colon M \times \mathbb{R} \to M$ as a family of maps $f_t\colon M \to M$ such that

(i) f_0 is the identity map on M,
(ii) $f_1(x) = r(x)$ for some retraction $r\colon M \to S$,
(iii) for every t the map $f_t\colon M \to M$ restricts to the identity on S.

We may rephrase condition (ii) in the definition as follows: there is a retraction $r\colon M \to S$ such that $f_1 = i \circ r$. Thus, a deformation retraction is a homotopy between the identity map $\mathbb{1}_M$ and $i \circ r$ for a retraction $r\colon M \to S$ such that this homotopy leaves S fixed for all time t.

Example. Any point p in a manifold M is a retract of M; simply take a retraction to be the constant map $r\colon M \to \{p\}$.

Example. The map F in Example 27.4 is a deformation retraction from the punctured plane $\mathbb{R}^2 - \{0\}$ to the unit circle S^1. The map F in Example 27.6 is a deformation retraction from $\mathbb{R}^n$ to a singleton $\{p\}$.

Generalizing Example 27.4, we prove the following theorem.

Proposition 27.9. *If $S \subset M$ is a deformation retract of M, then S and M have the same homotopy type.*

Proof. Let $F\colon M \times \mathbb{R} \to M$ be a deformation retraction and let $r(x) = f_1(x) = F(x,1)$ be the retraction. Because r is a retraction, the composite

$$S \xrightarrow{i} M \xrightarrow{r} S, \quad r \circ i = \mathbb{1}_S,$$

is the identity map on S. By the definition of a deformation retraction, the composite

$$M \xrightarrow{r} S \xrightarrow{i} M$$

is f_1 and the deformation retraction provides a homotopy

$$f_1 = i \circ r \sim f_0 = 1_M.$$

Therefore, $r\colon M \to S$ is a homotopy equivalence, with homotopy inverse $i\colon S \to M$.

$\square$

27.4 The Homotopy Axiom for de Rham Cohomology

We state here the homotopy axiom and derive a few consequences. The proof will be given in Section 29.

Theorem 27.10 (Homotopy axiom for de Rham cohomology). *Homotopic maps $f_0, f_1\colon M \to N$ induce the same map $f_0^* = f_1^* \colon H^*(N) \to H^*(M)$ in cohomology.*

Corollary 27.11. *If $f\colon M \to N$ is a homotopy equivalence, then the induced map in cohomology*

$$f^* \colon H^*(N) \to H^*(M)$$

is an isomorphism.

Proof (of Corollary). Let $g\colon N \to M$ be a homotopy inverse to f. Then

$$g \circ f \sim 1_M, \quad f \circ g \sim 1_N.$$

By the homotopy axiom,

$$(g \circ f)^* = 1_{H^*(M)}, \quad (f \circ g)^* = 1_{H^*(N)}.$$

By functoriality,

$$f^* \circ g^* = 1_{H^*(M)}, \quad g^* \circ f^* = 1_{H^*(N)}.$$

Therefore, f^* is an isomorphism in cohomology. $\square$

Corollary 27.12. *Suppose S is a submanifold of a manifold M and F is a deformation retraction from M to S. Let $r\colon M \to S$ be the retraction $r(x) = F(x, 1)$. Then r induces an isomorphism in cohomology*

$$r^* \colon H^*(S) \xrightarrow{\sim} H^*(M).$$

Proof. The proof of Proposition 27.9 shows that a retraction $r\colon M \to S$ is a homotopy equivalence. Apply Corollary 27.11. $\square$

Corollary 27.13 (Poincaré lemma). *Since $\mathbb{R}^n$ has the homotopy type of a point, the cohomology of $\mathbb{R}^n$ is*

$$H^k(\mathbb{R}^n) = \begin{cases} \mathbb{R} & \text{for } k = 0, \\ 0 & \text{for } k > 0. \end{cases}$$

More generally, any contractible manifold will have the same cohomology as a point.

Example 27.14 (*Cohomology of a punctured plane*). For any $p \in \mathbb{R}^2$, the translation $x \mapsto x - p$ is a diffeomorphism of $\mathbb{R}^2 - \{p\}$ with $\mathbb{R}^2 - \{0\}$. Because the punctured plane $\mathbb{R}^2 - \{0\}$ and the circle S^1 have the same homotopy type (Example 27.4), they have isomorphic cohomology. Hence, $H^k(\mathbb{R}^2 - \{p\}) \simeq H^k(S^1)$ for all $k \geq 0$.

Example. The central circle of an open Möbius band M is a deformation retract of M (Figure 27.3). Thus, the open Möbius band has the homotopy type of a circle. By the homotopy axiom,

$$H^k(M) = H^k(S^1) = \begin{cases} \mathbb{R} & \text{for } k = 0, 1, \\ 0 & \text{for } k > 1. \end{cases}$$

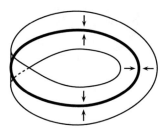

Fig. 27.3. The Möbius band deformation retracts to its central circle.

Problems

27.1. Homotopy equivalence
Let M, N, and P be manifolds. Prove that if M and N are homotopy equivalent and N and P are homotopy equivalent, then M and P are homotopy equivalent.

27.2. Contractibility and path-connectedness
Show that a contractible manifold is path-connected.

27.3. Deformation retraction from a cylinder to a circle
Show that the circle $S^1 \times \{0\}$ is a deformation retract of the cylinder $S^1 \times \mathbb{R}$.

§28 Computation of de Rham Cohomology

With the tools developed so far, we can compute the cohomology of many manifolds. This section is a compendium of some examples.

28.1 Cohomology Vector Space of a Torus

Cover a torus M with two open subsets U and V as shown in Figure 28.1.

Fig. 28.1. An open cover $\{U,V\}$ of a torus.

Both U and V are diffeomorphic to a cylinder and therefore have the homotopy type of a circle (Problem 27.3). Similarly, the intersection $U \cap V$ is the disjoint union of two cylinders A and B and has the homotopy type of a disjoint union of two circles. Our knowledge of the cohomology of a circle allows us to fill in many terms in the Mayer–Vietoris sequence:

$$
\begin{array}{c|ccc}
 & M & U \amalg V & U \cap V \\
\hline
H^2 & \xrightarrow{d_1^*} H^2(M) \to & 0 & \\
H^1 & \xrightarrow{d_0^*} H^1(M) \xrightarrow{i^*} & \mathbb{R} \oplus \mathbb{R} \xrightarrow{\beta} & \mathbb{R} \oplus \mathbb{R} \\
H^0 & 0 \to \quad \mathbb{R} & \to \mathbb{R} \oplus \mathbb{R} \xrightarrow{\alpha} & \mathbb{R} \oplus \mathbb{R}
\end{array}
\tag{28.1}
$$

Let $j_U : U \cap V \to U$ and $j_V : U \cap V \to V$ be the inclusion maps. Recall that H^0 of a connected manifold is the vector space of constant functions on the manifold (Proposition 24.1). If $a \in H^0(U)$ is the constant function with value a on U, then $j_U^* a = a|_{U \cap V} \in H^0(U \cap V)$ is the constant function with the value a on each component of $U \cap V$, that is,

$$
j_U^* a = (a, a).
$$

Therefore, for $(a,b) \in H^0(U) \oplus H^0(V)$,

$$\alpha(a,b) = b|_{U \cap V} - a|_{U \cap V} = (b,b) - (a,a) = (b-a,b-a).$$

Similarly, let us now describe the map

$$\beta \colon H^1(U) \oplus H^1(V) \to H^1(U \cap V) = H^1(A) \oplus H^1(B).$$

Since A is a deformation retract of U, the restriction $H^*(U) \to H^*(A)$ is an isomorphism, so if ω_U generates $H^1(U)$, then $j_U^* \omega_U$ is a generator of H^1 on A and on B. Identifying $H^1(U \cap V)$ with $\mathbb{R} \oplus \mathbb{R}$, we write $j_U^* \omega_U = (1,1)$. Let ω_V be a generator of $H^1(V)$. The pair of real numbers

$$(a,b) \in H^1(U) \oplus H^1(V) \simeq \mathbb{R} \oplus \mathbb{R}$$

stands for $(a\omega_U, b\omega_V)$. Then

$$\beta(a,b) = j_V^*(b\omega_V) - j_U^*(a\omega_U) = (b,b) - (a,a) = (b-a,b-a).$$

By the exactness of the Mayer–Vietoris sequence,

$$
\begin{aligned}
H^2(M) &= \operatorname{im} d_1^* && \text{(because } H^2(U) \oplus H^2(V) = 0) \\
&\simeq H^1(U \cap V)/\ker d_1^* && \text{(by the first isomorphism theorem)} \\
&\simeq (\mathbb{R} \oplus \mathbb{R})/\operatorname{im} \beta \\
&\simeq (\mathbb{R} \oplus \mathbb{R})/\mathbb{R} \simeq \mathbb{R}.
\end{aligned}
$$

Applying Problem 26.2 to the Mayer–Vietoris sequence (28.1), we get

$$1 - 2 + 2 - \dim H^1(M) + 2 - 2 + \dim H^2(M) = 0.$$

Since $\dim H^2(M) = 1$, this gives $\dim H^1(M) = 2$.

As a check, we can also compute $H^1(M)$ from the Mayer–Vietoris sequence using our knowledge of the maps α and β:

$$
\begin{aligned}
H^1(M) &\simeq \ker i^* \oplus \operatorname{im} i^* && \text{(by the first isomorphism theorem)} \\
&\simeq \operatorname{im} d_0^* \oplus \ker \beta && \text{(exactness of the M–V sequence)} \\
&\simeq (H^0(U \cap V)/\ker d_0^*) \oplus \ker \beta && \text{(first isomorphism theorem for } d_0^*) \\
&\simeq ((\mathbb{R} \oplus \mathbb{R})/\operatorname{im} \alpha) \oplus \mathbb{R} \\
&\simeq \mathbb{R} \oplus \mathbb{R}.
\end{aligned}
$$

28.2 The Cohomology Ring of a Torus

A torus is the quotient of $\mathbb{R}^2$ by the integer lattice $\Lambda = \mathbb{Z}^2$. The quotient map

$$\pi \colon \mathbb{R}^2 \to \mathbb{R}^2/\Lambda$$

induces a pullback map on differential forms,

$$\pi^*: \Omega^*(\mathbb{R}^2/\Lambda) \to \Omega^*(\mathbb{R}^2).$$

Since $\pi: \mathbb{R}^2 \to \mathbb{R}^2/\Lambda$ is a local diffeomorphism, its differential $\pi_*: T_q(\mathbb{R}^2) \to T_{\pi(q)}(\mathbb{R}^2/\Lambda)$ is an isomorphism at each point $q \in \mathbb{R}^2$. In particular, π is a submersion. By Problem 18.8, $\pi^*: \Omega^*(\mathbb{R}^2/\Lambda) \to \Omega^*(\mathbb{R}^2)$ is an injection.

For $\lambda \in \Lambda$, define $\ell_\lambda: \mathbb{R}^2 \to \mathbb{R}^2$ to be translation by λ,

$$\ell_\lambda(q) = q + \lambda, \quad q \in \mathbb{R}^2.$$

A differential form $\bar{\omega}$ on $\mathbb{R}^2$ is said to be *invariant under translation by* $\lambda \in \Lambda$ if $\ell_\lambda^* \bar{\omega} = \bar{\omega}$. The following proposition generalizes the description of differential forms on a circle given in Proposition 18.12, where Λ was the lattice $2\pi\mathbb{Z}$.

Proposition 28.1. *The image of the injection* $\pi^*: \Omega^*(\mathbb{R}^2/\Lambda) \to \Omega^*(\mathbb{R}^2)$ *is the subspace of differential forms on* $\mathbb{R}^2$ *invariant under translations by elements of* Λ.

Proof. For all $q \in \mathbb{R}^2$,

$$(\pi \circ \ell_\lambda)(q) = \pi(q + \lambda) = \pi(q).$$

Hence, $\pi \circ \ell_\lambda = \pi$. By the functoriality of the pullback,

$$\pi^* = \ell_\lambda^* \circ \pi^*.$$

Thus, for any $\omega \in \Omega^k(\mathbb{R}^2/\Lambda)$, $\pi^*\omega = \ell_\lambda^* \pi^*\omega$. This proves that $\pi^*\omega$ is invariant under all translations ℓ_λ, $\lambda \in \Lambda$.

Conversely, suppose $\bar{\omega} \in \Omega^k(\mathbb{R}^2)$ is invariant under translations ℓ_λ for all $\lambda \in \Lambda$. For $p \in \mathbb{R}^2/\Lambda$ and $v_1, \ldots, v_k \in T_p(\mathbb{R}^2/\Lambda)$, define

$$\omega_p(v_1, \ldots, v_k) = \bar{\omega}_{\bar{p}}(\bar{v}_1, \ldots, \bar{v}_k) \tag{28.2}$$

for any $\bar{p} \in \pi^{-1}(p)$ and $\bar{v}_1, \ldots, \bar{v}_k \in T_{\bar{p}}\mathbb{R}^2$ such that $\pi_* \bar{v}_i = v_i$. Note that once $\bar{p}$ is chosen, $\bar{v}_1, \ldots, \bar{v}_k$ are unique, since $\pi_*: T_{\bar{p}}(\mathbb{R}^2) \to T_p(\mathbb{R}^2/\Lambda)$ is an isomorphism. For ω to be well defined, we need to show that it is independent of the choice of $\bar{p}$. Now any other point in $\pi^{-1}(p)$ may be written as $\bar{p} + \lambda$ for some $\lambda \in \Lambda$. By invariance,

$$\bar{\omega}_{\bar{p}} = (\ell_\lambda^* \bar{\omega})_{\bar{p}} = \ell_\lambda^* (\bar{\omega}_{\bar{p}+\lambda}).$$

So

$$\bar{\omega}_{\bar{p}}(\bar{v}_1, \ldots, \bar{v}_k) = \ell_\lambda^* (\bar{\omega}_{\bar{p}+\lambda})(\bar{v}_1, \ldots, \bar{v}_k) = \bar{\omega}_{\bar{p}+\lambda}(\ell_{\lambda*}\bar{v}_1, \ldots, \ell_{\lambda*}\bar{v}_k). \tag{28.3}$$

Since $\pi \circ \ell_\lambda = \pi$, we have $\pi_*(\ell_{\lambda*}\bar{v}_i) = \pi_* \bar{v}_i = v_i$. Thus, (28.3) shows that ω_p is independent of the choice of $\bar{p}$, and $\omega \in \Omega^k(\mathbb{R}^2/\Lambda)$ is well defined. Moreover, by (28.2), for any $\bar{p} \in \mathbb{R}^2$ and $\bar{v}_1, \ldots, \bar{v}_k \in T_{\bar{p}}(\mathbb{R}^2)$,

$$\bar{\omega}_{\bar{p}}(\bar{v}_1, \ldots, \bar{v}_k) = \omega_{\pi(\bar{p})}(\pi_* \bar{v}_1, \ldots, \pi_* \bar{v}_k) = (\pi^*\omega)_{\bar{p}}(\bar{v}_1, \ldots, \bar{v}_k).$$

Hence, $\bar{\omega} = \pi^*\omega$. $\qquad\square$

Let x, y be the standard coordinates on $\mathbb{R}^2$. Since for any $\lambda \in \Lambda$,

$$\ell_\lambda^*(dx) = d(\ell_\lambda^* x) = d(x + \lambda) = dx,$$

by Proposition 28.1 the 1-form dx on $\mathbb{R}^2$ is π^* of a 1-form α on the torus $\mathbb{R}^2/\Lambda$. Similarly, dy is π^* of a 1-form β on the torus.

Note that

$$\pi^*(d\alpha) = d(\pi^*\alpha) = d(dx) = 0.$$

Since $\pi^* : \Omega^*(\mathbb{R}^2/\mathbb{Z}^2) \to \Omega^*(\mathbb{R}^2)$ is injective, $d\alpha = 0$. Similarly, $d\beta = 0$. Thus, both α and β are closed 1-forms on the torus.

Proposition 28.2. *Let M be the torus $\mathbb{R}^2/\mathbb{Z}^2$. A basis for the cohomology vector space $H^*(M)$ is represented by the forms* 1, α, β, $\alpha \wedge \beta$.

Proof. Let I be the closed interval $[0, 1]$, and $i : I^2 \hookrightarrow \mathbb{R}^2$ the inclusion map of the closed square I^2 into $\mathbb{R}^2$. The composite map $F = \pi \circ i : I^2 \hookrightarrow \mathbb{R}^2 \to \mathbb{R}^2/\mathbb{Z}^2$ represents the torus $M = \mathbb{R}^2/\mathbb{Z}^2$ as a parametrized set. Then $F^*\alpha = i^*(\pi^*\alpha) = i^*dx$, the restriction of dx to the square I^2. Similarly, $F^*\beta = i^*dy$.

As an integral over a parametrized set,

$$\int_M \alpha \wedge \beta = \int_{F(I^2)} \alpha \wedge \beta = \int_{I^2} F^*(\alpha \wedge \beta) = \int_{I^2} dx \wedge dy = \int_0^1 \int_0^1 dx\,dy = 1.$$

Thus, the closed 2-form $\alpha \wedge \beta$ represents a nonzero cohomology class on M. Since $H^2(M) = \mathbb{R}$ by the computation of Subsection 28.1, the cohomology class $[\alpha \wedge \beta]$ is a basis for $H^2(M)$.

Next we show that the cohomology classes of the closed 1-forms α, β on M constitute a basis for $H^1(M)$. Let $i_1, i_2 : I \to \mathbb{R}^2$ be given by $i_1(t) = (t, 0)$, $i_2(t) = (0, t)$. Define two closed curves C_1, C_2 in $M = \mathbb{R}^2/\mathbb{Z}^2$ as the images of the maps (Figure 28.2)

$$c_k : I \xrightarrow{i_k} \mathbb{R}^2 \xrightarrow{\pi} M = \mathbb{R}^2/\mathbb{Z}^2, \quad k = 1, 2,$$
$$c_1(t) = [(t, 0)], \quad c_2(t) = [(0, t)].$$

Each curve C_i is a smooth manifold and a parametrized set with parametrization c_i.

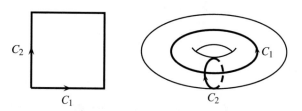

Fig. 28.2. Two closed curves on a torus.

Moreover,

$$c_1^*\alpha = (\pi \circ i_1)^*\alpha = i_1^*\pi^*\alpha = i_1^* dx = d i_1^* x = dt,$$
$$c_1^*\beta = (\pi \circ i_1)^*\beta = i_1^*\pi^*\beta = i_1^* dy = d i_1^* y = 0.$$

Similarly, $c_2^*\alpha = 0$ and $c_2^*\beta = dt$. Therefore,

$$\int_{C_1} \alpha = \int_{c_1(I)} \alpha = \int_I c_1^*\alpha = \int_0^1 dt = 1$$

and

$$\int_{C_1} \beta = \int_{c_1(I)} \beta = \int_I c_1^*\beta = \int_0^1 0 = 0.$$

In the same way, $\int_{C_2} \alpha = 0$ and $\int_{C_2} \beta = 1$.

Because $\int_{C_1} \alpha \neq 0$ and $\int_{C_2} \beta \neq 0$, neither α nor β is exact on M. Furthermore, the cohomology classes $[\alpha]$ and $[\beta]$ are linearly independent, for if $[\alpha]$ were a multiple of $[\beta]$, then $\int_{C_1} \alpha$ would have to be a nonzero multiple of $\int_{C_1} \beta = 0$. By Subsection 28.1, $H^1(M)$ is two-dimensional. Hence, $[\alpha], [\beta]$ is a basis for $H^1(M)$.

In degree 0, $H^0(M)$ has basis $[1]$, as is true for any connected manifold M. $\square$

The ring structure of $H^*(M)$ is clear from this proposition. Abstractly it is the algebra

$$\wedge(a,b) := \mathbb{R}[a,b]/(a^2, b^2, ab + ba), \quad \deg a = 1, \deg b = 1,$$

called the *exterior algebra* on two generators a and b of degree 1.

28.3 The Cohomology of a Surface of Genus g

Using the Mayer–Vietoris sequence to compute the cohomology of a manifold often leads to ambiguities, because there may be several unknown terms in the sequence. We can resolve these ambiguities if we can describe explicitly the maps occurring in the sequence. Here is an example of how this might be done.

Lemma 28.3. *Suppose p is a point in a compact oriented surface M without boundary, and $i: C \to M - \{p\}$ is the inclusion of a small circle around the puncture (Figure 28.3). Then the restriction map*

$$i^*: H^1(M - \{p\}) \to H^1(C)$$

is the zero map.

Proof. An element $[\omega] \in H^1(M - \{p\})$ is represented by a closed 1-form ω on $M - \{p\}$. Because the linear isomorphism $H^1(C) \simeq H^1(S^1) \simeq \mathbb{R}$ is given by integration over C, to identify $i^*[\omega]$ in $H^1(C)$, it suffices to compute the integral $\int_C i^*\omega$.

Fig. 28.3. Punctured surface.

If D is the open disk in M bounded by the curve C, then $M - D$ is a compact oriented surface with boundary C. By Stokes's theorem,

$$\int_C i^* \omega = \int_{\partial(M-D)} i^* \omega = \int_{M-D} d\omega = 0,$$

because $d\omega = 0$. Hence, $i^* : H^1(M - \{p\}) \to H^1(C)$ is the zero map. $\qquad\square$

Proposition 28.4. *Let M be a torus, p a point in M, and A the punctured torus $M - \{p\}$. The cohomology of A is*

$$H^k(A) = \begin{cases} \mathbb{R} & \text{for } k = 0, \\ \mathbb{R}^2 & \text{for } k = 1, \\ 0 & \text{for } k > 1. \end{cases}$$

Proof. Cover M with two open sets, A and a disk U containing p. Since A, U, and $A \cap U$ are all connected, we may start the Mayer–Vietoris sequence with the $H^1(M)$ term (Proposition 26.4(ii)). With $H^*(M)$ known from Section 28.1, the Mayer–Vietoris sequence becomes

		M	$U \amalg A$	$U \cap A \sim S^1$
H^2	$\overset{d_1^*}{\to}$	$\mathbb{R}$	$\to H^2(A) \to$	0
H^1	$0 \to$	$\mathbb{R} \oplus \mathbb{R}$	$\overset{\beta}{\to} H^1(A)$	$\overset{\alpha}{\to} H^1(S^1)$

Because $H^1(U) = 0$, the map $\alpha : H^1(A) \to H^1(S^1)$ is simply the restriction map i^*. By Lemma 28.3, $\alpha = i^* = 0$. Hence,

$$H^1(A) = \ker \alpha = \operatorname{im} \beta \simeq H^1(M) \simeq \mathbb{R} \oplus \mathbb{R}$$

and there is an exact sequence of linear maps

$$0 \to H^1(S^1) \overset{d_1^*}{\to} \mathbb{R} \to H^2(A) \to 0.$$

Since $H^1(S^1) \simeq \mathbb{R}$, it follows that $H^2(A) = 0$. $\qquad\square$

Proposition 28.5. *The cohomology of a compact orientable surface Σ_2 of genus 2 is*

$$H^k(\Sigma_2) = \begin{cases} \mathbb{R} & \text{for } k = 0, 2, \\ \mathbb{R}^4 & \text{for } k = 1, \\ 0 & \text{for } k > 2. \end{cases}$$

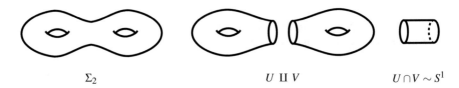

Σ_2 $U \amalg V$ $U \cap V \sim S^1$

Fig. 28.4. An open cover $\{U, V\}$ of a surface of genus 2.

Proof. Cover Σ_2 with two open sets U and V as in Figure 28.4. Since U, V, and $U \cap V$ are all connected, the Mayer–Vietoris sequence begins with

	M	$U \amalg V$	$U \cap V \sim S^1$
H^2	$\to H^2(\Sigma_2) \to$	0	
H^1	$0 \to H^1(\Sigma_2) \to$	$\mathbb{R}^2 \oplus \mathbb{R}^2$	$\xrightarrow{\alpha}$ $\mathbb{R}$

The map $\alpha \colon H^1(U) \oplus H^1(V) \to H^1(S^1)$ is the difference map

$$\alpha(\omega_U, \omega_V) = j_V^* \omega_V - j_U^* \omega_U,$$

where j_U and j_V are inclusions of an S^1 in $U \cap V$ into U and V, respectively. By Lemma 28.3, $j_U^* = j_V^* = 0$, so $\alpha = 0$. It then follows from the exactness of the Mayer–Vietoris sequence that

$$H^1(\Sigma_2) \simeq H^1(U) \oplus H^1(V) \simeq \mathbb{R}^4$$

and

$$H^2(\Sigma_2) \simeq H^1(S^1) \simeq \mathbb{R}. \qquad \square$$

A genus-2 surface Σ_2 can be obtained as the quotient space of an octagon with its edges identified following the scheme of Figure 28.5.

To see this, first cut Σ_2 along the circle e as in Figure 28.6.

Then the two halves A and B are each a torus minus an open disk (Figure 28.7), so that each half can be represented as a pentagon, before identification (Figure 28.8). When A and B are glued together along e, we obtain the octagon in Figure 28.5.

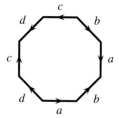

Fig. 28.5. A surface of genus 2 as a quotient space of an octagon.

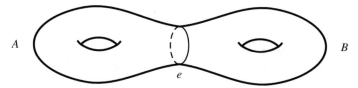

Fig. 28.6. A surface of genus 2 cut along a curve e.

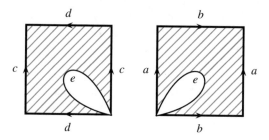

Fig. 28.7. Two halves of a surface of genus 2.

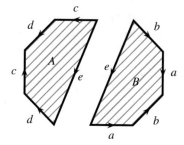

Fig. 28.8. Two halves of a surface of genus 2.

By Lemma 28.3, if $p \in \Sigma_2$ and $i \colon C \to \Sigma_2 - \{p\}$ is a small circle around p in Σ_2, then the restriction map

$$i^* \colon H^1(\Sigma_2 - \{p\}) \to H^1(C)$$

is the zero map. This allows us to compute inductively the cohomology of a compact orientable surface Σ_g of genus g.

Exercise 28.6 (Surface of genus 3). Compute the cohomology vector space of $\Sigma_2 - \{p\}$ and then compute the cohomology vector space of a compact orientable surface Σ_3 of genus 3.

Problems

28.1. Real projective plane
Compute the cohomology of the real projective plane $\mathbb{R}P^2$ (Figure 28.9).

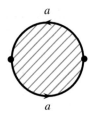

a

a

Fig. 28.9. The real projective plane.

28.2. The n-sphere
Compute the cohomology of the sphere S^n.

28.3. Cohomology of a multiply punctured plane

(a) Let p, q be distinct points in $\mathbb{R}^2$. Compute the de Rham cohomology of $\mathbb{R}^2 - \{p, q\}$.
(b) Let $p_1, \ldots, p_n$ be distinct points in $\mathbb{R}^2$. Compute the de Rham cohomology of $\mathbb{R}^2 - \{p_1, \ldots, p_n\}$.

28.4. Cohomology of a surface of genus g
Compute the cohomology vector space of a compact orientable surface Σ_g of genus g.

28.5. Cohomology of a 3-dimensional torus
Compute the cohomology ring of $\mathbb{R}^3/\mathbb{Z}^3$.

§29 Proof of Homotopy Invariance

In this section we prove the homotopy invariance of de Rham cohomology.

If $f: M \to N$ is a C^∞ map, the pullback maps on differential forms and on cohomology classes are normally both denoted by f^*. Since this might cause confusion in the proof of homotopy invariance, in this section we revert to our original convention of denoting the pullback of forms by

$$f^*: \Omega^k(N) \to \Omega^k(M)$$

and the induced map in cohomology by

$$f^\#: H^k(N) \to H^k(M).$$

The relation between these two maps is

$$f^\#[\omega] = [f^*\omega]$$

for $[\omega] \in H^k(N)$.

Theorem 29.1 (Homotopy axiom for de Rham cohomology). *Two smoothly homotopic maps $f, g: M \to N$ of manifolds induce the same map in cohomology:*

$$f^\# = g^\#: H^k(N) \to H^k(M).$$

We first reduce the problem to two special maps i_0 and $i_1: M \to M \times \mathbb{R}$, which are the 0-section and the 1-section, respectively, of the product line bundle $M \times \mathbb{R} \to M$:

$$i_0(x) = (x, 0), \quad i_1(x) = (x, 1).$$

Then we introduce the all-important technique of cochain homotopy. By finding a cochain homotopy between i_0^* and i_1^*, we prove that they induce the same map in cohomology.

29.1 Reduction to Two Sections

Suppose f and $g: M \to N$ are smoothly homotopic maps. Let $F: M \times \mathbb{R} \to N$ be a smooth homotopy from f to g. This means that

$$F(x, 0) = f(x), \quad F(x, 1) = g(x) \tag{29.1}$$

for all $x \in M$. For each $t \in \mathbb{R}$, define $i_t: M \to M \times \mathbb{R}$ to be the section $i_t(x) = (x, t)$. We can restate (29.1) as

$$F \circ i_0 = f, \quad F \circ i_1 = g.$$

By the functoriality of the pullback (Remark 24.7),

$$f^\# = i_0^\# \circ F^\#, \quad g^\# = i_1^\# \circ F^\#.$$

This reduces proving homotopy invariance to the special case

$$i_0^\# = i_1^\#.$$

The two maps $i_0, i_1 \colon M \to M \times \mathbb{R}$ are obviously smoothly homotopic via the identity map

$$1_{M \times \mathbb{R}} \colon M \times \mathbb{R} \to M \times \mathbb{R}.$$

29.2 Cochain Homotopies

The usual method for showing that two cochain maps $\varphi, \psi \colon \mathcal{A} \to \mathcal{B}$ induce the same map in cohomology is to find a linear map $K \colon \mathcal{A} \to \mathcal{B}$ of degree -1 such that

$$\varphi - \psi = d \circ K + K \circ d.$$

Such a map K is called a *cochain homotopy* from φ to ψ. Note that K is not assumed to be a cochain map. If a is a cocycle in $\mathcal{A}$, then

$$\varphi(a) - \psi(a) = dKa + Kda = dKa$$

is a coboundary, so that in cohomology

$$\varphi^\#[a] = [\varphi(a)] = [\psi(a)] = \psi^\#[a].$$

Thus, the existence of a cochain homotopy between φ and ψ implies that the induced maps $\varphi^\#$ and $\psi^\#$ in cohomology are equal.

Remark. Given two cochain maps $\varphi, \psi \colon \mathcal{A} \to \mathcal{B}$, if one could find a linear map $K \colon \mathcal{A} \to \mathcal{B}$ of degree -1 such that $\varphi - \psi = d \circ K$ on $\mathcal{A}$, then $\varphi^\#$ would be equal to $\psi^\#$ in cohomology. However, such a map almost never exists; it is necessary to have the term $K \circ d$ as well. The cylinder construction in homology theory [30, p. 65] shows why it is natural to consider $d \circ K + K \circ d$.

29.3 Differential Forms on $M \times \mathbb{R}$

Recall that a sum $\sum_\alpha \omega_\alpha$ of C^∞ differential forms on a manifold M is said to be *locally finite* if the collection $\{\operatorname{supp} \omega_\alpha\}$ of supports is locally finite. This means that every point p in M has a neighborhood V_p such that V_p intersects only finitely many of the sets $\operatorname{supp} \omega_\alpha$. If $\operatorname{supp} \omega_\alpha$ is disjoint from V_p, then $\omega_\alpha \equiv 0$ on V_p. Thus, on V_p the locally finite sum $\sum_\alpha \omega_\alpha$ is actually a finite sum. As an example, if $\{\rho_\alpha\}$ is a partition of unity, then the sum $\sum \rho_\alpha$ is locally finite.

Let $\pi \colon M \times \mathbb{R} \to M$ be the projection to the first factor. In this subsection we will show that every C^∞ differential form on $M \times \mathbb{R}$ is a locally finite sum of the following two types of forms:

(I) $f(x,t)\pi^*\eta$,

(II) $f(x,t)dt \wedge \pi^*\eta$,

where $f(x,t)$ is a C^∞ function on $M \times \mathbb{R}$ and η is a C^∞ form on M.

In general, a decomposition of a differential form on $M \times \mathbb{R}$ into a locally finite sum of type-I and type-II forms is far from unique. However, once we fix an atlas $\{(U_\alpha, \phi_\alpha)\}$ on M, a C^∞ partition of unity $\{\rho_\alpha\}$ subordinate to $\{U_\alpha\}$, and a collection $\{g_\alpha\}$ of C^∞ functions on M such that

$$g_\alpha \equiv 1 \text{ on } \operatorname{supp}\rho_\alpha \quad \text{and} \quad \operatorname{supp} g_\alpha \subset U_\alpha,$$

then there is a well-defined procedure to produce uniquely such a locally finite sum. The existence of the functions g_α follows from the smooth Urysohn lemma (Problem 13.3).

In the proof of the decomposition procedure, we will need the following simple but useful lemma on the extension of a C^∞ form by zero.

Lemma 29.2. *Let U be an open subset of a manifold M. If a smooth k-form $\tau \in \Omega^k(U)$ defined on U has support in a closed subset of M contained in U, then τ can be extended by zero to a smooth k-form on M.*

Proof. Problem 29.1.   □

Fix an atlas $\{(U_\alpha, \phi_\alpha)\}$, a partition of unity $\{\rho_\alpha\}$, and a collection $\{g_\alpha\}$ of C^∞ functions as above. Then $\{\pi^{-1}U_\alpha\}$ is an open cover of $M \times \mathbb{R}$, and $\{\pi^*\rho_\alpha\}$ is a partition of unity subordinate to $\{\pi^{-1}U_\alpha\}$ (Problem 13.6).

Let ω be any C^∞ k-form on $M \times \mathbb{R}$ and let $\omega_\alpha = (\pi^*\rho_\alpha)\omega$. Since $\sum \pi^*\rho_\alpha = 1$,

$$\omega = \sum_\alpha (\pi^*\rho_\alpha)\omega = \sum_\alpha \omega_\alpha. \tag{29.2}$$

Because $\{\operatorname{supp}\pi^*\rho_\alpha\}$ is locally finite, (29.2) is a locally finite sum. By Problem 18.4,

$$\operatorname{supp}\omega_\alpha \subset \operatorname{supp}\pi^*\rho_\alpha \cap \operatorname{supp}\omega \subset \operatorname{supp}\pi^*\rho_\alpha \subset \pi^{-1}U_\alpha.$$

Let $\phi_\alpha = (x^1, \ldots, x^n)$. Then on $\pi^{-1}U_\alpha$, which is homeomorphic to $U_\alpha \times \mathbb{R}$, we have coordinates $\pi^*x^1, \ldots, \pi^*x^n, t$. For the sake of simplicity, we sometimes write x^i instead of π^*x^i. On $\pi^{-1}U_\alpha$ the k-form ω_α may be written uniquely as a linear combination

$$\omega_\alpha = \sum_I a_I \, dx^I + \sum_J b_J \, dt \wedge dx^J, \tag{29.3}$$

where a_I and b_J are C^∞ functions on $\pi^{-1}U_\alpha$. This decomposition shows that ω_α is a finite sum of type-I and type-II forms on $\pi^{-1}U_\alpha$. By Problem 18.5, the supports of a_I and b_I are contained in $\operatorname{supp}\omega_\alpha$, hence in $\operatorname{supp}\pi^*\rho_\alpha$, a closed set in $M \times \mathbb{R}$. Therefore, by the lemma above, a_I and b_J can be extended by zero to C^∞ functions on $M \times \mathbb{R}$. Unfortunately, dx^I and dx^J make sense only on U_α and cannot be extended to M, at least not directly.

To extend the decomposition (29.3) to $M \times \mathbb{R}$, the trick is to multiply ω_α by $\pi^* g_\alpha$. Since $\operatorname{supp} \omega_\alpha \subset \operatorname{supp} \pi^* \rho_\alpha$ and $\pi^* g_\alpha \equiv 1$ on $\operatorname{supp} \pi^* \rho_\alpha$, we have the equality $\omega_\alpha = (\pi^* g_\alpha) \omega_\alpha$. Therefore,

$$\omega_\alpha = (\pi^* g_\alpha)\omega_\alpha = \sum_I a_I (\pi^* g_\alpha) \, dx^I + \sum_J b_J \, dt \wedge (\pi^* g_\alpha) \, dx^J,$$

$$= \sum_I a_I \pi^* (g_\alpha \, dx^I) + \sum_J b_J \, dt \wedge \pi^* (g_\alpha \, dx^J). \tag{29.4}$$

Now $\operatorname{supp} g_\alpha$ is a closed subset of M contained in U_α, so by Lemma 29.2 again, $g_\alpha \, dx^I$ can be extended by zero to M. Equations (29.2) and (29.4) prove that ω is a locally finite sum of type-I and type-II forms on $M \times \mathbb{R}$. Moreover, given $\{(U_\alpha, \phi_\alpha)\}$, $\{\rho_\alpha\}$, and $\{g_\alpha\}$, the decomposition in (29.4) is unique.

29.4 A Cochain Homotopy Between i_0^* and i_1^*

In the rest of the proof, fix an atlas $\{(U_\alpha, \phi_\alpha)\}$ for M, a C^∞ partition of unity $\{\rho_\alpha\}$ subordinate to $\{U_\alpha\}$, and a collection $\{g_\alpha\}$ of C^∞ functions on M as in Section 29.3. Let $\omega \in \Omega^k(M \times \mathbb{R})$. Using (29.2) and (29.4), we decompose ω into a locally finite sum

$$\omega = \sum_\alpha \omega_\alpha = \sum_{\alpha, I} a_I^\alpha \pi^* (g_\alpha \, dx_\alpha^I) + \sum_{\alpha, J} b_J^\alpha \, dt \wedge \pi^* (g_\alpha \, dx_\alpha^J),$$

where we now attach an index α to a_I, b_J, x^I, and x^J to indicate their dependence on α.

Define

$$K \colon \Omega^*(M \times \mathbb{R}) \to \Omega^{*-1}(M)$$

by the following rules:

(i) on type-I forms,

$$K(f \pi^* \eta) = 0;$$

(ii) on type-II forms,

$$K(f \, dt \wedge \pi^* \eta) = \left(\int_0^1 f(x, t) \, dt \right) \eta;$$

(iii) K is linear over locally finite sums.

Thus,

$$K(\omega) = K\left(\sum_\alpha \omega_\alpha \right) = \sum_{\alpha, J} \left(\int_0^1 b_J^\alpha(x, t) \, dt \right) g_\alpha \, dx_\alpha^J. \tag{29.5}$$

Given the data $\{(U_\alpha, \phi_\alpha)\}$, $\{\rho_\alpha\}$, $\{g_\alpha\}$, the decomposition $\omega = \omega_\alpha$ with ω_α as in (29.4) is unique. Therefore, K is well defined. It is not difficult to show that so defined, K is the unique linear operator $\Omega^*(M \times \mathbb{R}) \to \Omega^{*-1}(M)$ satisfying (i), (ii), and (iii) (Problem 29.3), so it is in fact independent of the data $\{(U_\alpha, \phi_\alpha)\}$, $\{\rho_\alpha\}$, and $\{g_\alpha\}$.

29.5 Verification of Cochain Homotopy

We check in this subsection that

$$d \circ K + K \circ d = i_1^* - i_0^*. \tag{29.6}$$

Lemma 29.3. (i) *The exterior derivative d is $\mathbb{R}$-linear over locally finite sums.*
(ii) *Pullback by a C^∞ map is $\mathbb{R}$-linear over locally finite sums.*

Proof. (i) Suppose $\sum \omega_\alpha$ is a locally finite sum of C^∞ k-forms. This implies that every point p has a neighborhood on which the sum is finite. Let U be such a neighborhood. Then

$$
\begin{aligned}
\left(d \sum \omega_\alpha \right) |_U &= d \left(\left(\sum \omega_\alpha \right) |_U \right) \quad \text{(Corollary 19.6)} \\
&= d \left(\sum \omega_\alpha |_U \right) \\
&= \sum d(\omega_\alpha |_U) \quad \left(\sum \omega_\alpha |_U \text{ is a finite sum} \right) \\
&= \sum (d\omega_\alpha)|_U \quad \text{(Corollary 19.6)} \\
&= \left(\sum d\omega_\alpha \right) |_U.
\end{aligned}
$$

Since M can be covered by such neighborhoods, $d\left(\sum \omega_\alpha \right) = \sum d\omega_\alpha$ on M. The homogeneity property $d(r\omega) = r\,d(\omega)$ for $r \in \mathbb{R}$ and $\omega \in \Omega^k(M)$ is trivial.
(ii) The proof is similar to (i) and is relegated to Problem 29.2. $\square$

By linearity of K, d, i_0^*, and i_1^* over locally finite sums, it suffices to check the equality (29.6) on any coordinate open set. Fix a coordinate open set $(U \times \mathbb{R}, \pi^* x^1, \ldots, \pi^* x^n, t)$ on $M \times \mathbb{R}$. On type-I forms,

$$Kd(f\,\pi^*\eta) = K \left(\frac{\partial f}{\partial t} dt \wedge \pi^* \eta + \sum_i \frac{\partial f}{\partial x^i} \pi^* dx^i \wedge \pi^* \eta + f\,\pi^* d\eta \right).$$

In the sum on the right-hand side, the second and third terms are type-I forms; they map to 0 under K. Thus,

$$
\begin{aligned}
Kd(f\,\pi^*\eta) &= K \left(\frac{\partial f}{\partial t} dt \wedge \pi^* \eta \right) = \left(\int_0^1 \frac{\partial f}{\partial t} dt \right) \eta \\
&= (f(x,1) - f(x,0))\eta = (i_1^* - i_0^*) \left(f(x,t)\,\pi^*\eta \right).
\end{aligned}
$$

Since $dK(f\,\pi^*\eta) = d(0) = 0$, on type-I forms,

$$d \circ K + K \circ d = i_1^* - i_0^*.$$

On type-II forms, by the antiderivation property of d,

$$
\begin{aligned}
dK(f\,dt \wedge \pi^*\eta) &= d \left(\left(\int_0^1 f(x,t)\,dt \right) \eta \right) \\
&= \sum \left(\frac{\partial}{\partial x^i} \int_0^1 f(x,t)\,dt \right) dx^i \wedge \eta + \left(\int_0^1 f(x,t)\,dt \right) d\eta \\
&= \sum \left(\int_0^1 \frac{\partial f}{\partial x^i}(x,t)\,dt \right) dx^i \wedge \eta + \left(\int_0^1 f(x,t)\,dt \right) d\eta.
\end{aligned}
$$

In the last equality, differentiation under the integral sign is permissible because $f(x,t)$ is C^∞. Furthermore,

$$Kd(f\,dt \wedge \pi^*\eta) = K\left(d(f\,dt) \wedge \pi^*\eta - f\,dt \wedge d\pi^*\eta\right)$$

$$= K\left(\sum_i \frac{\partial f}{\partial x^i}\,dx^i \wedge dt \wedge \pi^*\eta\right) - K(f\,dt \wedge \pi^*d\eta)$$

$$= -\sum_i \left(\int_0^1 \frac{\partial f}{\partial x^i}(x,t)\,dt\right)dx^i \wedge \eta - \left(\int_0^1 f(x,t)\,dt\right)d\eta.$$

Thus, on type-II forms,

$$d \circ K + K \circ d = 0.$$

On the other hand,

$$i_1^*(f(x,t)\,dt \wedge \pi^*\eta) = 0$$

because $i_1^*\,dt = di_1^*t = d(1) = 0$. Similarly, i_0^* also vanishes on type-II forms. Therefore,

$$d \circ K + K \circ d = 0 = i_1^* - i_0^*$$

on type-II forms.

This completes the proof that K is a cochain homotopy between i_0^* and i_1^*. The existence of the cochain homotopy K proves that the induced maps in cohomology $i_0^\#$ and $i_1^\#$ are equal. As we pointed out in Section 29.1,

$$f^\# = i_0^\# \circ F^\# = i_1^\# \circ F^\# = g^\#.$$

Problems

29.1. Extension by zero of a smooth k-form
Prove Lemma 29.2.

29.2. Linearity of pullback over locally finite sums
Let $h: N \to M$ be a C^∞ map, and $\sum \omega_\alpha$ a locally finite sum of C^∞ k-forms on M. Prove that $h^*\left(\sum \omega_\alpha\right) = \sum h^* \omega_\alpha$.

29.3. The cochain homotopy K
(a) Check that defined by (29.5), the linear map K satisfies the three rules in Section 29.4.
(b) Prove that a linear operator satisfying the three rules in Section 29.4 is unique if it exists.

Appendices

§A Point-Set Topology

Point-set topology, also called "general topology," is concerned with properties that remain invariant under homeomorphisms (continuous maps having continuous inverses). The basic development in the subject took place in the late nineteenth and early twentieth centuries. This appendix is a collection of basic results from point-set topology that are used throughout the book.

A.1 Topological Spaces

The prototype of a topological space is the Euclidean space $\mathbb{R}^n$. However, Euclidean space comes with many additional structures, such as a metric, coordinates, an inner product, and an orientation, that are extraneous to its topology. The idea behind the definition of a topological space is to discard all those properties of $\mathbb{R}^n$ that have nothing to do with continuous maps, thereby distilling the notion of continuity to its very essence.

In advanced calculus one learns several characterizations of a continuous map, among which is the following: a map f from an open subset of $\mathbb{R}^n$ to $\mathbb{R}^m$ is continuous if and only if the inverse image $f^{-1}(V)$ of any open set V in $\mathbb{R}^m$ is open in $\mathbb{R}^n$. This shows that continuity can be defined solely in terms of open sets.

To define open sets axiomatically, we look at properties of open sets in $\mathbb{R}^n$. Recall that in $\mathbb{R}^n$ the *distance* between two points p and q is given by

$$
d(p,q) = \left[\sum_{i=1}^{n} (p^i - q^i)^2 \right]^{1/2},
$$

and the *open ball* $B(p,r)$ with center $p \in \mathbb{R}^n$ and radius $r > 0$ is the set

$$
B(p,r) = \{ x \in \mathbb{R}^n \mid d(x,p) < r \}.
$$

A set U in $\mathbb{R}^n$ is said to be *open* if for every p in U, there is an open ball $B(p,r)$ with center p and radius r such that $B(p,r) \subset U$ (Figure A.1). It is clear that the union of an arbitrary collection $\{U_\alpha\}$ of open sets is open, but the same need not be true of the intersection of infinitely many open sets.

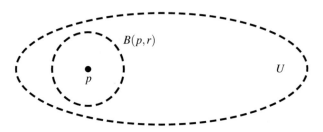

Fig. A.1. An open set in $\mathbb{R}^n$.

Example. The intervals $]-1/n, 1/n[$, $n = 1, 2, 3, \ldots$, are all open in $\mathbb{R}^1$, but their intersection $\bigcap_{n=1}^{\infty}]-1/n, 1/n[$ is the singleton set $\{0\}$, which is not open.

What is true is that the intersection of a *finite* collection of open sets in $\mathbb{R}^n$ is open. This leads to the definition of a topology on a set.

Definition A.1. A *topology* on a set S is a collection $\mathcal{T}$ of subsets containing both the empty set $\varnothing$ and the set S such that $\mathcal{T}$ is closed under arbitrary unions and finite intersections; i.e., if $U_\alpha \in \mathcal{T}$ for all α in an index set A, then $\bigcup_{\alpha \in A} U_\alpha \in \mathcal{T}$ and if $U_1, \ldots, U_n \in \mathcal{T}$, then $\bigcap_{i=1}^{n} U_i \in \mathcal{T}$.

The elements of $\mathcal{T}$ are called *open sets* and the pair $(S, \mathcal{T})$ is called a *topological space*. To simplify the notation, we sometimes simply refer to a pair $(S, \mathcal{T})$ as "the topological space S" when there is no chance of confusion. A *neighborhood* of a point p in S is an open set U containing p. If $\mathcal{T}_1$ and $\mathcal{T}_2$ are two topologies on a set S and $\mathcal{T}_1 \subset \mathcal{T}_2$, then we say that $\mathcal{T}_1$ is *coarser* than $\mathcal{T}_1$, or that $\mathcal{T}_2$ is *finer* than $\mathcal{T}_1$. A coarser topology has fewer open sets; conversely, a finer topology has more open sets.

Example. The open subsets of $\mathbb{R}^n$ as we understand them in advanced calculus form a topology on $\mathbb{R}^n$, the *standard topology* of $\mathbb{R}^n$. In this topology a set U is open in $\mathbb{R}^n$ if and only if for every $p \in U$, there is an open ball $B(p, \varepsilon)$ with center p and radius ε contained in U. Unless stated otherwise, $\mathbb{R}^n$ will always have its standard topology.

The criterion for openness in $\mathbb{R}^n$ has a useful generalization to a topological space.

Lemma A.2 (Local criterion for openness). *Let S be a topological space. A subset A is open in S if and only if for every $p \in A$, there is an open set V such that $p \in V \subset A$.*

Proof.

$(\Rightarrow)$ If A is open, we can take $V = A$.

$(\Leftarrow)$ Suppose for every $p \in A$ there is an open set V_p such that $p \in V_p \subset A$. Then

$$A \subset \bigcup_{p \in A} V_p \subset A,$$

so that equality $A = \bigcup_{p \in A} V_p$ holds. As a union of open sets, A is open. $\square$

Example. For any set S, the collection $\mathcal{T} = \{\varnothing, S\}$ consisting of the empty set $\varnothing$ and the entire set S is a topology on S, sometimes called the *trivial* or *indiscrete topology*. It is the coarsest topology on a set.

Example. For any set S, let $\mathcal{T}$ be the collection of all subsets of S. Then $\mathcal{T}$ is a topology on S, called the *discrete topology*. A *singleton set* is a set with a single element. The discrete topology can also be characterized as the topology in which every singleton subset $\{p\}$ is open. A topological space having the discrete topology is called a *discrete space*. The discrete topology is the finest topology on a set.

The complement of an open set is called a *closed set*. By de Morgan's laws from set theory, arbitrary intersections and finite unions of closed sets are closed (Problem A.3). One may also specify a topology by describing all the closed sets.

Remark. When we say that a topology is *closed* under arbitrary union and finite intersection, the word "closed" has a different meaning from that of a "closed subset."

Example A.3 (Finite-complement topology on $\mathbb{R}^1$). Let $\mathcal{T}$ be the collection of subsets of $\mathbb{R}^1$ consisting of the empty set $\varnothing$, the line $\mathbb{R}^1$ itself, and the complements of finite sets. Suppose F_α and F_i are finite subsets of $\mathbb{R}^1$ for $\alpha \in$ some index set A and $i = 1, \ldots, n$. By de Morgan's laws,

$$\bigcup_\alpha (\mathbb{R}^1 - F_\alpha) = \mathbb{R}^1 - \bigcap_\alpha F_\alpha \quad \text{and} \quad \bigcap_{i=1}^n (\mathbb{R}^1 - F_i) = \mathbb{R}^1 - \bigcup_{i=1}^n F_i.$$

Since the arbitrary intersection $\bigcap_{\alpha \in A} F_\alpha$ and the finite union $\bigcup_{i=1}^n F_i$ are both finite, $\mathcal{T}$ is closed under arbitrary unions and finite intersections. Thus, $\mathcal{T}$ defines a topology on $\mathbb{R}^1$, called the *finite-complement topology*.

For the sake of definiteness, we have defined the finite-complement topology on $\mathbb{R}^1$, but of course, there is nothing specific about $\mathbb{R}^1$ here. One can define in exactly the same way the finite-complement topology on any set.

Example A.4 (Zariski topology). One well-known topology is the *Zariski topology* from algebraic geometry. Let K be a field and let S be the vector space K^n. Define a subset of K^n to be *Zariski closed* if it is the zero set $Z(f_1, \ldots, f_r)$ of finitely many polynomials $f_1, \ldots, f_r$ on K^n. To show that these are indeed the closed subsets of a topology, we need to check that they are closed under arbitrary intersections and finite unions.

Let $I = (f_1, \ldots, f_r)$ be the ideal generated by $f_1, \ldots, f_r$ in the polynomial ring $K[x_1, \ldots, x_n]$. Then $Z(f_1, \ldots, f_r) = Z(I)$, the zero set of *all* the polynomials in the ideal I. Conversely, by the Hilbert basis theorem [11, §9.6, Th. 21], any ideal in $K[x_1, \ldots, x_n]$ has a finite set of generators. Hence, the zero set of finitely many polynomials is the same as the zero set of an ideal in $K[x_1, \ldots, x_n]$. If $I = (f_1, \ldots, f_r)$ and $J = (g_1, \ldots, g_s)$ are two ideals, then the *product ideal IJ* is the ideal in $K[x_1, \ldots, x_n]$ generated by all products $f_i g_j$, $1 \le i \le r$, $1 \le j \le s$. If $\{I_\alpha\}_{\alpha \in A}$ is a family of ideals in $K[x_1, \ldots, x_n]$, then their *sum* $\sum_\alpha I_\alpha$ is the smallest ideal in $K[x_1, \ldots, x_n]$ containing all the ideals I_α.

Exercise A.5 (Intersection and union of zero sets). Let I_α, I, and J be ideals in the polynomial ring $K[x_1, \ldots, x_n]$. Show that

(i)
$$\bigcap_\alpha Z(I_\alpha) = Z\left(\sum_\alpha I_\alpha\right)$$

and

(ii)
$$Z(I) \cup Z(J) = Z(IJ).$$

The complement of a Zariski-closed subset of K^n is said to be *Zariski open*. If $I = (0)$ is the zero ideal, then $Z(I) = K^n$, and if $I = (1) = K[x_1, \ldots, x_n]$ is the entire ring, then $Z(I)$ is the empty set $\emptyset$. Hence, both the empty set and K^n are Zariski open. It now follows from Exercise A.5 that the Zariski-open subsets of K^n form a topology on K^n, called the *Zariski topology* on K^n. Since the zero set of a polynomial on $\mathbb{R}^1$ is a finite set, the Zariski topology on $\mathbb{R}^1$ is precisely the finite-complement topology of Example A.3.

A.2 Subspace Topology

Let $(S, \mathcal{T})$ be a topological space and A a subset of S. Define $\mathcal{T}_A$ to be the collection of subsets
$$\mathcal{T}_A = \{U \cap A \mid U \in \mathcal{T}\}.$$

By the distributive property of union and intersection,
$$\bigcup_\alpha (U_\alpha \cap A) = \left(\bigcup_\alpha U_\alpha\right) \cap A$$

and
$$\bigcap_i (U_i \cap A) = \left(\bigcap_i U_i\right) \cap A,$$

which shows that $\mathcal{T}_A$ is closed under arbitrary unions and finite intersections. Moreover, $\emptyset, A \in \mathcal{T}_A$. So $\mathcal{T}_A$ is a topology on A, called the *subspace topology* or the *relative topology* of A in S, and elements of $\mathcal{T}_A$ are said to be *open in A*. To empha-size the fact that an open set U in A need not be open in S, we also say that U is *open relative to A* or *relatively open in A*. The subset A of S with the subspace topology $\mathcal{T}_A$ is called a *subspace* of S.

If A is an open subset of a topological space S, then a subset of A is relatively open in A if and only if it is open in S.

Example. Consider the subset $A = [0,1]$ of $\mathbb{R}^1$. In the subspace topology, the half-open interval $[0,1/2[$ is open relative to A, because

$$\left[0, \tfrac{1}{2}\right[\; = \; \left]-\tfrac{1}{2}, \tfrac{1}{2}\right[\cap A.$$

(See Figure A.2.)

$$-\tfrac{1}{2} \qquad 0 \qquad \tfrac{1}{2} \qquad 1$$

Fig. A.2. A relatively open subset of $[0, 1]$.

A.3 Bases

It is generally difficult to describe directly all the open sets in a topology $\mathcal{T}$. What one can usually do is to describe a subcollection $\mathcal{B}$ of $\mathcal{T}$ such that any open set is expressible as a union of open sets in $\mathcal{B}$.

Definition A.6. A subcollection $\mathcal{B}$ of a topology $\mathcal{T}$ on a topological space S is a *basis for the topology* $\mathcal{T}$ if given an open set U and point p in U, there is an open set $B \in \mathcal{B}$ such that $p \in B \subset U$. We also say that $\mathcal{B}$ *generates* the topology $\mathcal{T}$ or that $\mathcal{B}$ is a *basis for the topological space S*.

Example. The collection of all open balls $B(p,r)$ in $\mathbb{R}^n$, with $p \in \mathbb{R}^n$ and r a positive real number, is a basis for the standard topology of $\mathbb{R}^n$.

Proposition A.7. *A collection $\mathcal{B}$ of open sets of S is a basis if and only if every open set in S is a union of sets in $\mathcal{B}$.*

Proof.
($\Rightarrow$) Suppose $\mathcal{B}$ is a basis and U is an open set in S. For every $p \in U$, there is a basic open set $B_p \in \mathcal{B}$ such that $p \in B_p \subset U$. Therefore, $U = \bigcup_{p \in U} B_p$.

($\Leftarrow$) Suppose every open set in S is a union of open sets in $\mathcal{B}$. Given an open set U and a point p in U, since $U = \bigcup_{B_\alpha \in \mathcal{B}} B_\alpha$, there is a $B_\alpha \in \mathcal{B}$ such that $p \in B_\alpha \subset U$. Hence, $\mathcal{B}$ is a basis. $\qquad\square$

The following proposition gives a useful criterion for deciding whether a collection $\mathcal{B}$ of subsets is a basis for some topology.

Proposition A.8. *A collection $\mathcal{B}$ of subsets of a set S is a basis for some topology $\mathcal{T}$ on S if and only if*

(i) *S is the union of all the sets in $\mathcal{B}$, and*
(ii) *given any two sets B_1 and $B_2 \in \mathcal{B}$ and a point $p \in B_1 \cap B_2$, there is a set $B \in \mathcal{B}$ such that $p \in B \subset B_1 \cap B_2$ (Figure A.3).*

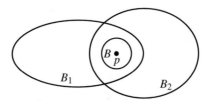

Fig. A.3. Criterion for a basis.

Proof.
($\Rightarrow$) (i) follows from Proposition A.7.

(ii) If $\mathcal{B}$ is a basis, then B_1 and B_2 are open sets and hence so is $B_1 \cap B_2$. By the definition of a basis, there is a $B \in \mathcal{B}$ such that $p \in B \subset B_1 \cap B_2$.

($\Leftarrow$) Define $\mathcal{T}$ to be the collection consisting of all sets that are unions of sets in $\mathcal{B}$. Then the empty set $\varnothing$ and the set S are in $\mathcal{T}$ and $\mathcal{T}$ is clearly closed under arbitrary union. To show that $\mathcal{T}$ is closed under finite intersection, let $U = \bigcup_\mu B_\mu$ and $V = \bigcup_\nu B_\nu$ be in $\mathcal{T}$, where $B_\mu, B_\nu \in \mathcal{B}$. Then

$$U \cap V = \left(\bigcup_\mu B_\mu \right) \cap \left(\bigcup_\nu B_\nu \right) = \bigcup_{\mu,\nu} (B_\mu \cap B_\nu).$$

Thus, any p in $U \cap V$ is in $B_\mu \cap B_\nu$ for some μ, ν. By (ii) there is a set B_p in $\mathcal{B}$ such that $p \in B_p \subset B_\mu \cap B_\nu$. Therefore,

$$U \cap V = \bigcup_{p \in U \cap V} B_p \in \mathcal{T}. \qquad \square$$

Proposition A.9. *Let $\mathcal{B} = \{B_\alpha\}$ be a basis for a topological space S, and A a subspace of S. Then $\{B_\alpha \cap A\}$ is a basis for A.*

Proof. Let U' be any open set in A and $p \in U'$. By the definition of subspace topology, $U' = U \cap A$ for some open set U in S. Since $p \in U \cap A \subset U$, there is a basic open set B_α such that $p \in B_\alpha \subset U$. Then

$$p \in B_\alpha \cap A \subset U \cap A = U',$$

which proves that the collection $\{B_\alpha \cap A \mid B_\alpha \in \mathcal{B}\}$ is a basis for A. $\qquad \square$

A.4 First and Second Countability

First and second countability of a topological space have to do with the countability of a basis. Before taking up these notions, we begin with an example. We say that a point in $\mathbb{R}^n$ is *rational* if all of its coordinates are rational numbers. Let $\mathbb{Q}$ be the set of rational numbers and $\mathbb{Q}^+$ the set of positive rational numbers. From real analysis, it is well known that every open interval in $\mathbb{R}$ contains a rational number.

Lemma A.10. *Every open set in $\mathbb{R}^n$ contains a rational point.*

Proof. An open set U in $\mathbb{R}^n$ contains an open ball $B(p,r)$, which in turn contains an open cube $\prod_{i=1}^n I_i$, where I_i is the open interval $]p^i - (r/\sqrt{n}), p^i + (r/\sqrt{n})[$ (see Problem A.4). For each i, let q^i be a rational number in I_i. Then $(q^1, \ldots, q^n)$ is a rational point in $\prod_{i=1}^n I_i \subset B(p,r) \subset U$. $\qquad\square$

Proposition A.11. *The collection $\mathcal{B}_{\mathrm{rat}}$ of all open balls in $\mathbb{R}^n$ with rational centers and rational radii is a basis for $\mathbb{R}^n$.*

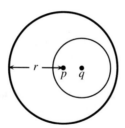

Fig. A.4. A ball with rational center q and rational radius $r/2$.

Proof. Given an open set U in $\mathbb{R}^n$ and point p in U, there is an open ball $B(p, r')$ with positive real radius r' such that $p \in B(p, r') \subset U$. Take a rational number r in $]0, r'[$. Then $p \in B(p, r) \subset U$. By Lemma A.10, there is a rational point q in the smaller ball $B(p, r/2)$. We claim that

$$p \in B\left(q, \frac{r}{2}\right) \subset B(p, r). \tag{A.1}$$

(See Figure A.4.) Since $d(p, q) < r/2$, we have $p \in B(q, r/2)$. Next, if $x \in B(q, r/2)$, then by the triangle inequality,

$$d(x, p) \le d(x, q) + d(q, p) < \frac{r}{2} + \frac{r}{2} = r.$$

So $x \in B(p, r)$. This proves the claim (A.1). Because $p \in B(q, r/2) \subset U$, the collection $\mathcal{B}_{\mathrm{rat}}$ of open balls with rational centers and rational radii is a basis for $\mathbb{R}^n$. $\qquad\square$

Both of the sets $\mathbb{Q}$ and $\mathbb{Q}^+$ are countable. Since the centers of the balls in $\mathcal{B}_{\text{rat}}$ are indexed by $\mathbb{Q}^n$, a countable set, and the radii are indexed by $\mathbb{Q}^+$, also a countable set, the collection $\mathcal{B}_{\text{rat}}$ is countable.

Definition A.12.   A topological space is said to be *second countable* if it has a countable basis.

Example A.13. Proposition A.11 shows that $\mathbb{R}^n$ with its standard topology is second countable. With the discrete topology, $\mathbb{R}^n$ would not be second countable. More generally, any uncountable set with the discrete topology is not second countable.

Proposition A.14. *A subspace A of a second-countable space S is second countable.*

Proof. By Proposition A.9, if $\mathcal{B} = \{B_i\}$ is a countable basis for S, then $\mathcal{B}_A := \{B_i \cap A\}$ is a countable basis for A. $\square$

Definition A.15. Let S be a topological space and p a point in S. A *basis of neighborhoods at p* or a *neighborhood basis at p* is a collection $\mathcal{B} = \{B_\alpha\}$ of neighborhoods of p such that for any neighborhood U of p, there is a $B_\alpha \in \mathcal{B}$ such that $p \in B_\alpha \subset U$. A topological space S is *first countable* if it has a countable basis of neighborhoods at every point $p \in S$.

Example. For $p \in \mathbb{R}^n$, let $B(p, 1/n)$ be the open ball of center p and radius $1/n$ in $\mathbb{R}^n$. Then $\{B(p, 1/n)\}_{n=1}^\infty$ is a neighborhood basis at p. Thus, $\mathbb{R}^n$ is first countable.

Example. An uncountable discrete space is first countable but not second countable. Every second-countable space is first countable (the proof is left to Problem A.18).

Suppose p is a point in a first-countable topological space and $\{V_i\}_{i=1}^\infty$ is a countable neighborhood basis at p. By taking $U_i = V_1 \cap \cdots \cap V_i$, we obtain a countable descending sequence

$$U_1 \supset U_2 \supset U_3 \supset \cdots$$

that is also a neighborhood basis at p. Thus, in the definition of first countability, we may assume that at every point the countable neighborhood basis at the point is a descending sequence of open sets.

A.5 Separation Axioms

There are various separation axioms for a topological space. The only ones we will need are the Hausdorff condition and normality.

Definition A.16. A topological space S is *Hausdorff* if given any two distinct points x, y in S, there exist disjoint open sets U, V such that $x \in U$ and $y \in V$. A Hausdorff space is *normal* if given any two disjoint closed sets F, G in S, there exist disjoint open sets U, V such that $F \subset U$ and $G \subset V$ (Figure A.5).

Fig. A.5. The Hausdorff condition and normality.

Proposition A.17. *Every singleton set (a one-point set) in a Hausdorff space S is closed.*

Proof. Let $x \in S$. For any $y \in S - \{x\}$, by the Hausdorff condition there exist an open set $U \ni x$ and an open set $V \ni y$ such that U and V are disjoint. In particular,

$$y \in V \subset S - U \subset S - \{x\}.$$

By the local criterion for openness (Lemma A.2), $S - \{x\}$ is open. Therefore, $\{x\}$ is closed. □

Example. The Euclidean space $\mathbb{R}^n$ is Hausdorff, for given distinct points x, y in $\mathbb{R}^n$, if $\varepsilon = \frac{1}{2} d(x, y)$, then the open balls $B(x, \varepsilon)$ and $B(y, \varepsilon)$ will be disjoint (Figure A.6).

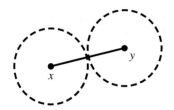

Fig. A.6. Two disjoint neighborhoods in $\mathbb{R}^n$.

Example A.18 (Zariski topology). Let $S = K^n$ be a vector space of dimension n over a field K, endowed with the Zariski topology. Every open set U in S is of the form $S - Z(I)$, where I is an ideal in $K[x_1, \ldots, x_n]$. The open set U is nonempty if and only if I is not the zero ideal. In the Zariski topology any two nonempty open sets intersect: if $U = S - Z(I)$ and $V = S - Z(J)$ are nonempty, then I and J are nonzero ideals and

$$\begin{aligned}
U \cap V &= (S - Z(I)) \cap (S - Z(J)) \\
&= S - (Z(I) \cup Z(J)) \qquad \text{(de Morgan's law)} \\
&= S - Z(IJ), \qquad\qquad \text{(Exercise A.5)}
\end{aligned}$$

which is nonempty because IJ is not the zero ideal. Therefore, K^n with the Zariski topology is not Hausdorff.

Proposition A.19. *Any subspace A of a Hausdorff space S is Hausdorff.*

Proof. Let x and y be distinct points in A. Since S is Hausdorff, there exist disjoint neighborhoods U and V of x and y respectively in S. Then $U \cap A$ and $V \cap A$ are disjoint neighborhoods of x and y respectively in A.   □

A.6 Product Topology

The *Cartesian product* of two sets A and B is the set $A \times B$ of all ordered pairs (a, b) with $a \in A$ and $b \in B$. Given two topological spaces X and Y, consider the collection $\mathcal{B}$ of subsets of $X \times Y$ of the form $U \times V$, with U open in X and V open in Y. We will call elements of $\mathcal{B}$ *basic open sets* in $X \times Y$. If $U_1 \times V_1$ and $U_2 \times V_2$ are in $\mathcal{B}$, then

$$(U_1 \times V_1) \cap (U_2 \times V_2) = (U_1 \cap U_2) \times (V_1 \cap V_2),$$

which is also in $\mathcal{B}$ (Figure A.7). From this, it follows easily that $\mathcal{B}$ satisfies the conditions of Proposition A.8 for a basis and generates a topology on $X \times Y$, called the *product topology*. Unless noted otherwise, this will always be the topology we assign to the product of two topological spaces.

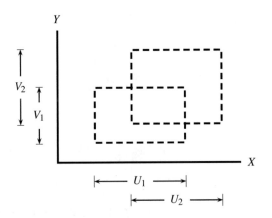

Fig. A.7. Intersection of two basic open subsets in $X \times Y$.

Proposition A.20. *Let $\{U_i\}$ and $\{V_j\}$ be bases for the topological spaces X and Y, respectively. Then $\{U_i \times V_j\}$ is a basis for $X \times Y$.*

Proof. Given an open set W in $X \times Y$ and point $(x, y) \in W$, we can find a basic open set $U \times V$ in $X \times Y$ such that $(x, y) \in U \times V \subset W$. Since U is open in X and $\{U_i\}$ is a basis for X,

$$x \in U_i \subset U$$

for some U_i. Similarly,

$$y \in V_j \subset V$$

for some V_j. Therefore,

$$(x,y) \in U_i \times V_j \subset U \times V \subset W.$$

By the definition of a basis, $\{U_i \times V_j\}$ is a basis for $X \times Y$.   □

Corollary A.21. *The product of two second-countable spaces is second countable.*

Proposition A.22. *The product of two Hausdorff spaces X and Y is Hausdorff.*

Proof. Given two distinct points $(x_1, y_1), (x_2, y_2)$ in $X \times Y$, without loss of generality we may assume that $x_1 \neq x_2$. Since X is Hausdorff, there exist disjoint open sets U_1, U_2 in X such that $x_1 \in U_1$ and $x_2 \in U_2$. Then $U_1 \times Y$ and $U_2 \times Y$ are disjoint neighborhoods of (x_1, y_1) and (x_2, y_2) (Figure A.8), so $X \times Y$ is Hausdorff.   □

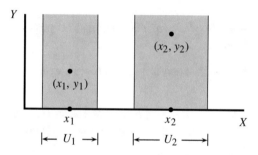

Fig. A.8. Two disjoint neighborhoods in $X \times Y$.

The product topology can be generalized to the product of an arbitrary collection $\{X_\alpha\}_{\alpha \in A}$ of topological spaces. Whatever the definition of the product topology, the projection maps $\pi_{\alpha_i} \colon \prod_\alpha X_\alpha \to X_{\alpha_i}$, $\pi_{\alpha_i}(\prod x_\alpha) = x_{\alpha_i}$ should all be continuous. Thus, for each open set U_{α_i} in X_{α_i}, the inverse image $\pi_{\alpha_i}^{-1}(U_{\alpha_i})$ should be open in $\prod_\alpha X_\alpha$. By the properties of open sets, a *finite* intersection $\bigcap_{i=1}^r \pi_{\alpha_i}^{-1}(U_{\alpha_i})$ should also be open. Such a finite intersection is a set of the form $\prod_{\alpha \in A} U_\alpha$, where U_α is open in X_α and $U_\alpha = X_\alpha$ for all but finitely many $\alpha \in A$. We define the *product topology* on the Cartesian product $\prod_{\alpha \in A} X_\alpha$ to be the topology with basis consisting of sets of this form. The product topology is the coarsest topology on $\prod_\alpha X_\alpha$ such that all the projection maps $\pi_{\alpha_i} \colon \prod_\alpha X_\alpha \to X_{\alpha_i}$ are continuous.

A.7 Continuity

Let $f \colon X \to Y$ be a function of topological spaces. Mimicking the definition from advanced calculus, we say that f is *continuous at a point p* in X if for every neighborhood V of $f(p)$ in Y, there is a neighborhood U of p in X such that $f(U) \subset V$. We say that f is *continuous on X* if it is continuous at every point of X.

Proposition A.23 (Continuity in terms of open sets). *A function $f: X \to Y$ is continuous if and only if the inverse image of any open set is open.*

Proof.
($\Rightarrow$) Suppose V is open in Y. To show that $f^{-1}(V)$ is open in X, let $p \in f^{-1}(V)$. Then $f(p) \in V$. Since f is assumed to be continuous at p, there is a neighborhood U of p such that $f(U) \subset V$. Therefore, $p \in U \subset f^{-1}(V)$. By the local criterion for openness (Lemma A.2), $f^{-1}(V)$ is open in X.

($\Leftarrow$) Let p be a point in X, and V a neighborhood of $f(p)$ in Y. By hypothesis, $f^{-1}(V)$ is open in X. Since $f(p) \in V$, $p \in f^{-1}(V)$. Then $U = f^{-1}(V)$ is a neighborhood of p such that $f(U) = f(f^{-1}(V)) \subset V$, so f is continuous at p. □

Example A.24 (*Continuity of an inclusion map*). If A is a subspace of X, then the inclusion map $i: A \to X$, $i(a) = a$ is continuous.

Proof. If U is open in X, then $i^{-1}(U) = U \cap A$, which is open in the subspace topology of A. □

Example A.25 (*Continuity of a projection map*). The projection $\pi: X \times Y \to X$, $\pi(x,y) = x$, is continuous.

Proof. Let U be open in X. Then $\pi^{-1}(U) = U \times Y$, which is open in the product topology on $X \times Y$. □

Proposition A.26. *The composition of continuous maps is continuous: if $f: X \to Y$ and $g: Y \to Z$ are continuous, then $g \circ f: X \to Z$ is continuous.*

Proof. Let V be an open subset of Z. Then

$$(g \circ f)^{-1}(V) = f^{-1}(g^{-1}(V)),$$

because for any $x \in X$,

$$x \in (g \circ f)^{-1}(V) \text{ iff } g(f(x)) \in V \text{ iff } f(x) \in g^{-1}(V) \text{ iff } x \in f^{-1}(g^{-1}(V)).$$

By Proposition A.23, since g is continuous, $g^{-1}(V)$ is open in Y. Similarly, since f is continuous, $f^{-1}(g^{-1}(V))$ is open in X. By Proposition A.23 again, $g \circ f: X \to Z$ is continuous. □

If A is a subspace of X and $f: X \to Y$ is a function, the *restriction* of f to A,

$$f|_A: A \to Y,$$

is defined by

$$(f|_A)(a) = f(a).$$

With $i: A \to X$ being the inclusion map, the restriction $f|_A$ is the composite $f \circ i$. Since both f and i are continuous (Example A.24) and the composition of continuous functions is continuous (Proposition A.26), we have the following corollary.

Corollary A.27. *The restriction $f|_A$ of a continuous function $f \colon X \to Y$ to a subspace A is continuous.*

Continuity may also be phrased in terms of closed sets.

Proposition A.28 (Continuity in terms of closed sets). *A function $f \colon X \to Y$ is continuous if and only if the inverse image of any closed set is closed.*

Proof. Problem A.9.   □

A map $f \colon X \to Y$ is said to be *open* if the image of every open set in X is open in Y; similarly, $f \colon X \to Y$ is said to be *closed* if the image of every closed set in X is closed in Y.

If $f \colon X \to Y$ is a bijection, then its inverse map $f^{-1} \colon Y \to X$ is defined. In this context, for any subset $V \subset Y$, the notation $f^{-1}(V)$ a priori has two meanings. It can mean either the inverse image of V under the map f,

$$f^{-1}(V) = \{x \in X \mid f(x) \in V\},$$

or the image of V under the map f^{-1},

$$f^{-1}(V) = \{f^{-1}(y) \in X \mid y \in V\}.$$

Fortunately, because $y = f(x)$ if and only if $x = f^{-1}(y)$, these two meanings coincide.

A.8 Compactness

While its definition may not be intuitive, the notion of compactness is of central importance in topology. Let S be a topological space. A collection $\{U_\alpha\}$ of open subsets of S is said to *cover S* or to be an *open cover* of S if $S \subset \bigcup_\alpha U_\alpha$. Of course, because S is the ambient space, this condition is equivalent to $S = \bigcup_\alpha U_\alpha$. A *subcover* of an open cover is a subcollection whose union still contains S. The topological space S is said to be *compact* if every open cover of S has a finite subcover.

With the subspace topology, a subset A of a topological space S is a topological space in its own right. The subspace A can be covered by open sets in A or by open sets in S. An *open cover of A in S* is a collection $\{U_\alpha\}$ of open sets in S that covers A. In this terminology, A is compact if and only if every open cover of A in A has a finite subcover.

Proposition A.29. *A subspace A of a topological space S is compact if and only if every open cover of A in S has a finite subcover.*

Proof.
($\Rightarrow$) Assume A compact and let $\{U_\alpha\}$ be an open cover of A in S. This means that $A \subset \bigcup_\alpha U_\alpha$. Hence,

$$A \subset \left(\bigcup_\alpha U_\alpha \right) \cap A = \bigcup_\alpha (U_\alpha \cap A).$$

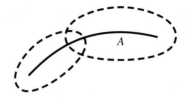

Fig. A.9. An open cover of A in S.

Since A is compact, the open cover $\{U_\alpha \cap A\}$ has a finite subcover $\{U_{\alpha_i} \cap A\}_{i=1}^r$. Thus,

$$A \subset \bigcup_{i=1}^r (U_{\alpha_i} \cap A) \subset \bigcup_{i=1}^r U_{\alpha_i},$$

which means that $\{U_{\alpha_i}\}_{i=1}^r$ is a finite subcover of $\{U_\alpha\}$.

($\Leftarrow$) Suppose every open cover of A in S has a finite subcover, and let $\{V_\alpha\}$ be an open cover of A in A. Then each V_α is equal to $U_\alpha \cap A$ for some open set U_α in S. Since

$$A \subset \bigcup_\alpha V_\alpha \subset \bigcup_\alpha U_\alpha,$$

by hypothesis there are finitely many sets U_{α_i} such that $A \subset \bigcup_i U_{\alpha_i}$. Hence,

$$A \subset \left(\bigcup_i U_{\alpha_i} \right) \cap A = \bigcup_i (U_{\alpha_i} \cap A) = \bigcup_i V_{\alpha_i}.$$

So $\{V_{\alpha_i}\}$ is a finite subcover of $\{V_\alpha\}$ that covers A. Therefore, A is compact. $\qquad\square$

Proposition A.30. *A closed subset F of a compact topological space S is compact.*

Proof. Let $\{U_\alpha\}$ be an open cover of F in S. The collection $\{U_\alpha, S - F\}$ is then an open cover of S. By the compactness of S, there is a finite subcover $\{U_{\alpha_i}, S - F\}$ that covers S, so $F \subset \bigcup_i U_{\alpha_i}$. This proves that F is compact. $\qquad\square$

Proposition A.31. *In a Hausdorff space S, it is possible to separate a compact subset K and a point p not in K by disjoint open sets; i.e., there exist an open set $U \supset K$ and an open set $V \ni p$ such that $U \cap V = \varnothing$.*

Proof. By the Hausdorff property, for every $x \in K$, there are disjoint open sets $U_x \ni x$ and $V_x \ni p$. The collection $\{U_x\}_{x \in K}$ is a cover of K by open subsets of S. Since K is compact, it has a finite subcover $\{U_{x_i}\}$.

Let $U = \bigcup_i U_{x_i}$ and $V = \bigcap_i V_{x_i}$. Then U is an open set of S containing K. Being the intersection of finitely many open sets containing p, V is an open set containing p. Moreover, the set

$$U \cap V = \bigcup_i (U_{x_i} \cap V)$$

is empty, since each $U_{x_i} \cap V$ is contained in $U_{x_i} \cap V_{x_i}$, which is empty. $\qquad\square$

Proposition A.32. *Every compact subset K of a Hausdorff space S is closed.*

Proof. By the preceding proposition, for every point p in $S - K$, there is an open set V such that $p \in V \subset S - K$. This proves that $S - K$ is open. Hence, K is closed. $\quad\square$

Exercise A.33 (Compact Hausdorff space).* Prove that a compact Hausdorff space is normal. (Normality was defined in Definition A.16.)

Proposition A.34. *The image of a compact set under a continuous map is compact.*

Proof. Let $f: X \to Y$ be a continuous map and K a compact subset of X. Suppose $\{U_\alpha\}$ is a cover of $f(K)$ by open subsets of Y. Since f is continuous, the inverse images $f^{-1}(U_\alpha)$ are all open. Moreover,

$$K \subset f^{-1}(f(K)) \subset f^{-1}\left(\bigcup_\alpha U_\alpha\right) = \bigcup_\alpha f^{-1}(U_\alpha).$$

So $\{f^{-1}(U_\alpha)\}$ is an open cover of K in X. By the compactness of K, there is a finite subcollection $\{f^{-1}(U_{\alpha_i})\}$ such that

$$K \subset \bigcup_i f^{-1}(U_{\alpha_i}) = f^{-1}\left(\bigcup_i U_{\alpha_i}\right).$$

Then $f(K) \subset \bigcup_i U_{\alpha_i}$. Thus, $f(K)$ is compact. $\quad\square$

Proposition A.35. *A continuous map $f: X \to Y$ from a compact space X to a Hausdorff space Y is a closed map.*

Proof. Let F be a closed subset of the compact space X. By Proposition A.30, F is compact. As the image of a compact set under a continuous map, $f(F)$ is compact in Y (Proposition A.34). As a compact subset of the Hausdorff space Y, $f(F)$ is closed (Proposition A.32). $\quad\square$

A continuous bijection $f: X \to Y$ whose inverse is also continuous is called a *homeomorphism*.

Corollary A.36. *A continuous bijection $f: X \to Y$ from a compact space X to a Hausdorff space Y is a homeomorphism.*

Proof. By Proposition A.28, to show that $f^{-1}: Y \to X$ is continuous, it suffices to prove that for every closed set F in X, the set $(f^{-1})^{-1}(F) = f(F)$ is closed in Y, i.e., that f is a closed map. The corollary then follows from Proposition A.35. $\quad\square$

Exercise A.37 (Finite union of compact sets). Prove that a finite union of compact subsets of a topological space is compact.

We mention without proof an important result. For a proof, see [29, Theorem 26.7, p. 167, and Theorem 37.3, p. 234].

Theorem A.38 (The Tychonoff theorem). *The product of any collection of compact spaces is compact in the product topology.*

A.9 Boundedness in $\mathbb{R}^n$

A subset A of $\mathbb{R}^n$ is said to be *bounded* if it is contained in some open ball $B(p,r)$; otherwise, it is *unbounded*.

Proposition A.39. *A compact subset of $\mathbb{R}^n$ is bounded.*

Proof. If A were an unbounded subset of $\mathbb{R}^n$, then the collection $\{B(0,i)\}_{i=1}^{\infty}$ of open balls with radius increasing to infinity would be an open cover of A in $\mathbb{R}^n$ that does not have a finite subcover.   □

By Propositions A.39 and A.32, a compact subset of $\mathbb{R}^n$ is closed and bounded. The converse is also true.

Theorem A.40 (The Heine–Borel theorem). *A subset of $\mathbb{R}^n$ is compact if and only if it is closed and bounded.*

For a proof, see for example [29].

A.10 Connectedness

Definition A.41. A topological space S is *disconnected* if it is the union $S = U \cup V$ of two disjoint nonempty open subsets U and V (Figure A.10). It is *connected* if it is not disconnected. A subset A of S is *disconnected* if it is disconnected in the subspace topology.

Fig. A.10. A disconnected space.

Proposition A.42. *A subset A of a topological space S is disconnected if and only if there are open sets U and V in S such that*

(i) $U \cap A \neq \emptyset$, $V \cap A \neq \emptyset$,
(ii) $U \cap V \cap A = \emptyset$,
(iii) $A \subset U \cup V$.

A pair of open sets in S with these properties is called a separation *of A (Figure A.11).*

Proof. Problem A.15.   □

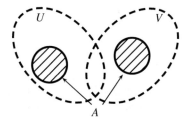

Fig. A.11. A separation of A.

Proposition A.43. *The image of a connected space X under a continuous map $f: X \to Y$ is connected.*

Proof. Suppose $f(X)$ is not connected. Then there is a separation $\{U,V\}$ of $f(X)$ in Y. By the continuity of f, both $f^{-1}(U)$ and $f^{-1}(V)$ are open in X. We claim that $\{f^{-1}(U), f^{-1}(V)\}$ is a separation of X.

(i) Since $U \cap f(X) \neq \varnothing$, the open set $f^{-1}(U)$ is nonempty.
(ii) If $x \in f^{-1}(U) \cap f^{-1}(V)$, then $f(x) \in U \cap V \cap f(X) = \varnothing$, a contradiction. Hence, $f^{-1}(U) \cap f^{-1}(V)$ is empty.
(iii) Since $f(X) \subset U \cup V$, we have $X \subset f^{-1}(U \cup V) = f^{-1}(U) \cup f^{-1}(V)$.

The existence of a separation of X contradicts the connectedness of X. This contradiction proves that $f(X)$ is connected. □

Proposition A.44. *In a topological space S, the union of a collection of connected subsets A_α having a point p in common is connected.*

Proof. Suppose $\bigcup_\alpha A_\alpha = U \cup V$, where U and V are disjoint open subsets of $\bigcup_\alpha A_\alpha$. The point $p \in \bigcup_\alpha A_\alpha$ belongs to U or V. Assume without loss of generality that $p \in U$.

For each α,

$$A_\alpha = A_\alpha \cap (U \cup V) = (A_\alpha \cap U) \cup (A_\alpha \cap V).$$

The two open sets $A_\alpha \cap U$ and $A_\alpha \cap V$ of A_α are clearly disjoint. Since $p \in A_\alpha \cap U$, $A_\alpha \cap U$ is nonempty. By the connectedness of A_α, $A_\alpha \cap V$ must be empty for all α. Hence,

$$V = \left(\bigcup_\alpha A_\alpha\right) \cap V = \bigcup_\alpha (A_\alpha \cap V)$$

is empty. So $\bigcup_\alpha A_\alpha$ must be connected. □

A.11 Connected Components

Let x be a point in a topological space S. By Proposition A.44, the union C_x of all connected subsets of S containing x is connected. It is called the *connected component* of S containing x.

Proposition A.45. *Let C_x be a connected component of a topological space S. Then a connected subset A of S is either disjoint from C_x or is contained entirely in C_x.*

Proof. If A and C_x have a point in common, then by Proposition A.44, $A \cup C_x$ is a connected set containing x. Hence, $A \cup C_x \subset C_x$, which implies that $A \subset C_x$. □

Accordingly, the connected component C_x is the largest connected subset of S containing x in the sense that it contains every connected subset of S containing x.

Corollary A.46. *For any two points x, y in a topological space S, the connected components C_x and C_y either are disjoint or coincide.*

Proof. If C_x and C_y are not disjoint, then by Proposition A.45, they are contained in each other. In this case, $C_x = C_y$. □

As a consequence of Corollary A.46, the connected components of S partition S into disjoint subsets.

A.12 Closure

Let S be a topological space and A a subset of S.

Definition A.47. The *closure* of A in S, denoted by $\overline{A}$, $\mathrm{cl}(A)$, or $\mathrm{cl}_S(A)$, is defined to be the intersection of all the closed sets containing A.

The advantage of the bar notation $\overline{A}$ is its simplicity, while the advantage of the $\mathrm{cl}_S(A)$ notation is its indication of the ambient space S. If $A \subset B \subset S$, then the closure of A in B and the closure of A in S need not be the same. In this case, it is useful to have the notations $\mathrm{cl}_B(A)$ and $\mathrm{cl}_M(A)$ for the two closures.

As an intersection of closed sets, $\overline{A}$ is a closed set. It is the smallest closed set containing A in the sense that any closed set containing A contains $\overline{A}$.

Proposition A.48 (Local characterization of closure). *Let A be a subset of a topological space S. A point $p \in S$ is in the closure $\mathrm{cl}(A)$ if and only if every neighborhood of p contains a point of A (Figure A.12).*

Here by "local," we mean a property satisfied by a basis of neighborhoods at a point.

Proof. We will prove the proposition in the form of its contrapositive:

$$p \notin \mathrm{cl}(A) \quad \Longleftrightarrow \quad \text{there is a neighborhood of } p \text{ disjoint from } A.$$

($\Rightarrow$) Suppose

$$p \notin \mathrm{cl}(A) = \bigcap \{F \text{ closed in } S \mid F \supset A\}.$$

Then $p \notin$ some closed set F containing A. It follows that $p \in S - F$, an open set disjoint from A.

($\Leftarrow$) Suppose $p \in$ an open set U disjoint from A. Then the complement $F := S - U$ is a closed set containing A and not containing p. Therefore, $p \notin \mathrm{cl}(A)$. □

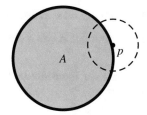

Fig. A.12. Every neighborhood of p contains a point of A.

Example. The closure of the open disk $B(\mathbf{0}, r)$ in $\mathbb{R}^2$ is the closed disk

$$\overline{B}(\mathbf{0}, r) = \{p \in \mathbb{R}^2 \mid d(p, \mathbf{0}) \leq r\}.$$

Definition A.49. A point p in S is an *accumulation point* of A if every neighborhood of p in S contains a point of A other than p. The set of all accumulation points of A is denoted by $\mathrm{ac}(A)$.

If U is a neighborhood of p in S, we call $U - \{p\}$ a *deleted neighborhood* of p. An equivalent condition for p to be an accumulation point of A is to require that every deleted neighborhood of p in S contain a point of A. In some books an accumulation point is called a *limit point*.

Example. If $A = [0, 1[\cup \{2\}$ in $\mathbb{R}^1$, then the closure of A is $[0, 1] \cup \{2\}$, but the set of accumulation points of A is only the closed interval $[0, 1]$.

Proposition A.50. *Let A be a subset of a topological space S. Then*

$$\mathrm{cl}(A) = A \cup \mathrm{ac}(A).$$

Proof.
($\supset$) By definition, $A \subset \mathrm{cl}(A)$. By the local characterization of closure (Proposition A.48), $\mathrm{ac}(A) \subset \mathrm{cl}(A)$. Hence, $A \cup \mathrm{ac}(A) \subset \mathrm{cl}(A)$.

($\subset$) Suppose $p \in \mathrm{cl}(A)$. Either $p \in A$ or $p \notin A$. If $p \in A$, then $p \in A \cup \mathrm{ac}(A)$. Suppose $p \notin A$. By Proposition A.48, every neighborhood of p contains a point of A, which cannot be p, since $p \notin A$. Therefore, every deleted neighborhood of p contains a point of A. In this case,

$$p \in \mathrm{ac}(A) \subset A \cup \mathrm{ac}(A).$$

So $\mathrm{cl}(A) \subset A \cup \mathrm{ac}(A)$.   □

Proposition A.51. *A set A is closed if and only if $A = \overline{A}$.*

Proof.
($\Leftarrow$) If $A = \overline{A}$, then A is closed because $\overline{A}$ is closed.

($\Rightarrow$) Suppose A is closed. Then A is a closed set containing A, so that $\overline{A} \subset A$. Because $A \subset \overline{A}$, equality holds.   □

Proposition A.52. *If $A \subset B$ in a topological space S, then $\overline{A} \subset \overline{B}$.*

Proof. Since $\overline{B}$ contains B, it also contains A. As a closed subset of S containing A, it contains $\overline{A}$ by definition. $\qquad\qquad\square$

Exercise A.53 (Closure of a finite union or finite intersection). Let A and B be subsets of a topological space S. Prove the following:

(a) $\overline{A \cup B} = \overline{A} \cup \overline{B}$,
(b) $\overline{A \cap B} \subset \overline{A} \cap \overline{B}$.

The example of $A =]a,0[$ and $B =]0,b[$ in the real line shows that in general, $\overline{A \cap B} \neq \overline{A} \cap \overline{B}$.

A.13 Convergence

Let S be a topological space. A *sequence* in S is a map from the set $\mathbb{Z}^+$ of positive integers to S. We write a sequence as $\langle x_i \rangle$ or $x_1, x_2, x_3, \ldots$.

Definition A.54. The sequence $\langle x_i \rangle$ *converges* to p if for every neighborhood U of p, there is a positive integer N such that for all $i \geq N$, $x_i \in U$. In this case we say that p is a *limit* of the sequence $\langle x_i \rangle$ and write $x_i \to p$ or $\lim_{i \to \infty} x_i = p$.

Proposition A.55 (Uniqueness of the limit). *In a Hausdorff space S, if a sequence $\langle x_i \rangle$ converges to p and to q, then $p = q$.*

Proof. Problem A.19. $\qquad\qquad\square$

Thus, in a Hausdorff space we may speak of *the* limit of a convergent sequence.

Proposition A.56 (The sequence lemma). *Let S be a topological space and A a subset of S. If there is a sequence $\langle a_i \rangle$ in A that converges to p, then $p \in \mathrm{cl}(A)$. The converse is true if S is first countable.*

Proof.
($\Rightarrow$) Suppose $a_i \to p$, where $a_i \in A$ for all i. By the definition of convergence, every neighborhood U of p contains all but finitely many of the points a_i. In particular, U contains a point in A. By the local characterization of closure (Proposition A.48), $p \in \mathrm{cl}(A)$.

($\Leftarrow$) Suppose $p \in \mathrm{cl}(A)$. Since S is first countable, we can find a countable basis of neighborhoods $\{U_n\}$ at p such that

$$U_1 \supset U_2 \supset \cdots.$$

By the local characterization of closure, in each U_i there is a point $a_i \in A$. We claim that the sequence $\langle a_i \rangle$ converges to p. If U is any neighborhood of p, then by the definition of a basis of neighborhoods at p, there is a U_N such that $p \in U_N \subset U$. For all $i \geq N$, we then have

$$U_i \subset U_N \subset U.$$

Therefore, for all $i \geq N$,

$$a_i \in U_i \subset U.$$

This proves that $\langle a_i \rangle$ converges to p. $\qquad\qquad\square$

Problems

A.1. Set theory
If U_1 and U_2 are subsets of a set X, and V_1 and V_2 are subsets of a set Y, prove that

$$(U_1 \times V_1) \cap (U_2 \times V_2) = (U_1 \cap U_2) \times (V_1 \cap V_2).$$

A.2. Union and intersection
Suppose $U_1 \cap V_1 = U_2 \cap V_2 = \varnothing$ in a topological space S. Show that the intersection $U_1 \cap U_2$ is disjoint from the union $V_1 \cup V_2$. (*Hint*: Use the distributive property of an intersection over a union.)

A.3. Closed sets
Let S be a topological space. Prove the following two statements.

(a) If $\{F_i\}_{i=1}^n$ is a finite collection of closed sets in S, then $\bigcup_{i=1}^n F_i$ is closed.
(b) If $\{F_\alpha\}_{\alpha \in A}$ is an arbitrary collection of closed sets in S, then $\bigcap_\alpha F_\alpha$ is closed.

A.4. Cubes versus balls
Prove that the open cube $]-a,a[^n$ is contained in the open ball $B(\mathbf{0}, \sqrt{n}a)$, which in turn is contained in the open cube $]-\sqrt{n}a, \sqrt{n}a[^n$. Therefore, open cubes with arbitrary centers in $\mathbb{R}^n$ form a basis for the standard topology on $\mathbb{R}^n$.

A.5. Product of closed sets
Prove that if A is closed in X and B is closed in Y, then $A \times B$ is closed in $X \times Y$.

A.6. Characterization of a Hausdorff space by its diagonal
Let S be a topological space. The diagonal Δ in $S \times S$ is the set

$$\Delta = \{(x,x) \in S \times S\}.$$

Prove that S is Hausdorff if and only if the diagonal Δ is closed in $S \times S$. (*Hint*: Prove that S is Hausdorff if and only if $S \times S - \Delta$ is open in $S \times S$.)

A.7. Projection
Prove that if X and Y are topological spaces, then the projection $\pi : X \times Y \to X$, $\pi(x,y) = x$, is an open map.

A.8. The ε-δ criterion for continuity
Prove that a function $f : A \to \mathbb{R}^m$ is continuous at $p \in A$ if and only if for every $\varepsilon > 0$, there exists a $\delta > 0$ such that for all $x \in A$ satisfying $d(x,p) < \delta$, one has $d(f(x), f(p)) < \varepsilon$.

A.9. Continuity in terms of closed sets
Prove Proposition A.28.

A.10. Continuity of a map into a product
Let X, Y_1, and Y_2 be topological spaces. Prove that a map $f = (f_1, f_2) : X \to Y_1 \times Y_2$ is continuous if and only if both components $f_i : X \to Y_i$ are continuous.

A.11. Continuity of the product map
Given two maps $f : X \to X'$ and $g : Y \to Y'$ of topological spaces, we define their *product* to be

$$f \times g : X \times Y \to X' \times Y', \quad (f \times g)(x,y) = (f(x), g(y)).$$

Note that if $\pi_1 : X \times Y \to X$ and $\pi_2 : X \times Y \to Y$ are the two projections, then $f \times g = (f \circ \pi_1, f \circ \pi_2)$. Prove that $f \times g$ is continuous if and only if both f and g are continuous.

A.12. Homeomorphism
Prove that if a continuous bijection $f\colon X \to Y$ is a closed map, then it is a homeomorphism (cf. Corollary A.36).

A.13.* The Lindelöf condition
Show that if a topological space is second countable, then it is Lindelöf; i.e., every open cover has a countable subcover.

A.14. Compactness
Prove that a finite union of compact sets in a topological space S is compact.

A.15.* Disconnected subset in terms of a separation
Prove Proposition A.42.

A.16. Local connectedness
A topological space S is said to be *locally connected at* $p \in S$ if for every neighborhood U of p, there is a connected neighborhood V of p such that $V \subset U$. The space S is *locally connected* if it is locally connected at every point. Prove that if S is locally connected, then the connected components of S are open.

A.17. Closure
Let U be an open subset and A an arbitrary subset of a topological space S. Prove that $U \cap \bar{A} \neq \varnothing$ if and only if $U \cap A \neq \varnothing$.

A.18. Countability
Prove that every second-countable space is first countable.

A.19.* Uniqueness of the limit
Prove Proposition A.55.

A.20.* Closure in a product
Let S and Y be topological spaces and $A \subset S$. Prove that

$$\mathrm{cl}_{S \times Y}(A \times Y) = \mathrm{cl}_S(A) \times Y$$

in the product space $S \times Y$.

A.21. Dense subsets
A subset A of a topological space S is said to be *dense* in S if its closure $\mathrm{cl}_S(A)$ equals S.

(a) Prove that A is dense in S if and only if for every $p \in S$, every neighborhood U of p contains a point of A.

(b) Let K be a field. Prove that a Zariski-open subset U of K^n is dense in K^n. (*Hint*: Example A.18.)

§B The Inverse Function Theorem on $\mathbb{R}^n$ and Related Results

This appendix reviews three logically equivalent theorems from real analysis: the inverse function theorem, the implicit function theorem, and the constant rank theorem, which describe the local behavior of a C^∞ map from $\mathbb{R}^n$ to $\mathbb{R}^m$. We will assume the inverse function theorem and from it deduce the other two in the simplest cases. In Section 11 these theorems are applied to manifolds in order to clarify the local behavior of a C^∞ map when the map has maximal rank at a point or constant rank in a neighborhood.

B.1 The Inverse Function Theorem

A C^∞ map $f: U \to \mathbb{R}^n$ defined on an open subset U of $\mathbb{R}^n$ is *locally invertible* or a *local diffeomorphism* at a point p in U if f has a C^∞ inverse in some neighborhood of p. The inverse function theorem gives a criterion for a map to be locally invertible. We call the matrix $Jf = [\partial f^i/\partial x^j]$ of partial derivatives of f the *Jacobian matrix* of f and its determinant $\det[\partial f^i/\partial x^j]$ the *Jacobian determinant* of f.

Theorem B.1 (Inverse function theorem). *Let $f: U \to \mathbb{R}^n$ be a C^∞ map defined on an open subset U of $\mathbb{R}^n$. At any point p in U, the map f is invertible in some neighborhood of p if and only if the Jacobian determinant $\det[\partial f^i/\partial x^j (p)]$ is not zero.*

For a proof, see for example [35, Theorem 9.24, p. 221]. Although the inverse function theorem apparently reduces the invertibility of f on an open set to a single number at p, because the Jacobian determinant is a continuous function, the non-vanishing of the Jacobian determinant at p is equivalent to its nonvanishing in a neighborhood of p.

Since the linear map represented by the Jacobian matrix $Jf(p)$ is the best linear approximation to f at p, it is plausible that f is invertible in a neighborhood of p if and only if $Jf(p)$ is also, i.e., if and only if $\det(Jf(p)) \neq 0$.

B.2 The Implicit Function Theorem

In an equation such as $f(x, y) = 0$, it is often impossible to solve explicitly for one of the variables in terms of the other. If we can show the existence of a function $y = h(x)$, which we may or may not be able to write down explicitly, such that $f(x, h(x)) = 0$, then we say that $f(x, y) = 0$ can be solved *implicitly* for y in terms of x. The implicit function theorem provides a sufficient condition on a system of equations $f^i(x^1, \ldots, x^n) = 0$, $i = 1, \ldots, m$, under which *locally* a set of variables can be solved implicitly as C^∞ functions of the other variables.

Example. Consider the equation

$$f(x,y) = x^2 + y^2 - 1 = 0.$$

The solution set is the unit circle in the xy-plane.

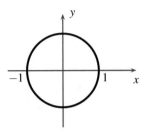

Fig. B.1. The unit circle.

From the picture we see that in a neighborhood of any point other than $(\pm 1, 0)$, y is a function of x. Indeed,

$$y = \pm\sqrt{1 - x^2},$$

and either function is C^∞ as long as $x \neq \pm 1$. At $(\pm 1, 0)$, there is no neighborhood on which y is a function of x.

On a smooth curve $f(x,y) = 0$ in $\mathbb{R}^2$,

> y can be expressed as a function of x in a neighborhood of a point (a,b)
> $\Longleftrightarrow$ the tangent line to $f(x,y) = 0$ at (a,b) is not vertical
> $\Longleftrightarrow$ the normal vector $\operatorname{grad} f := \langle f_x, f_y \rangle$ to $f(x,y) = 0$ at (a,b)
> is not horizontal
> $\Longleftrightarrow f_y(a,b) \neq 0$.

The implicit function theorem generalizes this condition to higher dimensions. We will deduce the implicit function theorem from the inverse function theorem.

Theorem B.2 (Implicit function theorem). *Let U be an open subset in $\mathbb{R}^n \times \mathbb{R}^m$ and $f: U \to \mathbb{R}^m$ a C^∞ map. Write $(x,y) = (x^1, \ldots, x^n, y^1, \ldots, y^m)$ for a point in U. At a point $(a,b) \in U$ where $f(a,b) = 0$ and the determinant $\det[\partial f^i / \partial y^j(a,b)]$ is nonzero, there exist a neighborhood $A \times B$ of (a,b) in U and a unique function $h: A \to B$ such that in $A \times B \subset U \subset \mathbb{R}^n \times \mathbb{R}^m$,*

$$f(x,y) = 0 \quad \Longleftrightarrow \quad y = h(x).$$

Moreover, h is C^∞.

Proof. To solve $f(x,y) = 0$ for y in terms of x using the inverse function theorem, we first turn it into an inverse problem. For this, we need a map between two open

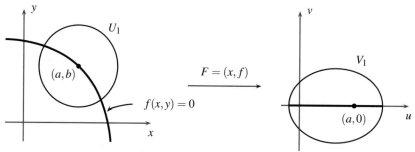

Fig. B.2. F^{-1} maps the u-axis to the zero set of f.

sets of the same dimension. Since $f(x,y)$ is a map from an open set U in $\mathbb{R}^{n+m}$ to $\mathbb{R}^m$, it is natural to extend f to a map $F \colon U \to \mathbb{R}^{n+m}$ by adjoining x to it as the first n components:

$$F(x,y) = (u,v) = (x, f(x,y)).$$

To simplify the exposition, we will assume in the rest of the proof that $n = m = 1$. Then the Jacobian matrix of F is

$$JF = \begin{bmatrix} 1 & 0 \\ \partial f/\partial x & \partial f/\partial y \end{bmatrix}.$$

At the point (a,b),

$$\det JF(a,b) = \frac{\partial f}{\partial y}(a,b) \neq 0.$$

By the inverse function theorem, there are neighborhoods U_1 of (a,b) and V_1 of $F(a,b) = (a,0)$ in $\mathbb{R}^2$ such that $F \colon U_1 \to V_1$ is a diffeomorphism with C^∞ inverse F^{-1} (Figure B.2). Since $F \colon U_1 \to V_1$ is defined by

$$u = x,$$
$$v = f(x,y),$$

the inverse map $F^{-1} \colon V_1 \to U_1$ must be of the form

$$x = u,$$
$$y = g(u,v)$$

for some C^∞ function $g \colon V_1 \to \mathbb{R}$. Thus, $F^{-1}(u,v) = (u, g(u,v))$.

The two compositions $F^{-1} \circ F$ and $F \circ F^{-1}$ give

$$(x,y) = (F^{-1} \circ F)(x,y) = F^{-1}(x, f(x,y)) = (x, g(x, f(x,y))),$$
$$(u,v) = (F \circ F^{-1})(u,v) = F(u, g(u,v)) = (u, f(u, g(u,v))).$$

Hence,

$$y = g(x, f(x,y)) \quad \text{for all } (x,y) \in U_1, \tag{B.1}$$
$$v = f(u, g(u,v)) \quad \text{for all } (u,v) \in V_1. \tag{B.2}$$

If $f(x,y) = 0$, then (B.1) gives $y = g(x,0)$. This suggests that we define $h(x) = g(x,0)$ for all $x \in \mathbb{R}^1$ for which $(x,0) \in V_1$. The set of all such x is homeomorphic to $V_1 \cap (\mathbb{R}^1 \times \{0\})$ and is an open subset of $\mathbb{R}^1$. Since g is C^∞ by the inverse function theorem, h is also C^∞.

Claim. For $(x,y) \in U_1$ such that $(x,0) \in V_1$,

$$f(x,y) = 0 \quad \Longleftrightarrow \quad y = h(x).$$

Proof (of Claim).
($\Rightarrow$) As we saw already, from (B.1), if $f(x,y) = 0$, then

$$y = g(x, f(x,y)) = g(x,0) = h(x). \tag{B.3}$$

($\Leftarrow$) If $y = h(x)$ and in (B.2) we set $(u,v) = (x,0)$, then

$$0 = f(x, g(x,0)) = f(x, h(x)) = f(x,y). \qquad \square$$

By the claim, in some neighborhood of $(a,b) \in U_1$, the zero set of $f(x,y)$ is precisely the graph of h. To find a product neighborhood of (a,b) as in the statement of the theorem, let $A_1 \times B$ be a neighborhood of (a,b) contained in U_1 and let $A = h^{-1}(B) \cap A_1$. Since h is continuous, A is open in the domain of h and hence in $\mathbb{R}^1$. Then $h(A) \subset B$,

$$A \times B \subset A_1 \times B \subset U_1, \quad \text{and} \quad A \times \{0\} \subset V_1.$$

By the claim, for $(x,y) \in A \times B$,

$$f(x,y) = 0 \quad \Longleftrightarrow \quad y = h(x).$$

Equation (B.3) proves the uniqueness of h. $\qquad\qquad\qquad\qquad\qquad\square$

Replacing a partial derivative such as $\partial f / \partial y$ with a Jacobian matrix $[\partial f^i / \partial y^j]$, we can prove the general case of the implicit function theorem in exactly the same way. Of course, in the theorem $y^1, \ldots, y^m$ need not be the last m coordinates in $\mathbb{R}^{n+m}$; they can be any set of m coordinates in $\mathbb{R}^{n+m}$.

Theorem B.3. *The implicit function theorem is equivalent to the inverse function theorem.*

Proof. We have already shown, at least for one typical case, that the inverse function theorem implies the implicit function theorem. We now prove the reverse implication.

So assume the implicit function theorem, and let $f : U \to \mathbb{R}^n$ be a C^∞ map defined on an open subset U of $\mathbb{R}^n$ such that at some point $p \in U$, the Jacobian determinant

$\det[\partial f^i/\partial x^j(p)]$ is nonzero. Finding a local inverse for $y = f(x)$ near p amounts to solving the equation

$$g(x,y) = f(x) - y = 0$$

for x in terms of y near $(p, f(p))$. Note that $\partial g^i/\partial x^j = \partial f^i/\partial x^j$. Hence,

$$\det\left[\frac{\partial g^i}{\partial x^j}(p, f(p))\right] = \det\left[\frac{\partial f^i}{\partial x^j}(p)\right] \neq 0.$$

By the implicit function theorem, x can be expressed in terms of y locally near $(p, f(p))$; i.e., there is a C^∞ function $x = h(y)$ defined in a neighborhood of $f(p)$ in $\mathbb{R}^n$ such that

$$g(x,y) = f(x) - y = f(h(y)) - y = 0.$$

Thus, $y = f(h(y))$. Since $y = f(x)$,

$$x = h(y) = h(f(x)).$$

Therefore, f and h are inverse functions defined near p and $f(p)$ respectively.   □

B.3   Constant Rank Theorem

Every C^∞ map $f: U \to \mathbb{R}^m$ on an open set U of $\mathbb{R}^n$ has a *rank* at each point p in U, namely the rank of its Jacobian matrix $[\partial f^i/\partial x^j(p)]$.

Theorem B.4 (Constant rank theorem). *If $f: \mathbb{R}^n \supset U \to \mathbb{R}^m$ has constant rank k in a neighborhood of a point $p \in U$, then after a suitable change of coordinates near p in U and $f(p)$ in $\mathbb{R}^m$, the map f assumes the form*

$$(x^1,\ldots,x^n) \mapsto (x^1,\ldots,x^k,0,\ldots,0).$$

More precisely, there are a diffeomorphism G of a neighborhood of p in U sending p to the origin in $\mathbb{R}^n$ and a diffeomorphism F of a neighborhood of $f(p)$ in $\mathbb{R}^m$ sending $f(p)$ to the origin in $\mathbb{R}^m$ such that

$$(F \circ f \circ G)^{-1}(x^1,\ldots,x^n) = (x^1,\ldots,x^k,0,\ldots,0).$$

Proof (for $n = m = 2$, $k = 1$). Suppose $f = (f^1, f^2): \mathbb{R}^2 \supset U \to \mathbb{R}^2$ has constant rank 1 in a neighborhood of $p \in U$. By reordering the functions f^1, f^2 or the variables x, y, we may assume that $\partial f^1/\partial x(p) \neq 0$. (Here we are using the fact that f has rank ≥ 1 at p.) Define $G: U \to \mathbb{R}^2$ by

$$G(x,y) = (u,v) = (f^1(x,y),y).$$

The Jacobian matrix of G is

$$JG = \begin{bmatrix} \partial f^1/\partial x & \partial f^1/\partial y \\ 0 & 1 \end{bmatrix}.$$

Since $\det JG(p) = \partial f^1/\partial x(p) \neq 0$, by the inverse function theorem there are neighborhoods U_1 of $p \in \mathbb{R}^2$ and V_1 of $G(p) \in \mathbb{R}^2$ such that $G\colon U_1 \to V_1$ is a diffeomorphism. By making U_1 a sufficiently small neighborhood of p, we may assume that f has constant rank 1 on U_1.

On V_1,

$$(u,v) = (G \circ G^{-1})(u,v) = (f^1 \circ G^{-1}, y \circ G^{-1})(u,v).$$

Comparing the first components gives $u = (f^1 \circ G^{-1})(u,v)$. Hence,

$$\begin{aligned}
(f \circ G^{-1})(u,v) &= (f^1 \circ G^{-1}, f^2 \circ G^{-1})(u,v) \\
&= (u, f^2 \circ G^{-1}(u,v)) \\
&= (u, h(u,v)),
\end{aligned}$$

where we set $h = f^2 \circ G^{-1}$.

Because $G^{-1}\colon V_1 \to U_1$ is a diffeomorphism and f has constant rank 1 on U_1, the composite $f \circ G^{-1}$ has constant rank 1 on V_1. Its Jacobian matrix is

$$J(f \circ G^{-1}) = \begin{bmatrix} 1 & 0 \\ \partial h/\partial u & \partial h/\partial v \end{bmatrix}.$$

For this matrix to have constant rank 1, $\partial h/\partial v$ must be identically zero on V_1. (Here we are using the fact that f has rank ≤ 1 in a neighborhood of p.) Thus, h is a function of u alone and we may write

$$(f \circ G^{-1})(u,v) = (u, h(u)).$$

Finally, let $F\colon \mathbb{R}^2 \to \mathbb{R}^2$ be the change of coordinates $F(x,y) = (x, y - h(x))$. Then

$$(F \circ f \circ G^{-1})(u,v) = F(u, h(u)) = (u, h(u) - h(u)) = (u, 0). \qquad \square$$

Example B.5. If a C^∞ map $f\colon \mathbb{R}^n \supset U \to \mathbb{R}^n$ defined on an open subset U of $\mathbb{R}^n$ has nonzero Jacobian determinant $\det(Jf(p))$ at a point $p \in U$, then by continuity it has nonzero Jacobian determinant in a neighborhood of p. Therefore, it has constant rank n in a neighborhood of p.

Problems

B.1.* The rank of a matrix
The *rank* of a matrix A, denoted by $\mathrm{rk}\,A$, is defined to be the number of linearly independent columns of A. By a theorem in linear algebra, it is also the number of linearly independent rows of A. Prove the following lemma.

Lemma. *Let A be an $m \times n$ matrix (not necessarily square), and k a positive integer. Then $\mathrm{rk}\,A \geq k$ if and only if A has a nonsingular $k \times k$ submatrix. Equivalently, $\mathrm{rk}\,A \leq k-1$ if and only if all $k \times k$ minors of A vanish. (A $k \times k$ minor of a matrix A is the determinant of a $k \times k$ submatrix of A.)*

B.2.* Matrices of rank at most r

For an integer $r \geq 0$, define D_r to be the subset of $\mathbb{R}^{m \times n}$ consisting of all $m \times n$ real matrices of rank at most r. Show that D_r is a closed subset of $\mathbb{R}^{m \times n}$. (*Hint*: Use Problem B.1.)

B.3.* Maximal rank

We say that the rank of an $m \times n$ matrix A is *maximal* if $\operatorname{rk} A = \min(m, n)$. Define $D_{\max}$ to be the subset of $\mathbb{R}^{m \times n}$ consisting of all $m \times n$ matrices of maximal rank r. Show that $D_{\max}$ is an open subset of $\mathbb{R}^{m \times n}$. (*Hint*: Suppose $n \leq m$. Then $D_{\max} = \mathbb{R}^{m \times n} - D_{n-1}$. Apply Problem B.2.)

B.4.* Degeneracy loci and maximal-rank locus of a map

Let $F: S \to \mathbb{R}^{m \times n}$ be a continuous map from a topological space S to the space $\mathbb{R}^{m \times n}$. The *degeneracy locus of rank r of F* is defined to be

$$D_r(F) := \{x \in S \mid \operatorname{rk} F(x) \leq r\}.$$

(a) Show that the degeneracy locus $D_r(F)$ is a closed subset of S. (*Hint*: $D_r(F) = F^{-1}(D_r)$, where D_r was defined in Problem B.2.)

(b) Show that the *maximal-rank locus of F*,

$$D_{\max}(F) := \{x \in S \mid \operatorname{rk} F(x) \text{ is maximal}\},$$

is an open subset of S.

B.5. Rank of a composition of linear maps

Suppose V, W, V', W' are finite-dimensional vector spaces.

(a) Prove that if the linear map $L: V \to W$ is surjective, then for any linear map $f: W \to W'$, $\operatorname{rk}(f \circ L) = \operatorname{rk} f$.

(b) Prove that if the linear map $L: V \to W$ is injective, then for any linear map $g: V' \to V$, $\operatorname{rk}(L \circ g) = \operatorname{rk} g$.

B.6. Constant rank theorem

Generalize the proof of the constant rank theorem (Theorem B.4) in the text to arbitrary n, m, and k.

B.7. Equivalence of the constant rank theorem and the inverse function theorem

Use the constant rank theorem (Theorem B.4) to prove the inverse function theorem (Theorem B.1). Hence, the two theorems are equivalent.

§C Existence of a Partition of Unity in General

This appendix contains a proof of Theorem 13.7 on the existence of a C^∞ partition of unity on a general manifold.

Lemma C.1. *Every manifold M has a countable basis all of whose elements have compact closure.*

Recall that if A is a subset of a topological space X, the notation $\overline{A}$ denotes the closure of A in X.

Proof (of Lemma C.1). Start with a countable basis $\mathcal{B}$ for M and consider the sub-collection $\mathcal{S}$ of elements in $\mathcal{B}$ that have compact closure. We claim that $\mathcal{S}$ is again a basis. Given an open subset $U \subset M$ and point $p \in U$, choose a neighborhood V of p such that $V \subset U$ and V has compact closure. This is always possible since M is locally Euclidean.

Since $\mathcal{B}$ is a basis, there is an open set $B \in \mathcal{B}$ such that

$$p \in B \subset V \subset U.$$

Then $\overline{B} \subset \overline{V}$. Because $\overline{V}$ is compact, so is the closed subset $\overline{B}$. Hence, $B \in \mathcal{S}$. Since for any open set U and any $p \in U$, we have found a set $B \in \mathcal{S}$ such that $p \in B \subset U$, the collection $\mathcal{S}$ of open sets is a basis.   □

Proposition C.2. *Every manifold M has a countable increasing sequence of subsets*

$$V_1 \subset \overline{V_1} \subset V_2 \subset \overline{V_2} \subset \cdots,$$

with each V_i open and $\overline{V_i}$ compact, such that M is the union of the V_i's (Figure C.1).

Proof. By Lemma C.1, M has a countable basis $\{B_i\}_{i=1}^{\infty}$ with each $\overline{B_i}$ compact. Any basis of M of course covers M. Set $V_1 = B_1$. By compactness, $\overline{V_1}$ is covered by finitely many of the B_i's. Define i_1 to be the smallest integer ≥ 2 such that

$$\overline{V_1} \subset B_1 \cup B_2 \cup \cdots \cup B_{i_1}.$$

Suppose open sets $V_1, \ldots, V_m$ have been defined, each with compact closure. As before, by compactness, $\overline{V_m}$ is covered by finitely many of the B_i's. If i_m is the smallest integer $\geq m+1$ and $\geq i_{m-1}$ such that

$$\overline{V_m} \subset B_1 \cup B_2 \cup \cdots \cup B_{i_m},$$

then we set

$$V_{m+1} = B_1 \cup B_2 \cup \cdots \cup B_{i_m}.$$

Since a finite union of compact sets is compact and

$$\overline{V_{m+1}} \subset \overline{B_1} \cup \overline{B_2} \cup \cdots \cup \overline{B_{i_m}}$$

is a closed subset of a compact set, $\overline{V_{m+1}}$ is compact. Since $i_m \ge m+1$, $B_{m+1} \subset V_{m+1}$. Thus,

$$M = \bigcup B_i \subset \bigcup V_i \subset M.$$

This proves that $M = \bigcup_{i=1}^{\infty} V_i$.   □

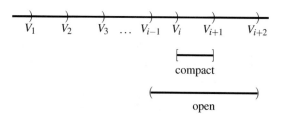

Fig. C.1. A nested open cover.

Define V_0 to be the empty set. For each $i \ge 1$, because $\overline{V_{i+1}} - V_i$ is a closed subset of the compact $\overline{V_{i+1}}$, it is compact. Moreover, it is contained in the open set $V_{i+2} - \overline{V_{i-1}}$.

Theorem 13.7 (Existence of a C^∞ partition of unity). *Let $\{U_\alpha\}_{\alpha \in A}$ be an open cover of a manifold M.*

(i) *There is a C^∞ partition of unity $\{\varphi_k\}_{k=1}^{\infty}$ with every φ_k having compact support such that for each k, $\operatorname{supp} \varphi_k \subset U_\alpha$ for some $\alpha \in A$.*
(ii) *If we do not require compact support, then there is a C^∞ partition of unity $\{\rho_\alpha\}$ subordinate to $\{U_\alpha\}$.*

Proof.
(i) Let $\{V_i\}_{i=0}^{\infty}$ be an open cover of M as in Proposition C.2, with V_0 the empty set. The idea of the proof is quite simple. For each i, we find finitely many smooth bump functions ψ_j^i on M, each with compact support in the open set $V_{i+2} - \overline{V_{i-1}}$ as well as in some U_α, such that their sum $\sum_j \psi_j^i$ is positive on the compact set $\overline{V_{i+1}} - V_i$. The collection $\{\operatorname{supp} \psi_j^i\}$ of supports over all i, j will be locally finite. Since the compact sets $\overline{V_{i+1}} - V_i$ cover M, the locally finite sum $\psi = \sum_{i,j} \psi_j^i$ will be positive on M. Then $\{\psi_j^i / \psi\}$ is a C^∞ partition of unity satisfying the conditions in (i).

We now fill in the details. Fix an integer $i \ge 1$. For each p in the compact set $\overline{V_{i+1}} - V_i$, choose an open set U_α containing p from the open cover $\{U_\alpha\}$. Then p is in the open set $U_\alpha \cap (V_{i+2} - \overline{V_{i-1}})$. Let ψ_p be a C^∞ bump function on M that is positive on a neighborhood W_p of p and has support in $U_\alpha \cap (V_{i+2} - \overline{V_{i-1}})$. Since $\operatorname{supp} \psi_p$ is a closed set contained in the compact set $\overline{V_{i+2}}$, it is compact.

The collection $\{W_p \mid p \in \overline{V_{i+1}} - V_i\}$ is an open cover of the compact set $\overline{V_{i+1}} - V_i$, and so there is a finite subcover $\{W_{p_1}, \ldots, W_{p_m}\}$, with associated bump functions

$\psi_{p_1}, \ldots, \psi_{p_m}$. Since m, W_{p_j}, and ψ_{p_j} all depend on i, we relabel them as $m(i)$, $W_1^i, \ldots, W_{m(i)}^i$, and $\psi_1^i, \ldots, \psi_{m(i)}^i$.

In summary, for each $i \geq 1$, we have found finitely many open sets $W_1^i, \ldots, W_{m(i)}^i$ and finitely many C^∞ bump functions $\psi_1^i, \ldots, \psi_{m(i)}^i$ such that

(1) $\psi_j^i > 0$ on W_j^i for $j = 1, \ldots, m(i)$;
(2) $W_1^i, \ldots, W_{m(i)}^i$ cover the compact set $\overline{V_{i+1}} - V_i$;
(3) $\operatorname{supp} \psi_j^i \subset U_{\alpha_{ij}} \cap (V_{i+2} - \overline{V_{i-1}})$ for some $\alpha_{ij} \in A$;
(4) $\operatorname{supp} \psi_j^i$ is compact.

As i runs from 1 to ∞, we obtain countably many bump functions $\{\psi_j^i\}$. The collection of their supports, $\{\operatorname{supp} \psi_j^i\}$, is locally finite, since only finitely many of these sets intersect any V_i. Indeed, since

$$\operatorname{supp} \psi_j^\ell \subset V_{\ell+2} - \overline{V_{\ell-1}}$$

for all ℓ, as soon as $\ell \geq i + 1$,

$$\left(\operatorname{supp} \psi_j^\ell \right) \cap V_i = \text{the empty set } \varnothing.$$

Any point $p \in M$ is contained in the compact set $\overline{V_{i+1}} - V_i$ for some i, and therefore $p \in W_j^i$ for some (i, j). For this (i, j), $\psi_j^i(p) > 0$. Hence, the sum $\psi := \Sigma_{i,j} \, \psi_j^i$ is locally finite and everywhere positive on M. To simplify the notation, we now relabel the countable set $\{\psi_j^i\}$ as $\{\psi_1, \psi_2, \psi_3, \ldots\}$. Define

$$\varphi_k = \frac{\psi_k}{\psi}.$$

Then $\Sigma \varphi_k = 1$ and

$$\operatorname{supp} \varphi_k = \operatorname{supp} \psi_k \subset U_\alpha$$

for some $\alpha \in A$. So $\{\varphi_k\}$ is a partition of unity with compact support such that for each k, $\operatorname{supp} \varphi_k \subset U_\alpha$ for some $\alpha \in A$.

(ii) For each $k = 1, 2, \ldots$, let $\tau(k)$ be an index in A such that

$$\operatorname{supp} \varphi_k \subset U_{\tau(k)}$$

as in the preceding paragraph. Group the collection $\{\varphi_k\}$ according to $\tau(k)$ and define

$$\rho_\alpha = \sum_{\tau(k)=\alpha} \varphi_k$$

if there is a k with $\tau(k) = \alpha$; otherwise, set $\rho_\alpha = 0$. Then

$$\sum_{\alpha \in A} \rho_\alpha = \sum_{\alpha \in A} \sum_{\tau(k)=\alpha} \varphi_k = \sum_{k=1}^\infty \varphi_k = 1.$$

By Problem 13.7,

$$\operatorname{supp} \rho_\alpha \subset \bigcup_{\tau(k)=\alpha} \operatorname{supp} \varphi_k \subset U_\alpha.$$

Hence, $\{\rho_\alpha\}$ is a C^∞ partition of unity subordinate to $\{U_\alpha\}$. $\square$

§D Linear Algebra

This appendix gathers together a few facts from linear algebra used throughout the book, especially in Sections 24 and 25.

The quotient vector space is a construction in which one reduces a vector space to a smaller space by identifying a subspace to zero. It represents a simplification, much like the formation of a quotient group or of a quotient ring. For a linear map $f: V \to W$ of vector spaces, the first isomorphism theorem of linear algebra gives an isomorphism between the quotient space $V / \ker f$ and the image of f. It is one of the most useful results in linear algebra.

We also discuss the direct sum and the direct product of a family of vector spaces, as well as the distinction between an internal and an external direct sum.

D.1 Quotient Vector Spaces

If V is a vector space and W is a subspace of V, a *coset* of W in V is a subset of the form

$$v + W = \{v + w \mid w \in W\}$$

for some $v \in V$.

Two cosets $v + W$ and $v' + W$ are equal if and only if $v' = v + w$ for some $w \in W$, or equivalently, if and only if $v' - v \in W$. This introduces an equivalence relation on V:

$$v \sim v' \iff v' - v \in W \iff v + W = v' + W.$$

A coset of W in V is simply an equivalence class under this equivalence relation. Any element of $v + W$ is called a *representative* of the coset $v + W$.

The set V/W of all cosets of W in V is again a vector space, with addition and scalar multiplication defined by

$$(u + W) + (v + W) = (u + v) + W,$$
$$r(v + W) = rv + W$$

for $u, v \in V$ and $r \in \mathbb{R}$. We call V/W the *quotient vector space* or *the quotient space of V by W*.

Example D.1. For $V = \mathbb{R}^2$ and W a line through the origin in $\mathbb{R}^2$, a coset of W in $\mathbb{R}^2$ is a line in $\mathbb{R}^2$ parallel to W. (For the purpose of this discussion, two lines in $\mathbb{R}^2$ are *parallel* if and only if they coincide or fail to intersect. This definition differs from the usual one in plane geometry in allowing a line to be parallel to itself.) The quotient space $\mathbb{R}^2/W$ is the collection of lines in $\mathbb{R}^2$ parallel to W (Figure D.1).

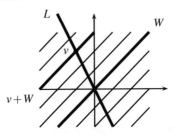

Fig. D.1. Quotient vector space of $\mathbb{R}^2$ by W.

D.2 Linear Transformations

Let V and W be vector spaces over $\mathbb{R}$. A map $f: V \to W$ is called a *linear transformation*, a *vector space homomorphism*, a *linear operator*, or a *linear map* over $\mathbb{R}$ if for all $u, v \in V$ and $r \in \mathbb{R}$,

$$f(u+v) = f(u) + f(v),$$
$$f(ru) = rf(u).$$

Example D.2. Let $V = \mathbb{R}^2$ and W a line through the origin in $\mathbb{R}^2$ as in Example D.1. If L is a line through the origin not parallel to W, then L will intersect each line in $\mathbb{R}^2$ parallel to W in one and only one point. This one-to-one correspondence

$$L \to \mathbb{R}^2/W,$$
$$v \mapsto v+W,$$

preserves addition and scalar multiplication, and so is an isomorphism of vector spaces. Thus, in this example the quotient space $\mathbb{R}^2/W$ can be identified with the line L.

If $f: V \to W$ is a linear transformation, the *kernel* of f is the set

$$\ker f = \{v \in V \mid f(v) = 0\}$$

and the *image* of f is the set

$$\operatorname{im} f = \{f(v) \in W \mid v \in V\}.$$

The kernel of f is a subspace of V and the image of f is a subspace of W. Hence, one can form the quotient spaces $V/\ker f$ and $W/\operatorname{im} f$. This latter space, $W/\operatorname{im} f$, denoted by $\operatorname{coker} f$, is called the *cokernel* of the linear map $f: V \to W$.

For now, denote by K the kernel of f. The linear map $f: V \to W$ induces a linear map $\bar{f}: V/K \to \operatorname{im} f$, by

$$\bar{f}(v+K) = f(v).$$

It is easy to check that $\bar{f}$ is linear and bijective. This gives the following fundamental result of linear algebra.

Theorem D.3 (The first isomorphism theorem). *Let* $f\colon V \to W$ *be a homomorphism of vector spaces. Then* f *induces an isomorphism*

$$\bar{f}\colon \frac{V}{\ker f} \xrightarrow{\sim} \operatorname{im} f.$$

D.3 Direct Product and Direct Sum

Let $\{V_\alpha\}_{\alpha \in I}$ be a family of real vector spaces. The *direct product* $\prod_\alpha V_\alpha$ is the set of all sequences (v_α) with $v_\alpha \in V_\alpha$ for all $\alpha \in I$, and the *direct sum* $\bigoplus_\alpha V_\alpha$ is the subset of the direct product $\prod_\alpha V_\alpha$ consisting of sequences (v_α) such that $v_\alpha = 0$ for all but finitely many $\alpha \in I$. Under componentwise addition and scalar multiplication,

$$(v_\alpha) + (w_\alpha) = (v_\alpha + w_\alpha),$$
$$r(v_\alpha) = (rv_\alpha), \quad r \in \mathbb{R},$$

both the direct product $\prod_\alpha V_\alpha$ and the direct sum $\bigoplus_\alpha V_\alpha$ are real vector spaces. When the index set I is finite, the direct sum coincides with the direct product. In particular, for two vector spaces A and B,

$$A \oplus B = A \times B = \{(a,b) \mid a \in A \text{ and } b \in B\}.$$

The *sum* of two subspaces A and B of a vector space V is the subspace

$$A + B = \{a + b \in V \mid a \in A,\ b \in B\}.$$

If $A \cap B = \{0\}$, this sum is called an *internal direct sum* and written $A \oplus_i B$. In an internal direct sum $A \oplus_i B$, every element has a representation as $a + b$ for a unique $a \in A$ and a unique $b \in B$. Indeed, if $a + b = a' + b' \in A \oplus_i B$, then

$$a - a' = b' - b \in A \cap B = \{0\}.$$

Hence, $a = a'$ and $b = b'$.

In contrast to the internal direct sum $A \oplus_i B$, the direct sum $A \oplus B$ is called the *external direct sum*. In fact, the two notions are isomorphic: the natural map

$$\varphi\colon A \oplus B \to A \oplus_i B,$$
$$(a,b) \mapsto a + b$$

is easily seen to be a linear isomorphism. For this reason, in the literature the internal direct sum is normally denoted by $A \oplus B$, just like the external direct sum.

If $V = A \oplus_i B$, then A is called a *complementary subspace to B in V*. In Example D.2, the line L is a complementary subspace to W, and we may identify the quotient vector space $\mathbb{R}^2/W$ with any complementary subspace to W.

In general, if W is a subspace of a vector space V and W' is a complementary subspace to W, then there is a linear map

$$\varphi\colon W' \to V/W,$$
$$w' \mapsto w' + W.$$

Exercise D.4. Show that $\varphi\colon W' \to V/W$ is an isomorphism of vector spaces.

Thus, the quotient space V/W may be identified with any complementary subspace to W in V. This identification is not canonical, for there are many complementary subspaces to a given subspace W and there is no reason to single out any one of them. However, when V has an inner product $\langle\,,\,\rangle$, one can single out a canonical complementary subspace, the *orthogonal complement* of W:

$$W^{\perp} = \{v \in V \mid \langle v, w\rangle = 0 \text{ for all } w \in W\}.$$

Exercise D.5. Check that $W^{\perp}$ is a complementary subspace to W.

In this case, there is a canonical identification $W^{\perp} \overset{\sim}{\to} V/W$.

Let $f\colon V \to W$ be a linear map of finite-dimensional vector spaces. It follows from the first isomorphism theorem and Problem D.1 that

$$\dim V - \dim(\ker f) = \dim(\operatorname{im} f).$$

Since the dimension is the only isomorphism invariant of a vector space, we therefore have the following corollary of the first isomorphism theorem.

Corollary D.6. *If* $f\colon V \to W$ *is a linear map of finite-dimensional vector spaces, then there is a vector space isomorphism*

$$V \simeq \ker f \oplus \operatorname{im} f.$$

(The right-hand side is an external direct sum because $\ker f$ and $\operatorname{im} f$ are not subspaces of the same vector space.)

Problems

D.1. Dimension of a quotient vector space
Prove that if $w_1,\ldots,w_m$ is a basis for W that extends to a basis $w_1,\ldots,w_m,v_1,\ldots,v_n$ for V, then $v_1 + W,\ldots,v_n + W$ is a basis for V/W. Therefore,

$$\dim V/W = \dim V - \dim W.$$

D.2. Dimension of a direct sum
Prove that if $a_1,\ldots,a_m$ is a basis for a vector space A and $b_1,\ldots,b_n$ is a basis for a vector space B, then $(a_i,0),(0,b_j)$, $i = 1,\ldots,m$, $j = 1,\ldots,n$, is a basis for the direct sum $A \oplus B$. Therefore,

$$\dim A \oplus B = \dim A + \dim B.$$

§E Quaternions and the Symplectic Group

First described by William Rowan Hamilton in 1843, *quaternions* are elements of the form

$$q = a + \mathbf{i}b + \mathbf{j}c + \mathbf{k}d, \quad a,b,c,d \in \mathbb{R},$$

that add componentwise and multiply according to the distributive property and the rules

$$\mathbf{i}^2 = \mathbf{j}^2 = \mathbf{k}^2 = -1,$$
$$\mathbf{ij} = \mathbf{k}, \quad \mathbf{jk} = \mathbf{i}, \quad \mathbf{ki} = \mathbf{j},$$
$$\mathbf{ij} = -\mathbf{ji}, \quad \mathbf{jk} = -\mathbf{kj}, \quad \mathbf{ki} = -\mathbf{ik}.$$

A mnemonic for the three rules $\mathbf{ij} = \mathbf{k}$, $\mathbf{jk} = \mathbf{i}$, $\mathbf{ki} = \mathbf{j}$ is that in going clockwise around the circle

the product of two successive elements is the next one. Under addition and multiplication, the quaternions satisfy all the properties of a field except the commutative property for multiplication. Such an algebraic structure is called a *skew field* or a *division ring*. In honor of Hamilton, the usual notation for the skew field of quaternions is $\mathbb{H}$.

A division ring that is also an algebra over a field K is called a *division algebra* over K. The real and complex fields $\mathbb{R}$ and $\mathbb{C}$ are commutative division algebras over $\mathbb{R}$. By a theorem of Ferdinand Georg Frobenius [13] from 1878, the skew field $\mathbb{H}$ of quaternions has the distinction of being the only (associative) division algebra over $\mathbb{R}$ other than $\mathbb{R}$ and $\mathbb{C}$.[1]

In this appendix we will derive the basic properties of quaternions and define the symplectic group in terms of quaternions. Because of the familiarity of complex matrices, quaternions are often represented by 2×2 complex matrices. Correspondingly, the symplectic group also has a description as a group of complex matrices.

One can define vector spaces and formulate linear algebra over a skew field, just as one would for vector spaces over a field. The only difference is that over a skew field it is essential to keep careful track of the order of multiplication. A vector space over $\mathbb{H}$ is called a *quaternionic* vector space. We denote by $\mathbb{H}^n$ the quaternionic vector space of n-tuples of quaternions. There are many potential pitfalls stemming from a choice of left and right, for example:

---

[1]If one allows an algebra to be nonassociative, then there are other division algebras over $\mathbb{R}$, for example Cayley's octonians.

(1) Should **i**, **j**, **k** be written on the left or on the right of a scalar?
(2) Should scalars multiply on the left or on the right of $\mathbb{H}^n$?
(3) Should elements of $\mathbb{H}^n$ be represented as column vectors or as row vectors?
(4) Should a linear transformation be represented by multiplication by a matrix on the left or on the right?
(5) In the definition of the quaternion inner product, should one conjugate the first or the second argument?
(6) Should a sesquilinear form on $\mathbb{H}^n$ be conjugate-linear in the first or the second argument?

The answers to these questions are not arbitrary, since the choice for one question may determine the correct choices for all the others. A wrong choice will lead to inconsistencies.

E.1 Representation of Linear Maps by Matrices

Relative to given bases, a linear map of vector spaces over a skew field will also be represented by a matrix. Since maps are written on the left of their arguments as in $f(x)$, we will choose our convention so that a linear map f corresponds to left multiplication by a matrix. In order for a vector in $\mathbb{H}^n$ to be multiplied on the left by a matrix, the elements of $\mathbb{H}^n$ must be column vectors, and for left multiplication by a matrix to be a linear map, scalar multiplication on $\mathbb{H}^n$ should be on the right. In this way, we have answered (1), (2), (3), and (4) above.

Let K be a skew field and let V and W be vector spaces over K, with scalar multiplication on the right. A map $f: V \to W$ is *linear* over K or *K-linear* if for all $x, y \in V$ and $q \in K$,

$$f(x+y) = f(x) + f(y),$$
$$f(xq) = f(x)q.$$

An *endomorphism* or a *linear transformation* of a vector space V over K is a K-linear map from V to itself. The endomorphisms of V over K form an algebra over K, denoted by $\mathrm{End}_K(V)$. An endomorphism $f: V \to V$ is *invertible* if it has a two-sided inverse, i.e., a linear map $g: V \to V$ such that $f \circ g = g \circ f = \mathbb{1}_V$. An invertible endomorphism of V is also called an *automorphism* of V. The *general linear group* $\mathrm{GL}(V)$ is by definition the group of all automorphisms of the vector space V. When $V = K^n$, we also write $\mathrm{GL}(n, K)$ for $\mathrm{GL}(V)$.

To simplify the presentation, we will discuss matrix representation only for endomorphisms of the vector space K^n. Let e_i be the column vector with 1 in the ith row and 0 everywhere else. The set $e_1, \ldots, e_n$ is called the *standard basis* for K^n. If $f: K^n \to K^n$ is K-linear, then

$$f(e_j) = \sum_i e_i a^i_j$$

for some matrix $A = [a^i_j] \in K^{n \times n}$, called the *matrix* of f (relative to the standard basis). Here a^i_j is the entry in the ith row and jth column of the matrix A. For $x = \sum_j e_j x^j \in K^n$,

$$f(x) = \sum_j f(e_j)x^j = \sum_{i,j} e_i a^i_j x^j.$$

Hence, the ith component of the column vector $f(x)$ is

$$(f(x))^i = \sum_j a^i_j x^j.$$

In matrix notation,

$$f(x) = Ax.$$

If $g \colon K^n \to K^n$ is another linear map and $g(e_j) = \sum_i e_i b^i_j$, then

$$(f \circ g)(e_j) = f\left(\sum_k e_k b^k_j\right) = \sum_k f(e_k)b^k_j = \sum_{i,k} e_i a^i_k b^k_j.$$

Thus, if $A = [a^i_j]$ and $B = [b^i_j]$ are the matrices representing f and g respectively, then the matrix product AB is the matrix representing the composite $f \circ g$. Therefore, there is an algebra isomorphism

$$\mathrm{End}_K(K^n) \overset{\sim}{\to} K^{n \times n}$$

between endomorphisms of K^n and $n \times n$ matrices over K. Under this isomorphism, the group $\mathrm{GL}(n, K)$ corresponds to the group of all invertible $n \times n$ matrices over K.

E.2 Quaternionic Conjugation

The *conjugate* of a quaternion $q = a + \mathbf{i}b + \mathbf{j}c + \mathbf{k}d$ is defined to be

$$\bar{q} = a - \mathbf{i}b - \mathbf{j}c - \mathbf{k}d.$$

It is easily shown that conjugation is an *antihomomorphism* from the ring $\mathbb{H}$ to itself: it preserves addition, but under multiplication,

$$\overline{pq} = \bar{q}\bar{p} \quad \text{for } p, q \in \mathbb{H}.$$

The *conjugate* of a matrix $A = [a^i_j] \in \mathbb{H}^{m \times n}$ is $\bar{A} = \left[\overline{a^i_j}\right]$, obtained by conjugating each entry of A. The *transpose* A^T of the matrix A is the matrix whose (i, j)-entry is the (j, i)-entry of A. In contrast to the case for complex matrices, when A and B are quaternion matrices, in general

$$\overline{AB} \neq \bar{A}\bar{B}, \quad \overline{AB} \neq \bar{B}\bar{A}, \quad \text{and } (AB)^T \neq B^T A^T.$$

However, it is true that

$$\overline{AB}^T = \bar{B}^T \bar{A}^T,$$

as one sees by a direct computation.

E.3 Quaternionic Inner Product

The *quaternionic inner product* on $\mathbb{H}^n$ is defined to be

$$\langle x,y \rangle = \sum_i \overline{x^i} y^i = \bar{x}^T y, \quad x, y \in \mathbb{H}^n,$$

with conjugation on the first argument $x = \langle x^1, \ldots, x^n \rangle$. For any $q \in \mathbb{H}$,

$$\langle xq, y \rangle = \bar{q}\langle x,y \rangle \quad \text{and} \quad \langle x, yq \rangle = \langle x,y \rangle q.$$

If conjugation were on the second argument, then the inner product would not have the correct linearity property with respect to scalar multiplication on the right.

For quaternion vector spaces V and W, we say that a map $f \colon V \times W \to \mathbb{H}$ is *sesquilinear* over $\mathbb{H}$ if it is conjugate-linear on the left in the first argument and linear on the right in the second argument: for all $v \in V$, $w \in W$, and $q \in \mathbb{H}$,

$$f(vq, w) = \bar{q} f(v,w),$$
$$f(v, wq) = f(v,w)q.$$

In this terminology, the quaternionic inner product is sesquilinear over $\mathbb{H}$.

E.4 Representations of Quaternions by Complex Numbers

A quaternion can be identified with a pair of complex numbers:

$$q = a + ib + jc + kd = (a+ib) + j(c - id) = u + jv \longleftrightarrow (u, v).$$

Thus, $\mathbb{H}$ is a vector space over $\mathbb{C}$ with basis 1, $\mathbf{j}$, and $\mathbb{H}^n$ is a vector space over $\mathbb{C}$ with basis $e_1, \ldots, e_n, \mathbf{j}e_1, \ldots, \mathbf{j}e_n$.

Proposition E.1. *Let q be a quaternion and let u, v be complex numbers.*

(i) *If $q = u + \mathbf{j}v$, then $\bar{q} = \bar{u} - \mathbf{j}v$.*
(ii) $\mathbf{j}u\mathbf{j}^{-1} = \bar{u}.$

Proof. Problem E.1.   □

By Proposition E.1(ii), for any complex vector $v \in \mathbb{C}^n$, one has $\mathbf{j}v = \bar{v}\mathbf{j}$. Although $\mathbf{j}e_i = e_i\mathbf{j}$, elements of $\mathbb{H}^n$ should be written as $u + \mathbf{j}v$, not as $u + v\mathbf{j}$, so that the map $\mathbb{H}^n \to \mathbb{C}^{2n}$, $u + \mathbf{j}v \mapsto (u, v)$, will be a complex vector space isomorphism.

For any quaternion $q = u + \mathbf{j}v$, left multiplication $\ell_q \colon \mathbb{H} \to \mathbb{H}$ by q is $\mathbb{H}$-linear and a fortiori $\mathbb{C}$-linear. Since

$$\ell_q(1) = u + \mathbf{j}v,$$
$$\ell_q(\mathbf{j}) = (u + \mathbf{j}v)\mathbf{j} = -\bar{v} + \mathbf{j}\bar{u},$$

the matrix of ℓ_q as a $\mathbb{C}$-linear map relative to the basis 1, $\mathbf{j}$ for $\mathbb{H}$ over $\mathbb{C}$ is the 2×2 complex matrix $\begin{bmatrix} u & -\bar{v} \\ v & \bar{u} \end{bmatrix}$. The map $\mathbb{H} \to \mathrm{End}_{\mathbb{C}}(\mathbb{C}^2)$, $q \mapsto \ell_q$ is an injective algebra homomorphism over $\mathbb{R}$, giving rise to a representation of the quaternions by 2×2 complex matrices.

E.5 Quaternionic Inner Product in Terms of Complex Components

Let $x = x_1 + \mathbf{j}x_2$ and $y = y_1 + \mathbf{j}y_2$ be in $\mathbb{H}^n$, with $x_1, x_2, y_1, y_2 \in \mathbb{C}^n$. We will express the quaternionic inner product $\langle x, y \rangle$ in terms of the complex vectors $x_1, x_2, y_1, y_2 \in \mathbb{C}^n$.

By Proposition E.1,

$$\langle x, y \rangle = \bar{x}^T y = \left(\bar{x}_1^T - \mathbf{j}x_2^T \right)(y_1 + \mathbf{j}y_2) \qquad \text{(since } \bar{x} = \bar{x}_1 - \mathbf{j}x_2\text{)}$$
$$= \left(\bar{x}_1^T y_1 + \bar{x}_2^T y_2 \right) + \mathbf{j} \left(x_1^T y_2 - x_2^T y_1 \right) \quad \text{(since } x_2^T \mathbf{j} = \mathbf{j}\bar{x}_2^T \text{ and } \bar{x}_1^T \mathbf{j} = \mathbf{j}x_1^T\text{)}.$$

Let

$$\langle x, y \rangle_1 = \bar{x}_1^T y_1 + \bar{x}_2^T y_2 = \sum_{i=1}^{n} \bar{x}_1^i y_1^i + \bar{x}_2^i y_2^i$$

and

$$\langle x, y \rangle_2 = x_1^T y_2 - x_2^T y_1 = \sum_{i=1}^{n} x_1^i y_2^i - x_2^i y_1^i.$$

So the quaternionic inner product $\langle \ , \ \rangle$ is the sum of a Hermitian inner product and $\mathbf{j}$ times a skew-symmetric bilinear form on $\mathbb{C}^{2n}$:

$$\langle \ , \ \rangle = \langle \ , \ \rangle_1 + \mathbf{j}\langle \ , \ \rangle_2.$$

Let $x = x_1 + \mathbf{j}x_2 \in \mathbb{H}^n$. By skew-symmetry, $\langle x, x \rangle_2 = 0$, so that

$$\langle x, x \rangle = \langle x, x \rangle_1 = \|x_1\|^2 + \|x_2\|^2 \geq 0.$$

The *norm* of a quaternionic vector $x = x_1 + \mathbf{j}x_2$ is defined to be

$$\|x\| = \sqrt{\langle x, x \rangle} = \sqrt{\|x_1\|^2 + \|x_2\|^2}.$$

In particular, the norm of a quaternion $q = a + \mathbf{i}b + \mathbf{j}c + \mathbf{k}d$ is

$$\|q\| = \sqrt{a^2 + b^2 + c^2 + d^2}.$$

E.6 ℍ-Linearity in Terms of Complex Numbers

Recall that an $\mathbb{H}$-linear map of quaternionic vector spaces is a map that is additive and commutes with right multiplication r_q for any quaternion q.

Proposition E.2. *Let V be a quaternionic vector space. A map $f \colon V \to V$ is $\mathbb{H}$-linear if and only if it is $\mathbb{C}$-linear and $f \circ r_{\mathbf{j}} = r_{\mathbf{j}} \circ f$.*

Proof. ($\Rightarrow$) Clear.

($\Leftarrow$) Suppose f is $\mathbb{C}$-linear and f commutes with $r_{\mathbf{j}}$. By $\mathbb{C}$-linearity, f is additive and commutes with r_u for any complex number u. Any $q \in \mathbb{H}$ can be written as $q = u + \mathbf{j}v$ for some $u, v \in \mathbb{C}$; moreover, $r_q = r_{u+\mathbf{j}v} = r_u + r_v \circ r_{\mathbf{j}}$ (note the order reversal in $r_{\mathbf{j}v} = r_v \circ r_{\mathbf{j}}$). Since f is additive and commutes with r_u, r_v, and $r_{\mathbf{j}}$, it commutes with r_q for any $q \in \mathbb{H}$. Therefore, f is $\mathbb{H}$-linear. □

Because the map $r_{\mathbf{j}}\colon \mathbb{H}^n \to \mathbb{H}^n$ is neither $\mathbb{H}$-linear nor $\mathbb{C}$-linear, it cannot be represented by left multiplication by a complex matrix. If $q = u + \mathbf{j}v \in \mathbb{H}^n$, where $u, v \in \mathbb{C}^n$, then

$$r_{\mathbf{j}}(q) = q\mathbf{j} = (u + \mathbf{j}v)\mathbf{j} = -\bar{v} + \mathbf{j}\bar{u}.$$

In matrix notation,

$$r_{\mathbf{j}}\left(\begin{bmatrix} u \\ v \end{bmatrix}\right) = \begin{bmatrix} -\bar{v} \\ \bar{u} \end{bmatrix} = c\left(\begin{bmatrix} 0 & -1 \\ 1 & 0 \end{bmatrix}\begin{bmatrix} u \\ v \end{bmatrix}\right) = -c\left(J\begin{bmatrix} u \\ v \end{bmatrix}\right), \tag{E.1}$$

where c denotes complex conjugation and J is the 2×2 matrix $\begin{bmatrix} 0 & 1 \\ -1 & 0 \end{bmatrix}$.

E.7 Symplectic Group

Let V be a vector space over a skew field K with conjugation, and let $B\colon V \times V \to K$ be a bilinear or sesquilinear function over K. Such a function is often called a bilinear or sesquilinear *form* over K. A K-linear automorphism $f\colon V \to V$ is said to *preserve* the form B if

$$B(f(x), f(y)) = B(x, y) \quad \text{for all } x, y \in V.$$

The set of these automorphisms is a subgroup of the general linear group $\mathrm{GL}(V)$.

When K is the skew field $\mathbb{R}$, $\mathbb{C}$, or $\mathbb{H}$, and B is the Euclidean, Hermitian, or quaternionic inner product respectively on K^n, the subgroup of $\mathrm{GL}(n, K)$ consisting of automorphisms of K^n preserving each of these inner products is called the *orthogonal*, *unitary*, or *symplectic group* and denoted by $\mathrm{O}(n)$, $\mathrm{U}(n)$, or $\mathrm{Sp}(n)$ respectively. Naturally, the automorphisms in these three groups are called *orthogonal*, *unitary*, or *symplectic* automorphisms.

In particular, the *symplectic group* is the group of automorphisms f of $\mathbb{H}^n$ such that

$$\langle f(x), f(y) \rangle = \langle x, y \rangle \quad \text{for all } x, y \in \mathbb{H}^n.$$

In terms of matrices, if A is the quaternionic matrix of such an f, then

$$\langle f(x), f(y) \rangle = \overline{Ax}^T Ay = \bar{x}^T \bar{A}^T Ay = \bar{x}^T y \quad \text{for all } x, y \in \mathbb{H}^n.$$

Therefore, $f \in \mathrm{Sp}(n)$ if and only if its matrix A satisfies $\bar{A}^T A = I$. Because $\mathbb{H}^n = \mathbb{C}^n \oplus \mathbf{j}\mathbb{C}^n$ is isomorphic to $\mathbb{C}^{2n}$ as a complex vector space and an $\mathbb{H}$-linear map is necessarily $\mathbb{C}$-linear, the group $\mathrm{GL}(n, \mathbb{H})$ is isomorphic to a subgroup of $\mathrm{GL}(2n, \mathbb{C})$ (see Problem E.2).

Example. Under the algebra isomorphisms $\mathrm{End}_{\mathbb{H}}(\mathbb{H}) \simeq \mathbb{H}$, elements of $\mathrm{Sp}(1)$ correspond to quaternions $q = a + \mathbf{i}b + \mathbf{j}c + \mathbf{k}d$ such that

$$\bar{q}q = a^2 + b^2 + c^2 + d^2 = 1.$$

These are precisely quaternions of norm 1. Therefore, under the chain of real vector space isomorphisms $\mathrm{End}_{\mathbb{H}}(\mathbb{H}) \simeq \mathbb{H} \simeq \mathbb{R}^4$, the group $\mathrm{Sp}(1)$ maps to S^3, the unit 3-sphere in $\mathbb{R}^4$.

The *complex symplectic group* $\mathrm{Sp}(2n, \mathbb{C})$ is the subgroup of $\mathrm{GL}(2n, \mathbb{C})$ consisting of automorphisms of $\mathbb{C}^{2n}$ preserving the skew-symmetric bilinear form $B \colon \mathbb{C}^{2n} \times \mathbb{C}^{2n} \to \mathbb{C}$,

$$B(x, y) = \sum_{i=1}^{n} x^i y^{n+i} - x^{n+i} y^i = x^T Jy, \quad J = \begin{bmatrix} 0 & I_n \\ -I_n & 0 \end{bmatrix},$$

where I_n is the $n \times n$ identity matrix. If $f \colon \mathbb{C}^{2n} \times \mathbb{C}^{2n} \to \mathbb{C}$ is given by $f(x) = Ax$, then

$$f \in \mathrm{Sp}(2n, \mathbb{C}) \iff B(f(x), f(y)) = B(x, y) \text{ for all } x, y \in \mathbb{C}^{2n}$$
$$\iff (Ax)^T JAy = x^T (A^T JA)y = x^T Jy \text{ for all } x, y \in \mathbb{C}^{2n}$$
$$\iff A^T JA = J.$$

Theorem E.3. *Under the injection* $\mathrm{GL}(n, \mathbb{H}) \hookrightarrow \mathrm{GL}(2n, \mathbb{C})$, *the symplectic group* $\mathrm{Sp}(n)$ *maps isomorphically to the intersection* $\mathrm{U}(2n) \cap \mathrm{Sp}(2n, \mathbb{C})$.

Proof.
$f \in \mathrm{Sp}(n)$

$\iff$ $f \colon \mathbb{H}^n \to \mathbb{H}^n$ is $\mathbb{H}$-linear and preserves the quaternionic inner product

$\iff$ $f \colon \mathbb{C}^{2n} \to \mathbb{C}^{2n}$ is $\mathbb{C}$-linear, $f \circ r_j = r_j \circ f$, and f preserves the Hermitian inner product and the standard skew-symmetric bilinear form on $\mathbb{C}^{2n}$ (by Proposition E.2 and Section E.5)

$\iff$ $f \circ r_j = r_j \circ f$ and $f \in \mathrm{U}(2n) \cap \mathrm{Sp}(2n, \mathbb{C})$.

We will now show that if $f \in \mathrm{U}(2n)$, then the condition $f \circ r_j = r_j \circ f$ is equivalent to $f \in \mathrm{Sp}(2n, \mathbb{C})$. Let $f \in \mathrm{U}(2n)$ and let A be the matrix of f relative to the standard basis in $\mathbb{C}^{2n}$. Then

$$(f \circ r_j)(x) = (r_j \circ f)(x) \text{ for all } x \in \mathbb{C}^{2n}$$
$$\iff -Ac(Jx) = -c(JAx) \text{ for all } x \in \mathbb{C}^{2n} \quad \text{(by (E.1))}$$
$$\iff c(\bar{A}Jx) = c(JAx) \text{ for all } x \in \mathbb{C}^{2n}$$
$$\iff \bar{A}Jx = JAx \text{ for all } x \in \mathbb{C}^{2n}$$
$$\iff J = \bar{A}^{-1}JA$$
$$\iff J = A^T JA \qquad \qquad \text{(since } A \in \mathrm{U}(2n)\text{)}$$
$$\iff f \in \mathrm{Sp}(2n, \mathbb{C}).$$

Therefore, the condition $f \circ r_j = r_j \circ f$ is redundant if $f \in \mathrm{U}(2n) \cap \mathrm{Sp}(2n, \mathbb{C})$. By the first paragraph of this proof, there is a group isomorphism $\mathrm{Sp}(n) \simeq \mathrm{U}(2n) \cap \mathrm{Sp}(2n, \mathbb{C})$. $\qquad \square$

Problems

E.1. Quaternionic conjugation
Prove Proposition E.1.

E.2. Complex representation of an $\mathbb{H}$-linear map

Suppose an $\mathbb{H}$-linear map $f \colon \mathbb{H}^n \to \mathbb{H}^n$ is represented relative to the standard basis $e_1, \ldots, e_n$ by the matrix $A = u + \mathbf{j}v \in \mathbb{H}^{n \times n}$, where $u, v \in \mathbb{C}^{n \times n}$. Show that as a $\mathbb{C}$-linear map, $f \colon \mathbb{H}^n \to \mathbb{H}^n$ is represented relative to the basis $e_1, \ldots, e_n, \mathbf{j}e_1, \ldots, \mathbf{j}e_n$ by the matrix $\begin{bmatrix} u & -\bar{v} \\ v & \bar{u} \end{bmatrix}$.

E.3. Symplectic and unitary groups of small dimension

For a field K, the *special linear group* $SL(n, K)$ is the subgroup of $GL(n, K)$ consisting of all automorphisms of K^n of determinant 1, and the *special unitary group* $SU(n)$ is the subgroup of $U(n)$ consisting of unitary automorphisms of $\mathbb{C}^n$ of determinant 1. Prove the following identifications or group isomorphisms.

(a) $Sp(2, \mathbb{C}) = SL(2, \mathbb{C})$.
(b) $Sp(1) \simeq SU(2)$. (*Hint:* Use Theorem E.3 and part (a).)
(c)

$$SU(2) \simeq \left\{ \begin{bmatrix} u & -\bar{v} \\ v & \bar{u} \end{bmatrix} \in \mathbb{C}^{2 \times 2} \;\middle|\; u\bar{u} + v\bar{v} = 1 \right\}.$$

(*Hint:* Use part (b) and the representation of quaternions by 2×2 complex matrices in Subsection E.4.)

Solutions to Selected Exercises Within the Text

3.6 Inversions

As a matrix, $\tau = \begin{bmatrix} 1 & 2 & 3 & 4 & 5 \\ 2 & 3 & 4 & 5 & 1 \end{bmatrix}$. Scanning the second row, we see that τ has four inversions: $(2,1)$, $(3,1)$, $(4,1)$, $(5,1)$. ◇

3.13 Symmetrizing operator

A k-linear function $h\colon V \to \mathbb{R}$ is symmetric if and only if $\tau h = h$ for all $\tau \in S_k$. Now

$$\tau(Sf) = \tau \sum_{\sigma \in S_k} \sigma f = \sum_{\sigma \in S_k} (\tau \sigma) f.$$

As σ runs over all elements of the permutation groups S_k, so does $\tau \sigma$. Hence,

$$\sum_{\sigma \in S_k} (\tau \sigma) f = \sum_{\tau \sigma \in S_k} (\tau \sigma) f = Sf.$$

This proves that $\tau(Sf) = Sf$. ◇

3.15 Alternating operator

$f(v_1, v_2, v_3) - f(v_1, v_3, v_2) + f(v_2, v_3, v_1) - f(v_2, v_1, v_3) + f(v_3, v_1, v_2) - f(v_3, v_2, v_1)$. ◇

3.20 Wedge product of two 2-covectors

$$(f \wedge g)(v_1, v_2, v_3, v_4)$$
$$= f(v_1, v_2) g(v_3, v_4) - f(v_1, v_3) g(v_2, v_4) + f(v_1, v_4) g(v_2, v_3)$$
$$+ f(v_2, v_3) g(v_1, v_4) - f(v_2, v_4) g(v_1, v_3) + f(v_3, v_4) g(v_1, v_2). \qquad ◇$$

3.22 Sign of a permutation

We can achieve the permutation τ from the initial configuration $1, 2, \ldots, k + \ell$ in k steps.

(1) First, move the element k to the very end across the ℓ elements $k+1, \ldots, k+\ell$. This requires ℓ transpositions.
(2) Next, move the element $k-1$ across the ℓ elements $k+1, \ldots, k+\ell$.
(3) Then move the element $k-2$ across the same ℓ elements, and so on.

Each of the k steps requires ℓ transpositions. In the end we achieve τ from the identity using ℓk transpositions.

Alternatively, one can count the number of inversions in the permutation τ. There are k inversions starting with $k+1$, namely, $(k+1,1),\ldots,(k+1,k)$. Indeed, for each $i=1,\ldots,\ell$, there are k inversions starting with $k+i$. Hence, the total number of inversions in τ is $k\ell$. By Proposition 3.8, $\operatorname{sgn}(\tau)=(-1)^{k\ell}$. ◇

4.3 A basis for 3-covectors
By Proposition 3.29, a basis for $A_3(T_p(\mathbb{R}^4))$ is $(dx^1 \wedge dx^2 \wedge dx^3)_p$, $(dx^1 \wedge dx^2 \wedge dx^4)_p$, $(dx^1 \wedge dx^3 \wedge dx^4)_p$, $(dx^2 \wedge dx^3 \wedge dx^4)_p$. ◇

4.4 Wedge product of a 2-form with a 1-form
The $(2,1)$-shuffles are $(1<2,3)$, $(1<3,2)$, $(2<3,1)$, with respective signs $+,\ -,\ +$. By equation (3.6),

$$(\omega \wedge \tau)(X,Y,Z) = \omega(X,Y)\tau(Z) - \omega(X,Z)\tau(Y) + \omega(Y,Z)\tau(X).$$ ◇

6.14 Smoothness of a map to a circle
Without further justification, the fact that both $\cos t$ and $\sin t$ are C^∞ proves only the smoothness of $(\cos t, \sin t)$ as a map from $\mathbb{R}$ to $\mathbb{R}^2$. To show that $F: \mathbb{R} \to S^1$ is C^∞, we need to cover S^1 with charts (U_i, ϕ_i) and examine in turn each $\phi_i \circ F: F^{-1}(U_i) \to \mathbb{R}$. Let $\{(U_i, \phi_i) \mid i=1,\ldots,4\}$ be the atlas of Example 5.16. On $F^{-1}(U_1)$, $\phi_1 \circ F(t) = (x \circ F)(t) = \cos t$ is C^∞. On $F^{-1}(U_3)$, $\phi_3 \circ F(t) = \sin t$ is C^∞. Similar computations on $F^{-1}(U_2)$ and $F^{-1}(U_4)$ prove the smoothness of F. ◇

6.18 Smoothness of a map to a Cartesian product
Fix $p \in N$, let (U,ϕ) be a chart about p, and let $(V_1 \times V_2, \psi_1 \times \psi_2)$ be a chart about $(f_1(p), f_2(p))$. We will be assuming either (f_1,f_2) smooth or both f_i smooth. In either case, (f_1,f_2) is continuous. Hence, by choosing U sufficiently small, we may assume $(f_1,f_2)(U) \subset V_1 \times V_2$. Then

$$(\psi_1 \times \psi_2) \circ (f_1,f_2) \circ \phi^{-1} = (\psi_1 \circ f_1 \circ \phi^{-1}, \psi_2 \circ f_2 \circ \phi^{-1})$$

maps an open subset of $\mathbb{R}^n$ to an open subset of $\mathbb{R}^{m_1+m_2}$. It follows that (f_1,f_2) is C^∞ at p if and only if both f_1 and f_2 are C^∞ at p. ◇

7.11 Real projective space as a quotient of a sphere
Define $\bar{f}: \mathbb{R}P^n \to S^n/\sim$ by $\bar{f}([x]) = [\frac{x}{\|x\|}] \in S^n/\sim$. This map is well defined because $\bar{f}([tx]) = [\frac{tx}{|t|x|}] = [\pm \frac{x}{\|x\|}] = [\frac{x}{\|x\|}]$. Note that if $\pi_1: \mathbb{R}^{n+1} - \{0\} \to \mathbb{R}P^n$ and $\pi_2: S^n \to S^n/\sim$ are the projection maps, then there is a commutative diagram

$$
\begin{array}{ccc}
\mathbb{R}^n - \{0\} & \xrightarrow{\ f\ } & S^n \\
{\scriptstyle \pi_1}\downarrow & & \downarrow{\scriptstyle \pi_2} \\
\mathbb{R}P^n & \xrightarrow[\ \bar{f}\ ]{} & S^n/\sim .
\end{array}
$$

By Proposition 7.1, $\bar{f}$ is continuous because $\pi_2 \circ f$ is continuous.

Next define $g: S^n \to \mathbb{R}^{n+1} - \{0\}$ by $g(x) = x$. This map induces a map $\bar{g}: S^n/\sim \to \mathbb{R}P^n$, $\bar{g}([x]) = [x]$. By the same argument as above, $\bar{g}$ is well defined and continuous. Moreover,

$$\bar{g} \circ \bar{f}([x]) = \left[\frac{x}{\|x\|}\right] = [x],$$

$$\bar{f} \circ \bar{g}([x]) = [x],$$

so $\bar{f}$ and $\bar{g}$ are inverses to each other. $\diamond$

8.14 Velocity vector versus the calculus derivative

As a vector at the point $c(t)$ in the real line, $c'(t)$ equals $a\,d/dx|_{c(t)}$ for some scalar a. Applying both sides of the equality to x, we get $c'(t)x = a\,dx/dx|_{c(t)} = a$. By the definition of $c'(t)$,

$$a = c'(t)x = c_*\left(\left.\frac{d}{dt}\right|_{c(t)}\right)x = \left.\frac{d}{dt}\right|_{c(t)} x \circ c = \left.\frac{d}{dt}\right|_{c(t)} c = \dot{c}(t).$$

Hence, $c'(t) = \dot{c}(t)\,d/dx|_{c(t)}$. $\diamond$

13.1 Bump function supported in an open set

Let (V, ϕ) be a chart centered at q such that V is diffeomorphic to an open ball $B(0, r)$. Choose real numbers a and b such that

$$\bar{B}(0, a) \subset B(0, b) \subset \bar{B}(0, b) \subset B(0, r).$$

With the σ given in (13.2), the function $\sigma \circ \phi$, extended by zero to M, gives the desired bump function. $\diamond$

15.2 Left multiplication

Let $i_a \colon G \to G \times G$ be the inclusion map $i_a(x) = (a, x)$. It is clearly C^∞. Then $\ell_a(x) = ax = (\mu \circ i_a)(x)$. Since $\ell_a = \mu \circ i_a$ is the composition of two C^∞ maps, it is C^∞. Moreover, because it has a two-sided C^∞ inverse $\ell_{a^{-1}}$, it is a diffeomorphism. $\diamond$

15.7 Space of symmetric matrices

Let

$$A = [a_{ij}] = \begin{bmatrix} a_{11} & a_{12} & \cdots & a_{1n} \\ * & a_{22} & \cdots & a_{2n} \\ \vdots & \vdots & \ddots & \vdots \\ * & * & \cdots & a_{nn} \end{bmatrix}$$

be a symmetric matrix. The symmetry condition $a_{ji} = a_{ij}$ implies that the entries below the diagonal are determined by the entries above the diagonal, and that there are no further conditions on the the entries above or on the diagonal. Thus, the dimension of S_n is equal to the number of entries above or on the diagonal. Since there are n such entries in the first row, $n-1$ in the second row, and so on,

$$\dim S_n = n + (n-1) + (n-2) + \cdots + 1 = \frac{n(n+1)}{2}.$$ $\diamond$

15.10 Induced topology versus the subspace topology

A basic open set in the induced topology on H is the image under f of an open interval in L. Such a set is not open in the subspace topology. A basic open set in the subspace topology on H is the intersection of H with the image of an open ball in $\mathbb{R}^2$ under the projection $\pi \colon \mathbb{R}^2 \to \mathbb{R}^2/\mathbb{Z}^2$; it is a union of infinitely many open intervals. Thus, the subspace topology is a subset of the induced topology, but not vice versa. $\diamond$

15.15 Distributivity over a convergent series

(i) We may assume $a \neq 0$, for otherwise there is nothing to prove. Let $\varepsilon > 0$. Since $s_m \to s$, there exists an integer N such that for all $m \geq N$, $\|s - s_m\| < \varepsilon/\|a\|$. Then for $m \geq N$,

$$\|as - as_m\| \leq \|a\|\,\|s - s_m\| < \|a\|\left(\frac{\varepsilon}{\|a\|}\right) = \varepsilon.$$

Hence, $as_m \to as$.

(ii) Set $s_m = \sum_{k=0}^m b_k$ and $s = \sum_{k=0}^\infty b_k$. The convergence of the series $\sum_{k=0}^\infty b_k$ means that $s_m \to s$. By (i), $as_m \to as$, which means that the sequence $as_m = \sum_{k=0}^m ab_k$ converges to $a\sum_{k=0}^\infty b_k$. Hence, $\sum_{k=0}^\infty ab_k = a\sum_{k=0}^\infty b_k$. ◇

18.5 Transition formula for a 2-form

$$
\begin{aligned}
a_{ij} = \omega(\partial/\partial x^i, \partial/\partial x^j) &= \sum_{k<\ell} b_{k\ell}\, dy^k \wedge dy^\ell (\partial/\partial x^i, \partial/\partial x^j)\\
&= \sum_{k<\ell} b_{k\ell} \left(dy^k(\partial/\partial x^i) dy^\ell(\partial/\partial x^j) - dy^k(\partial/\partial x^j) dy^\ell(\partial/\partial x^i) \right)\\
&= \sum_{k<\ell} b_{k\ell} \left(\frac{\partial y^k}{\partial x^i}\frac{\partial y^\ell}{\partial x^j} - \frac{\partial y^k}{\partial x^j}\frac{\partial y^\ell}{\partial x^i} \right) = \sum_{k<\ell} b_{k\ell} \frac{\partial(y^k, y^\ell)}{\partial(x^i, x^j)}.
\end{aligned}
$$
◇

Alternative solution: by Proposition 18.3,

$$
dy^k \wedge dy^\ell = \sum_{i<j} \frac{\partial(y^k, y^\ell)}{\partial(x^i, x^j)}\, dx^i \wedge dx^j.
$$

Hence,

$$
\sum_{i<j} a_{ij}\, dx^i \wedge dx^j = \sum_{k<\ell} b_{k\ell}\, dy^k \wedge dy^\ell = \sum_{i<j}\sum_{k<\ell} b_{k\ell} \frac{\partial(y^k, y^\ell)}{\partial(x^i, x^j)}\, dx^i \wedge dx^j.
$$

Comparing the coefficients of $dx^i \wedge dx^j$ gives

$$
a_{ij} = \sum_{k<\ell} b_{k\ell} \frac{\partial(y^k, y^\ell)}{\partial(x^i, x^j)}.
$$
◇

22.2 Smooth functions on a nonopen set

By definition, for each p in S there are an open set U_p in $\mathbb{R}^n$ and a C^∞ function $\tilde{f}_p : U_p \to \mathbb{R}^m$ such that $f = \tilde{f}_p$ on $U_p \cap S$. Extend the domain of $\tilde{f}_p$ to $\mathbb{R}^n$ by defining it to be zero on $\mathbb{R}^n - U_p$. Let $U = \bigcup_{p\in S} U_p$. Choose a partition of unity $\{\sigma_p\}_{p\in S}$ on U subordinate to the open cover $\{U_p\}_{p\in S}$ of U and define the function $\tilde{f} : U \to \mathbb{R}^m$ by

$$
\tilde{f} = \sum_{p\in S} \sigma_p \tilde{f}_p. \tag{$*$}
$$

Because $\operatorname{supp}\sigma_p \subset U_p$, the product $\sigma_p \tilde{f}_p$ is zero and hence smooth outside U_p; as a product of two C^∞ functions on U_p, $\sigma_p \tilde{f}_p$ is C^∞ on U_p. Therefore, $\sigma_p \tilde{f}_p$ is C^∞ on $\mathbb{R}^n$. Since the sum $(*)$ is locally finite, $\tilde{f}$ is well defined and C^∞ on $\mathbb{R}^n$ for the usual reason. (Every point $q \in U$ has a neighborhood W_q that intersects only finitely many of of the sets $\operatorname{supp}\sigma_p$, $p \in S$. Hence, the sum $(*)$ is a finite sum on W_q.)

Let $q \in S$. If $q \in U_p$, then $\tilde{f}_p(q) = f(q)$, and if $q \notin U_p$, then $\sigma_p(q) = 0$. Thus, for $q \in S$, one has

$$
\tilde{f}(q) = \sum_{p\in S} \sigma_p(q) \tilde{f}_p(q) = \sum_{p\in S} \sigma_p(q) f(q) = f(q).
$$
◇

25.5 Connecting homomorphism

The proof that the cohomology class of a is independent of the choice of b as preimage of c can be summarized in the commutative diagram

$$da'' = a - a' \;\; \xrightarrow{\;\;i\;\;} \;\; db - db'$$

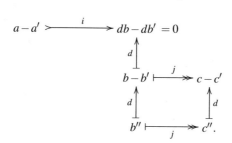

Suppose $b, b' \in B^k$ both map to c under j. Then $j(b - b') = jb - jb' = c - c = 0$. By the exactness at B^k, $b - b' = i(a'')$ for some $a'' \in A^k$.

With the choice of b as preimage, the element $d^*[c]$ is represented by a cocycle $a \in A^{k+1}$ such that $i(a) = db$. Similarly, with the choice of b' as preimage, the element $d^*[c]$ is represented by a cocycle $a' \in A^{k+1}$ such that $i(a') = db'$. Then $i(a - a') = d(b - b') = di(a'') = id(a'')$. Since i is injective, $a - a' = da''$, and thus $[a] = [a']$. This proves that $d^*[c]$ is independent of the choice of b.

The proof that the cohomology class of a is independent of the choice of c in the cohomology class $[c]$ can be summarized by the commutative diagram

$$a - a' \;\; \xrightarrow{\;\;i\;\;} \;\; db - db' = 0$$

Suppose $[c] = [c'] \in H^k(\mathcal{C})$. Then $c - c' = dc''$ for some $c'' \in C^{k-1}$. By the surjectivity of $j \colon B^{k-1} \to C^{k-1}$, there is a $b'' \in B^{k-1}$ that maps to c'' under j. Choose $b \in B^k$ such that $j(b) = c$ and let $b' = b - db'' \in B^k$. Then $j(b') = j(b) - jdb'' = c - dj(b'') = c - dc'' = c'$. With the choice of b as preimage, $d^*[c]$ is represented by a cocycle $a \in A^{k+1}$ such that $i(a) = db$. With the choice of b' as preimage, $d^*[c]$ is represented by a cocycle $a' \in A^{k+1}$ such that $i(a') = db'$. Then

$$i(a - a') = d(b - b') = ddb'' = 0.$$

By the injectivity of i, $a = a'$, so $[a] = [a']$. This shows that $d^*[c]$ is independent of the choice of c in the cohomology class $[c]$. ◇

A.33 Compact Hausdorff space

Let S be a compact Hausdorff space, and A, B two closed subsets of S. By Proposition A.30, A and B are compact. By Proposition A.31, for any $a \in A$ there are disjoint open sets $U_a \ni a$ and $V_a \supset B$. Since A is compact, the open cover $\{U_a\}_{a \in A}$ for A has a finite subcover $\{U_{a_i}\}_{i=1}^n$. Let $U = \bigcup_{i=1}^n U_{a_i}$ and $V = \bigcap_{i=1}^n V_{a_i}$. Then $A \subset U$ and $B \subset V$. The open sets U and V are disjoint because if $x \in U \cap V$, then $x \in U_{a_i}$ for some i and $x \in V_{a_i}$ for the same i, contradicting the fact that $U_{a_i} \cap V_{a_i} = \varnothing$. ◇

Hints and Solutions to Selected End-of-Section Problems

Problems with complete solutions are starred (*). Equations are numbered consecutively within each problem.

1.2* A C^∞ function very flat at 0

(a) Assume $x > 0$. For $k = 1$, $f'(x) = (1/x^2)e^{-1/x}$. With $p_2(y) = y^2$, this verifies the claim. Now suppose $f^{(k)}(x) = p_{2k}(1/x)e^{-1/x}$. By the product rule and the chain rule,

$$f^{(k+1)}(x) = p_{2k-1}\left(\frac{1}{x}\right)\cdot\left(-\frac{1}{x^2}\right)e^{-1/x} + p_{2k}\left(\frac{1}{x}\right)\cdot\frac{1}{x^2}e^{-1/x}$$

$$= \left(q_{2k+1}\left(\frac{1}{x}\right) + q_{2k+2}\left(\frac{1}{x}\right)\right)e^{-1/x}$$

$$= p_{2k+2}\left(\frac{1}{x}\right)e^{-1/x},$$

where $q_n(y)$ and $p_n(y)$ are polynomials of degree n in y. By induction, the claim is true for all $k \geq 1$. It is trivially true for $k = 0$ also.

(b) For $x > 0$, the formula in (a) shows that $f(x)$ is C^∞. For $x < 0$, $f(x) \equiv 0$, which is trivially C^∞. It remains to show that $f^{(k)}(x)$ is defined and continuous at $x = 0$ for all k.

Suppose $f^{(k)}(0) = 0$. By the definition of the derivative,

$$f^{(k+1)}(0) = \lim_{x\to 0}\frac{f^{(k)}(x) - f^{(k)}(0)}{x} = \lim_{x\to 0}\frac{f^{(k)}(x)}{x}.$$

The limit from the left is clearly 0. So it suffices to compute the limit from the right:

$$\lim_{x\to 0^+}\frac{f^{(k)}(x)}{x} = \lim_{x\to 0^+}\frac{p_{2k}\left(\frac{1}{x}\right)e^{-1/x}}{x} = \lim_{x\to 0^+}p_{2k+1}\left(\frac{1}{x}\right)e^{-1/x} \qquad (1.2.1)$$

$$= \lim_{y\to\infty}\frac{p_{2k+1}(y)}{e^y} \quad \left(\text{replacing }\frac{1}{x}\text{ by }y\right).$$

Applying l'Hôpital's rule $2k+1$ times, we reduce this limit to 0. Hence, $f^{(k+1)}(0) = 0$. By induction, $f^{(k)}(0) = 0$ for all $k \geq 0$.

A similar computation as (1.2.1) for $\lim_{x\to 0}f^{(k)}(x) = 0$ proves that $f^{(k)}(x)$ is continuous at $x = 0$. ◊

1.3 (b) $h(t) = (\pi/(b-a))(t-a) - (\pi/2)$.

1.5

(a) The line passing through $(0,0,1)$ and (a,b,c) has a parametrization

$$x = at, \quad y = bt, \quad z = (c-1)t + 1.$$

This line intersects the xy-plane when

$$z = 0 \Leftrightarrow t = \frac{1}{1-c} \Leftrightarrow (x,y) = \left(\frac{a}{1-c}, \frac{b}{1-c}\right).$$

To find the inverse of g, write down a parametrization of the line through $(u,v,0)$ and $(0,0,1)$ and solve for the intersection of this line with S.

1.6* Taylor's theorem with remainder to order 2
To simplify the notation, we write $\mathbf{0}$ for $(0,0)$. By Taylor's theorem with remainder, there exist C^∞ functions g_1, g_2 such that

$$f(x,y) = f(\mathbf{0}) + xg_1(x,y) + yg_2(x,y). \tag{1.6.1}$$

Applying the theorem again, but to g_1 and g_2, we obtain

$$g_1(x,y) = g_1(\mathbf{0}) + xg_{11}(x,y) + yg_{12}(x,y), \tag{1.6.2}$$
$$g_2(x,y) = g_2(\mathbf{0}) + xg_{21}(x,y) + yg_{22}(x,y). \tag{1.6.3}$$

Since $g_1(\mathbf{0}) = \partial f/\partial x(\mathbf{0})$ and $g_2(\mathbf{0}) = \partial f/\partial y(\mathbf{0})$, substituting (1.6.2) and (1.6.3) into (1.6.1) gives the result. $\diamond$

1.7* A function with a removable singularity
In Problem 1.6, set $x = t$ and $y = tu$. We obtain

$$f(t,tu) = f(\mathbf{0}) + t\frac{\partial f}{\partial x}(\mathbf{0}) + tu\frac{\partial f}{\partial y}(\mathbf{0}) + t^2(\cdots),$$

where

$$(\cdots) = g_{11}(t,tu) + ug_{12}(t,tu) + u^2 g_{22}(t,tu)$$

is a C^∞ function of t and u. Since $f(\mathbf{0}) = \partial f/\partial x(\mathbf{0}) = \partial f/\partial y(\mathbf{0}) = 0$,

$$\frac{f(t,tu)}{t} = t(\cdots),$$

which is clearly C^∞ in t, u and agrees with g when $t = 0$. $\diamond$

1.8 See Example 1.2(ii).

3.1 $f = \sum g_{ij}\alpha^i \otimes \alpha^j$.

3.2

(a) Use the formula $\dim \ker f + \dim \operatorname{im} f = \dim V$.
(b) Choose a basis $e_1, \ldots, e_{n-1}$ for $\ker f$, and extend it to a basis $e_1, \ldots, e_{n-1}, e_n$ for V. Let $\alpha^1, \ldots, \alpha^n$ be the dual basis for $V^\vee$. Write both f and g in terms of this dual basis.

3.3 We write temporarily α^I for $\alpha^{i_1} \otimes \cdots \otimes \alpha^{i_k}$ and e_J for $(e_{j_1}, \ldots, e_{j_k})$.

(a) Prove that $f = \sum f(e_I)\alpha^I$ by showing that both sides agree on all (e_J). This proves that the set $\{\alpha^I\}$ spans.

(b) Suppose $\sum c_I \alpha^I = 0$. Applying both sides to e_J gives $c_J = \sum c_I \alpha^I(e_J) = 0$. This proves that the set $\{\alpha^I\}$ is linearly independent.

3.9 To compute $\omega(v_1,\ldots,v_n)$ for any $v_1,\ldots,v_n \in V$, write $v_j = \sum_i e_i a_j^i$ and use the fact that ω is multilinear and alternating.

3.10* Linear independence of covectors

($\Rightarrow$) If $\alpha^1,\ldots,\alpha^k$ are linearly dependent, then one of them is a linear combination of the others. Without loss of generality, we may assume that

$$\alpha^k = \sum_{i=1}^{k-1} c_i \alpha^i.$$

In the wedge product $\alpha^1 \wedge \cdots \wedge \alpha^{k-1} \wedge (\sum_{i=1}^{k-1} c_i \alpha^i)$, every term has a repeated α^i. Hence, $\alpha^1 \wedge \cdots \wedge \alpha^k = 0$.

($\Leftarrow$) Suppose $\alpha^1,\ldots,\alpha^k$ are linearly independent. Then they can be extended to a basis $\alpha^1,\ldots,\alpha^k,\ldots,\alpha^n$ for $V^\vee$. Let $v_1,\ldots,v_n$ be the dual basis for V. By Proposition 3.27,

$$(\alpha^1 \wedge \cdots \wedge \alpha^k)(v_1,\ldots,v_k) = \det[\alpha^i(v_j)] = \det[\delta_j^i] = 1.$$

Hence, $\alpha^1 \wedge \cdots \wedge \alpha^k \neq 0$. $\Diamond$

3.11* Exterior multiplication

($\Leftarrow$) Clear because $\alpha \wedge \alpha = 0$.

($\Rightarrow$) Suppose $\alpha \wedge \gamma = 0$. Extend α to a basis $\alpha^1,\ldots,\alpha^n$ for $V^\vee$, with $\alpha^1 = \alpha$. Write $\gamma = \sum c_J \alpha^J$, where J runs over all strictly ascending multi-indices $1 \leq j_1 < \cdots < j_k \leq n$. In the sum $\alpha \wedge \gamma = \sum c_J \alpha \wedge \alpha^J$, all the terms $\alpha \wedge \alpha^J$ with $j_1 = 1$ vanish, since $\alpha = \alpha^1$. Hence,

$$0 = \alpha \wedge \gamma = \sum_{j_1 \neq 1} c_J \alpha \wedge \alpha^J.$$

Since $\{\alpha \wedge \alpha^J\}_{j_1 \neq 1}$ is a subset of a basis for $A_{k+1}(V)$, it is linearly independent, and so all c_J are 0 if $j_1 \neq 1$. Thus,

$$\gamma = \sum_{j_1=1} c_J \alpha^J = \alpha \wedge \left(\sum_{j_1=1} c_J \alpha^{j_2} \wedge \cdots \wedge \alpha^{j_k} \right).$$ $\Diamond$

4.1 $\omega(X) = yz,\ d\omega = -dx \wedge dz$.

4.2 Write $\omega = \sum_{i<j} c_{ij} dx^i \wedge dx^j$. Then $c_{ij}(p) = \omega_p(e_i,e_j)$, where $e_i = \partial/\partial x^i$. Calculate $c_{12}(p)$, $c_{13}(p)$, and $c_{23}(p)$. The answer is $\omega_p = p^3 dx^1 \wedge dx^2$.

4.3 $dx = \cos\theta\, dr - r\sin\theta\, d\theta,\ dy = \sin\theta\, dr + r\cos\theta\, d\theta,\ dx \wedge dy = r\, dr \wedge d\theta$.

4.4 $dx \wedge dy \wedge dz = \rho^2 \sin\phi\, d\rho \wedge d\phi \wedge d\theta$.

4.5 $\alpha \wedge \beta = (a_1 b_1 + a_2 b_2 + a_3 b_3)\, dx^1 \wedge dx^2 \wedge dx^3$.

5.3 The image $\phi_4(U_{14}) = \left\{ (x,z) \mid -1 < z < 1,\ 0 < x < \sqrt{1-z^2} \right\}$.

The transition function $(\phi_1 \circ \phi_4^{-1})(x,z) = \phi_1(x,y,z) = (y,z) = \left(-\sqrt{1-x^2-z^2}, z \right)$ is a C^∞ function of x,z.

5.4* Existence of a coordinate neighborhood

Let U_β be any coordinate neighborhood of p in the maximal atlas. Any open subset of U_β is again in the maximal atlas, because it is C^∞ compatible with all the open sets in the maximal atlas. Thus $U_\alpha := U_\beta \cap U$ is a coordinate neighborhood such that $p \in U_\alpha \subset U$.

6.3* Group of automorphisms of a vector space

The manifold structure $GL(V)_e$ is the maximal atlas on $GL(V)$ containing the coordinate chart $(GL(V), \phi_e)$. The manifold structure $GL(V)_u$ is the maximal atlas on $GL(V)$ containing the coordinate chart $(GL(V), \phi_u)$. The two maps $\phi_e \colon GL(V) \to \mathbb{R}^{n \times n}$ and $\phi_u \colon GL(V) \to \mathbb{R}^{n \times n}$ are C^∞ compatible, because $\phi_e \circ \phi_u^{-1} \colon GL(n, \mathbb{R}) \to GL(n, \mathbb{R})$ is conjugation by the change-of-basis matrix from u to e. Therefore, the two maximal atlases are in fact the same.   ◇

7.4* Quotient space of a sphere with antipodal points identified

(a) Let U be an open subset of S^n. Then $\pi^{-1}(\pi(U)) = U \cup a(U)$, where $a \colon S^n \to S^n$, $a(x) = -x$, is the antipodal map. Since the antipodal map is a homeomorphism, $a(U)$ is open, and hence $\pi^{-1}(\pi(U))$ is open. By the definition of quotient topology, $\pi(U)$ is open. This proves that π is an open map.

(b) The graph R of the equivalence relation $\sim$ is

$$R = \{(x,x) \in S^n \times S^n\} \cup \{(x,-x) \in S^n \times S^n\} = \Delta \cup (\mathbb{1} \times a)(\Delta).$$

By Corollary 7.8, because S^n is Hausdorff, the diagonal Δ in $S^n \times S^n$ is closed. Since $\mathbb{1} \times a \colon S^n \times S^n \to S^n \times S^n$, $(x,y) \mapsto (x,-y)$ is a homeomorphism, $(\mathbb{1} \times a)(\Delta)$ is also closed. As a union of the two closed sets Δ and $(\mathbb{1} \times a)(\Delta)$, R is closed in $S^n \times S^n$. By Theorem 7.7, $S^n/\!\!\sim$ is Hausdorff.   ◇

7.5* Orbit space of a continuous group action

Let U be an open subset of S. For each $g \in G$, since right multiplication by g is a homeomorphism $S \to S$, the set Ug is open. But

$$\pi^{-1}(\pi(U)) = \cup_{g \in G} Ug,$$

which is a union of open sets, hence is open. By the definition of the quotient topology, $\pi(U)$ is open.   ◇

7.9* Compactness of real projective space

By Exercise 7.11 there is a continuous surjective map $\pi \colon S^n \to \mathbb{R}P^n$. Since the sphere S^n is compact, and the continuous image of a compact set is compact (Proposition A.34), $\mathbb{R}P^n$ is compact.   ◇

8.1* Differential of a map

To determine the coefficient a in $F_*(\partial/\partial x) = a\partial/\partial u + b\partial/\partial v + c\partial/\partial w$, we apply both sides to u to get

$$F_*\left(\frac{\partial}{\partial x}\right)u = \left(a\frac{\partial}{\partial u} + b\frac{\partial}{\partial v} + c\frac{\partial}{\partial w}\right)u = a.$$

Hence,

$$a = F_*\left(\frac{\partial}{\partial x}\right)u = \frac{\partial}{\partial x}(u \circ F) = \frac{\partial}{\partial x}(x) = 1.$$

Similarly,

$$b = F_*\left(\frac{\partial}{\partial x}\right)v = \frac{\partial}{\partial x}(v \circ F) = \frac{\partial}{\partial x}(y) = 0$$

and

$$c = F_* \left(\frac{\partial}{\partial x} \right) w = \frac{\partial}{\partial x} (w \circ F) = \frac{\partial}{\partial x} (xy) = y.$$

So $F_*(\partial/\partial x) = \partial/\partial u + y\partial/\partial w$.   ◇

8.3 One can directly calculate $a = F_*(X)u$ and $b = F_*(X)v$ or more simply, one can apply Problem 8.2. The answer is $a = -(\sin\alpha)x - (\cos\alpha)y$, $b = (\cos\alpha)x - (\sin\alpha)y$.

8.5* Velocity of a curve in local coordinates
We know that $c'(t) = \sum a^j \partial/\partial x^j$. To compute a^i, evaluate both sides on x^i:

$$a^i = \left(\sum a^j \frac{\partial}{\partial x^j} \right) x^i = c'(t)x^i = c_* \left(\frac{d}{dt} \right) x^i = \frac{d}{dt} (x^i \circ c) = \frac{d}{dt} c^i = \dot{c}^i(t). \qquad ◇$$

8.6 $c'(0) = -2y\partial/\partial x + 2x\partial/\partial y.$

8.7* Tangent space to a product
If $(U, \phi) = (U, x^1, \ldots, x^m)$ and $(V, \psi) = (V, y^1, \ldots, y^n)$ are charts about p in M and q in N respectively, then by Proposition 5.18, a chart about (p, q) in $M \times N$ is

$$(U \times V, \phi \times \psi) = (U \times V, (\pi_1^*\phi, \pi_2^*\psi)) = (U \times V, \bar{x}^1, \ldots, \bar{x}^n, \bar{y}^1, \ldots, \bar{y}^n),$$

where $\bar{x}^i = \pi_1^* x^i$ and $\bar{y}^i = \pi_2^* y^i$. Let $\pi_{1*}(\partial/\partial\bar{x}^j) = \sum a_j^i \partial/\partial x^i$. Then

$$a_j^i = \pi_{1*} \left(\frac{\partial}{\partial\bar{x}^j} \right) x^i = \frac{\partial}{\partial\bar{x}^j} (x^i \circ \pi_1) = \frac{\partial\bar{x}^i}{\partial\bar{x}^j} = \delta_j^i.$$

Hence,

$$\pi_{1*} \left(\frac{\partial}{\partial\bar{x}^j} \right) = \sum_i \delta_j^i \frac{\partial}{\partial x^i} = \frac{\partial}{\partial x^j}.$$

This really means that

$$\pi_{1*} \left(\frac{\partial}{\partial\bar{x}^j} \bigg|_{(p,q)} \right) = \frac{\partial}{\partial x^j} \bigg|_p. \qquad (8.7.1)$$

Similarly,

$$\pi_{1*} \left(\frac{\partial}{\partial\bar{y}^j} \right) = 0, \quad \pi_{2*} \left(\frac{\partial}{\partial\bar{x}^j} \right) = 0, \quad \pi_{2*} \left(\frac{\partial}{\partial\bar{y}^j} \right) = \frac{\partial}{\partial y^j}. \qquad (8.7.2)$$

A basis for $T_{(p,q)}(M \times N)$ is

$$\frac{\partial}{\partial\bar{x}^1} \bigg|_{(p,q)}, \ldots, \frac{\partial}{\partial\bar{x}^m} \bigg|_{(p,q)}, \frac{\partial}{\partial\bar{y}^1} \bigg|_{(p,q)}, \ldots, \frac{\partial}{\partial\bar{y}^n} \bigg|_{(p,q)}.$$

A basis for $T_pM \times T_qN$ is

$$\left(\frac{\partial}{\partial x^1} \bigg|_p, 0 \right), \ldots, \left(\frac{\partial}{\partial x^m} \bigg|_p, 0 \right), \left(0, \frac{\partial}{\partial y^1} \bigg|_q \right), \ldots, \left(0, \frac{\partial}{\partial y^n} \bigg|_q \right).$$

By (8.7.1) and (8.7.2), the linear map (π_{1*}, π_{2*}) maps a basis of $T_{(p,q)}(M \times N)$ to a basis of $T_pM \times T_qN$. It is therefore an isomorphism.   ◇

8.8 (a) Let $c(t)$ be a curve starting at e in G with $c'(0) = X_e$. Then $\alpha(t) = (c(t), e)$ is a curve starting at (e, e) in $G \times G$ with $\alpha'(0) = (X_e, 0)$. Compute $\mu_{*, (e,e)}$ using $\alpha(t)$.

8.9* Transforming vectors to coordinate vectors

Let $(V, y^1, \ldots, y^n)$ be a chart about p. Suppose $(X_j)_p = \sum_i a_j^i \, \partial/\partial y^i|_p$. Since $(X_1)_p, \ldots, (X_n)_p$ are linearly independent, the matrix $A = [a_j^i]$ is nonsingular.

Define a new coordinate system $x^1, \ldots, x^n$ by

$$y^i = \sum_{j=1}^n a_j^i x^j \quad \text{for } i = 1, \ldots, n. \tag{8.9.1}$$

By the chain rule,

$$\frac{\partial}{\partial x^j} = \sum_i \frac{\partial y^i}{\partial x^j} \frac{\partial}{\partial y^i} = \sum a_j^i \frac{\partial}{\partial y^i}.$$

At the point p,

$$\left. \frac{\partial}{\partial x^j} \right|_p = \sum a_j^i \left. \frac{\partial}{\partial y^i} \right|_p = (X_j)_p.$$

In matrix notation,

$$\begin{bmatrix} y^1 \\ \vdots \\ y^n \end{bmatrix} = A \begin{bmatrix} x^1 \\ \vdots \\ x^n \end{bmatrix}, \quad \text{so} \quad \begin{bmatrix} x^1 \\ \vdots \\ x^n \end{bmatrix} = A^{-1} \begin{bmatrix} y^1 \\ \vdots \\ y^n \end{bmatrix}.$$

This means that (8.9.1) is equivalent to $x^j = \sum_{i=1}^n (A^{-1})_i^j y^i$.   ◊

8.10 (a) For all $x \le p$, $f(x) \le f(p)$. Hence,

$$f'(p) = \lim_{x \to p^-} \frac{f(x) - f(p)}{x - p} \ge 0. \tag{8.10.1}$$

Similarly, for all $x \ge p$, $f(x) \le f(p)$, so that

$$f'(p) = \lim_{x \to p^+} \frac{f(x) - f(p)}{x - p} \le 0. \tag{8.10.2}$$

The two inequalities (8.10.1) and (8.10.2) together imply that $f'(p) = 0$.

9.1 $c \in \mathbb{R} - \{0, -108\}$.

9.2 Yes, because it is a regular level set of the function $f(x, y, z, w) = x^5 + y^5 + z^5 + w^5$.

9.3 Yes; see Example 9.12.

9.4* Regular submanifolds

Let $p \in S$. By hypothesis there is an open set U in $\mathbb{R}^2$ such that on $U \cap S$ one of the coordinates is a C^∞ function of the other. Without loss of generality, we assume that $y = f(x)$ for some C^∞ function $f \colon A \subset \mathbb{R} \to B \subset \mathbb{R}$, where A and B are open sets in $\mathbb{R}$ and $V := A \times B \subset U$. Let $F \colon V \to \mathbb{R}^2$ be given by $F(x, y) = (x, y - f(x))$. Since F is a diffeomorphism onto its image, it can be used as a coordinate map. In the chart $(V, x, y - f(x))$, $V \cap S$ is defined by the vanishing of the coordinate $y - f(x)$. This proves that S is a regular submanifold of $\mathbb{R}^2$.   ◊

9.5 $(\mathbb{R}^3, x, y, z - f(x, y))$ is an adapted chart for $\mathbb{R}^3$ relative to $\Gamma(f)$.

9.6 Differentiate (9.3) with respect to t.

9.10* The transversality theorem

(a) $f^{-1}(U) \cap f^{-1}(S) = f^{-1}(U \cap S) = f^{-1}(g^{-1}(0)) = (g \circ f)^{-1}(0)$.

(b) Let $p \in f^{-1}(U) \cap f^{-1}(S) = f^{-1}(U \cap S)$. Then $f(p) \in U \cap S$. Because S is a fiber of g, the pushforward $g_*(T_{f(p)}S)$ equals 0. Because $g: U \to \mathbb{R}^k$ is a projection, $g_*(T_{f(p)}M) = T_0(\mathbb{R}^k)$. Applying g_* to the transversality equation (9.4), we get

$$g_* f_*(T_p N) = g_*(T_{f(p)}M) = T_0(\mathbb{R}^k).$$

Hence, $g \circ f: f^{-1}(U) \to \mathbb{R}^k$ is a submersion at p. Since p is an arbitrary point of $f^{-1}(U) \cap f^{-1}(S) = (g \circ f)^{-1}(0)$, this set is a regular level set of $g \circ f$.

(c) By the regular level set theorem, $f^{-1}(U) \cap f^{-1}(S)$ is a regular submanifold of $f^{-1}(U) \subset N$. Thus every point $p \in f^{-1}(S)$ has an adapted chart relative to $f^{-1}(S)$ in N. $\lozenge$

10.7 Let $e_1, \ldots, e_n$ be a basis for V and $\alpha^1, \ldots, \alpha^n$ the dual basis for $V^\vee$. Then a basis for $A_n(V)$ is $\alpha^1 \wedge \cdots \wedge \alpha^n$ and $L^*(\alpha^1 \wedge \cdots \wedge \alpha^n) = c\alpha^1 \wedge \cdots \wedge \alpha^n$ for some constant c. Suppose $L(e_j) = \sum_i a^i_j e_i$. Compute c in terms of a^i_j.

11.1 Let $c(t) = (x^1(t), \ldots, x^{n+1}(t))$ be a curve in S^n with $c(0) = p$ and $c'(0) = X_p$. Differentiate $\sum_i (x^i)^2(t) = 1$ with respect to t. Let H be the plane $\{(a^1, a^2, a^3) \in \mathbb{R}^3 \mid \sum a^i p^i = 0\}$. Show that $T_p(S^2) \subset H$. Because both sets are linear spaces and have the same dimension, equality holds.

11.3* Critical points of a smooth map on a compact manifold

First Proof. Suppose $f: N \to \mathbb{R}^m$ has no critical point. Then it is a submersion. The projection to the first factor, $\pi: \mathbb{R}^m \to \mathbb{R}$, is also a submersion. It follows that the composite $\pi \circ f: N \to \mathbb{R}$ is a submersion. This contradicts the fact that as a continuous function from a compact manifold to $\mathbb{R}$, the function $\pi \circ f$ has a maximum and hence a critical point (see Problem 8.10).

Second Proof. Suppose $f: N \to \mathbb{R}^m$ has no critical point. Then it is a submersion. Since a submersion is an open map (Corollary 11.6), the image $f(N)$ is open in $\mathbb{R}^m$. But the continuous image of a compact set is compact and a compact subset of $\mathbb{R}^m$ is closed and bounded. Hence, $f(N)$ is a nonempty proper closed subset of $\mathbb{R}^m$. This is a contradiction, because being connected, $\mathbb{R}^m$ cannot have a nonempty proper subset that is both open and closed. $\lozenge$

11.4 At $p = (a, b, c)$, $i_*(\partial/\partial u|_p) = \partial/\partial x - (a/c)\partial/\partial z$, and $i_*(\partial/\partial v|_p) = \partial/\partial y - (b/c)\partial/\partial z$.

11.5 Use Proposition A.35 to show that f is a closed map. Then apply Problem A.12.

12.1* The Hausdorff condition on the tangent bundle

Let (p, v) and (q, w) be distinct points of the tangent bundle TM.

Case 1: $p \neq q$. Because M is Hausdorff, p and q can be separated by disjoint neighborhoods U and V. Then TU and TV are disjoint open subsets of TM containing (p, v) and (q, w), respectively.

Case 2: $p = q$. Let (U, ϕ) be a coordinate neighborhood of p. Then (p, v) and (p, w) are distinct points in the open set TU. Since TU is homeomorphic to the open subset $\phi(U) \times \mathbb{R}^n$ of $\mathbb{R}^{2n}$, and any subspace of a Hausdorff space is Hausdorff, TU is Hausdorff. Therefore, (p, v) and (p, w) can be separated by disjoint open sets in TU. $\lozenge$

13.1* Support of a finite sum

Let A be the set where $\sum \rho_i$ is not zero and A_i the set where ρ_i is not zero:

$$A = \{x \in M \mid \sum \rho_i(x) \neq 0\}, \quad A_i = \{x \in M \mid \rho_i(x) \neq 0\}.$$

If $\sum \rho_i(x) \neq 0$, then at least one $\rho_i(x)$ must be nonzero. This implies that $A \subset \bigcup A_i$. Taking the closure of both sides gives $\mathrm{cl}(A) \subset \overline{\bigcup A_i}$. For a finite union, $\overline{\bigcup A_i} = \bigcup \overline{A_i}$ (Exercise A.53). Hence,

$$\mathrm{supp}\left(\sum \rho_i\right) = \mathrm{cl}(A) \subset \overline{\bigcup A_i} = \bigcup \overline{A_i} = \bigcup \mathrm{supp}\,\rho_i. \qquad \Diamond$$

13.2* Locally finite family and compact set

For each $p \in K$, let W_p be a neighborhood of p that intersects only finitely many of the sets A_α. The collection $\{W_p\}_{p \in K}$ is an open cover of K. By compactness, K has a finite subcover $\{W_{p_i}\}_{i=1}^r$. Since each W_{p_i} intersects only finitely many of the A_α, the finite union $W := \bigcup_{i=1}^r W_{p_i}$ intersects only finitely many of the A_α. $\qquad \Diamond$

13.3 (a) Take $f = \rho_{M-B}$.

13.5* Support of the pullback by a projection

Let $A = \{p \in M \mid f(p) \neq 0\}$. Then $\mathrm{supp}\,f = \mathrm{cl}_M(A)$. Observe that

$$(\pi^* f)(p,q) = f(p) \neq 0 \quad \text{iff } p \in A.$$

Hence,

$$\{(p,q) \in M \times N \mid (\pi^* f)(p,q) \neq 0\} = A \times N.$$

So

$$\mathrm{supp}(\pi^* f) = \mathrm{cl}_{M \times N}(A \times N) = \mathrm{cl}_M(A) \times N = (\mathrm{supp}\,f) \times N$$

by Problem A.20. $\qquad \Diamond$

13.7* Closure of a locally finite union

($\supset$) Since $A_\alpha \subset \bigcup A_\alpha$, taking the closure of both sides gives

$$\overline{A_\alpha} \subset \overline{\bigcup A_\alpha}.$$

Hence, $\bigcup \overline{A_\alpha} \subset \overline{\bigcup A_\alpha}$.

($\subset$) Instead of proving $\overline{\bigcup A_\alpha} \subset \bigcup \overline{A_\alpha}$, we will prove the contrapositive: if $p \notin \bigcup \overline{A_\alpha}$, then $p \notin \overline{\bigcup A_\alpha}$. Suppose $p \notin \bigcup \overline{A_\alpha}$. By local finiteness, p has a neighborhood W that meets only finitely many of the A_α's, say $A_{\alpha_1}, \ldots, A_{\alpha_m}$ (see the figure below).

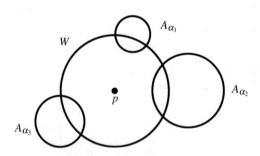

Since $p \notin \overline{A_\alpha}$ for any α, $p \notin \bigcup_{i=1}^m \overline{A_{\alpha_i}}$. Note that W is disjoint from A_α for all $\alpha \neq \alpha_i$, so $W - \bigcup_{i=1}^m \overline{A_{\alpha_i}}$ is disjoint from A_α for all α. Because $\bigcup_{i=1}^m \overline{A_{\alpha_i}}$ is closed, $W - \bigcup_{i=1}^m \overline{A_{\alpha_i}}$ is an open set containing p disjoint from $\bigcup A_\alpha$. By the local characterization of closure (Proposition A.48), $p \notin \overline{\bigcup A_\alpha}$. Hence, $\overline{\bigcup A_\alpha} \subset \bigcup \overline{A_\alpha}$. $\qquad \Diamond$

14.1* Equality of vector fields

The implication in the direction $(\Rightarrow)$ is obvious. For the converse, let $p \in M$. To show that $X_p = Y_p$, it suffices to show that $X_p[h] = Y_p[h]$ for any germ $[h]$ of C^∞ functions in $C_p^\infty(M)$. Suppose $h \colon U \to \mathbb{R}$ is a C^∞ function that represents the germ $[h]$. We can extend it to a C^∞ function $\tilde{h} \colon M \to \mathbb{R}$ by multiplying it by a C^∞ bump function supported in U that is identically 1 in a neighborhood of p. By hypothesis, $X\tilde{h} = Y\tilde{h}$. Hence,

$$X_p\tilde{h} = (X\tilde{h})_p = (Y\tilde{h})_p = Y_p\tilde{h}. \tag{14.1.1}$$

Because $\tilde{h} = h$ in a neighborhood of p, we have $X_p h = X_p\tilde{h}$ and $Y_p h = Y_p\tilde{h}$. It follows from (14.1.1) that $X_p h = Y_p h$. Thus, $X_p = Y_p$. Since p is an arbitrary point of M, the two vector fields X and Y are equal. $\diamondsuit$

14.6 Integral curve starting at a zero of a vector field

(a)* Suppose $c(t) = p$ for all $t \in \mathbb{R}$. Then

$$c'(t) = 0 = X_p = X_{c(t)}$$

for all $t \in \mathbb{R}$. Thus, the constant curve $c(t) = p$ is an integral curve of X with initial point p. By the uniqueness of an integral curve with a given initial point, this is the maximal integral curve of X starting at p.

14.8 $c(t) = 1/((1/p) - t)$ on $(-\infty, 1/p)$.

14.10 Show that both sides applied to a C^∞ function h on M are equal. Then use Problem 14.1.

14.11 $-\partial/\partial y$.

14.12 $c^k = \sum_i \left(a^i \dfrac{\partial b^k}{\partial x^i} - b^i \dfrac{\partial a^k}{\partial x^i} \right).$

14.14 Use Example 14.15 and Proposition 14.17.

15.3

(a) Apply Proposition A.43.

(b) Apply Proposition A.43.

(c) Apply Problem A.16.

(d) By (a) and (b), the subset G_0 is a subgroup of G. By (c), it is an open submanifold.

15.4* Open subgroup of a connected Lie group

For any $g \in G$, left multiplication $\ell_g \colon G \to G$ by g maps the subgroup H to the left coset gH. Since H is open and ℓ_g is a homeomorphism, the coset gH is open. Thus, the set of cosets gH, $g \in G$, partitions G into a disjoint union of open subsets. Since G is connected, there can be only one coset. Therefore, $H = G$. $\diamondsuit$

15.5 Let $c(t)$ be a curve in G with $c(0) = a$, $c'(0) = X_a$. Then $(c(t), b)$ is a curve through (a, b) with initial velocity $(X_a, 0)$. Compute $\mu_{*,(a,b)}(X_a, 0)$ using this curve (Proposition 8.18). Compute similarly $\mu_{*,(a,b)}(0, Y_b)$.

15.7* Differential of the determinant map

Let $c(t) = Ae^{tX}$. Then $c(0) = A$ and $c'(0) = AX$. Using the curve $c(t)$ to calculate the differential yields

$$\det_{A,*}(AX) = \frac{d}{dt}\Big|_{t=0} \det(c(t)) = \frac{d}{dt}\Big|_{t=0} (\det A) \det e^{tX}$$

$$= (\det A) \frac{d}{dt}\Big|_{t=0} e^{t \operatorname{tr} X} = (\det A) \operatorname{tr} X. \qquad \Diamond$$

15.8* Special linear group
If $\det A = 1$, then Exercise 15.7 gives

$$\det_{*,A}(AX) = \operatorname{tr} X.$$

Since $\operatorname{tr} X$ can assume any real value, $\det_{*,A} : T_A GL(n,\mathbb{R}) \to \mathbb{R}$ is surjective for all $A \in \det^{-1}(1)$. Hence, 1 is a regular value of $\det$. $\qquad \Diamond$

15.10
(a) $O(n)$ is defined by polynomial equations.
(b) If $A \in O(n)$, then each column of A has length 1.

15.11 Write out the conditions $A^T A = I$, $\det A = 1$. If $a^2 + b^2 = 1$, then (a,b) is a point on the unit circle, and so $a = \cos \theta$, $b = \sin \theta$ for some $\theta \in [0, 2\pi]$.

15.14
$$\begin{bmatrix} \cosh 1 & \sinh 1 \\ \sinh 1 & \cosh 1 \end{bmatrix},$$

where $\cosh t = (e^t + e^{-t})/2$ and $\cosh t = (e^t - e^{-t})/2$ are hyperbolic cosine and sine, respectively.

15.16 The correct target space for f is the vector space $K_{2n}(\mathbb{C})$ of $2n \times 2n$ skew-symmetric complex matrices.

16.4 Let $c(t)$ be a curve in $Sp(n)$ with $c(0) = I$ and $c'(0) = X$. Differentiate $c(t)^T J c(t) = J$ with respect to t.

16.5 Mimic Example 16.6. The left-invariant vector fields on $\mathbb{R}^n$ are the constant vector fields $\sum_{i=1}^n a^i \, \partial/\partial x^i$, where $a^i \in \mathbb{R}$.

16.9 A basis $X_{1,e}, \ldots, X_{n,e}$ for the tangent space $T_e(G)$ at the identity gives rise to a frame consisting of left-invariant vector fields $X_1, \ldots, X_n$.

16.10* The pushforward of left-invariant vector fields
Under the isomorphisms $\varphi_H : T_e H \overset{\sim}{\to} L(H)$ and $\varphi_G : T_e G \overset{\sim}{\to} L(G)$, the Lie brackets correspond and the pushforward maps correspond. Thus, this problem follows from Proposition 16.14 by the correspondence.

A more formal proof goes as follows. Since X and Y are left-invariant vector fields, $X = \tilde{A}$ and $Y = \tilde{B}$ for $A = X_e$ and $B = Y_e \in T_e H$. Then

$$
\begin{aligned}
F_*[X,Y] = F_*[\tilde{A}, \tilde{B}] = F_*([A,B]\tilde{\ }) & \quad \text{(Proposition 16.10)} \\
= (F_*[A,B])\tilde{\ } & \quad \text{(definition of } F_* \text{ on } L(H)) \\
= [F_* A, F_* B]\tilde{\ } & \quad \text{(Proposition 16.14)} \\
= [(F_* A)\tilde{\ }, (F_* B)\tilde{\ }] & \quad \text{(Proposition 16.10)} \\
= [F_* \tilde{A}, F_* \tilde{B}] & \quad \text{(definition of } F_* \text{ on } L(H)) \\
= [F_* X, F_* Y]. &
\end{aligned}
$$

$\qquad \Diamond$

16.11 (b) Let $(U, x^1, \ldots, x^n)$ be a chart about e in G. Relative to this chart, the differential c_{a*} at e is represented by the Jacobian matrix $[\partial(x^i \circ c_a)/\partial x^j|_e]$. Since $c_a(x) = axa^{-1}$ is a C^∞ function of x and a, all the partial derivatives $\partial(x^i \circ c_a)/\partial x^j|_e$ are C^∞ and therefore $\mathrm{Ad}(a)$ is a C^∞ function of a.

17.1 $\omega = (x\,dx + y\,dy)/(x^2 + y^2)$.

17.2 $a_j = \sum_i b_i\, \partial y^i / \partial x^j$.

17.4 (a) Suppose $\omega_p = \sum c_i\, dx^i|_p$. Then

$$\lambda_{\omega_p} = \pi^*(\omega_p) = \sum c_i \pi^*(dx^i|_p) = \sum c_i (\pi^* dx^i)_{\omega_p} = \sum c_i (d\pi^* x^i)_{\omega_p} = \sum c_i (d\tilde{x}^i)_{\omega_p}.$$

Hence, $\lambda = \sum c_i\, d\tilde{x}^i$.

18.4* Support of a sum or product

(a) If $(\omega + \tau)(p) \neq 0$, then $\omega(p) \neq 0$ or $\tau(p) \neq 0$. Hence,

$$Z(\omega + \tau)^c \subset Z(\omega)^c \cup Z(\tau)^c.$$

Taking the closure of both sides and using the fact that $\overline{A \cup B} = \overline{A} \cup \overline{B}$, we get

$$\operatorname{supp}(\omega + \tau) \subset \operatorname{supp}\omega \cup \operatorname{supp}\tau.$$

(b) Suppose $(\omega \wedge \tau)_p \neq 0$. Then $\omega_p \neq 0$ and $\tau_p \neq 0$. Hence,

$$Z(\omega \wedge \tau)^c \subset Z(\omega)^c \cap Z(\tau)^c.$$

Taking the closure of both sides and using the fact that $\overline{A \cap B} \subset \overline{A} \cap \overline{B}$, we get

$$\operatorname{supp}(\omega \wedge \tau) \subset \overline{Z(\omega)^c \cap Z(\tau)^c} \subset \operatorname{supp}\omega \cap \operatorname{supp}\tau. \qquad \diamondsuit$$

18.6* Locally finite supports

Let $p \in \operatorname{supp}\omega$. Since $\{\operatorname{supp}\rho_\alpha\}$ is locally finite, there is a neighborhood W_p of p in M that intersects only finitely many of the sets $\operatorname{supp}\rho_\alpha$. The collection $\{W_p \mid p \in \operatorname{supp}\omega\}$ covers $\operatorname{supp}\omega$. By the compactness of $\operatorname{supp}\omega$, there is a finite subcover $\{W_{p_1}, \ldots, W_{p_m}\}$. Since each W_{p_i} intersects only finitely many $\operatorname{supp}\rho_\alpha$, $\operatorname{supp}\omega$ intersects only finitely many $\operatorname{supp}\rho_\alpha$.

By Problem 18.4,

$$\operatorname{supp}(\rho_\alpha \omega) \subset \operatorname{supp}\rho_\alpha \cap \operatorname{supp}\omega.$$

Thus, for all but finitely many α, $\operatorname{supp}(\rho_\alpha \omega)$ is empty; i.e., $\rho_\alpha \omega \equiv 0$. $\qquad \diamondsuit$

18.8* Pullback by a surjective submersion

The fact that $\pi^* \colon \Omega^*(M) \to \Omega^*(\tilde{M})$ is an algebra homomorphism follows from Propositions 18.9 and 18.11.

Suppose $\omega \in \Omega^k(M)$ is a k-form on M for which $\pi^*\omega = 0$ in $\Omega^k(\tilde{M})$. To show that $\omega = 0$, pick an arbitrary point $p \in M$, and arbitrary vectors $v_1, \ldots, v_k \in T_p M$. Since π is surjective, there is a point $\tilde{p} \in \tilde{M}$ that maps to p. Since π is a submersion at $\tilde{p}$, there exist $\tilde{v}_1, \ldots, \tilde{v}_k \in T_{\tilde{p}}\tilde{M}$ such that $\pi_{*,\tilde{p}}\tilde{v}_i = v_i$. Then

$$\begin{aligned}
0 &= (\pi^*\omega)_{\tilde{p}}(\tilde{v}_1, \ldots, \tilde{v}_k) && \text{(because } \pi^*\omega = 0) \\
&= \omega_{\pi(\tilde{p})}(\pi_* \tilde{v}_1, \ldots, \pi_* \tilde{v}_k) && \text{(definition of } \pi^*\omega) \\
&= \omega_p(v_1, \ldots, v_k).
\end{aligned}$$

Since $p \in M$ and $v_1, \ldots, v_k \in T_p M$ are arbitrary, this proves that $\omega = 0$. Therefore, $\pi^* : \Omega^*(M) \to \Omega^*(\tilde{M})$ is injective.   ◇

18.9 (c) Because $f(a)$ is the pullback by $\mathrm{Ad}(a^{-1})$, we have $f(a) = \det(\mathrm{Ad}(a^{-1}))$ by Problem 10.7. According to Problem 16.11, $\mathrm{Ad}(a^{-1})$ is a C^∞ function of a.

19.1 $F^*(dx \wedge dy \wedge dz) = d(x \circ F) \wedge d(y \circ F) \wedge d(z \circ F)$. Apply Corollary 18.4(ii).

19.2 $F^*(u\,du + v\,dv) = (2x^3 + 3xy^2)dx + (3x^2 y + 2y^3)dy$.

19.3 $c^* \omega = dt$.

19.5* Coordinates and differential forms

Let $(V, x^1, \ldots, x^n)$ be a chart about p. By Corollary 18.4(ii),

$$df^1 \wedge \cdots \wedge df^n = \det \left[\frac{\partial f^i}{\partial x^j} \right] dx^1 \wedge \cdots \wedge dx^n.$$

So $(df^1 \wedge \cdots \wedge df^n)_p \neq 0$ if and only if $\det[\partial f^i / \partial x^j(p)] \neq 0$. By the inverse function theorem, this condition is equivalent to the existence of a neighborhood W on which the map $F := (f^1, \ldots, f^n) : W \to \mathbb{R}^n$ is a C^∞ diffeomorphism onto its image. In other words, $(W, f^1, \ldots, f^n)$ is a chart.   ◇

19.7 Mimic the proof of Proposition 19.3.

19.9* Vertical plane

Since $ax + by$ is the zero function on the vertical plane, its differential is identically zero:

$$a\,dx + b\,dy = 0.$$

Thus, at each point of the plane, dx is a multiple of dy or vice versa. In either case, $dx \wedge dy = 0$.   ◇

19.11

(a) Mimic Example 19.8. Define

$$U_x = \{(x, y) \in M \mid f_x \neq 0\} \quad \text{and} \quad U_y = \{(x, y) \in M \mid f_y \neq 0\},$$

where f_x, f_y are the partial derivatives $\partial f / \partial x$, $\partial f / \partial y$ respectively. Because 0 is a regular value of f, every point in M satisfies $f_x \neq 0$ or $f_y \neq 0$. Hence, $\{U_x, U_y\}$ is an open cover of M. Define $\omega = dy/f_x$ on U_x and $-dx/f_y$ on U_y. Show that ω is globally defined on M. By the implicit function theorem, in a neighborhood of a point $(a, b) \in U_x$, x is a C^∞ function of y. It follows that y can be used as a local coordinate and the 1-form dy/f_x is C^∞ at (a, b). Thus, ω is C^∞ on U_x. A similar argument shows that ω is C^∞ on U_y.

(b) On M, $df = f_x\,dx + f_y\,dy + f_z\,dz \equiv 0$.

(c) Define $U_i = \{p \in \mathbb{R}^{n+1} \mid \partial f / \partial x^i(p) \neq 0\}$ and

$$\omega = (-1)^{i-1} \frac{dx^1 \wedge \cdots \wedge \widehat{dx^i} \wedge \cdots \wedge dx^{n+1}}{\partial f / \partial x^i} \quad \text{on } U_i.$$

19.13 $\nabla \times \mathbf{E} = -\partial \mathbf{B} / \partial t$ and $\mathrm{div}\,\mathbf{B} = 0$.

20.3* Derivative of a smooth family of vector fields

Let $(V, y^1, \ldots, y^n)$ be another coordinate neighborhood of p such that

$$X_t = \sum_j b^j(t,q)\frac{\partial}{\partial y^j} \quad \text{on } V.$$

On $U \cap V$,

$$\frac{\partial}{\partial x^i} = \sum_j \frac{\partial y^j}{\partial x^i}\frac{\partial}{\partial y^j}.$$

Substituting this into (20.2) in the text and comparing coefficients with the expression for X_t above, we get

$$b^j(t,q) = \sum_i a^i(t,q)\frac{\partial y^j}{\partial x^i}.$$

Since $\partial y^j/\partial x^i$ does not depend on t, differentiating both sides of this equation with respect to t gives

$$\frac{\partial b^j}{\partial t} = \sum_i \frac{\partial a^i}{\partial t}\frac{\partial y^j}{\partial x^i}.$$

Hence,

$$\sum_j \frac{\partial b^j}{\partial t}\frac{\partial}{\partial y^j} = \sum_{i,j} \frac{\partial a^i}{\partial t}\frac{\partial y^j}{\partial x^i}\frac{\partial}{\partial y^j} = \sum_i \frac{\partial a^i}{\partial t}\frac{\partial}{\partial x^i}. \qquad \diamond$$

20.6* Global formula for the exterior derivative

By Theorem 20.12,

$$(\mathcal{L}_{Y_0}\omega)(Y_1,\ldots,Y_k) = Y_0\left(\omega(Y_1,\ldots,Y_k)\right) - \sum_{j=1}^{k} \omega(Y_1,\ldots,[Y_0,Y_j],\ldots,Y_k)$$

$$= Y_0\left(\omega(Y_1,\ldots,Y_k)\right) + \sum_{j=1}^{k} (-1)^j \omega([Y_0,Y_j],Y_1,\ldots,\widehat{Y}_j,\ldots,Y_k). \quad (20.6.1)$$

By the induction hypothesis, Theorem 20.14 is true for $(k-1)$-forms. Hence,

$$-(d\iota_{Y_0}\omega)(Y_1,\ldots,Y_k) = -\sum_{i=1}^{k}(-1)^{i-1}Y_i\left((\iota_{Y_0}\omega)(Y_1,\ldots,\widehat{Y}_i,\ldots,Y_k)\right)$$

$$- \sum_{1\le i<j\le k}(-1)^{i+j}(\iota_{Y_0}\omega)([Y_i,Y_j],Y_1,\ldots,\widehat{Y}_i,\ldots,\widehat{Y}_j,\ldots,Y_k)$$

$$= \sum_{i=1}^{k}(-1)^i Y_i\left(\omega(Y_0,Y_1,\ldots,\widehat{Y}_i,\ldots,Y_k)\right)$$

$$+ \sum_{1\le i<j\le k}(-1)^{i+j}\omega([Y_i,Y_j],Y_0,Y_1,\ldots,\widehat{Y}_i,\ldots,\widehat{Y}_j,\ldots,Y_k). \quad (20.6.2)$$

Adding (20.6.1) and (20.6.2) gives

$$\sum_{i=0}^{k}(-1)^i Y_i\left(\omega(Y_0,\ldots,\widehat{Y}_i,\ldots,Y_k)\right) + \sum_{j=1}^{k}(-1)^j \omega([Y_0,Y_j],\widehat{Y}_0,Y_1,\ldots,\widehat{Y}_j,\ldots,Y_k)$$

$$+ \sum_{1\le i<j\le k}(-1)^{i+j}\omega([Y_i,Y_j],Y_0,\ldots,\widehat{Y}_i,\ldots,\widehat{Y}_j,\ldots,Y_k),$$

which simplifies to the right-hand side of Theorem 20.14. $\qquad \diamond$

21.1* Locally constant map on a connected space

We first show that for every $y \in Y$, the inverse $f^{-1}(y)$ is an open set. Suppose $p \in f^{-1}(y)$. Then $f(p) = y$. Since f is locally constant, there is a neighborhood U of p such that $f(U) = \{y\}$. Thus, $U \subset f^{-1}(y)$. This proves that $f^{-1}(y)$ is open.

The equality $S = \bigcup_{y \in Y} f^{-1}(y)$ exhibits S as a disjoint union of open sets. Since S is connected, this is possible only if there is just one such open set $S = f^{-1}(y_0)$. Hence, f assumes the constant value y_0 on S. $\Diamond$

21.5 The map F is orientation-preserving.

21.6 Use Problem 19.11(c) and Theorem 21.5.

21.9 See Problem 12.2.

22.1 The topological boundary $\mathrm{bd}(M)$ is $\{0, 1, 2\}$; the manifold boundary ∂M is $\{0\}$.

22.3* Inward-pointing vectors at the boundary

($\Leftarrow$) Suppose $(U, \phi) = (U, x^1, \ldots, x^n)$ is a chart for M centered at p such that $X_p = \sum a^i \, \partial/\partial x^i|_p$ with $a^n > 0$. Then the curve $c(t) = \phi^{-1}(a^1 t, \ldots, a^n t)$ in M satisfies

$$c(0) = p, \quad c(]0, \varepsilon[) \subset M^\circ, \quad \text{and} \quad c'(0) = X_p. \tag{22.3.1}$$

So X_p is inward-pointing.

($\Rightarrow$) Suppose X_p is inward-pointing. Then $X_p \notin T_p(\partial M)$ and there is a curve $c \colon [0, \varepsilon[\to M$ such that (22.3.1) holds. Let $(U, \phi) = (U, x^1, \ldots, x^n)$ be a chart centered at p. On $U \cap M$, we have $x^n \geq 0$. If $(\phi \circ c)(t) = (c^1(t), \ldots, c^n(t))$, then $c^n(0) = 0$ and $c^n(t) > 0$ for $t > 0$. Therefore, the derivative of c^n at $t = 0$ is

$$\dot{c}^n(0) = \lim_{t \to 0^+} \frac{c^n(t) - c^n(0)}{t} = \lim_{t \to 0^+} \frac{c^n(t)}{t} \geq 0.$$

Since $X_p = \sum_{i=1}^n \dot{c}^i(0) \, \partial/\partial x^i|_p$, the coefficient of $\partial/\partial x^n|_p$ in X_p is $\dot{c}^n(0)$. In fact, $\dot{c}^n(0) > 0$ because if $\dot{c}^n(0)$ were 0, then X_p would be tangent to ∂M at p. $\Diamond$

22.4* Smooth outward-pointing vector field along the boundary

Let $p \in \partial M$ and let $(U, x^1, \ldots, x^n)$ be a coordinate neighborhood of p. Write

$$X_{\alpha, p} = \sum_{i=1}^n a^i(X_{\alpha, p}) \left. \frac{\partial}{\partial x^i} \right|_p .$$

Then

$$X_p = \sum_\alpha \rho_\alpha(p) X_{\alpha, p} = \sum_{i=1}^n \sum_\alpha \rho_\alpha(p) a^i(X_{\alpha, p}) \left. \frac{\partial}{\partial x^i} \right|_p .$$

Since $X_{\alpha, p}$ is outward-pointing, the coefficient $a^n(X_{\alpha, p})$ is negative by Problem 22.3. Because $\rho_\alpha(p) \geq 0$ for all α with $\rho_\alpha(p)$ positive for at least one α, the coefficient $\sum_\alpha \rho_\alpha(p) a^i(X_{\alpha, p})$ of $\partial/\partial x^n|_p$ in X_p is negative. Again by Problem 22.3, this proves that X_p is outward-pointing.

The smoothness of the vector field X follows from the smoothness of the partition of unity ρ_α and of the coefficient functions $a^i(X_{\alpha, p})$ as functions of p. $\Diamond$

22.6* Induced atlas on the boundary

Let $r^1, \ldots, r^n$ be the standard coordinates on the upper half-space $\mathcal{H}^n$. As a shorthand, we write $a = (a^1, \ldots, a^{n-1})$ for the first $n-1$ coordinates of a point in $\mathcal{H}^n$. Since the transition function

$$\psi \circ \phi^{-1} : \phi(U \cap V) \to \psi(U \cap V) \subset \mathcal{H}^n$$

takes boundary points to boundary points and interior points to interior points,

(i) $\left(r^n \circ \psi \circ \phi^{-1}\right)(a,0) = 0$, and

(ii) $\left(r^n \circ \psi \circ \phi^{-1}\right)(a,t) > 0$ for $t > 0$,

where $(a,0)$ and (a,t) are points in $\phi(U \cap V) \subset \mathcal{H}^n$.

Let $x^j = r^j \circ \phi$ and $y^i = r^i \circ \psi$ be the local coordinates on the charts (U,ϕ) and (V,ψ) respectively. In particular, $y^n \circ \phi^{-1} = r^n \circ \psi \circ \phi^{-1}$. Differentiating (i) with respect to r^j gives

$$\left.\frac{\partial y^n}{\partial x^j}\right|_{\phi^{-1}(a,0)} = \left.\frac{\partial (y^n \circ \phi^{-1})}{\partial r^j}\right|_{(a,0)} = \left.\frac{\partial (r^n \circ \psi \circ \phi^{-1})}{\partial r^j}\right|_{(a,0)} = 0 \qquad \text{for } j = 1,\ldots,n-1.$$

From (i) and (ii),

$$\begin{aligned}
\left.\frac{\partial y^n}{\partial x^n}\right|_{\phi^{-1}(a,0)} &= \left.\frac{\partial \left(y^n \circ \phi^{-1}\right)}{\partial r^n}\right|_{(a,0)} \\
&= \lim_{t \to 0^+} \frac{\left(y^n \circ \phi^{-1}\right)(a,t) - \left(y^n \circ \phi^{-1}\right)(a,0)}{t} \\
&= \lim_{t \to 0^+} \frac{\left(y^n \circ \phi^{-1}\right)(a,t)}{t} \geq 0,
\end{aligned}$$

since both t and $\left(y^n \circ \phi^{-1}\right)(a,t)$ are positive.

The Jacobian matrix of $J = [\partial y^i / \partial x^j]$ of the overlapping charts U and V at a point $p = \phi^{-1}(a,0)$ in $U \cap V \cap \partial M$ therefore has the form

$$J = \begin{pmatrix}
\dfrac{\partial y^1}{\partial x^1} & \cdots & \dfrac{\partial y^1}{\partial x^{n-1}} & \dfrac{\partial y^1}{\partial x^n} \\
\vdots & \ddots & \vdots & \vdots \\
\dfrac{\partial y^{n-1}}{\partial x^1} & \cdots & \dfrac{\partial y^{n-1}}{\partial x^{n-1}} & \dfrac{\partial y^{n-1}}{\partial x^n} \\
0 & \cdots & 0 & \dfrac{\partial y^n}{\partial x^n}
\end{pmatrix} = \begin{pmatrix} A & * \\ 0 & \dfrac{\partial y^n}{\partial x^n} \end{pmatrix},$$

where the upper left $(n-1) \times (n-1)$ block $A = [\partial y^i / \partial x^j]_{1 \leq i,j \leq n-1}$ is the Jacobian matrix of the induced charts $U \cap \partial M$ and $V \cap \partial M$ on the boundary. Since $\det J(p) > 0$ and $\partial y^n / \partial x^n(p) > 0$, we have $\det A(p) > 0$. $\qquad \diamondsuit$

22.7* Boundary orientation of the left half-space

Because a smooth outward-pointing vector field along ∂M is $\partial/\partial x^1$, by definition an orientation form of the boundary orientation on ∂M is the contraction

$$\iota_{\partial/\partial x^1}(dx^1 \wedge dx^2 \wedge \cdots \wedge dx^n) = dx^2 \wedge \cdots \wedge dx^n. \qquad \diamondsuit$$

22.8 Viewed from the top, C_1 is clockwise and C_0 is counterclockwise.

22.9 Compute $\iota_X(dx^1 \wedge \cdots \wedge dx^{n+1})$.

22.10 (a) An orientation form on the closed unit ball is $dx^1 \wedge \cdots \wedge dx^{n+1}$ and a smooth outward-pointing vector field on U is $\partial/\partial x^{n+1}$. By definition, an orientation form on U is the contraction

$$\iota_{\partial/\partial x^{n+1}}(dx^1 \wedge \cdots \wedge dx^{n+1}) = (-1)^n dx^1 \wedge \cdots \wedge dx^n.$$

22.11 (a) Let ω be the orientation form on the sphere in Problem 22.9. Show that $a^*\omega = (-1)^{n+1}\omega$.

23.1 Let $x = au$ and $y = bv$.

23.2 Use the Heine–Borel theorem (Theorem A.40).

23.3* Integral under a diffeomorphism

Let $\{(U_\alpha, \phi_\alpha)\}$ be an oriented atlas for M that specifies the orientation of M, and $\{\rho_\alpha\}$ a partition of unity on M subordinate to the open cover $\{U_\alpha\}$. Assume that $F: N \to M$ is orientation-preserving. By Problem 21.4, $\{(F^{-1}(U_\alpha), \phi_\alpha \circ F)\}$ is an oriented atlas for N that specifies the orientation of N. By Problem 13.6, $\{F^*\rho_\alpha\}$ is a partition of unity on N subordinate to the open cover $\{F^{-1}(U_\alpha)\}$.

By the definition of the integral,

$$
\begin{aligned}
\int_N F^*\omega &= \sum_\alpha \int_{F^{-1}(U_\alpha)} (F^*\rho_\alpha)\,(F^*\omega) \\
&= \sum_\alpha \int_{F^{-1}(U_\alpha)} F^*(\rho_\alpha\omega) \\
&= \sum_\alpha \int_{(\phi_\alpha \circ F)(F^{-1}(U_\alpha))} (\phi_\alpha \circ F)^{-1*} F^*(\rho_\alpha\omega) \\
&= \sum_\alpha \int_{\phi_\alpha(U_\alpha)} (\phi_\alpha^{-1})^*(\rho_\alpha\omega) \\
&= \sum_\alpha \int_{U_\alpha} \rho_\alpha\omega = \int_M \omega.
\end{aligned}
$$

If $F: N \to M$ is orientation-reversing, then $\{(F^{-1}(U_\alpha), \phi_\alpha \circ F)\}$ is an oriented atlas for N that gives the opposite orientation of N. Using this atlas to calculate the integral as above gives $-\int_N F^*\omega$. Hence in this case $\int_M \omega = -\int_N F^*\omega$. $\Diamond$

23.4* Stokes's theorem for $\mathbb{R}^n$ and for $\mathcal{H}^n$

An $(n-1)$-form ω with compact support on $\mathbb{R}^n$ or $\mathcal{H}^n$ is a linear combination

$$\omega = \sum_{i=1}^n f_i\, dx^1 \wedge \cdots \wedge \widehat{dx^i} \wedge \cdots \wedge dx^n. \tag{23.4.1}$$

Since both sides of Stokes's theorem are $\mathbb{R}$-linear in ω, it suffices to check the theorem for just one term of the sum (23.4.1). So we may assume

$$\omega = f\, dx^1 \wedge \cdots \wedge \widehat{dx^i} \wedge \cdots \wedge dx^n,$$

where f is a C^∞ function with compact support in $\mathbb{R}^n$ or $\mathcal{H}^n$. Then

$$
\begin{aligned}
d\omega &= \frac{\partial f}{\partial x^i}\, dx^i \wedge dx^1 \wedge \cdots \wedge dx^{i-1} \wedge \widehat{dx^i} \wedge \cdots \wedge dx^n \\
&= (-1)^{i-1}\frac{\partial f}{\partial x^i}\, dx^1 \wedge \cdots \wedge dx^i \wedge \cdots \wedge dx^n.
\end{aligned}
$$

Since f has compact support in $\mathbb{R}^n$ or $\mathcal{H}^n$, we may choose $a > 0$ large enough that $\operatorname{supp} f$ lies in the interior of the cube $[-a, a]^n$.

Stokes's theorem for $\mathbb{R}^n$

By Fubini's theorem, one can first integrate with respect to x^i:

$$\int_{\mathbb{R}^n} d\omega = \int_{\mathbb{R}^n} (-1)^{i-1} \frac{\partial f}{\partial x^i} dx^1 \cdots dx^n$$

$$= (-1)^{i-1} \int_{\mathbb{R}^{n-1}} \left(\int_{-\infty}^{\infty} \frac{\partial f}{\partial x^i} dx^i \right) dx^1 \cdots \widehat{dx^i} \cdots dx^n$$

$$= (-1)^{i-1} \int_{\mathbb{R}^{n-1}} \left(\int_{-a}^{a} \frac{\partial f}{\partial x^i} dx^i \right) dx^1 \cdots \widehat{dx^i} \cdots dx^n.$$

But

$$\int_{-a}^{a} \frac{\partial f}{\partial x^i} dx^i = f(\ldots, x^{i-1}, a, x^{i+1}, \ldots) - f(\ldots, x^{i-1}, -a, x^{i+1}, \ldots)$$

$$= 0 - 0 = 0,$$

because the support of f lies in the interior of $[-a,a]^n$. Hence, $\int_{\mathbb{R}^n} d\omega = 0$.

The right-hand side of Stokes's theorem is $\int_{\partial \mathbb{R}^n} \omega = \int_{\varnothing} \omega = 0$, because $\mathbb{R}^n$ has empty boundary. This verifies Stokes's theorem for $\mathbb{R}^n$.

Stokes's theorem for $\mathcal{H}^n$

Case 1: $i \neq n$.

$$\int_{\mathcal{H}^n} d\omega = (-1)^{i-1} \int_{\mathcal{H}^n} \frac{\partial f}{\partial x^i} dx^1 \cdots dx^n$$

$$= (-1)^{i-1} \int_{\mathcal{H}^{n-1}} \left(\int_{-\infty}^{\infty} \frac{\partial f}{\partial x^i} dx^i \right) dx^1 \cdots \widehat{dx^i} \cdots dx^n$$

$$= (-1)^{i-1} \int_{\mathcal{H}^{n-1}} \left(\int_{-a}^{a} \frac{\partial f}{\partial x^i} dx^i \right) dx^1 \cdots \widehat{dx^i} \cdots dx^n$$

$$= 0 \quad \text{for the same reason as the case of } \mathbb{R}^n.$$

As for $\int_{\partial \mathcal{H}^n} \omega$, note that $\partial \mathcal{H}^n$ is defined by the equation $x^n = 0$. Hence, on $\partial \mathcal{H}^n$, the 1-form dx^n is identically zero. Since $i \neq n$, $\omega = f dx^1 \wedge \cdots \wedge \widehat{dx^i} \wedge \cdots \wedge dx^n \equiv 0$ on $\partial \mathcal{H}^n$, so $\int_{\partial \mathcal{H}^n} \omega = 0$. Thus, Stokes's theorem holds in this case.

Case 2: $i = n$.

$$\int_{\mathcal{H}^n} d\omega = (-1)^{n-1} \int_{\mathcal{H}^n} \frac{\partial f}{\partial x^n} dx^1 \cdots dx^n$$

$$= (-1)^{n-1} \int_{\mathbb{R}^{n-1}} \left(\int_{0}^{\infty} \frac{\partial f}{\partial x^n} dx^n \right) dx^1 \cdots dx^{n-1}.$$

In this integral

$$\int_{0}^{\infty} \frac{\partial f}{\partial x^n} dx^n = \int_{0}^{a} \frac{\partial f}{\partial x^n} dx^n$$

$$= f(x^1, \ldots, x^{n-1}, a) - f(x^1, \ldots, x^{n-1}, 0)$$

$$= -f(x^1, \ldots, x^{n-1}, 0).$$

Hence,

$$\int_{\mathcal{H}^n} d\omega = (-1)^n \int_{\mathbb{R}^{n-1}} f(x^1,\dots,x^{n-1},0)\,dx^1 \cdots dx^{n-1} = \int_{\partial \mathcal{H}^n} \omega$$

because $(-1)^n \mathbb{R}^{n-1}$ is precisely $\partial \mathcal{H}^n$ with its boundary orientation. So Stokes's theorem also holds in this case.   ◇

23.5 Take the exterior derivative of $x^2 + y^2 + z^2 = 1$ to obtain a relation among the 1-forms dx, dy, and dz on S^2. Then show for example that for $x \neq 0$, one has $dx \wedge dy = (z/x)dy \wedge dz$.

24.1 Assume $\omega = df$. Derive a contradiction using Problem 8.10(b) and Proposition 17.2.

25.4* The snake lemma
If we view each column of the given commutative diagram as a cochain complex, then the diagram is a short exact sequence of cochain complexes

$$0 \to \mathcal{A} \to \mathcal{B} \to \mathcal{C} \to 0.$$

By the zig-zag lemma, it gives rise to a long exact sequence in cohomology. In the long exact sequence, $H^0(\mathcal{A}) = \ker \alpha$, $H^1(\mathcal{A}) = A^1/\operatorname{im}\alpha = \operatorname{coker}\alpha$, and similarly for $\mathcal{B}$ and $\mathcal{C}$.   ◇

26.2 Define $d_{-1} = 0$. Then the given exact sequence is equivalent to a collection of short exact sequences

$$0 \to \operatorname{im}d_{k-1} \to A^k \xrightarrow{d_k} \operatorname{im}d_k \to 0, \qquad k = 0,\dots,m-1.$$

By the rank–nullity theorem,

$$\dim A^k = \dim(\operatorname{im}d_{k-1}) + \dim(\operatorname{im}d_k).$$

When we compute the alternating sum of the left-hand side, the right-hand side will cancel to 0.   ◇

28.1 Let U be the punctured projective plane $\mathbb{R}P^2 - \{p\}$ and V a small disk containing p. Because U can be deformation retracted to the boundary circle, which after identification is in fact $\mathbb{R}P^1$, U has the homotopy type of $\mathbb{R}P^1$. Since $\mathbb{R}P^1$ is homeomorphic to S^1, $H^*(U) \simeq H^*(S^1)$. Apply the Mayer–Vietoris sequence. The answer is $H^0(\mathbb{R}P^2) = \mathbb{R}$, $H^k(\mathbb{R}P^2) = 0$ for $k > 0$.

28.2 $H^k(S^n) = \mathbb{R}$ for $k = 0, n$, and $H^k(S^n) = 0$ otherwise.

28.3 One way is to apply the Mayer–Vietoris sequence to $U = \mathbb{R}^2 - \{p\}$, $V = \mathbb{R}^2 - \{q\}$.

A.13* The Lindelöf condition
Let $\{B_i\}_{i \in I}$ be a countable basis and $\{U_\alpha\}_{\alpha \in A}$ an open cover of the topological space S. For every $p \in U_\alpha$, there exists a B_i such that

$$p \in B_i \subset U_\alpha.$$

Since this B_i depends on p and α, we write $i = i(p,\alpha)$. Thus,

$$p \in B_{i(p,\alpha)} \subset U_\alpha.$$

Now let J be the set of all indices $j \in I$ such that $j = i(p,\alpha)$ for some p and some α. Then $\bigcup_{j \in J} B_j = S$ because every p in S is contained in some $B_{i(p,\alpha)} = B_j$.

   For each $j \in J$, choose an $\alpha(j)$ such that $B_j \subset U_{\alpha(j)}$. Then $S = \bigcup_j B_j \subset \bigcup_j U_{\alpha(j)}$. So $\{U_{\alpha(j)}\}_{j \in J}$ is a countable subcover of $\{U_\alpha\}_{\alpha \in A}$.   ◇

A.15* Disconnected subset in terms of a separation

($\Rightarrow$) By (iii),

$$A = (U \cap V) \cap A = (U \cap A) \cup (V \cap A).$$

By (i) and (ii), $U \cap A$ and $V \cap A$ are disjoint nonempty open subsets of A. Hence, A is disconnected.

($\Leftarrow$) Suppose A is disconnected in the subspace topology. Then $A = U' \cup V'$, where U' and V' are two disjoint nonempty open subsets of A. By the definition of the subspace topology, $U' = U \cap A$ and $V' = V \cap A$ for some open sets U, V in S.

  (i) holds because U' and V' are nonempty.
  (ii) holds because U' and V' are disjoint.
  (iii) holds because $A = U' \cup V' \subset U \cup V$. $\Diamond$

A.19* Uniqueness of the limit

Suppose $p \neq q$. Since S is Hausdorff, there exist disjoint open sets U_p and U_q such that $p \in U_p$ and $q \in U_q$. By the definition of convergence, there are integers N_p and N_q such that for all $i \geq N_p$, $x_i \in U_p$ and for all $i \geq N_q$, $x_i \in U_q$. This is a contradiction, since $U_p \cap U_q$ is the empty set. $\Diamond$

A.20* Closure in a product

($\subset$) By Problem A.5, $\mathrm{cl}(A) \times Y$ is a closed set containing $A \times Y$. By the definition of closure, $\mathrm{cl}(A \times Y) \subset \mathrm{cl}(A) \times Y$.

($\supset$) Conversely, suppose $(p, y) \in \mathrm{cl}(A) \times Y$. If $p \in A$, then $(p, y) \in A \times Y \subset \mathrm{cl}(A \times Y)$. Suppose $p \notin A$. By Proposition A.50, p is an accumulation point of A. Let $U \times V$ be any basis open set in $S \times Y$ containing (p, y). Because $p \in \mathrm{ac}(A)$, the open set U contains a point $a \in A$ with $a \neq p$. So $U \times V$ contains the point $(a, y) \in A \times Y$ with $(a, y) \neq (p, y)$. This proves that (p, y) is an accumulation point of $A \times Y$. By Proposition A.50 again, $(p, y) \in \mathrm{ac}(A \times Y) \subset \mathrm{cl}(A \times Y)$. This proves that $\mathrm{cl}(A) \times Y \subset \mathrm{cl}(A \times Y)$. $\Diamond$

B.1* The rank of a matrix

($\Rightarrow$) Suppose $\mathrm{rk}\, A \geq k$. Then one can find k linearly independent columns, which we call a_1, ..., a_k. Since the $m \times k$ matrix $[a_1 \cdots a_k]$ has rank k, it has k linearly independent rows b^1, ..., b^k. The matrix B whose rows are b^1, ..., b^k is a $k \times k$ submatrix of A, and $\mathrm{rk}\, B = k$. In other words, B is a nonsingular $k \times k$ submatrix of A.

($\Leftarrow$) Suppose A has a nonsingular $k \times k$ submatrix B. Let $a_1, ..., a_k$ be the columns of A such that the submatrix $[a_1 \cdots a_k]$ contains B. Since $[a_1 \cdots a_k]$ has k linearly independent rows, it also has k linearly independent columns. Thus, $\mathrm{rk}\, A \geq k$. $\Diamond$

B.2* Matrices of rank at most r

Let A be an $m \times n$ matrix. By Problem B.1, $\mathrm{rk}\, A \leq r$ if and only if all $(r+1) \times (r+1)$ minors $m_1(A), \ldots, m_s(A)$ of A vanish. As the common zero set of a collection of continuous functions, D_r is closed in $\mathbb{R}^{m \times n}$. $\Diamond$

B.3* Maximal rank

For definiteness, suppose $n \leq m$. Then the maximal rank is n and every matrix $A \in \mathbb{R}^{m \times n}$ has rank $\leq n$. Thus,

$$D_{\max} = \{A \in \mathbb{R}^{m \times n} \mid \mathrm{rk}\, A = n\} = \mathbb{R}^{m \times n} - D_{n-1}.$$

Since D_{n-1} is a closed subset of $\mathbb{R}^{m \times n}$ (Problem B.2), $D_{\max}$ is open in $\mathbb{R}^{m \times n}$. $\Diamond$

B.4* Degeneracy loci and maximal-rank locus of a map

(a) Let D_r be the subset of $\mathbb{R}^{m \times n}$ consisting of matrices of rank at most r. The degeneracy locus of rank r of the map $F : S \to \mathbb{R}^{m \times n}$ may be described as

$$D_r(F) = \{x \in S \mid F(x) \in D_r\} = F^{-1}(D_r).$$

Since D_r is a closed subset of $\mathbb{R}^{m \times n}$ (Problem B.2) and F is continuous, $F^{-1}(D_r)$ is a closed subset of S.

(b) Let $D_{\max}$ be the subset of $\mathbb{R}^{m \times n}$ consisting of all matrices of maximal rank. Then $D_{\max}(F) = F^{-1}(D_{\max})$. Since $D_{\max}$ is open in $\mathbb{R}^{m \times n}$ (Problem B.3) and F is continuous, $F^{-1}(D_{\max})$ is open in S.   ◊

B.7 Use Example B.5.

List of Notations

$\mathbb{R}^n$	Euclidean space of dimension n (p. 4)
$p = (p^1, \ldots, p^n)$	point in $\mathbb{R}^n$ (p. 4)
C^∞	smooth or infinitely differentiable (p. 4)
$\partial f / \partial x^i$	partial derivative with respect to x^i (pp. 4, 67)
$f^{(k)}(x)$	the kth derivative of $f(x)$ (p. 5)
$B(p, r)$	open ball in $\mathbb{R}^n$ with center p and radius r (pp. 7, 317)
$]a, b[$	open interval in $\mathbb{R}^1$ (p. 8)
$T_p(\mathbb{R}^n)$ or $T_p R^n$	tangent space to $\mathbb{R}^n$ at p (p. 10)
$v = \begin{bmatrix} v^1 \\ v^2 \\ v^3 \end{bmatrix} = \langle v^1, \ldots, v^n \rangle$	column vector (p. 11)
$\{e_1, \ldots, e_n\}$	standard basis for $\mathbb{R}^n$ (p. 11)
$D_v f$	directional derivative of f in the direction of v at p (p. 11)
$x \sim y$	equivalence relation (p. 11)
C_p^∞	algebra of germs of C^∞ functions at p in $\mathbb{R}^n$ (p. 12)
$\mathcal{D}_p(\mathbb{R}^n)$	vector space of derivations at p in $\mathbb{R}^n$ (p. 13)
$\mathfrak{X}(U)$	vector space of C^∞ vector fields on U (p. 15)
$\mathrm{Der}(A)$	vector space of derivations of an algebra A (p. 17)
δ_j^i	Kronecker delta (p. 13)
$\mathrm{Hom}(V, W)$	vector space of linear maps $f : V \to W$ (p. 19)
$V^\vee = \mathrm{Hom}(V, \mathbb{R})$	dual of a vector space (p. 19)
V^k	Cartesian product $V \times \cdots \times V$ of k copies of V (p. 22)
$L_k(V)$	vector space of k-linear functions on V (p. 22)

$(a_1\, a_2\, \cdots\, a_r)$	cyclic permutation, r-cycle (p. 20)
(ab)	transposition (p. 20)
S_k	group of permutations of k objects (p. 20)
$\operatorname{sgn}(\sigma)$ or $\operatorname{sgn}\sigma$	sign of a permutation (p. 20)
$A_k(V)$	vector space of alternating k-linear functions on V (p. 23)
σf	a function f acted on by a permutation σ (p. 23)
e	identity element of a group (p. 24)
$\sigma \cdot x$	left action of σ on x (p. 24)
$x \cdot \sigma$	right action of σ on x (p. 24)
Sf	symmetrizing operator applied to f (p. 24)
Af	alternating operator applied to f (p. 24)
$f \otimes g$	tensor product of multilinear functions f and g (p. 25)
$f \wedge g$	wedge product of multicovectors f and g (p. 26)
$B = [b^i_j]$ or $[b_{ij}]$	matrix whose (i,j)-entry is b^i_j or b_{ij} (p. 30)
$\det[b^i_j]$ or $\det[b_{ij}]$	determinant of the matrix $[b^i_j]$ or $[b_{ij}]$ (p. 30)
$\bigwedge(V)$	exterior algebra of a vector space (p. 30)
$I = (i_1,\ldots,i_k)$	multi-index (p. 31)
e_I	k-tuple $(e_{i_1},\ldots,e_{i_k})$ (p. 31)
α^I	k-covector $\alpha^{i_1} \wedge \cdots \wedge \alpha^{i_k}$ (p. 31)
$T^*_p(\mathbb{R}^n)$ or $T^*_p\mathbb{R}^n$	cotangent space to $\mathbb{R}^n$ (p. 34)
df	differential of a function (pp. 34, 191)
dx^I	$dx^{i_1} \wedge \cdots \wedge dx^{i_k}$ (p. 36)
$\Omega^k(U)$	vector space of C^∞ k-forms on U (pp. 36, 203)
$\Omega^*(U)$	direct sum $\bigoplus_{k=0}^n \Omega^k(U)$ (p. 37, 206)
$\omega(X)$	the function $p \mapsto \omega_p(X_p)$ (p. 37)
$\mathcal{F}(U)$ or $C^\infty(U)$	ring of C^∞ functions on U (p. 38)
$d\omega$	exterior derivative of ω (p. 38)
f_x	$\partial f/\partial x$, partial derivative of f with respect to x (p. 38)
$\bigoplus_{k=0}^\infty A^k$	direct sum of $A^0, A^1, \ldots$ (p. 30)
$\operatorname{grad} f$	gradient of a function f (p. 41)
$\operatorname{curl} \mathbf{F}$	curl of a vector field $\mathbf{F}$ (p. 41)
$\operatorname{div} \mathbf{F}$	divergence of a vector field $\mathbf{F}$ (p. 41)
$H^k(U)$	kth de Rham cohomology of U (p. 43)

$\{U_\alpha\}_{\alpha \in A}$	open cover (p. 48)
$(U,\phi),(U,\phi:U \to \mathbb{R}^n)$	chart or coordinate open set (p. 48)
$\mathbb{1}_U$	identity map on U (p. 49)
$U_{\alpha\beta}$	$U_\alpha \cap U_\beta$ (p. 50)
$U_{\alpha\beta\gamma}$	$U_\alpha \cap U_\beta \cap U_\gamma$ (p. 50)
$\mathfrak{U} = \{(U_\alpha, \phi_\alpha)\}$	atlas (p. 50)
$\mathbb{C}$	complex plane (p. 50)
$\coprod$	disjoint union (pp. 51, 129)
$\phi_\alpha\|_{U_\alpha \cap V}$	restriction of ϕ_α to $U_\alpha \cap V$ (p. 54)
$\Gamma(f)$	graph of f (p. 54)
$K^{m \times n}$	vector space of $m \times n$ matrices with entries in K (p. 54)
$\mathrm{GL}(n,K)$	general linear group over a field K (p. 54)
$M \times N$	product manifold (p. 55)
$f \times g$	Cartesian product of two maps (p. 55)
S^n	unit sphere in $\mathbb{R}^{n+1}$ (p. 58)
F^*h	pullback of a function h by a map F (p. 60)
$J(f) = [\partial F^i/\partial x^j]$	Jacobian matrix (p. 68)
$\det[\partial F^i/\partial x^j]$	Jacobian determinant (p. 68)
$\dfrac{\partial(F^1,\ldots,F^n)}{\partial(x^1,\ldots,x^n)}$	Jacobian determinant (p. 68)
$\mu: G \times G \to G$	multiplication on a Lie group (p. 66)
$\iota: G \to G$	inverse map of a Lie group (p. 66)
$K^\times$	nonzero elements of a field K (p. 66)
S^1	unit circle in $\mathbb{C}^\times$ (p. 66)
$A = [a_{ij}], [a^i_j]$	matrix whose (i,j)-entry is a_{ij} or a^i_j (p. 67)
$S/\sim$	quotient (p. 71)
$[x]$	equivalence class of x (p. 71)
$\pi^{-1}(U)$	inverse image of U under π (p. 71)
$\mathbb{R}P^n$	real projective space of dimension n (p. 76)
$\|x\|$	modulus of x (p. 77)
$a^1 \wedge \cdots \wedge \widehat{a^i} \wedge \cdots \wedge a^n$	the caret $\widehat{}$ means to omit a^i (p. 80)
$G(k,n)$	Grassmannian of k-planes in $\mathbb{R}^n$ (p. 82)
$\mathrm{rk}\,A$	rank of a matrix A (p. 82 (p. 344)
$C^\infty_p(M)$	germs of C^∞ functions at p in M (p. 87)

$T_p(M)$ or T_pM	tangent space to M at p (p. 87)
$\partial/\partial x^i\|_p$	coordinate tangent vector at p (p. 87)
$d/dt\|_p$	coordinate tangent vector of a 1-dimensional manifold (p. 87)
$F_{*,p}$ or F_*	differential of F at p (p. 87)
$c(t)$	curve in a manifold (p. 92)
$c'(t) := c_*\left(\left.\dfrac{d}{dt}\right\|_{t_0}\right)$	velocity vector of a curve (p. 92)
$\dot{c}(t)$	derivative of a real-valued function (p. 92)
ϕ_S	coordinate map on a submanifold S (p. 100)
$f^{-1}(\{c\})$ or $f^{-1}(c)$	level set (p. 103)
$Z(f) = f^{-1}(0)$	zero set (p. 103)
$\mathrm{SL}(n,K)$	special linear group over a field K (pp. 107, 109)
m_{ij} or $m_{ij}(A)$	(i,j)-minor of a matrix A (p. 107)
$\mathrm{Mor}(A,B)$	the set of morphisms from A to B (p. 110)
$\mathbb{1}_A$	identity map on A (p. 110)
(M,q)	pointed manifold (p. 111)
$\simeq$	isomorphism (p. 111)
$\mathcal{F}, \mathcal{G}$	functors (p. 111)
$\mathcal{C}, \mathcal{D}$	categories (p. 111)
$\{e_1,\ldots,e_n\}$	basis for a vector space V (p. 113)
$\{\alpha^1,\ldots,\alpha^n\}$	dual basis for $V^\vee$ (p. 113)
$L^\vee$	dual of linear map L (p. 113)
$\mathrm{O}(n)$	orthogonal group (p. 117)
A^T	transpose of a matrix A (p. 117)
ℓ_g	left multiplication by g (p. 117)
r_g	right multiplication by g (p. 117)
$D_{\max}(F)$	maximal rank locus of $F: S \to \mathbb{R}^{m \times n}$ (pp. 118, 345)
$i: N \to M$	inclusion map (p. 123)
TM	tangent bundle (p. 129)
$\tilde{\phi}$	coordinate map on the tangent bundle (p. 130)
$E_p := \pi^{-1}(p)$	fiber at p of a vector bundle (p. 133)
X	vector field (p. 136)
X_p	tangent vector at p (p. 136)
$\Gamma(U,E)$	vector space of C^∞ sections of E over U (p. 137)

$\Gamma(E) := \Gamma(M,E)$	vector space of C^∞ sections of E over M (p. 137)
$\mathrm{supp}\, f$	support of a function f (p. 140)
$\overline{B}(p,r)$	closed ball in $\mathbb{R}^n$ with center p and radius r (p. 143)
$\overline{A}$, $\mathrm{cl}(A)$, or $\mathrm{cl}_S(A)$	closure of a set A in S (pp. 148, 334)
$c_t(p)$	integral curve through p (p. 152)
$\mathrm{Diff}(M)$	group of diffeomorphisms of M (p. 153)
$F_t(q) = F(t,q)$	local flow (p. 156)
$[X,Y]$	Lie bracket of vector fields, bracket in a Lie algebra (pp. 157, 158)
$\mathfrak{X}(M)$	Lie algebra of C^∞ vector fields on M (p. 158)
S_n	vector space of $n \times n$ real symmetric matrices (p. 166)
$\mathbb{R}^2/\mathbb{Z}^2$	torus (p. 167)
$\|X\|$	norm of a matrix (p. 169
$\exp(X)$ or e^X	exponential of a matrix X (p. 170)
$\mathrm{tr}(X)$	trace (p. 171)
$Z(G)$	center of a group G (p. 176)
$\mathrm{SO}(n)$	special orthogonal group (p. 176)
$\mathrm{U}(n)$	unitary group (p. 176)
$\mathrm{SU}(n)$	special unitary group (p. 177)
$\mathrm{Sp}(n)$	compact symplectic group (p. 177)
J	the matrix $\begin{bmatrix} 0 & I_n \\ -I_n & 0 \end{bmatrix}$ (p. 177)
I_n	$n \times n$ identity matrix (p. 177)
$\mathrm{Sp}(2n,\mathbb{C})$	complex symplectic group (p. 177)
K_n	space of $n \times n$ real skew-symmetric matrices (p. 179)
$\tilde{A}$	left-invariant vector field generated by $A \in T_e G$ (p. 180)
$L(G)$	Lie algebra of left-invariant vector fields on G (p. 180)
$\mathfrak{g}$	Lie algebra (p. 182)
$\mathfrak{h} \subset \mathfrak{g}$	Lie subalgebra (p. 182)
$\mathfrak{gl}(n,\mathbb{R})$	Lie algebra of $\mathrm{GL}(n,\mathbb{R})$ (p. 183)
$\mathfrak{sl}(n,\mathbb{R})$	Lie algebra of $\mathrm{SL}(n,\mathbb{R})$ (p. 186)
$\mathfrak{o}(n)$	Lie algebra of $\mathrm{O}(n)$ (p. 186)
$\mathfrak{u}(n)$	Lie algebra of $\mathrm{U}(n)$ (p. 186)
$(df)_p$, $df|_p$	value of a 1-form at p (p. 191)

$T_p^*(M)$ or T_p^*M	cotangent space at p (p. 190)
T^*M	cotangent bundle (p. 192)
$F^*: T_{F(p)}^*M \to T_p^*N$	codifferential (p. 196)
$F^*\omega$	pullback of a differential form ω by F (pp. 196, 205)
$\bigwedge^k(V^\vee) = A_k(V)$	k-covectors on a vector space V (p. 200)
ω_p	value of a differential form ω at p (p. 200)
$\mathfrak{J}_{k,n}$	the set of strictly ascending multi-indices $1 \le i_1 < \cdots < i_k \le n$ (p. 201)
$\bigwedge^k(T^*M)$	kth exterior power of the cotangent bundle (p. 203)
$\Omega^k(G)^G$	left-invariant k-forms on a Lie group G (p. 208)
$\operatorname{supp}\omega$	support of a k-form (p. 208)
$d\omega$	exterior derivative of a differential form ω (p. 213)
$\omega|_S$	restriction of a differential from ω to a submanifold S (p. 216)
$\hookrightarrow$	inclusion map (p. 216)
$\mathcal{L}_X Y$	the Lie derivative of a vector field Y along X (p. 224)
$\mathcal{L}_X \omega$	the Lie derivative of a differential form ω along X (p. 226)
$\iota_v \omega$	interior multiplication of ω by v (p. 227)
$(v_1, \ldots, v_n)$	ordered basis (p. 237)
$[v_1, \ldots, v_n]$	ordered basis as a matrix (p. 238)
$(M, [\omega])$	oriented manifold with orientation $[\omega]$ (p. 244)
$-M$	the oriented manifold having the opposite orientation as M (p. 246)
$\mathcal{H}^n$	closed upper half-space (p. 248)
M°	interior of a manifold with boundary (pp. 248, 252)
∂M	boundary of a manifold with boundary (pp. 248, 251)
$\mathcal{L}^1$	left half-line (p. 251)
$\operatorname{int}(A)$	topological interior of a subset A (p. 252)
$\operatorname{ext}(A)$	exterior of a subset A (p. 252)
$\operatorname{bd}(A)$	topological boundary of a subset A (p. 252)
$\{p_0, \ldots, p_n\}$	partition of a closed interval (p. 260)
$P = \{P_1, \ldots, P_n\}$	partition of a closed rectangle (p. 260)
$L(f, P)$	lower sum of f with respect to a partition P (p. 260)
$U(f, P)$	upper sum of f with respect to a partition P (p. 260)

$\overline{\int}_R f$	upper integral of f over a closed rectangle R (p. 261)
$\underline{\int}_R f$	lower integral of f over a closed rectangle R (p. 261)
$\int_R f(x)\,\lvert dx^1 \cdots dx^n\rvert$	Riemann integral of f over a closed rectangle R (p. 261)
$\int_U \omega$	Riemann integral of a differential form ω over U (p. 263)
$\mathrm{vol}(A)$	volume of a subset A of $\mathbb{R}^n$ (p. 262)
$\mathrm{Disc}(f)$	set of discontinuities of a function f (p. 262)
$\Omega_c^k(M)$	vector space of C^∞ k-forms with compact support on M (p. 265)
$Z^k(M)$	vector space of closed k-forms on M (p. 275)
$B^k(M)$	vector space of exact k-forms on M (p. 275)
$H^k(M)$	de Rham cohomology of M in degree k (p. 275)
$[\omega]$	cohomology class of ω (p. 275)
$F^\#$ or F^*	induced map in cohomology (p. 278)
$H^*(M)$	the cohomology ring $\oplus_{k=0}^n H^k(M)$ (p. 279)
$\mathcal{C} = (\{C^k\}_{k \in \mathbb{Z}}, d)$	cochain complex (p. 281)
$(\Omega^*(M), d)$	de Rham complex (p. 281)
$H^k(\mathcal{C})$	kth cohomology of $\mathcal{C}$ (p. 283)
$Z^k(\mathcal{C})$	subspace of k-cocycles (p. 283)
$B^k(\mathcal{C})$	subspace of k-coboundaries (p. 283
$d^*\colon H^k(\mathcal{C}) \to H^{k+1}(\mathcal{A})$	connecting homomorphism (p. 284)
$\rightarrowtail$	injection or maps to under an injection (p. 285)
$\longmapsto\!\!\!\!\to$	maps to under a surjection (p. 285)
$i_U\colon U \to M$	inclusion map of U in M (p. 288)
$j_U\colon U \cap V \to U$	inclusion map of $U \cap V$ in U (p. 288)
$\longrightarrow\!\!\!\!\to$	surjection (p. 291)
$\chi(M)$	Euler characteristic of M (p. 295)
$f \sim g$	f is homotopic to g (p. 296)
Σ_g	compact orientable surface of genus g (p. 310)
$d(p,q)$	distance between p and q (p. 317)
(a,b)	open interval (p. 318)
$(S, \mathcal{T})$	a set S with a topology $\mathcal{T}$ (p. 318)
$Z(f_1, \ldots, f_r)$	zero set of $f_1, \ldots, f_r$ (p. 319)
$Z(I)$	zero set of all the polynomials in an ideal I (p. 320)

IJ	the product ideal (p. 320)
$\sum_\alpha I_\alpha$	sum of ideals (p. 320)
$\mathcal{T}_A$	subspace topology or relative topology of A (p. 320)
$\mathbb{Q}$	the set of rational numbers (p. 323)
$\mathbb{Q}^+$	the set of positive rational numbers (p. 323)
$A \times B$	Cartesian product of two sets A and B (p. 326)
C_x	connected component of a point x (p. 333)
$\mathrm{ac}(A)$	the set of accumulation points of A (p. 335)
$\mathbb{Z}^+$	the set of positive integers (p. 336)
D_r	the set of matrices of rank $\leq r$ in $\mathbb{R}^{m \times n}$ (p. 345)
$D_{\max}$	the set of matrices of maximal rank in $\mathbb{R}^{m \times n}$ (p. 345)
$D_r(F)$	degeneracy locus of rank r of a map $F : S \to \mathbb{R}^{m \times n}$ (p. 345)
$\ker f$	kernel of a homomorphism f (p. 350)
$\operatorname{im} f$	image of a map f (p. 350)
$\operatorname{coker} f$	cokernel of a homomorphism f (p. 350)
$v + W$	coset of a subspace W (p. 349)
V/W	quotient vector space of V by W (p. 349)
$\prod_\alpha V_\alpha, A \times B$	direct product (p. 351)
$\bigoplus_\alpha V_\alpha, A \oplus B$	direct sum (p. 351)
$A + B$	sum of two vector subspaces (p. 351)
$A \oplus_i B$	internal direct sum (p. 351)
$W^\perp$	W "perp," orthogonal complement of W (p. 352)
$\mathbb{H}$	skew field of quaternions (p. 353)
$\operatorname{End}_K(V)$	algebra of endomorphisms of V over K (p. 354)

References

[1] V. I. Arnold, *Mathematical Methods of Classical Mechanics*, 2nd ed., Springer, New York, 1989.

[2] P. Bamberg and S. Sternberg, *A Course in Mathematics for Students of Physics*, Vol. 2, Cambridge University Press, Cambridge, UK, 1990.

[3] W. Boothby, *An Introduction to Differentiable Manifolds and Riemannian Geometry*, 2nd ed., Academic Press, Boston, 1986.

[4] R. Bott and L. W. Tu, *Differential Forms in Algebraic Topology*, 3rd corrected printing, Graduate Texts in Mathematics, Vol. 82, Springer, New York, 1995.

[5] É. Cartan, Sur certaines expressions différentielles et le problème de Pfaff, Ann. E.N.S. (3), vol. XVI (1899), pp. 239–332 (= *Oeuvres complètes*, vol. II, Gauthier-Villars, Paris, 1953, pp. 303–396).

[6] C. Chevalley, *Theory of Lie Groups*, Princeton University Press, Princeton, 1946.

[7] L. Conlon, *Differentiable Manifolds*, 2nd ed., Birkhäuser Boston, Cambridge, MA, 2001.

[8] G. de Rham, Sur l'analysis situs des variétés à n-dimension, Journal Math. Pure et Appl. (9) 10 (1931), pp. 115–200.

[9] G. de Rham, Über mehrfache Integrale, Abhandlungen aus dem Mathematischen Hansischen Universität (Universität Hamburg) 12 (1938), pp. 313–339.

[10] G. de Rham, *Variétés différentiables*, Hermann, Paris, 1960 (in French); *Differentiable Manifolds*, Springer, New York, 1984 (in English).

[11] D. Dummit and R. Foote, Abstract Algebra, 3rd ed., John Wiley and Sons, Hoboken, NJ, 2004.

[12] T. Frankel, *The Geometry of Physics: An Introduction*, 2nd ed., Cambridge University Press, Cambridge, UK, 2003.

[13] F. G. Frobenius, Über lineare Substitutionen und bilineare Formen, Journal für die reine und angewandte Mathematik (Crelle's Journal) 84 (1878), pp. 1–63. Reprinted in *Gesammelte Abhandlungen*, Band I, pp. 343–405.

[14] C. Godbillon, *Géométrie différentielle et mécanique analytique*, Hermann, Paris, 1969.

[15] E. Goursat, Sur certains systèmes d'équations aux différentielles totales et sur une généralisation du problème de Pfaff, Ann. Fac. Science Toulouse (3) 7 (1917), pp. 1–58.

[16] M. J. Greenberg, *Lectures on Algebraic Topology*, W. A. Benjamin, Menlo Park, CA, 1966.

[17] V. Guillemin and A. Pollack, *Differential Topology*, Prentice-Hall, Englewood Cliffs, NJ, 1974.

[18] A. Hatcher, *Algebraic Topology*, Cambridge University Press, Cambridge, UK, 2002.

[19] I. N. Herstein, *Topics in Algebra*, 2nd ed., John Wiley and Sons, New York, 1975.

[20] M. Karoubi and C. Leruste, *Algebraic Topology via Differential Geometry*, Cambridge University Press, Cambridge, UK, 1987.

[21] V. Katz, The history of Stokes' theorem, Mathematics Magazine 52 (1979), pp. 146–156.

[22] V. Katz, Differential forms, in *History of Topology*, edited by I. M. James, Elsevier, Amsterdam, 1999, pp. 111–122.

[23] M. Kervaire, A manifold which does not admit any differentiable structure, Commentarii Mathematici Helvetici 34 (1960), pp. 257–270.

[24] M. Kervaire and J. Milnor, Groups of homotopy spheres: I, Annals of Mathematics 77 (1963), pp. 504–537.

[25] J. M. Lee, *Introduction to Smooth Manifolds*, Graduate Texts in Mathematics, Vol. 218, Springer, New York, 2003.

[26] J. E. Marsden and M. J. Hoffman, *Elementary Classical Analysis*, 2nd ed., W. H. Freeman, New York, 1993.

[27] J. Milnor, On manifolds homeomorphic to the 7-sphere, Annals of Mathematics 64 (1956), pp. 399–405.

[28] J. Milnor, *Topology from the Differentiable Viewpoint*, University Press of Virginia, Charlottesville, VA, 1965.

[29] J. Munkres, *Topology*, 2nd ed., Prentice–Hall, Upper Saddle River, NJ, 2000.

[30] J. Munkres, *Elements of Algebraic Topology*, Perseus Publishing, Cambridge, MA, 1984.

[31] J. Munkres, *Analysis on Manifolds*, Addison–Wesley, Menlo Park, CA, 1991.

[32] H. Poincaré, Sur les résidus des intégrales doubles, Acta Mathematica 9 (1887), pp. 321–380. *Oeuvres*, Tome III, pp. 440–489.

[33] H. Poincaré, Analysis situs, Journal de l'École Polytechnique 1 (1895), pp. 1–121; Oeuvres 6, pp. 193–288.

[34] H. Poincaré, *Les méthodes nouvelles de la mécanique céleste*, vol. III, Gauthier-Villars, Paris, 1899, Chapter XXII.

[35] W. Rudin, *Principles of Mathematical Analysis*, 3rd ed., McGraw–Hill, New York, 1976.

[36] M. Spivak, *A Comprehensive Introduction to Differential Geometry*, Vol. 1, 3rd ed., Publish or Perish, Houston, 2005.

[37] O. Veblen and J. H. C. Whitehead, A set of axioms for differential geometry, Proceedings of National Academy of Sciences 17 (1931), pp. 551–561. *The Mathematical Works of J. H. C. Whitehead*, Vol. 1, Pergamon Press, 1962, pp. 93–104.

[38] F. Warner, *Foundations of Differentiable Manifolds and Lie Groups*, Springer, New York, 1983.

[39] M. Zisman, Fibre bundles, fibre maps, in *History of Topology*, edited by I. M. James, Elsevier, Amsterdam, 1999, pp. 605–629.

Index

 **Loring W. Tu** was born in Taipei, Taiwan, and grew up in Taiwan, Canada, and the United States. He attended McGill University and Princeton University as an undergraduate, and obtained his Ph.D. from Harvard University under the supervision of Phillip A. Griffiths. He has taught at the University of Michigan, Ann Arbor, and at Johns Hopkins University, and is currently Professor of Mathematics at Tufts University in Massachusetts.

An algebraic geometer by training, he has done research at the interface of algebraic geometry, topology, and differential geometry, including Hodge theory, degeneracy loci, moduli spaces of vector bundles, and equivariant cohomology. He is the coauthor with Raoul Bott of *Differential Forms in Algebraic Topology*.

Printed in the United States
By Bookmasters